ASTRONOMY AND ASTROPHYSICS LIBRARY

Series Editors: M. Harwit, R. Kippenhahn, J.-P. Zahn

Martin Harwit

Astrophysical Concepts

Second Edition

With 175 Illustrations

Springer-Verlag

New York Berlin Heidelberg London Paris
Tokyo Hong Kong Barcelona Budapest

Martin Harwit, National Air and Space Museum, Smithsonian Institution, Washington, D.C. 20560, USA

Series Editors

Martin Harwit

National Air and Space Museum
Smithsonian Institution
Washington, D.C. 20560, USA

Rudolf Kippenhahn

Max-Planck-Institut für
Physik und Astrophysik
Institut für Astrophysik
Karl-Schwarzschild-Straße 1
D-8046 Garching,
Fed. Rep. of Germany

Jean-Paul Zahn

Université Paul Sabatier
Observatoires du Pic-du-Midi
et de Toulouse
14, Avenue Edouard-Belin
F-31400 Toulouse, France

Cover photo: M51, Whirlpool Galaxy. Courtesy U.S. Naval Observatory.

Library of Congress Cataloging-in-Publication Data
Harwit, Martin, 1931–
   Astrophysical concepts.
   (Astronomy and astrophysics library)
   Bibliography: p.
   Includes index.
   1. Astrophysics.   I. Title.   II. Series.
QB461.H37   1988   523.01   87-32387

The first edition was published in 1973 by John Wiley and Sons, Inc., and was reprinted and published in 1982 by Concepts.

Typeset by Asco Trade Typesetting Ltd., Hong Kong.
Printed and bound by Edwards Brothers, Inc., Ann Arbor, Michigan.
Printed in the United States of America.

9 8 7 6 5 4 3

ISBN 0-387-96683-8 Springer-Verlag New York Berlin Heidelberg
ISBN 3-540-96683-8 Springer-Verlag Berlin Heidelberg New York

Preface to the Second Edition

My principal aim in writing this book was to present a wide range of astrophysical topics in sufficient depth to give the reader a general quantitative understanding of the subject. The book outlines cosmic events but does not portray them in detail—it provides a series of astrophysical sketches. I think this approach befits the present uncertainties and changing views in astrophysics.

The material is based on notes I prepared for a course aimed at seniors and beginning graduate students in physics and astronomy at Cornell. This course defined the level at which the book is written.

For readers who are versed in physics but are unfamiliar with astronomical terminology, Appendix A is included. It gives a brief background of astronomical concepts and should be read before starting the main text.

The first few chapters outline the scope of modern astrophysics and deal with elementary problems concerning the size and mass of cosmic objects. However, it soon becomes apparent that a broad foundation in physics is needed to proceed. This base is developed in Chapters 4 to 7 by using, as examples, specific astronomical situations. Chapters 8 to 10 enlarge on the topics first outlined in Chapter 1 and show how we can obtain quantitative insights into the structure and evolution of stars, the dynamics of cosmic gases, and the large-scale behavior of the universe. Chapter 11 discusses life in the universe, while Chapter 12 attempts a synthesis of everything discussed before, by tracing the history of the universe from its beginnings to the formation of the sun and the planets.

Throughout the book I emphasize astrophysical concepts. This means that objects such as asteroids, stars, supernovae, or quasars are not described in individual chapters or sections. Instead, they are mentioned throughout the text whenever relevant physical principles are discussed. Thus the common features of many astronomical situations are underlined, but there is a partition of information about specific astronomical objects. For example, different aspects of neutron stars and pulsars are discussed in Chapters 5, 6, 8, Appendix A, and elsewhere. To compensate for this treatment, a comprehensive index is included.

I have sketched no more than the outlines of several traditional astronomical topics, such as the theories of radiative transfer, stellar atmospheres, and polytropic gas spheres, because a complete presentation would have required extensive mathematical development to be genuinely useful. However, the main physical concepts of these subjects are worked into the text, often as remarks without specific mention. In addition I refer, where appropriate, to other sources that treat these topics in greater detail.

The list of references is designed for readers who wish to cover any given area in greater depth. I have cited only authors who actively have contributed to a field and whose views bring the reader closer to the subject. Although some of the cited articles are popular, the writing is accurate.

In this Second Edition, I have tried to update the book throughout, revising at innumerable points, page by page, to make up for the passage of 15 years since the book was first published. I have also added a totally new final chapter, which attempts to trace our current overview of how various features of the universe, large and small, are structurally related.

A book that covers a major portion of astrophysics must be guided by the many excellent monographs and review articles that exist today. It is impossible to acknowledge all of them properly and to give credit to the astrophysicists whose viewpoints strongly influenced my writing. I am also grateful for the many suggestions for improvements offered by my colleagues at Cornell and by several generations of students who saw this book evolve from a series of informal lecture notes. Finally, I would like to thank Barbara L. Boettcher for preparing the drawings.

Martin Harwit

Contents

CHAPTER 10. STRUCTURE OF THE UNIVERSE   431

1

An Approach to Astrophysics

In a sense each of us has been inside a star; in a sense each of us has been in the vast empty spaces between the stars; and—if the universe ever had a beginning—each of us was there!

Every molecule in our bodies contains matter that once was subjected to the tremendous temperatures and pressures at the center of a star. This is where the iron in our red blood cells originated. The oxygen we breathe, the carbon and nitrogen in our tissues, and the calcium in our bones, also were formed through the fusion of smaller atoms at the center of a star.

Terrestrial ores containing uranium, plutonium, lead, and many other massive atoms must have been formed in a supernova explosion—the self-destruction of a star in which a sun's mass is hurled into space at huge velocity. In fact, most of the matter on earth and in our bodies must have gone through such a catastrophic event!

The elements lithium, beryllium, and boron, which we find in traces on earth, seem to have originated through cosmic ray bombardment in interstellar space. At that epoch the earth we now walk on was distributed so tenuously that a gram of soil would have occupied a volume the size of the entire planet.

To account for the deuterium, the heavy hydrogen isotope found on earth, we may have to go back to a cosmic explosion signifying the birth of the entire universe.

How do we know all this? And how sure are we of this knowledge?

This book was written to answer such questions and to provide a means for making astrophysical judgments.

We are just beginning a long and exciting journey into the universe. There is much to be learned, much to be discarded, and much to be revised. We have excellent theories, but theories are guides for understanding the truth. They are not truth itself. We must therefore continually revise them if they are to keep leading us in the right direction.

In going through the book, just as in devising new theories, we will find ourselves baffled by choices between the real and the apparent. We will have to learn that it may still be too early to make such choices, that reality in astrophysics has often been short-lived, and that—disturbing though it would be—we may some day have to reconcile ourselves to the realization that our theories had recognized only superficial effects—not the deeper, truly motivating, factors.

We may therefore do well to avoid an immediate preoccupation with astrophysical "reality." We should take the longer view and look closely at those physical concepts likely to play a role in the future evolution of our understanding. We may reason this way:

The development of astrophysics in the last few decades has been revolutionary. We have discarded what had appeared to be our most reliable theories, replaced them, and frequently found even the replacements lacking. The only constant in this revolution has been the pool of astrophysical concepts. It has not substantially changed, and it has provided a continuing source of material for our evolving theories.

This pool contained the neutron stars 35 years before their discovery, and it contained black holes three decades before astronomers started searching for them. The best investment of our efforts may lie in a deeper exploration of these concepts.

In astrophysics we often worry whether we should organize our thinking around individual objects—planets, stars, pulsars, and galaxies—or whether we should divide the subject according to physical principles common to the various astrophysical processes.

Our emphasis on concepts will make the second approach more appropriate. It will, however, also raise some problems: Much of the information about individual types of objects will be distributed throughout the book, and can be gathered only through use of the index. This leads to a certain unevenness in the presentation.

The unevenness is made even more severe by the varied mathematical treatment: No astrophysical picture is complete if we cannot assign a numerical value to its scale. In this book, we will therefore consistently aim at obtaining a rough order of magnitudes characteristic of the different phenomena. In some cases, this aim leads to no mathematical difficulties. In other problems, we will have to go through rather complex mathematical preparations before even the crudest answers emerge. The estimates of the curvature of the universe in Chapter 10 are an example of these more complex approaches.

Given these difficulties, which appear to be partly dictated by the nature of modern astrophysics, let us examine the most effective ways to use this book.

For those who have no previous background in astronomy, Appendix A may provide a good starting point. It briefly describes the astronomical objects we will study and introduces astronomical notation. This notation will be used throughout the book and is generally not defined in other chapters. Those who have previously studied astronomy will be able to start directly with the present chapter that presents the current searches going on in astrophysics—the questions that we seek to answer. Chapters 2 and 3 show that, while some of the rough dimensions of the universe can be measured by conceptually simple means, a deeper familiarity with physics is required to understand the cosmic sources of energy and the nature of cosmic evolution. The physical tools we need are therefore presented in the intermediate Chapters 4 to 7. We then gather these tools to work our way through theories of the synthesis of chemical elements mentioned right at the start of this section, the formation and evolution of stars, the processes that take place in interstellar space, the evolution of the universe, and the astrophysical setting for the origins of life.

    This is an exciting, challenging venture; but we have a long way to go.

    Let us start.

1:1 CHANNELS FOR ASTRONOMICAL INFORMATION

Imagine a planet inhabited by a blind civilization. One day an inventor discovers an instrument sensitive to visible light and this device is found to be useful for many purposes, particularly for astronomy.

    Human beings can see light and we would expect to have a big headstart in astronomy compared to any civilization that was just discovering methods for detecting visible radiation. Think then of an even more advanced culture that could detect not only visible light but also all other electromagnetic radiation and that had telescopes and detectors sensitive to *cosmic rays, neutrinos*, and *gravitational waves*. Clearly that civilization's knowledge of astronomy could be far greater than ours.

    Four entirely independent channels are known to exist by means of which information can reach us from distant parts of the universe.

    (a) Electromagnetic radiation: γ-rays, X-rays, ultraviolet, visible, infrared, and radio waves.

    (b) Cosmic ray particles: These comprise high energy electrons, protons, and heavier nuclei as well as the (unstable) neutrons and mesons. Some cosmic ray particles consist of antimatter.

    (c) Neutrinos and antineutrinos: There are two known types of neutrinos and antineutrinos; those associated with electrons and others associated with μ-mesons. A third type associated with τ-mesons is being sought.

    (d) Gravitational waves.

Most of us are familiar with channel *a*, currently the channel through which we obtain the bulk of astronomical information. However, let us briefly describe channels (b), (c), and (d).

(b) There are fundamental differences between cosmic ray particles and the other three information carriers: (i) cosmic ray particles move at very nearly the speed of light, while electromagnetic and gravitational waves—as well as neutrinos, if they are found to have no rest mass—move at precisely the speed of light; (ii) cosmic rays have a positive rest mass; and (iii) when electrically charged the particles can be deflected by cosmic magnetic fields so that the direction from which a cosmic ray particle arrives at the earth often is not readily related to the actual direction of the source.

Cosmic ray astronomy is far more advanced than either neutrino or gravitational wave work. Detectors and detector arrays exist, but the technical difficulties still are great. Nonetheless, through cosmic ray studies we hope to learn a great deal about the chemistry of the universe on a large scale and we hope, eventually, to single out regions of the universe in which as yet unknown, grandiose accelerators produce these highly energetic particles. We do not yet know how or where the cosmic ray particles gain their high energies; we merely make guesses, expressed in the form of different theories on the origin of cosmic rays (Ro64a, Go69, Gu69, Hi84).

(c) Neutrinos, have zero or at least low rest mass. They have one great advantage in that they can traverse great depths of matter without being absorbed. Neutrino astronomy could give us a direct look at the interior of stars, much as X-rays can be used to examine a metal block for internal flaws or a medical patient for lung ailments. Neutrinos could also convey information about past ages of the universe because, except for a systematic energy loss due to the expansion of the universe, the neutrinos are preserved in almost unmodified form over many aeons.* Much of the history of the universe must be recorded in the ambient neutrino flux, but so far we do not know how to tap this information (We62).

A first serious search for solar neutrinos has been conducted and has shown that there are fewer emitted neutrinos than had been predicted (Da68). This has led to a re-examination of theories on the nuclear reactions taking place in the interior of the sun. First direct evidence for copious generation of neutrinos in the explosion of a supernova has given neutrino astronomy a promising new start (Hi87, Bi87).

(d) Gravitational waves, when reliably detected, will yield information on the motion of very massive bodies. Gravitational waves have not yet been directly detected, though their existence is indirectly inferred from observations on closely spaced pairs of compact stars and changes in their orbital motions about a common

---

* One aeon = 10^9 y.

center of mass. We seem therefore to be on the threshold of important discoveries that are sure to have a significant influence on astronomy (We70).

It is clear that astronomy cannot be complete until techniques are developed to detect all of the four principal means by which information can reach us. Until that time astrophysical theories must remain provisory.

Not only must we be able to detect these information carriers, but we will also have to develop detectors that cover the entire spectral range for each type of carrier. The importance of this is shown by the great contribution made by radio astronomy. Until two or three decades ago, all our astronomical information was obtained in the visible, near infrared, or near ultraviolet regions; no one at that time suspected that a wealth of information was available in the radio, infrared, X-ray or gamma-ray spectrum. Yet, today the only complete maps we have of our own Galaxy lie in these spectral ranges. They show, respectively, the distributions of pulsars and molecular, atomic or ionized gas; clouds of dust; bright, hot X-ray emitting stars and X-ray binaries; and gamma rays produced through the interaction of cosmic rays with gas in Galactic clouds.

Just as we have made our first astrophysically significant neutrino observations and are reaching for gravitational wave detection, a variety of new carriers of information have been proposed. We now speak of axions, photinos, magnetic monopoles, tachyons, and other carriers of information which—should they exist—could serve as further channels for communication through which we could gather astrophysical information. All these hypothesized entities arise from an extension of known theory into domains where we still lack experimental data. Theoretically, they are plausible, but there is no evidence that they exist in nature. Axions, photinos, and magnetic monopoles could, however, be making themselves felt through their gravitational attraction, even though otherwise unobserved. We infer from the hot, massive gaseous haloes around giant elliptical galaxies that these galaxies contain far more matter than is observed in stars and interstellar gases. The same inference is drawn from the surprisingly high speeds at which stars orbit the centers of spiral galaxies even when located at the extreme periphery of the galaxies' disks. Could this dark matter consist of such exotic particles? Tachyons, in turn, are interesting because they would travel at speeds exceeding the speed of light. Should intelligent life exist elsewhere in the universe, tachyons might provide a preferred method of rapid communication.

1:2 X-RAY ASTRONOMY: DEVELOPMENT OF A NEW FIELD

The development of a new branch of astronomy often follows a general pattern: Vague theoretical thinking tells us that no new development is to be expected at all. Consequently, it is not until some chance observation focuses attention onto a new area that serious preliminary measurements are undertaken. Many of these initial findings later have to be discarded as techniques improve.

These awkward developmental stages are always exciting; let us outline the evolution of X-ray astronomy, as an example, to convey the sense of advances that should take place in astronomy and astrophysics in the next few years.

Until 1962 only solar X-ray emission had been observed. This flux is so weak that no one expected a large X-ray flux from sources outside the solar system. Then, in June 1962, R. Giacconi, H. Gursky, and F. Paolini of the American Science and Engineering Corporation (ASE) and B. Rossi of M.I.T. (Gi62) flew a set of large area Geiger counters in an Aerobee rocket. The increased area of these counters was designed to permit detection of X-rays scattered by the moon, but originating from the sun. The counters were sensitive in the wavelength region from 2 to 8 Å.

No lunar X-ray flux could be detected. However, a source of X-rays was discovered in a part of the sky not far from the center of the Galaxy and a diffuse background flux of X-ray counts was evident from all portions of the sky. Various arguments showed that this flux probably was not emitted in the outer layers of the earth's atmosphere, and therefore should be cosmic in origin. Later flights by the same group verified their first results.

At this point a team of researchers at the U.S. Naval Research Laboratory became interested. They had experience with solar X-ray observations and were able to construct an X-ray counter some 10 times more sensitive than that flown by Giacconi's group. Instead of the very wide field of view used by that group, the NRL team limited their field of view to 10 degrees of arc so that their map of the sky could show somewhat finer detail (Bo64a).

An extremely powerful source was located in the constellation Scorpius about 20 degrees of arc from the Galactic center. At first this source remained unidentified. Photographic plates showed no unusual objects in that part of the sky. The NRL group also discovered a second source, some eight times weaker than the Scorpio source. This was identified as the Crab Nebula, a remnant of a supernova explosion observed by Chinese astronomers in 1054 A.D. The NRL team, whose members were Bowyer, Byram, Chubb, and Friedman, believed that these two sources accounted for most of the emission observed by Giaconni's group.

Many explanations were advanced about the possible nature of these sources. Arguments were given in favor of emission by a new breed of highly dense stars whose cores consisted of neutrons. Other theories suggested that the emission might come from extremely hot interstellar gas clouds. No decision could be made on the basis of observations because none of the apparatus flown had fine enough angular resolving power. Nor did the NRL team expect to attain such instrumental resolving power for some years to come.

Then, early in 1964, Herbert Friedman at NRL heard that the moon would occult the Crab Nebula some seven weeks later. Here was a great opportunity to test whether at least one cosmic X-ray source was extended or stellar. For, as the edge of the moon passes over a well-defined point source, all the radiation is suddenly cut off. On the other hand, a diffuse source is slowly covered as the moon moves across the celestial sphere; accordingly, the radiation should be cut off gradually.

No other lunar occultation of either the Scorpio source or the Crab Nebula was expected for many years; so the NRL group went into frenzied preparations and seven weeks later a payload was ready. The flight had to be timed to within seconds since the Aerobee rocket to be used only gave 5 minutes of useful observing time at altitude. Two possible flight times were available; one at the beginning of the eclipse, the other at the end. Because of limited flight duration it was not possible to observe both the initial immersion and subsequent egress from behind the moon.

The first flight time was set for 22:42:30 Universal Time on July 7, 1964. That time would allow the group to observe immersion of the central 2 minutes of arc of the Nebula. Launch took place within half a second of the prescribed time. At altitude, an attitude control system oriented the geiger counters. At 160 seconds after launch, the control system locked on the Crab. By 200 seconds a noticeable decrease in flux could be seen and by 330 seconds the X-ray count was down to normal background level. The slow eclipse had shown that the Crab Nebula is an extended source. One could definitely state that at least one of the cosmic X-ray sources was diffuse. Others might be due to stars. But this one was not (Bo64b).

Roughly seven weeks after this NRL flight the ASE-MIT group was also ready to test angular sizes of X-ray sources. Their experiment was more general in that any source could be viewed. Basically it made use of a collimator that had been designed by the Japanese physicist, M. Oda (Od65). This device consisted of two wire grids separated by a distance D that was large compared to the open space between wires, which was slightly less than the wire diameter d.

The principle on which this collimator works is illustrated in Fig. 1.1. When the angular diameter of the source is small compared to d/D, alternating strong and weak signals are detected as the collimator aperture is swept across the source. If $\theta \gg d/D$ virtually no change in signal strength is detected as a function of orientation.

In their first flight the MIT-ASE group found the Scorpio source to have an angular diameter small compared to $1/2°$. Two months later a second flight confirmed that the source diameter was small, in fact, less than $1/8°$. A year and a half later this group found that the source must be far smaller yet, less than $20''$ in diameter. On this flight two collimators with different wire spacing were used. This meant that the transmission peaks for the two collimators coincided only

1:2

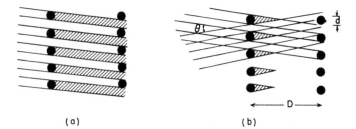

(a)   (b)

Fig. 1.1 Principle of operation of an X-ray astronomical wire-grid collimator. (*a*) For parallel light the front grid casts a sharp shadow on the rear grid. As the collimator is rotated, light is alternately transmitted and stopped depending on whether the shadow is cast on the wires of the rear grid or between them. (*b*) For light from a source whose angular dimension $\theta \gg d/D$, the shadow cast by the front grid is washed out. Rotation of the collimator assembly then does not give rise to a strong variation of the transmitted X-ray flux.

for normal incidence and, in this way, yielded an accurate position of the Scorpio source (Gu66). An optical identification was then obtained at the Tokyo Observatory and subsequently confirmed at Mount Palomar (Sa66a). It showed an intense ultraviolet object that flickered on a time scale less than one minute. These are characteristics associated with old novae near their minimum phase.

The brightness and color of neighbouring stars in the vicinity of Sco XR-1 showed that these stars were at a distance of a few hundred light years from the sun, and this gave us a good first estimate of the total energy output of the source. A search on old plates showed that the mean photographic brightness of the object had not changed much since 1896.

Interestingly, the 1969 discovery that the Crab Nebula contains a pulsar sent X-ray astronomers back to data previously collected. Some of these records showed the characteristic 33 millisecond pulsations, and showed that an appreciable fraction of the flux—10 to 15%—comes from a point source—now believed to be a neutron star formed in the supernova explosion (Fr69). Our views of the Crab as a predominantly diffuse X-ray source had to be revised.

Many other Galactic X-ray sources have by now been located and identified; and frequently they have a violet, stellar (pointlike) appearance similar to Sco XR-1. These objects sometimes suddenly increase in brightness by many magnitudes within hours. Others pulsate regularly, somewhat like the Crab Nebula pulsar. The range of X-ray energies at which the observations have by now been carried out, is quite wide too, and both visual and X-ray spectra are available for many sources.

By now, hundreds of extragalactic X-ray sources, many of them quasars or galaxies exhibiting violently active nuclei, have also been seen. The first of these was M87, a galaxy known to be a bright radio source (By67). It is a peculiar galaxy consisting of a spherical distribution of stars from which a jet of gas seems to be ejected. The jet is bluish in visible light, and probably gives off light by virtue of highly relativistic electrons spiraling about magnetic lines of force and emitting radiation by the synchrotron mechanism (see Chapter 6)—a mechanism by which highly accelerated energetic particles lose energy in a synchroton.

Theorists have now proposed a variety of explanations for cosmic X-ray sources and also for the continuum X-ray background that appears to permeate the universe. Many experiments are being planned to test these theories. X-ray, visual, infrared, and radio astronomers compare their results to see if a common explanation can be found. Progress is rapid and perhaps in a few years this field will no longer be quite as exciting. But by then another branch of astronomy will have opened up and the excitement will be renewed.

The fundamental nature of astrophysical discoveries being made—or remaining to be made—leaves little room for doubt that a large part of current theory will be drastically revised over the next decades. Much of what is known today must be regarded as tentative and all parts of the field have to be viewed with healthy skepticism.

We expect that much will still be learned using the methods that have been so successful in the past. However, there are parts of astrophysics—notably cosmology—in which the very way in which we think and our whole way of approaching scientific problems may be a hindrance. It is therefore useful to describe the starting point from which we always embark.

1:3 THE APPROPRIATE SET OF PHYSICAL LAWS

Nowadays *astrophysics* and *astronomy* have come to mean almost the same thing. In earlier days it was not clear at all that the study of stars had anything in common with physics. But physical explanations for the observations not only of stars, but of interstellar matter and of processes that take place on the scale of galaxies, have been so successful that we confidently assume all astronomical processes to be subject to physical reasoning.

Several points must, however, be kept in mind. First, the laws of physics that we apply to astrophysical processes are largely based on experiments that we can carry out with equipment in a very confined range of sizes. For example, we measure the speed of light over regions that maximally have dimensions of the order of 10^{14} cm, the size of the inner solar system. Our knowledge of large-scale dynamics also is based on detailed studies of the solar system. We then extrapolate the dynamical laws gained on such a small scale to processes that go on, on a cosmic scale of $\sim 10^{18}$ to 10^{28} cm; but we have no guarantee that this extrapolation is warranted.

It may well be true that these local laws do in fact hold over the entire range of cosmic mass and distance scales; but we only have to recall that the laws of quantum mechanics, which hold on a scale of 10^{-8} cm, are quite different from the laws we would have expected on the basis of classical measurements carried out with objects 1 cm in size.

A second point, similar in vein, is the question of the constancy of the "constants of nature." We do not know, in observing a distant galaxy from which light has traveled many *aeons*, whether the electrons and atomic nuclei carried the same charge in the past as they do now. If the charge was different, then perhaps the energy of the emitted light would be different too, and our interpretation of the observed spectra would have to be changed.

A third point concerns the uniqueness of the universe.

Normal questions of physics are answered by experiment. We alter one feature of our apparatus and note the effect on another. Cosmic questions, however, do not permit this kind of approach. The universe is unique. We cannot alter phenomena on a very large scale, at least not at our current level of technological development; and if we did, it is not clear that we would be able to discern real changes. There would simply be no available apparatus that would not in itself become affected by the experiment—no reference frame against which to detect the change. In short, we may not be asking questions that can be answered in physical terms, because the methods of physics, and more generally of all science, depend on our ability to conduct experiments; truly cosmic problems may just not permit such an approach.

This then is the current situation: We know a great deal about some as yet apparently unrelated astronomical events. We feel that an interconnection must exist, but we are not sure. Not knowing, we divide our knowledge into a number of different "areas": cosmology, galactic structure, stellar evolution, cosmic rays, and so on. We do this with misgivings, but the strategy is to seek a connection by solving individual small problems. All the time we expect to widen the areas of understanding, until some day contact is made between them and a firm bridge of knowledge is established between previously separated domains.

How far will this approach work for us? How soon will the philosophical difficulties connected with the uniqueness of the universe arise? We do not yet know; but we expect to face the problem when we get there.

In the meantime we can address ourselves to a number of concrete problems which, although still unsolved, nevertheless are expected to have solutions that can be reached using the laws of physics as we know them. Among these are questions concerning the origin and evolution of stars, of galaxies, of planetary systems. There are also questions concerning the origins of the various chemical elements;

been entirely *anaerobic*. As the atmosphere slowly became rich in oxygen, and life changed to take advantage of oxygen as a source of energy, some anaerobes remained and sought refuge where oxygen could not penetrate and where competition from the *aerobes*, or oxygen-metabolizing organisms, was not severe (Op61a, b; Sh66).

One of the interesting problems of astrophysics, then, is to try to understand the chemistry of the primitive earth. By noting the overall composition of solar surface material and the chemical composition of the atmospheres of other planets where conditions may have always remained stable, we may be able to understand what changes have taken place on earth. As already mentioned, the chemistry of comets may also help to produce an understanding of the initial conditions that existed on the young earth.

Is life, even as we know it, abundant in the universe? The probabilities involved in an answer to that question are still thoroughly speculative. If we estimated conservatively, we might suggest that life exists only on planets around stars having the same general characteristics as the sun; and even then, we might need to postulate the existence of a planet just at the distance where water neither freezes nor boils. Unfortunately, we do not know enough about the formation of planets to be able to estimate the likelihood for the occurrence of such a combination of star and planet. Once this kind of information is available, however, we will still face the problem of estimating the likelihood of life spontaneously catching on, on such a planet.

Increasingly sophisticated laboratory tests now are possible. They seek to establish the kind of lifelike molecule that could occur under conditions assumed to have held on the primitive earth. In time, these experiments might go far enough to synthesize lifelike primitive organisms. Once that is done, the probability of the formation of life could be estimated more realistically.

There are other possibilities too. Perhaps life is more in the nature of an infection that, having started in a given planetary system, is then able to spread from one system to another, either through natural causes, or through the intervention of intelligent beings who would like to see life propagate over wide regions.

If this second situation is true, then life would have had to be formed only once, and from then on no further spontaneous formation would have been necessary. The study of the spontaneous origin of life on a primitive earth could then lead to considerable error if extrapolated to deal with the probability for the occurrence of life elsewhere.

The assumption of intelligent life existing in other regions of the Galaxy or universe is of course fascinating. Can we contact such life? How would we communicate? If such an intelligent civilization, far more advanced than ours, exists, is it trying to communicate with us? Is there some unique best way of communicating, which a better understanding of physics and astrophysics will some day provide? Do we have to communicate by means of electromagnetic signals, or are there perhaps faster than light particles—tachyons—that we will discover later on and that would almost certainly be used by an intelligent civilization bent on saving time?

If other civilizations exist, should we visit them, or is that even possible outside our solar system? After all, the purpose of a visit is to see, talk, and touch; all of that could be done by improved communication techniques provided only that the distant civilization is able, and also willing, to communicate. There are relatively few things that cannot be settled that way, although without some actual exchange of mass, we probably could not decide whether a given civilization was made of matter or antimatter.

There are many fundamental questions of life on which astrophysics can throw new light and the interest of astrophysicists in biological problems is bound to increase in the next few years.

1:10 UNOBSERVED ASTRONOMICAL OBJECTS

In Appendix A we list a wide variety of astronomical objects; and we might think that we know enough to form a world picture with some reasonable assurance of—at least—being sensible.

To avoid this trap of complacency, we should complete our list of astronomical objects by citing those that have not yet been observed. We might think that this would be difficult; but it is not. To illustrate this, we first restrict ourselves to photographic observations of diffuse objects. An extension to other techniques will become obvious later.

We produce a plot comparing the absolute photographic magnitude and the logarithm of the diameter of different objects (Fig. 1.11). This was first conceived by Halton Arp (Ar65).

We note that all objects normally discovered on photographic plates have to lie on a strip between the two slanted lines on the Arp plot. Objects lying to the left or above this strip appear stellar. But since there are about 10^{11} stars that can appear on photographs taken of the Galaxy, abnormal or highly compact objects with a stellar appearance could never be separated from bona fide stars, without an inordinate amount of labor.

To detect something unusual about objects falling into this upper region on the chart, some other peculiar earmark must be found. For example, quasars lie above the strip. They were first discovered by virtue of their radio emission and only later identified as distant objects by means of individually obtained spectra.

To the right and below the strip, the surface brightness of a diffuse object is so low that the foreground glow emitted by the night sky outshines the object,

1:10

Fig. 1.11 Diagram showing the diameter-brightness strip onto which extended objects observed through the atmosphere tend to fall. Objects in the upper left-hand corner are very compact, and are not readily distinguished from ordinary stars. In the lower right-hand corner, atmospheric night sky emission interferes with observations. The upper and lower crosses, respectively, represent the quasars 3C273 and 3C48. Their diameters are quite uncertain. The upper and lower filled circles represent the Fornax and Draco galaxies—minor members of the Local Group of galaxies (Based on a drawing by Arp Ar65.)

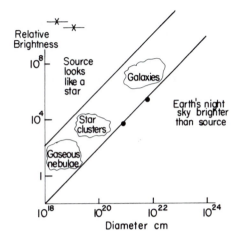

making it undetectable. Exceptions to this are the Local Group minor galaxies Fornax and Draco in which individual stars can be counted. If these objects were more distant, individual stars would not be detected and the objects would not be visible.

We note that the strip of observable objects covers only a small portion of the available area on the plot. This means that we have not yet had the opportunity to see many different varieties of objects that probably occur in nature. For it would be too much of a coincidence to expect all classes of objects in the universe to fit neatly into a pattern defined by our own instrumental capabilities—and to fall onto the strip of observables in the Arp diagram.

By taking instruments above the atmosphere, in rockets and satellites, we are able to get above much of the night sky emission. The demarcation line on the right can therefore be moved downward and further to the right. It also is possible to obtain a higher image resolving power above the atmosphere, since the main limitation to the resolving power on the ground is due to atmospheric scintillation. This feature would move the line on the left of the strip upward and to the left. The combined effect is to widen the strip altogether and to allow us to identify a larger variety of objects than is accessible from the ground. This is one reason for launching an observatory into an orbit high above the atmosphere, and we may expect new discoveries in astronomy when such observations become possible.

Of course, not all objects emit visible radiation, and so we cannot expect to find out all there is to know in astronomy simply by making visual observations. Table 1.5 gives 10 different observable entities, chosen to be roughly representative of the various kinds of astronomical objects we know. We note that only three

Table 1.5 Comparison of Objects Seen in the Visual and Radio parts of the Electromagnetic Spectrum.[a]

	VISIBLE	RADIO	CORRELATION	NEW
1. COMETS	√	X	0	
2. PLANETS	√	√	I	
3. STARS	√	X	0	
4. IONIZED REGIONS	√	√	I	
5. INTERSTELLAR MASERS	X	√	0	□
6. GLOBULAR CLUSTERS	√	X	0	
7. GALAXIES	√	√	I	
8. PULSARS	X	√	0	□
9. QUASARS	X	√	0	□
10. BACKGROUND	X	√	0	□
			3/10	4/10

[a]We find that many of the objects are more readily apparent observed with one technique than with the other, even though some emission may be present in both parts of the spectrum.

objects could have been discovered equally well through visual or through radio observations. Four of the objects emit only radio waves, or are most easily distinguished as unusual objects through their radio emission. *Quasars,* for example, do emit in the visual domain; but they look like ordinary stellar objects until we painstakingly observe their spectra. As stated above, any stellar object is easily lost among the 10^{11} ordinary stars in our galaxy; and it is therefore only through their radio emission that *quasistellar objects* were discovered as early as they were.

We note from the highly subjective Fig. 1.12 that most of our knowledge about the universe still comes from visual observations mainly because more observations have been carried out in the visible than in other parts of the spectrum.

A few more decades of radio observations may however reverse this state of affairs and, certainly, continued infrared, X-ray, and γ-ray observations will also follow that trend.

Table 1.5 shows vividly that observations made with new techniques help to discover new phenomena. They do not merely tell us something additional about phenomena we have already observed. We can therefore expect that an entirely new set of astronomical objects will become uncovered as we perfect observations throughout

(a) The entire electromagnetic spectrum, going all the way from the lowest frequencies in the hundred kilocycle radio band up to the highest energy gamma rays.

(b) The entire modulation frequency spectrum, going up to megacycle frequencies: Pulsars would never have been discovered had it not been for electronic innovations that permitted observations of intensity changes over millisecond

Fig. 1.12 Subjective drawing indicating the amount of information that has been gained through observations in the various portions of the electromagnetic spectrum. We could equally well have plotted "total time spent making observations throughout the history of astronomy", and come up with a similar plot. The ordinate has a quite arbitrary scale, probably more nearly logarithmic than linear. The peak *V* represents observations in the visual part of the spectrum.

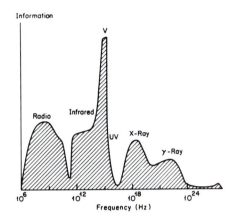

time intervals. Using photographic plates, where exposure times of the order of an hour are representative, one could not expect to discover objects with periodic brightness undulations much shorter than an hour. At the other extreme, analysis of old photographic plates cannot yield discernible variations for phenomena whose period is much longer than a few decades.

(c) The entire spatial frequency domain. As already indicated, many observing techniques are good for stellar or at least highly compact objects, but are not capable of detecting a uniform background. Other techniques permit background measurements, but not the observation of faint compact objects. This is particularly true of infrared observations. Until we have observed the entire range of possible angular sizes, from the lowest limits of angular resolution, all the way up to a uniform background, there must remain potentially interesting unobserved astronomical objects.

(d) The entire set of communication channels: electromagnetic and gravitational radiation, cosmic rays, neutrinos, and, if they exist, any others. These channels again can be expected to exhibit the existence of new phenomena, in a universe rich far beyond our most adventurous speculation.

(e) All of the above should lead to progress. However, there are many possible astronomical objects that simply are beyond observation through any currently planned telescopes. For example, if 10 percent of the mass of our galaxy consisted of snowballs (fist-sized chunks of frozen water freely floating through interstellar space) we would never know it. The amount of light scattered from these objects would be too low to make them detectable. They would not be able to penetrate the solar system to become visible as meteors because the sun would evaporate them away long before they ever approached the earth's orbit. Snowballs would therefore have to remain undetected until spaceships traveled beyond the solar system. After that they could become a major nuisance, since a spaceship moving nearly at the speed of light could be completely destroyed on colliding with one of these miniature icebergs. Similarly black dwarf stars, or unaccompanied planets, would have been difficult to detect thus far.

When we look at the unfinished work implied by the points (a) to (e), we must be prepared to accept the thought that astronomers may have seen no more than a few percent by number of all observable important phenomena characterizing the universe. From this point of view, it almost seems premature to construct sophisticated cosmological theories and cosmic models.

On the other hand, these theories and models often suggest novel observations that produce new results. We should therefore think of astrophysical theory not so much as a structure that summarizes all we know about the universe. Rather, it is a continually changing pattern of thought that permits us to grope in the right directions.

2

The Cosmic
Distance Scale

2:1 SIZE OF THE SOLAR SYSTEM

A first requirement for the establishment of a cosmic distance scale is the correct measurement of distances within the solar system. The basic step in this procedure is the measurement of the distance to Venus. The most precise way of obtaining this distance is through the use of radar techniques. Another method is described in problems 2–2 and 2–3 at the end of this chapter.

A radar pulse is sent out in the direction of Venus, and the time between its transmission and reception is measured. Since time measurements can be made with great accuracy, the distance to Venus and the dimensions of its orbit can be established within a kilometer.

Once the distance to Venus is known at closest approach a, and most distant separation b, and these measurements are repeated over a number of years, the diameter and eccentricity of both the earth's and Venus's orbit can be computed. The mean distance from the earth to the sun is then directly available as the mean value of $(a + b)/2$ (Fig. 2.1). This distance is called the *astronomical unit*. A check on the earth-Venus distance is obtained from trajectories of space vehicles sent to Venus.

2:2 TRIGONOMETRIC PARALLAX

When observations are made from opposite extremes in the earth's orbit about the sun, a nearby star will appear displaced relative to more distant stars in the same part of the sky. The *parallax*, p, is defined as half the apparent angular displacement measured in this way. The distance d to the star is then

$$d = \frac{\text{astronomical unit}}{\tan p} \qquad (2-1)$$

or

$$d = 1.5 \times 10^{13} (\tan p)^{-1} \text{ cm}$$

A star whose parallax is one second of arc is at a distance of 3×10^{18} cm, since $\tan 1'' = 5 \times 10^{-6}$. This distance forms a convenient astrophysical unit length, and is called the parsec, pc. 1 pc $= 3 \times 10^{18}$ cm.

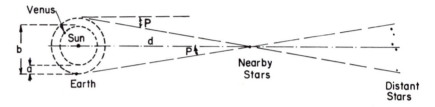

Fig. 2.1   Measurement of the astronomical unit and trigonometric parallax.

The trigonometric parallax can be reliably determined out to distances of about 50 pc, where the parallax is 0.02 sec of arc.

2:3 SPECTROSCOPIC PARALLAX

Once the distance to nearby stars has been determined, we can correlate absolute brightness with spectral type. Bright stars of recognizable spectral type then become distance indicators across large distances where only the brightest stars are individually recognized and trigonometric parallax cannot be used.

2:4 THE MOVING CLUSTER METHOD

As explained below this method has thus far been used only for the Hyades star cluster, whose members are close together and move through space as a group. Three measurements need to be made:

(a) The *radial velocity v*, of the cluster stars along the line of sight is measured from spectral displacement—Dopler shifts.

2:4

(b) and (c) The *proper motion* (apparent motion perpendicular to the line of sight) of individual stars in the cluster is measured by intercomparing their positions on photographic plates taken many decades apart. This motion has the form of an apparent contraction of the cluster angular diameter θ. Its rate of decrease, $\dot{\theta}$, can be computed and gives the fractional rate of increase of the cluster's distance from the sun (Fig. 2.2).

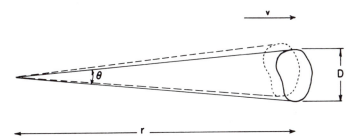

Fig. 2.2 The Hyades cluster at distance r recedes from us with a radial velocity component v. If the cluster diameter D remains constant, θ must decrease. Equation 2-2 shows how r can be determined from a measurement of v, θ, and $\dot{\theta}$, the rate of change of θ with time.

The quantities θ, $\dot{\theta}$, and v can be related to the cluster's distance r, since $\theta = D/r$ and the time derivative is $\dot{\theta} = -D\dot{r}/r^2$ for a constant cluster diameter D. Since $\dot{r} = v$,

$$\dot{\theta} = -\frac{v\theta}{r} \qquad r = -\frac{v\theta}{\dot{\theta}} \qquad (2\text{-}2)$$

The distance r to the Hyades cluster is therefore available in terms of three directly measured quantities. This simple scheme is restricted to use with the Hyades stars because no other cluster is near enough to yield accurate proper motions. Clearly use of the method depends on the dynamical stability of the Hyades group. If the stars were not gravitationally bound but, say, were expanding away from a common origin, the method would yield a false measure of the distance, because the time derivative $\dot{D}$ would then contribute to $\dot{\theta}$.

2:5 METHOD OF WILSON AND BAPPU

The H and K lines of singly ionized calcium appear in the spectra of most late-type stars. The shape of these lines is rather complex. First, there is a broad absorption line. In its center we see a thinner central emission feature and super-posed on this is an even finer central dark absorption band. The broad absorption labeled H_1 and K_1, respectively, is due to cool gas in the outer atmosphere of a star. The lines are broad because the absorption of calcium is so strong that it extinguishes radiation even in the wings of the spectral lines where the absorption coefficient is relatively low. The insert in Figure 2.3 sketches the line shape.

The emission lines H_2 and K_2 are due to re-emission higher in the atmosphere. The emission from these lines can be absorbed a second time—and produces fine lines respectively labeled H_3 and K_3. This absoprtion line is produced by cool gas lying even higher in the star's atmosphere than the atoms responsible for the emission lines H_2 and K_2.

Bappu and Wilson (Wi57) noticed that there exists a correlation of the width of the H_2 and K_2 components with the stellar brightness:

$$\frac{dM_v}{d\log(W_2)} = \text{constant} \qquad (2\text{--}3)$$

W_2 is the line width measured in cycles per second. There is no clear explanation for this linear relationship between the visual magnitude and the logarithm of the line width; it simply is empirical.

Using the brightness of the sun as one data point and the brightness of four Hyades stars as the other, the slope of the line can be determined. Hence, the measured brightness of a star and its H_2 and K_2 line widths establish its distance. When properly used, the uncertainty of the distance measured by this method is of the order of 10%. Some care must, however, be taken since the Wilson–Bappu relationship does not appear to hold for stars of all spectral types. Figure 2.3 shows a calibration of the line width W_2 in terms of stars whose trigonometric parallax is known. The relation (2–3) is illustrated by this plot. This is not a fundamental method for calibrating distances. It does, however, serve as a useful cross check, and gives us additional confidence in our other methods of judging the distances of stars.

2:6 SUPERPOSITION OF MAIN SEQUENCES

This method is based on the assumption that main sequence stars have identical properties in all galactic clusters. This means that the slope of the main sequence is the same for all such clusters and, moreover, it requires that main sequence stars of a given spectral type or color have the same absolute brightness in all clusters (see Fig. 1.4). On this assumption we can compare the brightness of the main sequence of the Hyades and any other galactic cluster. The vertical shift necessary to bring the two main sequences into superposition gives the relative distances of the clusters.

2:6

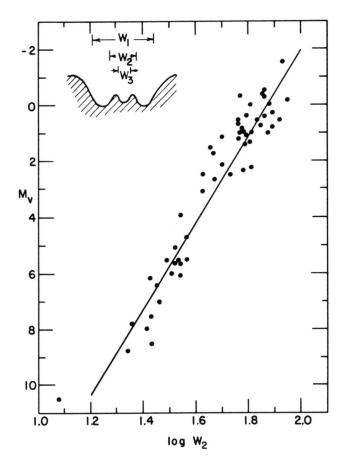

Fig. 2.3 Plot of the logarithm of the ionized calcium CaII emission line width W_2 against absolute magnitudes derived from trigonometric parallaxes. The insert explains the labeling of line widths W_1, W_2, and W_3. (After Wilson and Bappu, Wi57.)

PROBLEM 2-1. If the shift in apparent magnitudes is $\Delta m = m_{GC} - m_{Hya}$ show that the relative distances are

$$\Delta m = 5 \log \frac{r_{GC}}{r_{Hya}} + A' \qquad (2\text{-}4)$$

where A' is a correction for the difference in interstellar reddening of the galactic cluster, GC, and the Hyades. The derivation is analogous to the work leading to equation A-2.

The factor A' can be determined through use of stellar line spectra as explained in section A:6.

To obtain the distance to the globular clusters we can proceed on one of three different assumptions:

(a) The Hertzsprung–Russell diagram of the globular cluster has a segment that runs essentially parallel to the galactic cluster main sequence. We can assume that this segment coincides with the main sequence of the Hyades cluster. The distance of the globular cluster can then be calculated in terms of equation 2–4.

(b) Alternatively we can assume that the segment coincides with the main sequence defined by a group of dwarf stars in the sun's immediate neighborhood. The distance to these dwarf stars is determined by trigonometric parallax.

(c) Finally we can assume that the mean absolute magnitude for short-period variables (RR Lyrae variables) is the same in globular clusters and in the solar neighborhood (see section 2:7, below).

None of these three choices is safe in itself. However, when applied to the globular cluster M3, the third entry in the Messier catalog, all three methods give distance values in fair agreement with each other. This verifies that the main sequences of different groupings of stars coincide reasonably well, and can be used as distance indicators.

2:7 BRIGHTNESS OF RR LYRAE VARIABLES

We find that the apparent brightness of all RR Lyrae variables in a given globular cluster is the same regardless of the variable's period. Since these stars are intrinsically bright, and since their short period makes them stand out, they serve as ideal distance indicators. We assume that the absolute brightness of these stars is the same not only within a given cluster, but also elsewhere. The relative distance of two clusters can then be determined by the inverse square law corrected for interstellar extinction (equation 2–4).

2:8 BRIGHTNESS OF CEPHEID VARIABLES

At the turn of the century, cepheid variables in the Magellanic clouds were found to have periods that are a function of brightness. The Magellanic clouds are dwarf companions to the Galaxy. They are small galaxies in their own right and are compact enough so that all their stars can be taken to be at essentially the same distance from the sun. By comparing the brightness of cepheids in the

Magellanic clouds to that in globular clusters, one was able to obtain relative distances to these objects.

However, there was a pitfall in this comparison. The cepheids in the Clouds are population I stars, normally found in the disk of a galaxy. Globular clusters, on the other hand, belong to the halo component that is more or less spherically distributed about the center of a galaxy. Generally, *population I* consists of bright early stars, and dwarf late stars that lie on the main sequence. *Population II* is characterized by dwarf late stars and late-type giants.

In 1952 Baade analyzed the brightness of cepheids in the Andromeda Nebula, comparing population I with population II regions. He found that population I cepheids were about 1.5 magnitudes brighter than population II cepheids. The *distance modulus* of M31 had previously been derived by comparison of these brighter cepheids with type II cepheids in clusters within our own Galaxy. The distance to M31 had therefore been erroneously underestimated by a factor of two. Baade's measurements showed that this distance, in fact, the distance to all galaxies had to be doubled.

2:9 THE BRIGHTNESS OF NOVAE AND HII REGIONS

Novae have an absolute brightness that is related to the decay rate of the brightness after an outburst. The great intrinsic brightness of a nova makes it a very useful distance indicator for nearby galaxies.

The diameters of bright HII (ionized hydrogen) regions also form good yardsticks by which to judge the distances of such galaxies.

2:10 DISTANCE–RED-SHIFT RELATION

The distances of various galaxies can be compared by making a comparison of bright objects within the galaxies. Suitable candidates are O stars, novae, cepheid variables, and HII regions. These individual objects can be detected out to distances about as far as the nearer Virgo cluster galaxies. By comparing the distance estimated from the apparent brightness of such stars and the sizes of HII regions, it is possible to show that the spectral red shift of light from these galaxies is linearly related to distance: $\Delta\lambda/\lambda \propto r$.

We can also compare the brightness of individual galaxies to estimate relative distances. Here we must be careful to compare galaxies of the same general type. To minimize errors due to statistical variation in brightness, we sometimes compare not the brightest, but rather, say, the 10th brightest galaxy in two different clusters. By this device we hope to avoid selecting galaxies that are unusually

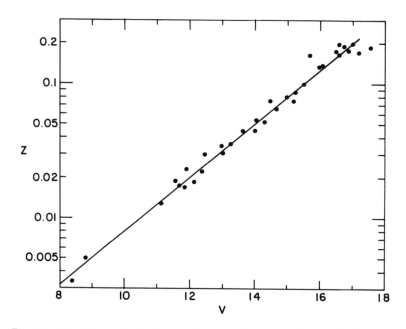

Fig. 2.4 Red-shift–magnitude diagram for the brightest members of 38 clusters of galaxies. After J. V. Peach (Pe69) based on data obtained by A. Sandage. z is the red shift $\Delta\lambda/\lambda$. V is the visual magnitude. (Reprinted with permission from *Nature*, Vol. 223, p. 1141, copyright © 1969 Macmillan Magazines Ltd.)

bright. Figure 2.4, however, is a plot of the brightest cluster galaxies as a function of red shift.

The data show a linear distance–red-shift relation. It is not clear how far this linearity persists, but for many cosmological purposes we use the red shift as a reliable indicator of a galaxy's distance. This procedure may, however, not be appropriate for quasars.

We should still note that distance measurements are not easy and that errors cannot always be avoided. In 1958, Sandage (Sa58) discovered that previous observers had mistaken ionized hydrogen regions for bright stars. This had led them to underestimate the distance to galaxies by a factor of ~ 3 beyond the error previously unearthed by Baade. Within a space of five years the dimensions of the universe therefore had to be revised upward by a total factor of ~ 6. It is not unlikely that, from time to time, other corrections may lead to further revisions of the cosmic distance scale. However, Fig. 2.5 shows that we can frequently check astronomical distances by several different methods, and eventually we should be able to derive a reliable distance scale. At present, a red-shift velocity of 75 km sec^{-1} is estimated to indicate that a galaxy is at a distance of 1 Mpc.

Dimensions   Methods

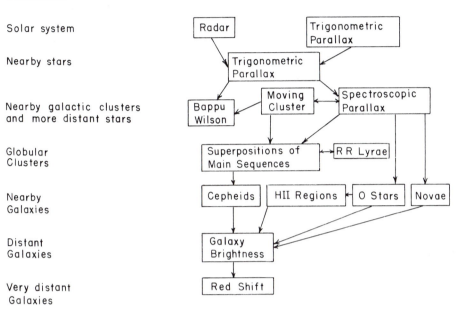

Fig. 2.5 Flowchart of distance indicators.

The velocity-distance proportionality constant—the Hubble constant H—has a value $H = 75$ km sec^{-1} Mpc^{-1}.

Once we know the distance to the various galaxies, we can estimate typical intergalactic distances and typical number densities of galaxies for cosmological purposes. The variation of number density with distance, or more accurately, with spectral red shift, can in principle be used to determine the geometric properties of the universe. By such means we may hope to determine whether the universe is open or closed, and whether it is finite or infinitely large. We will return to such questions in Chapter 10, but a simple argument based on Euclidean geometry is given in the next section.

2:11 SEELIGER'S THEOREM AND NUMBER COUNTS IN COSMOLOGY

If a set of emitting objects is homogeneously distributed in space, then N_m/N_{m-1}, the ratio of objects whose apparent magnitude is less than m to those whose apparent magnitude is less than $m - 1$ is 3.98. This is called Seeliger's theorem. Let us see how this result is obtained.

2:11

Let $m - 1$ be the apparent magnitude of a given star at distance r_1 (see Fig 2.6). Then the distance r_0 at which its apparent magnitude would be m is $r_0 = (2.512)^{1/2}r_1$. At that distance its apparent brightness is reduced by $(r_0/r_1)^2 = 2.512$. All this follows directly from our definition of the magnitude scale in section A:5.

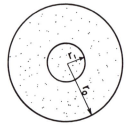

Fig. 2.6

If stars are uniformly distributed in space and have a fixed brightness, they will appear brighter than apparent magnitude m out to a distance r_0, but brighter than $m - 1$ only out to a distance r_1. The ratio of the number of stars brighter than a certain magnitude, N_m/N_{m-1}, is proportional to the volume occupied.

$$\frac{N_m}{N_{m-1}} = \frac{r_0^3}{r_1^3} = (2.512)^{3/2} = 3.98 \tag{2-5}$$

Since this is true for stars of any given brightness, it will also be true for any homogeneous distribution of stars, regardless of their luminosities. Equation 2–5 states that the flux obtained from a source is proportional to r^{-2}, while the number of sources observed down to a given flux limit is proportional to r^3. Hence the number of sources observed brighter than a certain strength (flux density) $S(v)$ at a given spectral frequency v is

$$N \propto S(v)^{-3/2} \quad \text{since} \quad N \propto r^3 \text{ and } S(v) \propto r^{-2} \tag{2-6}$$

This proportionality, which already was of interest in classical stellar astronomy, has become even more important in modern cosmology, where it is usually found in a somewhat different form. If we take the logarithm of both sides of equation 2–6 we find

$$\log N \propto -\tfrac{3}{2}\log S(v) \tag{2-7}$$

In radioastronomy a comparison of $\log N$ and $\log S$, often called the $\log N$–$\log S$ plot, means this: If the logarithm of the number of sources brighter than a given level is plotted against the logarithm of the brightness, at the spectral frequency at which the instrument operates, then the slope of the plot should be constant,

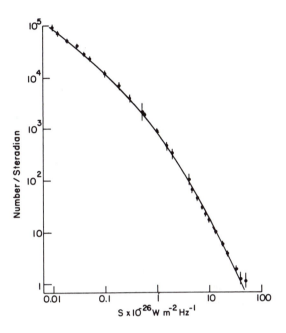

Fig. 2.7 Plot of log N against log S, where N is the number of
radio sources per unit solid angle with flux density, measured in
watts per square meter per Hertz, greater than S. The measurements were made by Pooley and Ryle (Po68) and have been
discussed in the context of other studies by Ryle (Ry68).

with a value of $-3/2$, provided (a) the sources are homogeneously distributed in
space, (b) space is Euclidean, and (c) we compensate for any cosmic red shift in
apparent brightness. This latter requirement comes about because observations
are made at one given frequency v. If, say, the source intrinsically is very bright
at high frequencies, then red shift to lower frequencies would make it look de-
ceptively bright. A correction for the spectral shape for radio sources is therefore
required. A correction for the red shift is required, in any case, because a cosmically
red shifted source already appears weaker just from the time dilation effect,
that is, from the increased apparent spacing between the emission times of radio
frequency photons (Ke68). How these corrections are to be made is discussed in
section 10:6.

If these red shift effects are ignored, say at small cosmic distances, or else corrected
for on a statistical basis, then the homogeneity and Euclidean geometry required
both by an inflationary and by a steady state universe (10:2, 10:5) predict a
log N–log S slope of -1.5. Actual observations indicate a slope not precisely of this

value, but near enough to keep such theories within contention, at least by this criterion.

If true deviations from this slope eventually become established, we will have evidence to show either that space is non-Euclidean, or that the radio sources are inhomogeneously distributed. Either of these consequences would imply an evolving cosmos (see Chapter 10). Figure 2.7 shows the current state of observational results. The slope at high flux densities is -1.85; at low densities, it is -0.8. These results are partly dependent on the frequency at which observations are carried out. At 1400 MHz, a slope of -1.5 appears to hold for flux densities $S \geq 0.5 \times 10^{-26}\,\mathrm{Wm^{-2}\,Hz^{-1}}$, except that for the very few strong sources a steeper slope exists. This may just be a local inhomogeneity characteristic of the Galaxy's neighborhood (Br72).

PROBLEMS DEALING WITH THE SIZE OF ASTRONOMICAL OBJECTS

The methods here described are not those normally used by astronomers. However, they allow us to obtain insight into the dimensions of planetary and stellar systems without recourse to the more sophisticated methods covered in this chapter. The first six steps were already known to Newton (Ne00).

2–2. The distance, R, to Venus can be accurately obtained by triangulation when Venus is at its point of closest approach. Two observers separated by 10^4 km along a line perpendicular to the direction of Venus find the position of Venus on the star background to differ by 49″ of arc. Calculate the distance of Venus at closest approach.

2–3. At this distance, the angular diameter of Venus is 64″, while at greatest separation its angular diameter is 10″. Assuming both the earth's and Venus's orbit to be circular, compute the two orbital radii. Assume the orbits to be concentric.

2–4. The mean angular diameter of Saturn at smallest separation is about 1.24 times as great as at the largest separation. (These mean angular diameters have to be averaged over many orbital revolutions, because Saturn's orbit about the sun is appreciably eccentric). Calculate the semimajor axis, a, of Saturn's orbit about the sun.

2–5. Both the sun and the moon subtend an angular diameter of $1/2°$ at the earth. The lunar disk at full moon is only about 2×10^{-6} as bright as the sun's disk. Knowing that the moon is much nearer the earth than the sun, compute the reflection coefficient K of the lunar surface, assuming that the light is reflected isotropically, into 2π sterad. Show that this reflection coefficient is appreciably lower than that of terrestrial surface material (which is estimated to have a mean reflection coefficient of order 0.3). Actually the moon scatters light mainly in the backward direction, so the result obtained here gives an artificially elevated value for K.

2–6. Assume that Saturn has an angular diameter of $\sim 17''$ at the sun. Let its distance from both the earth and the sun be considered to be 9.5 AU. If the light received from Saturn is 0.86×10^{-11} that received from the sun, compute the reflection coefficient of Saturn's surface. Note that Saturn is known to shine primarily by reflection, since its moons cast a shadow on the surface when they pass between the planet and the sun.

2–7. Saturn appears to emit 0.86×10^{-11} as much light as the sun. How far would the sun have to be removed from the earth to appear with a luminosity identical to that of Saturn, that is, to appear like a first magnitude star?

2–8. Assuming the sun to be a typical star, we conclude that the nearest stars are of the order of 5.2×10^{18} cm distant. We further assume that this is the characteristic distance between stars in the disk of the Andromeda spiral galaxy M31. We note that M31 appears to be a system viewed more or less perpendicular to the disk containing the spiral arms. Other spiral galaxies viewed in profile indicate that the thickness of the disk is about $0.003L$, where L is the diameter of the galaxy. In terms of the distance D of M31, show that the flux received would be

$$\sim \left(\frac{0.003\, SL^3}{D^2} \right) \times \frac{\pi}{4} \frac{10^{-18}}{5.2}$$

where S is the flux we would expect to receive from the sun if it were 5.2×10^{18} cm from the earth.

2–9. If the brighter regions of M31 subtend an angular diameter of $3°$ at the earth, and if the galaxy is a 5th magnitude object, calculate the distance of the galaxy. Show that its diameter is ~ 6 kpc. (Note that the actual diameter of M31 is about half an order of magnitude larger.)

2–10. Find the distance of the smallest resolved galaxies, on the assumption that all spiral galaxies are of the size of M31. The smallest resolved objects for currently available telescopes are of the order of $2''$ of arc in diameter.

2–11. We note that the light from distant galaxies appears red shifted in proportion to their distance as judged by their angular diameters. If the smallest resolved objects are red shifted by 30% of the spectral frequency, that is, $\Delta v/v \sim 0.3$, calculate at what distance the linear distance–red-shift law would require galaxies to attain the speed of light. This distance is sometimes called the effective radius of the universe.

2–12.   Olbers' Paradox: Let there be n stars per unit volume throughout the universe.

(a) What is the number of stars seen at distances r to $r + dr$ within a solid angle Ω?

(b) How much light from these stars is incident on unit area at the observer's position, assuming each star to be as bright as the sun?

(c) Integrating out to $r = \infty$ how much light is incident on unit detector area at the observer?

This problem will be discussed at length in Chapter 10.

ANSWERS TO SELECTED PROBLEMS

2–2. $R = 4.2 \times 10^7$ km.

2–3. $R_e = 1.5 \times 10^8$ km, $R_v = 1.1 \times 10^8$ km.

2–4. $(a + 1)/(a - 1) = 1.24$. Hence $a = 9.5$ AU.

2–5.   If $L_\odot$ is the solar luminosity, r is the radius of the moon, and R is the distance of the moon—and the earth—from the sun, then $S = (\pi r^2/4\pi R^2) L_\odot$ is the radiation accepted by the moon. This light is spread into 2π solid angle so that, at the distance D of the earth, the flux per unit area is $(K \cdot S)/2\pi D^2$, which has to be compared with $L_\odot/4\pi R^2$ coming directly from the sun.

$$\therefore \frac{Kr^2}{2D^2} = 2 \times 10^{-6} \quad \text{and} \quad K \sim 0.2.$$

2–6.   Saturn's diameter is $2r \sim 7.8 \times 10^{-4}$ AU

$$\therefore \frac{\dfrac{\pi r^2 L_\odot}{4\pi(9.5)^2} \cdot \dfrac{K}{2\pi(9.5)^2}}{\dfrac{L_\odot}{4\pi(1)^2}} = 0.86 \times 10^{-11}$$

Hence $K \sim 0.90$.

2–7.   The distance at which the sun would appear to be a first magnitude star is $r = 5.2 \times 10^{18}$ cm

2–8.   If $L_\odot$ is the sun's luminosity, and D is the distance, the flux from the galaxy is

$$\frac{\text{(volume of galaxy) (number density) } L_\odot}{4\pi D^2}$$

$$\sim \frac{\pi/4(L^2)\,(.003L)\cdot\left(\dfrac{1}{5.2\times10^{18}}\right)^3\cdot L_\odot}{4\pi D^2}$$

$$= \frac{\pi/4(L^2)(.003L)}{D^2}\cdot\frac{S}{(5.2\times10^{18})}.$$

2–9.   Comparing the magnitude of M31 to a first magnitude star, and taking $\theta = 3/57 = L/D$ we see from Problem 2–8 that $D \sim 0.1$ Mpc, $L \sim 6$ kpc. The actual distance to M31 is given in Table 1.4.

2–10. Distance $= 2 \times 10^{27}$ cm

2–11. Distance $= 7 \times 10^{27}$ cm.

2–12. (a) $\Omega\, nr^2\, dr.$

(b) $\Omega nr^2\, dr\dfrac{L_\odot}{4\pi r^2} = \dfrac{\Omega n}{4\pi}\,L_\odot\, dr.$

(c) The integral (b) would diverge if distant stars were not eclipsed by nearer stars. When eclipses are taken into account, the flux at the observer is finite and is equal to the flux at the surface of the sun.

3

Dynamics and Masses of Astronomical Bodies

The motion of astronomical bodies was first correctly analyzed by Newton, in the second half of the 17th century. He saw that a variety of apparently unrelated observations all had common features and should form part of a single theory of gravitational interaction. To formulate the theory, he had to invent mathematical techniques that described the observations and showed their interrelationship. His struggles with the mathematical problems are recorded in his book *Principia Mathematica* (Ne00).

The intervening three centuries since Newton's discoveries have allowed his mathematical formulation to be streamlined, so that it can now be presented in brief form; but the underlying astrophysics remains unchanged.

The aim of this chapter will be to show how astronomical observations lead to the conclusions reached by Newton (1642–1727). We will then show the importance of Newtonian dynamics in determining the masses of all astronomical objects. It is interesting that a correct evaluation of these masses was not obtained until more than a century after Newton's work. We will discuss the gravitational interaction of matter with antimatter and finally mention some of the limitations of Newton's work.

3:1 UNIVERSAL GRAVITATIONAL ATTRACTION

A number of astronomical observations and experimental results were known to Newton when he first tried to understand the dynamics of bodies. Many of the experimental results dealing with the motion of falling bodies had been found by Galileo (1564–1642) (Ga00). The astronomical observations, which treated the motions of planets, had been gathered over many years by Tycho Brahe (1546–1601). Johannes Kepler (1571–1630) had then analyzed these data and summarized them in terms of three empirical laws. Newton postulated that the work of Kepler and of Galileo was related. We will not retrace his reasoning here, but rather will outline the evidence with some of the advantages of three centuries of hindsight.

We know from experiments with sets of identical springs and sets of identical masses that a single mass accelerated through the release of, say, two stretched springs, mounted side by side, is accelerated at twice the rate experienced by the same mass when impelled by one spring only (Fig. 3.1). Of course, the springs have to be stretched to the same length. Measurements of this kind lead us to assert that an acceleration always is associated with a force, and is directly proportional to it.

$$F \propto \ddot{r} \qquad (3\text{--}1)$$

This is a brief way of stating Newton's first and second laws.

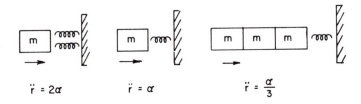

$$\ddot{r} = 2\alpha \qquad \ddot{r} = \alpha \qquad \ddot{r} = \frac{\alpha}{3}$$

Fig. 3.1 Definition of inertial mass.

In a related experiment three interconnected masses accelerated by releasing a single spring would be accelerated at only one third the rate experienced by one mass acted on by the same spring. This second type of measurement shows that the acceleration produced is inversely proportional to the mass of the impelled body:

$$\ddot{r} \propto \frac{1}{m} \qquad (3\text{--}2)$$

Combining relations (3–1) and (3–2) we obtain the proportionality relation

$$\ddot{r} \propto F/m \qquad (3\text{--}3)$$

The acceleration produced on a body is proportional to the force acting on it and inversely proportional to its mass. When the impelling force is zero, the body remains unaccelerated; its velocity stays constant and may be zero.

We can go one step further and say that the force is equal to the mass times the acceleration. This defines the unit of force in terms of the other two quantities:

$$F = m\ddot{r} \qquad (3\text{--}4)$$

With these ideas in mind we can draw a significant conclusion from Galileo's experiments that showed that two bodies placed at identical points near the earth fall (are accelerated) at equal rates, even though their masses may be quite different. This independence of mass, interpreted in terms of the proportionality relation (3–3), shows that the accelerating force is proportional to the mass of the falling body. We will need to make use of this point in the arguments that follow.

We can now consider Galileo's work that concerned itself with projectiles. A projectile fired at a given angle falls to earth at a greater distance if its initial velocity is large. We can ask what would happen if the initial velocity was increased indefinitely. The projectile would keep falling to earth at larger and larger distances and, neglecting atmospheric effects, it could presumably circle the earth if given enough initial velocity. If the projectile still retained its original velocity on returning to its initial position, the circling motion would continue. The projectile would orbit the earth much like the moon.

Newton already knew a number of facts about the motion of the moon and he performed calculations to show that the moon behaves in every way just like a projectile placed into an orbit around the earth.

In addition to the experiments of Galileo, Newton also was aware of the observational results summarized by Kepler. There were three principal observations that are summarized in *Kepler's laws*:

(i) The orbits along which planets move about the sun are ellipses.

(ii) The area swept out by the radius vector joining sun and planet is the same in equal time intervals. This means that the angular velocity about the sun is small when the planet is distant, and is large when the planet is close to the sun. The moon shows the same behavior as it orbits about the earth.

(iii) The period a planet requires to describe a complete elliptical orbit about the sun is related to the length of the semimajor axis of the ellipse: The square of the period P is proportional to the cube of the semimajor axis a (Figure 3.2). This law also describes the motion of satellites (moons) about their parent planets.

Newton therefore had three pieces of information:

(i) He knew that projectiles fall because they are gravitationally attracted toward the earth.

(ii) He knew that there are certain similarities between the motions of projectiles and the motion of the moon about the earth.

(iii) He knew that the motion of the moon is similar to that of Jupiter's and Saturn's satellites and that those motions are governed by the same laws that described the motions of planets about the sun.

These ideas led him to attempt an explanation of all these phenomena in terms of accelerations produced by gravitational attraction.

3:1

He already suspected that in the interaction of two bodies equal but oppositely directed forces act on both bodies (Newton's third law). The fact that a planet is attracted by the sun, but can also attract a satellite by gravitational means, indicates that there is no real difference between the *attracting* and the *falling* body. If the force acting on one of Galileo's falling bodies was proportional to its own mass—as stated above—then the force must also be proportional to the mass of the earth. The gravitational force of attraction between two bodies must then be proportional to the product of their masses m_a and m_b:

$$F \propto m_a m_b \qquad (3\text{--}5)$$

Since the acceleration of distant planets is smaller than that of planets lying close to the sun, this force must also be inversely dependent on the distance between the bodies. Similarly, the distance and orbital period of the moon show it to have an acceleration toward the earth, much smaller than that of objects at the earth's surface. Quantitatively it indicates that $F \propto r^{-2}$. In any case, F must drop faster than $F \propto r^{-1}$, because otherwise the effects of distant stars would influence a planet's orbital motion more strongly than the sun.* As seen from Problems 2–2 and 2–7, Newton knew the distances to other stars, and knew that there were a large number of stars surrounding the sun. As a reasonable choice of distance dependence he tried an inverse square relationship. We will show in the next section that a force law of the form

$$F \propto m_a m_b r^{-2} \qquad (3\text{--}6)$$

allows us to derive Kepler's laws of motions. To turn this proportionality relation into the form of an equation, we write

$$F = \frac{m_a m_b}{r^2} G \qquad (3\text{--}7)$$

where the proportionality constant G is the *gravitational constant*. Sometimes called the *Newtonian gravitational constant*, G is a universal constant of nature whose value must be experimentally determined, as discussed in section 3:6 below.

3:2 ELLIPSES AND CONIC SECTIONS

Since the planets are known to describe elliptical orbits about the sun, it is convenient to start the discussion of their motions by defining a set of parameters in terms of which the elliptical paths can be described.

---

* If *differential* acceleration of the sun and earth is considered, the effect of the distant stars is not so striking. However, with a r^{-1} force, the sun would still rob the earth of its moon.

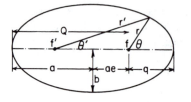

Fig. 3.2   The terminology of conic sections:

f, f' The two foci of the ellipse.
a Semimajor axis.
b Semiminor axis.
e Eccentricity: the focus of the ellipse is displaced from the center by a distance ae.
q Distance of the pericenter; we can see that $q = a(1 - e)$.
Q Distance of the apocenter; $Q = a(1 + e)$.
θ True anomaly, the angle between the radius vector r, and the major axis as seen from the focus f.
r The radius vector from the focus f.
r' The radius vector from the focus f'.

We can define an ellipse as the set of all points, the sum of whose distances from the two foci is constant:

$$r + r' = \text{constant}$$

Since the ellipse is symmetrical about the two foci, we can see from Fig. 3.2 that this constant must have the value $2a$:

$$r + r' = 2a \tag{3-8}$$

Hence $b = \sqrt{a^2 - a^2 e^2}$ by the theorem of Pythagoras.
The figure also shows that

$$r \sin \theta = r' \sin \theta' \tag{3-9}$$

and that

$$r \cos \theta - r' \cos \theta' = -2ae \tag{3-10}$$

These two equations, respectively, represent the laws of sines and of cosines for plane triangles. Squaring (3–9) and (3–10) and adding these expressions gives

$$r^2 + 4aer \cos \theta + 4a^2 e^2 = r'^2 \tag{3-11}$$

Substituting from (3–8) then gives

$$r = \frac{a(1 - e^2)}{1 + e \cos \theta} \tag{3-12}$$

an equation that we will need below. Actually equation (3–12) is more general than shown here; it describes any conic section. When the eccentricity is $0 < e < 1$, the figure described is an ellipse. If $e = 0$, we retrieve the expression for a circle of radius a. If $e = 1$, a becomes infinite, the product $a(1 - e^2)$ can remain finite, and the equation describes a parabola. If $e > 1$, equation (3–12) describes a hyperbola.

3:2

3:3 CENTRAL FORCE

From Newton's laws and Kepler's second law, a simple but important deduction can be drawn at once. Kepler's law can be stated in vector form as

$$\mathbf{r} \wedge \dot{\mathbf{r}} = 2 A \mathbf{n} \qquad (3\text{--}13)$$

Here $\mathbf{r}$ is the radius vector from the sun to the planet, $\dot{\mathbf{r}}$ is the planet's velocity with respect to the sun. A is a constant and the symbol $\wedge$ stands for the *vector*, or *cross*, product. The product of r, $\dot{r}$, and the sine of the angle between these two vectors is twice the area swept out by the radius vector in unit time. $\mathbf{n}$ is a unit vector whose direction is normal to the plane in which the planet moves.

We see that the time derivative of equation 3–13 is

$$\frac{d}{dt}(\mathbf{r} \wedge \dot{\mathbf{r}}) = \mathbf{r} \wedge \ddot{\mathbf{r}} = 0 \qquad (3\text{--}14)$$

since both A and $\mathbf{n}$ are constant. Multiplying this expression by the mass of the planet m, and using equation 3–14 we find that

$$\mathbf{F} \wedge \mathbf{r} = 0 \qquad (3\text{--}15)$$

Since neither the force nor the radius vector vanishes in elliptical motion, it is clear that the force and radius vectors must be colinear. Whatever the nature of the force acting on the planet may be, it is clear that this force acts along the radius vector: Such a force is called a *central force*. A planet is pulled toward the sun at all times; and the components of a binary star always are mutually attracted.

3:4 TWO-BODY PROBLEM WITH ATTRACTIVE FORCE

Define a coordinate system whose origin lies at the *center of mass* of bodies a and b. The positions and masses of the bodies are related (Fig. 3.3) by

$$\mathbf{r}_a = -\frac{m_b}{m_a}\mathbf{r}_b \qquad (3\text{--}16)$$

Fig. 3.3 Center of mass (CM) of two bodies a and b.

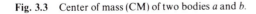
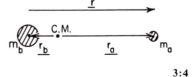

We know that in planetary motion we are dealing with a central attractive force, and that the force should decrease more rapidly than the inverse first power of the distance between attracting bodies. We postulate that the attractive force is an inverse square law force. If this postulate is correct, we should obtain the correct laws of motion as given by Kepler's laws. We will show below that this indeed is true.

For a central force decreasing as the square of the distance between two attracting bodies, we write the force F_a on body a due to body b as

$$\mathbf{F}_a = m_a\ddot{\mathbf{r}}_a = -\frac{m_a m_b G}{r^3}\mathbf{r} \tag{3-17}$$

where m_a and m_b are the masses of the two bodies. From the definition of $\mathbf{r}$ and the center of mass, we have

$$\mathbf{r} = \mathbf{r}_a - \mathbf{r}_b = \left(1 + \frac{m_a}{m_b}\right)\mathbf{r}_a \tag{3-18}$$

Combining (3–17) and (3–18) we have

$$\ddot{\mathbf{r}}_a = -\frac{GM}{r^3}\mathbf{r}_a \qquad M \equiv m_a + m_b \tag{3-19}$$

where M is the total mass of the two bodies. Subtracting a similar expression for r_b we obtain

$$\ddot{\mathbf{r}} = -\frac{GM}{r^3}\mathbf{r} \tag{3-20}$$

We see that the acceleration of each body relative to the other is influenced only by the total mass of the system and the separation of the bodies. If equation 3–20 is multiplied by a mass term μ, we obtain a force term that is a function only of r, M, μ, and the gravitational constant:

$$\mathbf{F}(\mu,M,r) = -\frac{GM\mu}{r^3}\mathbf{r} = \frac{-Gm_a m_b \mathbf{r}}{r^3} \tag{3-21}$$

If this force is to be equal to the force acting between the two masses, we must satisfy equation 3–7 which means that

$$\mu = \frac{m_a m_b}{m_a + m_b} \tag{3-22}$$

μ is called the *reduced mass*.

3:4

The equation of motion (3–20) taken together with equation 3–21, shows that the orbit of each mass about the other is equivalent to the orbit of a mass μ about a mass M that is fixed—or moves in unaccelerated motion. There is a great advantage to this reformulation. Newton's laws of motion only hold when referred to certain reference frames, for example, a stationary coordinate system, or one in uniform unaccelerated motion (see also sections 3:8 and 5:1). It was for this reason that the motion of each mass a and b was initially referred to the center of mass. This procedure, however, required us to keep separate accounts of the time evolution of $\mathbf{r}_a$ and $\mathbf{r}_b$. The separation $\mathbf{r}$ was only determined subsequently by adding r_a and r_b. This two-step procedure is avoided if equations 3–20 and 3–21 are used, since $\mathbf{r}$ can then be determined directly.

3:5 KEPLER'S LAWS

Consider a polar coordinate system with unit vectors $\boldsymbol{\varepsilon}_r$ and $\boldsymbol{\varepsilon}_\theta$ (Fig. 3.4) A particle is placed at position $\mathbf{r} = r\boldsymbol{\varepsilon}_r$. Since the rate of change of the unit vectors can be expressed as

$$\dot{\boldsymbol{\varepsilon}}_r = \dot{\theta}\boldsymbol{\varepsilon}_\theta \tag{3–23}$$

defining the rate of change (rotation) of the radial direction, and

$$\dot{\boldsymbol{\varepsilon}}_\theta = -\dot{\theta}\boldsymbol{\varepsilon}_r \tag{3–24}$$

giving the rate of change for the tangential direction, we can write the first and second time derivatives of r as

$$\dot{\mathbf{r}} = \dot{r}\boldsymbol{\varepsilon}_r + r\dot{\theta}\boldsymbol{\varepsilon}_\theta \tag{3–25}$$

$$\ddot{\mathbf{r}} = (\ddot{r} - r\dot{\theta}^2)\boldsymbol{\varepsilon}_r + (2\dot{r}\dot{\theta} + r\ddot{\theta})\boldsymbol{\varepsilon}_\theta \tag{3–26}$$

From equations 3–20 and 3–26 we obtain two separate equations, respectively, for the components along and perpendicular to the radius vector

$$\ddot{r} = -\frac{GM}{r^2} + r\dot{\theta}^2 \tag{3–27}$$

and

$$2\dot{r}\dot{\theta} + r\ddot{\theta} = 0 = 2\dot{r}r\dot{\theta} + r^2\ddot{\theta} \tag{3–28}$$

Equation 3–28 integrates to

$$r^2\dot{\theta} = h \tag{3–29}$$

where h is a constant that is twice the area swept out by the radius vector per unit time. This relationship has a superficial resemblance to the law of conservation

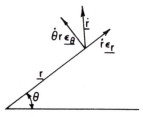

Fig. 3.4 Vector components of the velocity $\dot{r}$.

of angular momentum (per unit mass). But that law would involve the distances r_a and r_b, instead of r. Equation 3–29 does state Kepler's second law, however, and that is satisfactory.

Combining equations 3–27 and 3–29 we have

$$\ddot{r} - \frac{h^2}{r^3} + \frac{MG}{r^2} = 0 \qquad (3\text{–}30)$$

PROBLEM 3-1. Choose a substitution of variables

$$y = r^{-1}, \qquad \dot{\theta}\frac{d}{d\theta} = \frac{d}{dt} \qquad (3\text{–}31)$$

to rewrite equation 3–30 in the form

$$\frac{d^2 y}{d\theta^2} + y = \frac{MG}{h^2} \qquad (3\text{–}32)$$

Show that this has the solution

$$y = B\cos(\theta - \theta_0) + \frac{MG}{h^2} \qquad (3\text{–}33)$$

This leads to

$$r = \frac{1}{B\cos(\theta - \theta_0) + \dfrac{MG}{h^2}} \qquad (3\text{–}34)$$

This is the expression for a conic section (see equation 3–12). It therefore represents a generalization of Kepler's first law. Gravitationally attracted bodies move along conic sections which, in the case of planets, are ellipses. We see this if we set

$$a(1 - e^2) = \frac{h^2}{MG} \qquad (3\text{–}35)$$

3:5

and

$$e = \frac{Bh^2}{MG} \tag{3-36}$$

The minimum value of r occurs for $\theta = \theta_0$.

Let r_m be a relative maximum or minimum distance between the two bodies. Then the entire velocity at separation r_m must be transverse to the radius vector and by equation 3-29

$$\frac{(r_m\dot\theta)^2}{2} = \frac{h^2}{2r_m{}^2} \tag{3-37}$$

is the kinetic energy per unit mass. The *total energy* per unit mass is the sum of *kinetic* and *potential energy* per unit mass.

$$\mathscr{E} = \frac{h^2}{2r_m{}^2} - \frac{MG}{r_m} \tag{3-38}$$

Solving for r_m we have

$$r_m = \left(\frac{MG}{h^2} \pm \sqrt{\frac{M^2G^2}{h^4} + \frac{2\mathscr{E}}{h^2}}\right)^{-1} \tag{3-39}$$

Hence the quantity B in equation 3-34 has the value

$$B = +\sqrt{\frac{M^2G^2}{h^4} + \frac{2\mathscr{E}}{h^2}} \tag{3-40}$$

the sign being determined by the condition that the minimum r value occur at $\theta - \theta_0 = 0$.

Equations 3-12 and 3-35 show that the minimum value of r is

$$q = \frac{h^2}{MG(1 + e)} \tag{3-41}$$

Substituting this into equation 3-38 we then have an expression for the energy in terms of the semimajor axis a,

$$\mathscr{E} = (e^2 - 1)\frac{M^2G^2}{2h^2} = -\frac{MG}{2a} \tag{3-42}$$

where we have made use of expression 3-35. Since the total energy per unit mass is the sum of kinetic and potential energy, also per unit mass, we see that

$$\mathscr{E} = \frac{v^2}{2} - \frac{MG}{r} \tag{3-43}$$

and from (3–42) we obtain the orbital speed as

$$v^2 = MG\left(\frac{2}{r} - \frac{1}{a}\right)$$

(3–44)

We now can make a number of useful statements:

(i) If S is the area swept out by the radius vector

$$\frac{dS}{dt} = \frac{1}{2}h, \quad S - S_0 = \frac{1}{2}ht$$

(3–45)

For an ellipse, the total area is

$$S - S_0 = \pi ab = \pi a^2 (1 - e^2)^{1/2}$$

so that from equation 3–35 the period of the orbit is

(3–46)

$$P = \frac{2}{h}\pi a^2 (1 - e^2)^{1/2} = \frac{2\pi a^{3/2}}{\sqrt{MG}}$$

(3–47)

Equation 3–47 is a statement of Kepler's third law.

(ii) If the eccentricity is $e = 1$, the total energy is zero, by equation 3–42 and the motion is parabolic. Astronomical observations have shown that some comets approaching the sun from very large distances have orbits that are practically parabolic, although they may be slightly elliptical or slightly hyperbolic. At best, these comets therefore are only loosely bound to the sun. A small gravitational perturbation by a passing star evidently can make the total energy of some of these comets slightly positive, and they escape from the solar system to wander about in interstellar space.

We should still note that one of the big advances brought about by Newton's theory was the realization that both cometary and planetary orbits could be understood in terms of one and the same theory of gravitation. Prior to that no such connection was known.

(iii) If the eccentricity $e > 1$, the total energy is positive, and the motion of the two masses is unbound. After one near approach the bodies recede from each other indefinitely.

(iv) If the eccentricity is zero, the motion is circular with some radius R and the energy obtained from equation 3–42 is $- MG/2R$ per unit mass. Equation 3–44 then states that v^2 equals MG/R or that the gravitational attractive force MG/R^2 must equal v^2/R, which sometimes is called the *centrifugal force*,—a fictitious force that is supposed to "keep the orbiting mass at constant radius R despite the attractive pull of M."

Thus far we have shown that the motion of one mass about another describes a conic section. In addition, we can show that the orbit of each mass about the common center of mass is a conic section also. Equation 3–19 can be rewritten as

$$\ddot{\mathbf{r}}_a = - \frac{GM}{(1 + m_a/m_b)^3} \frac{\mathbf{r}_a}{r_a^3} \qquad (3\text{–}19a)$$

This is of the same form as equation 3–20 and we can, therefore, readily obtain equations similar in form to expressions 3–27, 3–28, 3–29, and finally 3–34. This argument also holds true if we were to talk about the vector $\mathbf{r}_b$ instead of $\mathbf{r}_a$. Hence both masses m_a and m_b are orbiting about the center of mass along paths that describe conic sections.

Let us still see how we can determine the masses of the components of a spectroscopic binary. This is the most important means we have for determining stellar masses. For such binaries we can measure the radial velocities of both stars throughout their orbits (Fig 3.5).

It is relatively easy to determine the period of such a binary by looking at the repeating shifts of the superposed spectral lines. Equation 3–47 then gives the ratio (a^3/M) of the semimajor orbital axis cubed and the sum of the masses. If the binary, in addition, is an eclipsing binary, so that the line of sight is known to lie close to the orbital plane, then the semimajor axes of the orbits of the two components about the common center of mass can be found; and this gives the individual component masses if use is made of component equations derived from (3–19a).

For a few visual binaries that are close enough to permit accurate observations, the motion of the individual components relative to distant background stars again permits computation of the individual semimajor axes, provided the trigonometric parallax also is known. The orbital period then allows us to compute the individual masses through Kepler's third law and equation 3–19a.

We note that expressions such as 3–35, 3–36, 3–44, and 3–47, which connect measurable orbital characteristics to M and G always depend on the product MG and, hence, permit a determination neither of the system's total mass, nor

Fig. 3.5 Binary star orbits and the individual semimajor axes for the two stars orbiting their common center of mass.

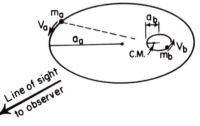

of the gravitational constant. For a long time this presented a serious difficulty. However:

PROBLEM 3–2. Show how a rough measure of G can be obtained from falling mass experiments when the known size of the earth and some estimate of its density are used to determine the earth's mass. In section 3:6 we show how G was eventually measured by Cavendish. Note that for an accurate determination of the earth's density, G has to be accurately known.

3:6 DETERMINATION OF THE GRAVITATIONAL CONSTANT

Henry Cavendish (1731–1810) an English chemist, discovered a means of measuring the gravitational constant G, late in the 18 century, more than a 100 years after Newton had first shown how the motion of the planets depends on the mass the sun. Until Cavendish performed his experiment, the absolute masses of celestial objects could not be accurately determined; there were only relative values of, say, planetary masses as judged by orbits of their moons.

In the Cavendish experiment a torsion balance is used. Typically such a device may consist of a fine quartz fiber to which a rod bearing masses m_1 and m_2 is attached, as shown in Figure 3.6a. Each mass is at some distance L from the fiber. We can calibrate the balance by noting the torsion that can be induced in the fiber when a small measurable torque is applied to the system. We can apply this torque by having a spring with known force constant exert a horizontal force at the position of m_2.

If masses M_1 and M_2, respectively, are placed at a small horizontal distance from masses m_1 and m_2, we may observe a twist of the fiber, in the sense shown in Figure 3.6b. We can determine the distances r_1 and r_2, respectively, between m_1 and M_1, and m_2 and M_2 to find the horizontal forces acting on the ends of the bar and, hence, establish the torque acting on the quartz fiber. That torque N is

$$N = L\left(\frac{m_1 M_1 G}{r_1^2} + \frac{m_2 M_2 G}{r_2^2}\right) \tag{3–48}$$

From the measured deflection of the masses we can determine the value of N, in terms of the calibration previously obtained on the twisted quartz wire. Since we can measure N, L, r_1, r_2, and the masses of all the different bodies, we now have the values of all quantities in (3–48) except for G, which can then be directly determined from the equation. That value is $G = 6.7 \times 10^{-8}$ dyn cm^2 g^{-2}.

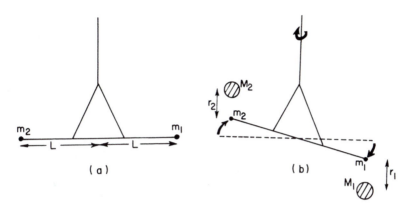

Fig. 3.6 The Cavendish experiment to determine the gravitational constant, *G*.

Once the value of *G* is known, the mass of the sun is readily determined using Kepler's third law, equation 3–47. That law actually involves the total mass *M* of the sun and planet, but by performing the calculations for a number of different planets we can verify that the mass of the sun is very nearly equal to *M* and that the mass of the planets $\sim 0.0013 \, M_\odot$ can be neglected to a good approximation. The approximate mass of the earth can be derived in a similar way, making use of the known orbit of the moon.

3:7 THE CONCEPT OF MASS

If we examine what we have said about the measurement of masses, we find that there really are two quite distinct ways of determining the mass of a body. (i) We can measure its acceleration in response to a measured force (equation 3–4), or (ii) we can measure the force acting on the body when a given mass is placed at a specified distance—this is what we do when we weigh the body with a spring balance.

The first of these is a dynamic measurement, the second can be static. The mass of a body measured in the first way is called its *inertial mass*, while the mass measured by means of the second method is called the *gravitational mass*.

Suppose now that we take a steel ball whose gravitational mass is m_1. We take a wooden ball that is slightly too heavy, and slowly file away excess material until its gravitational mass is also equal to m_1. If the two balls now are placed on a pan balance, they should leave the balance arm in a horizontal position, because the earth attracts both masses equally.

3:7

The question now is whether the inertial mass of these two bodies is always the same. Will the wooden ball be accelerated at the same rate as the steel ball in response to a given force? Until we try the experiment we cannot be sure.

This question intrigued the Hungarian baron Roland von Eötvös around the turn of the century. He suspended two weights of different composition but identical weight on a torsion balance, with the horizontal bar along the East-West direction (Fig. 3.7a). As the earth rotated, two forces acted on each mass, (i) a gravitational attraction that is equal, since the weights of the masses are equal and (ii) a *centrifugal force* due to the earth's rotation. If the centrifugal force on mass A was greater or less than on mass B, this would indicate that their inertial masses differed. The bar would rotate until the torsional force in the suspending wire compensated for the inequality in the centrifugal forces.* Eötvös never observed such a rotation of the bar and concluded that the inertial and gravitational masses of the bodies were identical to within one part in about 10^8.

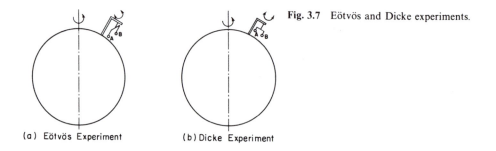

Fig. 3.7 Eötvös and Dicke experiments.

(a) Eötvös Experiment   (b) Dicke Experiment

This experiment now has been refined by R. H. Dicke and his co-workers. They suspended their weights with the bar in a North-South direction. As the earth turns about its axis, mass A might be attracted more or less strongly toward the sun than mass B. One should then observe a diurnal effect with the balance arm first swinging in one direction, then in the other. No such effect was observed and the conclusion is that the gravitational and inertial masses are identical to within about one part in 10^{11}. The advantage of this kind of experiment is that it is dynamic, not static, and has a definite expected periodicity for which we can look (Ro64b).

We may now ask whether the gravitational mass of matter is the same as that of antimatter. If there existed galaxies composed of antimatter, would they attract or repel a galaxy consisting of matter? L. I. Schiff (Sc58a) has given a tentative

---

* This experiment works best, roughly halfway between the equator and poles, say, Budapest or Princeton.

answer to such questions. He points out that many atomic nuclei emit virtual positron electron pairs. This means that part of the time a fraction of the total nuclear energy is to be found in the form of an electron and a matched positron. Such a pair of particles is continually formed and reassimilated and is never actually emitted.

Schiff computes that if the positrons had a negative gravitational mass, then the ratio of inertial to gravitational mass would be affected for a number of substances for which the electron-positron virtual pair formation is a major effect. The ratio of the two kinds of masses would then be different from unity by about 1, 2, and 4 parts in 10^7 for aluminum, copper, and platinum, respectively. Experiments with such substances have been performed, and inequalities of this size are ruled out by the Eötvös and Dicke experiments. It follows that matter and antimatter ought to have gravitational masses of the same sign and that galaxies interacting gravitationally with antigalaxies cannot be distinguished on dynamical grounds.

Note that we have really only shown that the inertial and gravitational mass have the same sign for positrons as for electrons. But we actually know from dynamical experiments in magnetic fields that the inertial mass of the positron equals that of the electron, so that our previous conclusion should follow at once: Matter and antimatter both have positive mass.

There is one difficulty with this argument. As Schiff himself recognized, we are not absolutely certain that virtual electron-positron pairs behave exactly like real pairs. Could it be that the gravitational mass of a real positron differs from that of a virtual positron? Unfortunately, we will not be absolutely sure until we make a direct measurement of the positron's motion in a gravitational field.

3:8 INERTIAL FRAMES OF REFERENCE—THE EQUIVALENCE PRINCIPLE

We noted earlier that Newton's laws of motion hold only when the motion is described in coordinates that refer to a frame of reference that is either fixed, or is moving at constant velocity with respect to the distant galaxies. Such a reference frame is known as an *inertial coordinate system.*

Several perplexing questions arise when we try to understand the significance of these frames of reference. They can be described by some simple experiments.

(1) Suppose that a man were blindfolded and placed on a merry-go-round. He could determine quite accurately whether he was being spun around because he would be able to feel the centrifugal force acting on him when the merry-go-round was moving. If he adjusted the mechanism until he felt no centrifugal force

he would find, on taking off his blindfold, that the merry-go-round was stationary with respect to the distant galaxies.

(2) A blindfolded man placed in a rocket in interstellar space could adjust his controls until he felt no forces on himself. On taking a closer look, he would find that he had adjusted the engine to give zero thrust. He might find that he was moving at constant velocity with respect to the distant galaxies. Alternately, however, he might find that he had strayed into the vicinity of a star and was freely falling toward it! Einstein first postulated that freely falling, nonrotating coordinate frames are fully equivalent to Newton's inertial frames that move at constant velocity with respect to the distant galaxies. All laws of physics have precisely the same form in both types of frames. This *equivalence principle* will prove to be very useful in sections 3:9 and 5:13.

When we talk about a motion with respect to the distant galaxies, we really mean a motion with respect to the mean velocity of all galaxies at very large distances. Galaxies are receding in all directions but, as far as we can tell, there always exists a local frame of reference in which the motions of distant galaxies appear symmetrical, no matter which direction we look.

This suggests that perhaps the local frame of zero acceleration is determined by the distribution of the galaxies in the universe. Just how this determination comes about is a basic unanswered question of the theory of gravitation. The thought that the overall distribution of mass within the universe should determine a local inertial framework is due to E. Mach and is sometimes called *Mach's principle*. There are many related questions all of which are involved in the same basic thought: "Is the inertial mass of a body determined by the distribution of matter in the universe? Is the gravitational constant determined by the distribution of the distant galaxies? As a result, would the value of the gravitational constant change with time, as the galaxies recede from each other? Are the atomic constants of physics related to the large-scale structure of the universe?" As yet there are no answers to these fundamental questions.

3:9 GRAVITATIONAL RED-SHIFT AND TIME DILATION

Einstein's *principle of relativity* (5:1) states that mass and energy are related in such a way that any stationary mass m has an equivalent energy mc^2 associated with it (Ei07). Einstein showed that the separate laws of conservation of mass and of energy merged into a more general *conservation of mass-energy*. This predicts a gravitational red shift for radiation emitted at the surface of a star. Consider two particles, an electron and a positron, at rest at a very great distance from a star. The rest mass of each particle is m_0. If the particles fall in toward the star's

3:9

surface each one acquires a total mass-energy

$$E \equiv m_r c^2 = \left(m_0 c^2 + \frac{m_0 MG}{r} \right) = m_0 c^2 \left(1 + \frac{MG}{rc^2} \right) \qquad (3\text{–}49)$$

at distance r from the star. The second term in the parantheses represents the conversion of potential into kinetic energy. Now let the two particles be deflected without loss of energy or momentum, so that they collide head-on, and annihilate. Two photons, each with frequency

$$\nu_r = \frac{m_r c^2}{h} \qquad (3\text{–}50)$$

will be formed in this process. These photons are now permitted to escape from r, but through reflections from stationary mirrors—which produce no frequency shifts—we can make them collide again at a large distance from the star.

In this collision they can form an electron-positron pair. If energy is conserved, then the frequency ν_0 at a large distance from the star must again be

$$\nu_0 = \frac{m_0 c^2}{h} \qquad (3\text{–}51)$$

Otherwise there would be either too much or too little energy to recreate a positron-electron pair at rest. Hence

$$\nu_0 = \frac{\nu_r}{1 + MG/rc^2} \qquad (3\text{–}52)$$

The frequency at a large distance from the star is less than the emitted frequency. For a star like the sun $M \sim 2 \times 10^{33}$ g, the radius $R \sim 7 \times 10^{10}$ cm, and $MG/Rc^2 \sim 2 \times 10^{-6}$ at the sun's surface. For a neutron star whose mass would be about the same, but whose radius would be 10^5 times less, the fractional frequency shift

$$\frac{\Delta \nu}{\nu_r} = \frac{\nu_0 - \nu_r}{\nu_r} = -\frac{MG}{rc^2} \left[1 + \frac{MG}{rc^2} \right]^{-1} \qquad (3\text{–}53)$$

becomes comparable to unity, and the frequency shift $\Delta \nu$ becomes comparable to the frequency itself.

We will see in the next section that the frequency of electromagnetic waves can give a very accurate measure of time and can therefore be used as a clock. Such a clock, placed in a strong gravitational field, would therefore run more slowly. Quite generally, the rate at which a clock runs is determined by the potential $\mathbb{V}(r)$ at the position r, of the clock. The period P of this clock measured by an observer outside the potential field, that is by an observer located at $\mathbb{V} = 0$, appears to be

$$P_0 = P_r\left(1 - \frac{\mathbb{V}}{c^2}\right) \tag{3–54}$$

In section 3:11 we outline an experiment that has been designed to measure this *time dilation*, which leads to a delay in the arrival, at the earth, of pulsar pulses that have passed close to the sun.

3:10 MEASURES OF TIME

In describing the orbital motions of planets about the sun, we have obtained expressions for position as a function of time. But how is this parameter, time, actually measured?

There are a number of ways (see D. H. Sadler Sa68a) of measuring time and it is interesting to see how these methods interrelate—some rather basic questions of physics are involved. Let us first describe some imaginary clocks that could be constructed. They may not be practical but they should work in principle.

First Clock

Take an amount of tritium H^3 that beta decays into the helium isotope He^3. If the tritium is kept at a temperature around $10°K$, the helium will diffuse out as it is formed. We weigh the tritium. When the mass has dropped to half its initial value, we say that a time of one unit, NT, has elapsed. We could set up a clock that struck each time the remaining mass was reduced by a factor of two.

Second Clock

Take a quantity of the cesium isotope Cs^{133}. It has a transition between two hyperfine levels of the ground state. We measure the frequency of the radiation (radio wave) emitted in this transition. The period of this electromagnetic wave can serve as a unit of time, AT.

Third Clock

We set up a telescope that is always pointed at the local zenith. Each time a given distant galaxy reappears exactly in the center of the telescope's field of view, we say that one unit of time, UT, has elapsed.

Fourth Clock

We note the plane described by Jupiter as it rotates about the sun. We mark the instant that the earth crosses this plane in its motion about the sun. The earth crosses the plane twice per orbit; once it does so from North to South, and the next time from South to North. If we define the interval between successive N to S crossings as one unit of time, ET, we have still another means of measuring time.

We call these measures—NT, AT, UT and ET—*nuclear time, atomic time, universal time,* and *ephemeris time.* As we have chosen to define them here, the units of time, respectively, correspond to ~ 12 years, $(9,192,631,770)^{-1}$ seconds, 1 day, and 1 year.

The basic differences between the clocks are these: The first clock uses beta decay, a weak interaction, as its basic mechanism. The second clock uses an electromagnetic process to measure time. The third clock uses the earth's rotation to measure time; this is an inertial process. Finally, the fourth clock makes use of a gravitational force to measure time.

Since each of these clocks depends on quite different physical processes, we worry that they might not measure the same "kinds" of time. There is no reason, for example, why atomic time and ephemeris time as defined above should describe intervals having a constant ratio. At present the ratio of these time units is about $3 \times 10^{17}:1$. Will this ratio be the same some 10^9 y from now? Or does the strength of the gravitational field, or of the weak interactions, change after years in such a way that one of these clocks becomes accelerated relative to the other?

We can test this question experimentally and such tests have, in fact, been proposed. Their results would be of great importance to cosmology. For, in order to understand the nuclear history of the universe and the formation of chemical elements, we have to know how nuclear reaction rates in stars may have been affected by the overall evolution of the universe over the past aeons. This will appear more clearly after the synthesis of nuclei in stars has been discussed in Chapter 8.

The important point to realize is that we have enumerated four quite different ways of defining time.* The last two are related if the general theory of relativity holds true; and their interrelationship becomes a test for theories of gravitation.

In practice, the intercomparison of these clocks is difficult. Planetary perturbations make the earth's orbit about the sun irregular. Earthquakes and other

---

* Actually, there are five ways. Nuclear β and α decay rates are based on weak and strong nuclear forces, respectively. We will show in Chapter 10 that these two kinds of clock apparently have run at identical rates over the past few æons.

disturbances affect the rotation rate of the earth. An incomplete understanding of these effects makes it hard to compare the UT and ET rates with time measured by atomic clocks. Eventually, however, such practical difficulties should be overcome and an intercomparison of time scales will be possible.

3:11 USES OF PULSAR TIME

Many pulsars emit signals with a periodicity that appears to vary by less than one part in 10^8 over an interval of a year. These signals can therefore be used to define a time scale. The mechanism of the clock is not yet understood, but is thought to involve the rotational period of neutron stars. In any case, its regularity allows us to put it to scientific use. For many purposes we do not need to know how a clock works as long as its accuracy can be verified.

Counselman and Shapiro have listed a number of interesting gravitational effects that can be studied using pulsars (Co68).

(a) The orbit of the earth could be determined more precisely than it is known now. The pulsar emission would act like a "one-way" radar. Counting pulse rates from different pulsars would allow us to measure the instantaneous velocity of the earth, relative to some arbitrarily defined inertial frame. Integrating these velocities over a series of time intervals would yield the earth's position as a function of time, that is, the shape of the orbit and its orientation.

Such measurements can also yield data on the positions and masses of the outermost planets. Their motions affect the position of the solar system's *barycenter* and hence also the orbit of the earth. The periodicity of these effects is determined by the orbital periods of the planets and we should find corresponding periodic variations in the pulse counts (As71).

(b) A pulsar located near the ecliptic plane will appear close to the sun once each year. When the light pulses pass very close to the limb of the sun, they should be slowed down because all clocks are slowed by the presence of a strong gravitational field and because the speed of light measured locally at the sun would still appear to be c. The arrival time of pulses at the earth will therefore be delayed by an amount of order 100 μsec, depending only on how close to the sun the radiation passes. By keeping track of the arrival times we can compute the delay and see whether the measured delay agrees with the predictions of relativity theory. To do this, we have to first correct for the time delay due to the relatively high index of refraction of the solar corona. This is possible because the delay due to refraction is proportional to v^{-2}, while the gravitational delay is independent of frequency v. Several pulsars pass within $1°$ of the sun and hence are suitable objects for such tests.

3:11

(c) Since pulsars are located within the galaxy, the shear motion of stars in the galaxy would yield an acceleration relative to the sun that could be detected by keeping track of pulse arrival times. The differential rotation of the galaxy could therefore be mapped very accurately.

3:12 GALACTIC ROTATION

The mass of the Galaxy is not distributed evenly. The density is greatest near the nucleus. For this reason, stars near the Galactic center tend to have angular velocities $\dot\theta(r)$ appreciably larger than stars at greater distances, that is, $d\dot\theta/dr < 0$. Suppose for simplicity that all stars have idealized circular orbits about the Galactic center. Let the sun be at distance r_s from the center. Relative to the sun, matter at Galactic longitude l, and at distance r from the center C has an approach velocity $v(r, l)$ along the line of sight (Fig. 3.8).

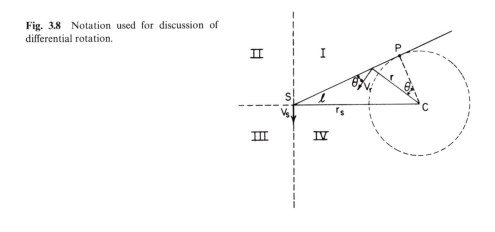

Fig. 3.8 Notation used for discussion of differential rotation.

$$v(r,l) = [- r_s\dot\theta(r_s)\sin l + r\dot\theta(r)\cos\theta] = [\dot\theta(r) - \dot\theta(r_s)]\, r_s \sin l \qquad (3\text{--}55)$$

where the simple form of the expression on the extreme right is due to the fact that $r\cos\theta = r_s\sin l$, as evident from Fig. 3.8. We note from (3–55) and from the fact that $d\dot\theta/dr < 0$, that $v(r,l)$ is positive in the quadrants I and III so that stars and gas in these directions should appear to approach, and their spectra should be blue-shifted. In quadrants II and IV stellar spectra should appear red-shifted.

This, in fact, is what is observed. In 1927 the Dutch astronomer Oort was able to use this evidence to prove that stars in our Galaxy are in *differential rotation* about the Galactic center (Oo27a,b).

At any given Galactic longitude l, the highest velocity should be observed at point P, where the line of sight is tangential. By noting the maximum velocity at any given elongation l, we can construct a model of the Galaxy giving both its mass distribution and distance of the sun from the Galactic center. We will adopt $r_s \sim 9.5$ kpc ± 1.5 kpc (vdBe68). However, recent results suggest a 20% lower value.

Differential rotation tends to shear aggregates of gas and dust as they orbit the Galactic center. For some time this effect was considered responsible for the appearance of spiral arms in some galaxies. However, more recently, Lin (Li67) has suggested that the spiral structure represents a local increase in density and that this enhanced density spiral travels around the galaxy like a wave, at a "pattern" velocity different from that of the speed of the stars involved. For the Galaxy this speed is about 13.5 km sec^{-1} times the distance from the center measured in kiloparsecs. At our distance from the center, this would be 135 km sec^{-1}, while the Galactic rotation (velocity of the stars) is ~ 250 km sec^{-1}. In contrast to this, the stars in the bar of a barred spiral galaxy do appear to move with the pattern, that is, similar to a solid body (see Problem 3–13).

3:13 SCATTERING IN AN INVERSE SQUARE LAW FIELD

When a meteorite approaches the earth, its orbit can become appreciably changed. Similarly, a comet passing close to Jupiter can be given enough energy to escape the solar system. In both cases the smaller object is scattered or deflected by the larger body. For a particle initially approaching from direction $\theta_\infty - \theta_0$ (Fig. 3.9), the orbital trajectory is given by equations 3–34 and 3–40.

$$\frac{1}{r} = \frac{MG}{h^2}\left[1 + \sqrt{1 + \frac{2\mathscr{E}h^2}{M^2G^2}} \cos(\theta - \theta_0) \right] \tag{3-56}$$

At large distances from the scatterer, the asymptotic motion is along directions (see equation 3–42)

$$\cos(\theta_\infty - \theta_0) = -\left(1 + \frac{2\mathscr{E}h^2}{M^2G^2} \right)^{-1/2} = -\frac{1}{e}, \quad r \to \infty \tag{3-57}$$

This has solutions for two values of $|\theta_\infty - \theta_0|$, one corresponding to the incoming and the other to the scattered asymptotic direction. The angle through which the object is deflected is $\Theta = 2(\theta_\infty - \theta_0) - \pi$. We see that

$$\sin\frac{\Theta}{2} = -\cos(\theta_\infty - \theta_0) = \frac{1}{e} \tag{3-58}$$

3:13

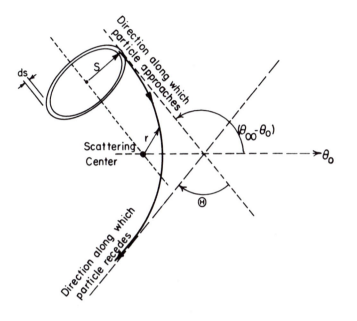

Fig. 3.9 Scattering in an attractive inverse square law field.

We note that h, twice the area swept out per unit time, is

$$h = v_0 s$$

where s is the *impact parameter* (Fig. 3.9) and v_0 is the *approach velocity* of the scattered particle at a large distance, $r \to \infty$.

$$\text{Since } \mathscr{E} = \frac{v_0^2}{2}, \quad h^2 = 2\mathscr{E}s^2 \tag{3-59}$$

and

$$\sin \frac{\Theta}{2} = \left[1 + \left(\frac{2s\mathscr{E}}{MG} \right)^2 \right]^{-1/2} \tag{3-60}$$

This leads to

$$\cot \frac{\Theta}{2} = \frac{2\mathscr{E}s}{MG} \tag{3-61}$$

If any object having an impact parameter between s and $s + ds$ is scattered into an angle between Θ and $\Theta + d\Theta$, we say that the *differential cross section*

3:13

$\sigma(\Theta)$ for scattering is given by

$$2\pi s \, ds \equiv - \sigma(\Theta) \, d\Omega = - 2\pi \, \sigma(\Theta) \sin \Theta \, d\Theta \qquad (3\text{--}62)$$

In this equation the expression on the left represents the area of a ring through which all particles approaching from a given direction have to flow if they are to be scattered into the solid angle $d\Omega$ enclosed between two cones having half angles Θ and $\Theta + d\Theta$, respectively. The expression on the right gives the solid angle between these two cones multiplied by the differential cross section. The differential cross section is therefore just a parameter that assures conservation of scattered particles. The negative sign appears because an increase in the impact parameter s results in a decreasing scattering angle Θ. The differential cross section is proportional to the probability for scattering into an angle between Θ and $\Theta + d\Theta$, since $2\pi s \, ds$ (see equation 3–62) is the probability for encounter at impact parameter values between s and $s + ds$.

We now can rewrite (3–62) in the form

$$\sigma(\Theta) = \frac{s \, ds}{\sin \Theta \, d\Theta} \qquad (3\text{--}63)$$

which, together with expressions 3–60 for s yields

$$\sigma(\Theta) = \frac{1}{4} \left(\frac{MG}{2\mathscr{E}} \right)^2 \frac{1}{\sin^4(\Theta/2)} \qquad (3\text{--}64)$$

3:14 STELLAR DRAG

If a high velocity star moves through a surrounding field of low velocity stars, it experiences a drag because it is slightly deflected in each *distant encounter*. We can compute this drag in an elementary way through the use of the scattering theory derived above.

First, we note that the star's velocity loss along the initial direction of approach to a scattering mass—which we will take here to be another star—is

$$\Delta v = v_0 \, (1 - \cos \Theta)$$

where v_0 is the approach velocity relative to the scattering center at large distances. This is not an overall velocity loss; just a decrease in the component along the direction of approach. The change in momentum is $\mu \Delta v$ where, again, μ is the reduced mass. The force on the high velocity star, opposite to its initial direction of motion, therefore, is

$$F = \sum_i \frac{\mu_i \Delta v_i}{\Delta t} \tag{3-65}$$

where Δt is the time during which a change Δv_i takes place and the summation is taken over all stars i that are being encountered during this time interval. This summation can be replaced by considering a gas of stars with number density n. In terms of the probability or cross section for scattering into angle Θ, at any given encounter, the force becomes

$$F = 2\pi \mu v_0^2 n \int_{\Theta_{\text{max}}}^{\Theta_{\text{min}}} (1 - \cos \Theta) \sigma(\Theta) \sin \Theta \, d\Theta \tag{3-66}$$

This assumes that all the deflections due to interactions with individual stars are small, and that the forces along the direction of motion add linearly. Using equation 3–62 we have

$$F = 2\pi \mu v_0^2 n \int_{s_{\text{min}}}^{s_{\text{max}}} (1 - \cos \Theta) s \, ds \tag{3-67}$$

Instead of integrating for all stars, we integrate the impact parameter s over all possible values for a single star and then multiply by the stellar density n. This is an equivalent procedure because the probability of encountering a star at impact parameter s is proportional to s and to n. The extra factor v_0 that appears in the expression takes account of the increasing number of encounters, per unit time, at large velocities. If we set $\theta_0 \equiv 0$ and $\theta_\infty \equiv \theta$, then

$$- \cos \Theta = \cos 2\theta \equiv \frac{1 - \tan^2 \theta}{1 + \tan^2 \theta} \tag{3-68}$$

but (see Fig. 3.9),

$$\cot \frac{\Theta}{2} \equiv \tan \theta = \frac{2\mathscr{E} s}{MG} = \frac{s v_0^2}{MG} \equiv \alpha s \tag{3-69}$$

so that

$$F = 2\pi \mu v_0^2 n \int s \frac{2}{1 + \tan^2 \theta} \, ds \tag{3-70}$$

and

$$F = 4\pi \mu v_0^2 n \int \frac{s \, ds}{1 + \alpha^2 s^2} \tag{3-71}$$

$$F = \frac{4\pi \mu v_0^2 n}{2\alpha^2} \ln(1 + \alpha^2 s^2) \Bigg]_{s_{\text{min}}}^{s_{\text{max}}} \tag{3-72}$$

We define a slowing down time, or *relaxation time*, τ

$$\tau \equiv \frac{\mu v_0}{F} \qquad (3\text{-}73)$$

In this calculation we have assumed that the star is moving through an assembly of stationary "field" stars. As long as the random motion of these stars is low compared to v_0, equation 3–72 holds quite well. However, when the random stellar velocities approach v_0, the particle can alternately be accelerated or slowed down by collisions and the above derivation no longer holds. For the sun, moving with a velocity of $v_0 = 20\,\mathrm{km}\,\mathrm{sec}^{-1}$ through the ambient star field, $\alpha = 3 \times 10^{-14}\,\mathrm{cm}^{-1}$, $n \sim 10^{-56}\,\mathrm{cm}^{-3}$, and $\mu \sim 10^{33}$ g. If $s_{max} \sim 10^{19}$ cm, roughly the mean separation of stars,* and s_{min} is much smaller, then $F \sim 10^{18}\mathrm{dyn}$. However, even with this large force, $\tau \sim 10^{21}\mathrm{sec}$, that is, far greater than the estimated age of the universe. This large value of τ is disconcerting because it is symptomatic of a general problem in stellar dynamics. We find such aggregates as globular clusters to be in configurations close to those we would expect in thermodynamic equilibrium. This would mean that the stars must interact quite strongly to transfer energy to each other; and yet the above mechanism will not accomplish this at anywhere near a satisfactory rate. Neither will other mechanisms of the same general class. The interaction of these stars must be dominated by some other process that we do not yet understand. We will discuss this again in section 3:16. However, we might note that interaction of stars with gas clouds or clouds of stars produces a larger effect than that of individual stars' encounters (Sp51a). If the mass of the cloud is $M \sim 10^6\,M_\odot$ and $n \sim 10^{-65}\,\mathrm{cm}^{-3}$, F increases by 10^3, and τ decreases by 10^3. Here s_{max} might be chosen $\sim 10^{22}$ cm.

Collisions need not always act to slow particles down. When stars in the plane of the Galaxy interact with the much more massive clouds of gas, they can actually become accelerated to high velocities. In Table A.6 we show that, relative to the sun, older stars have higher root mean squared random velocities than younger stars. This may be due to collisions with such clouds. As will be shown in Chapter 4 an assembly of bodies tends to arrange itself in such a way that translational energies are equal (equipartition of energy). The massive clouds, therefore, tend to pass some of their energy on to the less massive stars and, in so doing, accelerate them to velocities higher than v_c, the velocity of the clouds.

A quite different class of problems in which the above calculations are useful deals with charged particles. The inverse square law electrostatic forces allow us to derive equations quite similar to (3–72) and (3–73), and we can compute the

---

* When s_{max} is much larger than the mean separation, encounters begin to overlap in time, and (3–72) becomes an overestimate of F because the effects of individual stars will tend to cancel through symmetry in their distribution (Ch43).

$$\therefore s_{max} \sim n^{-1/3} \qquad (3\text{-}74)$$

electrostatic drag on fast electrons traveling through the interstellar medium and on charged interstellar or interplanetary dust grains moving at typical velocities of ~ 10 km sec^{-1} through a partially ionized medium. This effect plays a major role especially in the dynamics of dust grains moving through the interstellar medium. In section 6:16 we will also see that the distant collisions of electrons and ions are described by equations like (3–67) and that the opacity or emissivity of an ionized plasma can be computed making use of these equations. The radio emission from hot ionized interstellar gas can then be directly related to the plasma density, or rather to the collision frequency in a line-of-sight column through the cloud.

3:15 VIRIAL THEOREM

The theorem we will prove here again is statistical. It describes the overall dynamic behavior of a large assembly of bodies, rather than the precise behavior of any given individual body belonging to the assembly.

Consider a system of masses m_j at positions $\mathbf{r}_j$. Let the force on m_j be $\mathbf{F}_j$. We now write the identity

$$\frac{d}{dt}\sum_j \mathbf{p}_j \cdot \mathbf{r}_j = \sum_j \mathbf{p}_j \cdot \dot{\mathbf{r}}_j + \sum_j \dot{\mathbf{p}}_j \cdot \mathbf{r}_j \tag{3–75}$$

$$= 2\mathbb{T} + \sum_j \mathbf{F}_j \cdot \mathbf{r}_j \tag{3–76}$$

where $\mathbb{T}$ is the kinetic energy of the entire system and the time derivative of the momentum $\dot{\mathbf{p}}_j$ is equal to the force $\mathbf{F}_j$. For the moment we do not identify the left side of the equation with any physically interesting quantity. Taking the time average of both sides, we obtain

$$\frac{1}{\tau}\int_0^\tau \frac{d}{dt}\sum_j \mathbf{p}_j \cdot \mathbf{r}_j \, dt = \langle\, 2\mathbb{T} + \sum_j \mathbf{F}_j \cdot \mathbf{r}_j \,\rangle \tag{3–77}$$

where the brackets denote a time average. A particularly interesting situation concerns a bound system in which each member of the assembly remains a member for all time. In this situation all the $\mathbf{r}_j$ values must remain finite since no particle escapes from the system, and all $\mathbf{p}_j$ values must remain finite because the total energy of the system is finite.

Since $\sum_j \mathbf{p}_j \cdot \mathbf{r}_j$ remains finite, the integral of its derivative must also remain finite for all time. This means that the left side of equation 3–77 consists of a finite quantity divided by τ that can be made arbitrarily large—if we average over a very large or infinite period. The left side of equation 3–77 therefore approaches zero and we can set

$$\langle 2\mathbb{T} \rangle + \langle \sum_j \mathbf{F}_j \cdot \mathbf{r}_j \rangle = 0 \tag{3–78}$$

If the force is derivable from a potential, this equation becomes

$$\langle 2\mathbb{T} \rangle - \langle \sum_j \nabla \mathbb{V}(r_j) \cdot \mathbf{r}_j \rangle = 0 \tag{3–79}$$

where $\mathbb{V}(r_j)$ is the potential energy of mass m_j at position r_j. The force in such a situation is a function of position only and can be written as the negative gradient ∇ of the potential energy:

$$\mathbf{F}_j = -\nabla \mathbb{V}(r_j) \tag{3–80}$$

If the potential is proportional to r^n, the gradient lies along the radial direction and

$$\sum_j \nabla \mathbb{V}(r_j) \cdot \mathbf{r}_j = \sum_j \frac{\partial \mathbb{V}(r_j)}{\partial r_j} r_j \tag{3–81}$$

Calling the total potential energy of the entire assembly $\mathbb{V}$, we obtain

$$\langle \mathbb{T} \rangle = \frac{n}{2} \langle \mathbb{V} \rangle , \quad \mathbb{V} \equiv \sum_j \mathbb{V}(r_j) , \quad -2 < n \tag{3–82}$$

This relation runs into difficulty for $n < -2$, since the total energy $\langle \mathbb{T} \rangle + \langle \mathbb{V} \rangle$ would then be positive, indicating that the system would no longer be bound. For an inverse square law force, as in gravitation or electrostatics, the potential goes as the inverse first power, $n = -1$, and

$$\langle \mathbb{T} \rangle = -\frac{1}{2} \langle \mathbb{V} \rangle \tag{3–83}$$

This theorem is of great importance and finds many applications in astrophysics. For example, it provides the only current estimate for the mass of clusters of galaxies. That estimate is obtained by observing the spread in radial *Doppler velocities* among different galaxies in the cluster, which gives the mean kinetic energy per unit mass. Equation 3–83 then yields the mean potential energy per unit mass, and if a typical cluster diameter is known from the cluster's distance and from the angle it subtends in the sky, we can obtain a rough estimate of the total cluster mass on the assumption that

$$\frac{\mathbb{V}}{M} \sim \frac{MG}{R} \tag{3–84}$$

Here M is the cluster mass and R is some weighted cluster radius, somewhat smaller than the observed radius of the cluster. The estimated cluster mass would normally be in error (that is, too high) by a factor less than ~ 2 if the actually observed cluster radius is used in equation 3–84.

An interesting problem occurs when we measure the masses of clusters of galaxies making use of the virial theorem. The masses of individual galaxies within the cluster can be determined from their rotations, Problem 3–9. From these we can compute the potential energy of the entire cluster, if the cluster dimensions are computed from the apparent diameter and red-shift distance. An independent estimate of the potential energy, however, is obtained from (3–83) if the random velocities of the individual galaxies are taken to compute $\mathbb{T}$. To do this, we note the variations in red shifts from galaxy to galaxy and estimate the actual random velocities. Strangely the results of using (3–83) always give values of $\langle \mathbb{T} \rangle$ and, hence, $\langle \mathbb{V} \rangle$ that are about an order of magnitude higher than the total potential energy computed on the basis of individual galactic masses. We conclude that either (i) there is a lot of undetected *dark matter* in clusters, or (ii) the clusters are breaking up, or (iii) we do not understand dynamics on such a large scale. For example, we might ask whether galaxies in a cluster could not participate in the overall cosmic expansion. Could this account for the apparent disruption of the clusters? The answer seems to depend on factors we do not yet know. If the bulk of the mass-energy in the universe is in the form of dark matter, cosmic expansion seems to play a minor role. The problem of dark matter, particularly as it relates to galaxy mass-distribution will be discussed further in section 12:19. *Dark matter* is a generic term used for any type of matter making itself felt gravitationally, but not detectable, at least to date, by direct observational means. Problem 4–5 and the discussion following it treat the cluster problem from an observational viewpoint.

3:16 STABILITY AGAINST TIDAL DISRUPTION

When a swarm of gravitationally bound particles, having a total mass m, approaches too close to a massive object M, the swarm tends to be torn apart. The same thing can happen to a solid body held together by gravitational forces, when it approaches a much more massive object.

The reason for this is quite simple. If we consider that the center of mass of the swarm is at a distance r from the mass M, and is falling straight toward it, then its acceleration toward M is $- MG/r^2$. Let r' be the swarm radius. A particle P_0 (Fig. 3.10), at the surface of the swarm nearest to M, would be accelerated at a rate $- MG/(r - r')^2$ towards M, if it were not for the fact that a gravitational attraction from the center of the swarm tends to accelerate it away from M at a rate mG/r'^2. In order for the particle to be pulled steadily away from the swarm, we must have the condition

$$MG\left(-\frac{1}{r^2} + \frac{1}{(r - r')^2} \right) > \frac{mG}{r'^2} \tag{3–85}$$

3:16

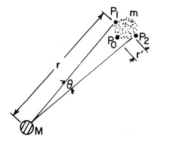

Fig. 3.10 A swarm of gravitationally bound particles—stars, atoms, molecules—can be disrupted through close encounter with a massive object M.

Expanding the expression on the left and keeping only terms down to first order in r', we obtain

$$\frac{2M}{r^3} > \frac{m}{r'^3} \tag{3-86}$$

Similarly for a swarm in a perfectly circular orbit about M, disruption occurs when

$$\frac{3M}{r^3} > \frac{m}{r'^3} \tag{3-87}$$

PROBLEM 3-3.   Derive the result (3-87). In doing this, it is helpful to think of the swarm as moving without rotation about its center, and to consider its center of mass as having a *centrifugal repulsion*

$$F_c = r\,\dot\theta^2 \tag{3-88}$$

away from M. This is different from the "repulsion" $(r - r')\,\dot\theta^2$ at P_0.

The precise ratios of the masses M and m will therefore vary with differing orbits and the rotation of m will also play a role in determining its stability. What is important to note, however, is that the density of the swarm is more important a consideration than its actual mass, or size, taken individually.

There is a second effect that also plays an important role. Again, consider a direct infall. Here points P_1 and P_2 (Fig. 3.10) would be accelerated radially toward M and would tend to converge. The effective acceleration of P_1 and P_2 relative to each other would be roughly

$$2\,\frac{MG}{r^2}\,\frac{r'}{r} = \frac{2MGr'}{r^3}$$

due to this effect taken by itself. This is important whenever it is larger than the acceleration mG/r'^2 due to the mass of the swarm itself, that is, when (3–86) holds. There exists a lateral compression, therefore, that accompanies the tidal disruption and tends to concentrate the swarm, while the tidal forces attempt to tear it apart. What actually happens under these combined effects will be better understood in terms of the Liouville theorem in Chapter 4.

It is now worth noting that there are two particular conditions where this kind of tidal disruption seems to play a leading role. First, comets that approach too close to the sun or even too close to the massive planet Jupiter have been observed to break up into two or more fragments, and the general nature of the tidal theory seems to be borne out.

Equally interesting is the effect that tidal disruption seems to have on globular clusters. Von Hoerner (vHo57) has examined the orbits of these clusters statistically, and finds that they have orbits that draw them very close to the Galactic center. The massive Galactic nucleus then seems able to rob such clusters of loosely bound outer members. At the center of the cluster, where the density is largest, disruptive effects are then relatively small, while at the periphery, where m/r'^3 is low, stars can be pulled away more readily.

We can now also see why the interaction of stars within a globular cluster may only play a limited role in determining the ultimate velocity distribution of stars in the cluster. The treatment of section 3:14, and the very long star encounter relaxation time τ predicted by equation 3–73 may not give a true picture of the actual evolution of clusters into the well-defined, compact, spherical aggregates we observe. Interaction with the Galactic nucleus must have an appreciable, perhaps even a dominant, influence on this distribution. We will touch on this problem again in section 4:21.

In section 1:8 we had said that some of the dwarf galaxies of the Local Group can never have been very close to either the Galaxy or M31. We can conclude this directly from criteria like (3–86) and (3–87).

PROBLEMS

3–4. The orbital period for the earth moving about the sun is given by equation 3–47. The distance of the sun can be obtained most accurately by the radar method described in section 2:1. Averaged over the earth's eccentric orbit it has a mean value of 1.5×10^{13} cm. Assuming the earth's mass, $m_E \ll M_\odot$, show that the sun's mass $M_\odot = 2.0 \times 10^{33}$ g.

3–5. A radar signal reflected from the moon returns 2.56 sec after transmission. The speed of light is 3.00×10^{10} cm sec^{-1}. Assume the period of the moon to be roughly 27.3 days. Find the mass of the earth assuming the moon's mass is small compared to the earth.

Note: In this way we can determine the mass of any planet with a moon. When a planet has no moon, its mass is determined by the perturbations it produces on the orbits of nearby planets. Such a calculation is quite time-consuming, but introduces no essentially new physical concepts. The calculations proceed within the framework of Newtonian dynamics.

3–6. Since the moon and the earth revolve about a common center of mass, the apparent motion of Mars has a periodicity of one month superposed on its normal orbit. The distance of the moon is $D \sim 3.8 \times 10^5$ km. The distance of Mars at closest approach is $L \sim 5.6 \times 10^7$ km. The apparent displacement of Mars over a period of a half month is ~ 34 sec of arc. What is the mass of the moon?

3–7. A meteor approaches the earth with a speed v_0, when it is at a very large distance from the earth. Show that the meteor will strike the earth, at least at grazing incidence if its impact parameter s is given by

$$s \leq [R^2 + 2MGR \, v_0^{-2}]^{1/2}$$

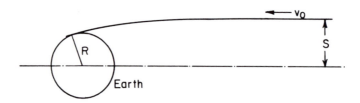

Fig. 3.11   Impact of a meteorite or a cloud of meteors on the earth's atmosphere.

3–8. If a cloud of meteors approaches the earth at relative speed v_0, show that the rate of mass capture is $\pi(R^2 v_0 + 2MGR/v_0)\rho$, where ρ is the mass density of the cloud. Both Problems 3–7 and 3–8 neglect the sun's influence on the meteors. This problem might be more appropriately done after reading section 4:5.

3–9. A disk-shaped rotating galaxy is seen edge on. By Doppler-shift spectroscopic measurements we can determine the speed V with which the stars near the edge of the galaxy rotate about its center. Show that the mass of the galaxy in terms of the observed velocity is $\sim V^2 R/G$. State the assumptions made. R is the radius of the galaxy.

3–10. In the vicinity of young clusters we occasionally see *runaway stars*, O or B stars that evidently were part of the cluster until recently but are receding rapidly. Blaauw (Bl61) has suggested that the runaways initially may have been part of binaries in which the companion exploded as a supernova, leaving only part of its mass behind. Suppose that the initial motion was circular, with initial orbital velocity v for the surviving star. If the initial mass of the companion was M, and the final mass after the explosion is only $M/10$, what will be the final velocity V of the runaway star at large distance from the explosion? The mass of the surviving star is m. Refer v and V to the system's center of mass.

3–11. A gravitationally bound body spins rapidly (but not at relativistic velocities). At what rotational velocity will it break up if its mass is m and radius is r? Assume the body remains spherical until breakup—even though this assumption will not normally hold.

3–12. Observations on the compact radio source 3C279, which is occulted by the sun once a year, show that radio waves are bent as they pass very close to the sun (Hi71). Show that this bending is a consequence of the equivalence principle.

3–13. In the barred spiral galaxy NGC 7479, Doppler-shift velocities indicate that the bar rotates as a solid body (Bu60), that is, with constant angular velocity ω along its entire length. Show that such a motion can take place when the distribution of mass is actually spherical (but only the bar consists of luminous stars) and when the mass $M(r)$ enclosed by a sphere of radius r increases sufficiently rapidly with increasing distance r from the galaxy's center (Fig. 1.10). Show that $dM(r)/dr$ is proportional to r^2 and to ω^2, in that case. Barred spirals may, however, be set up in ways altogether different from this process. We do not yet know. Aarseth (As60, 61) has discussed the stability of an actual cylindrical bar of stars.

ANSWERS TO SELECTED PROBLEMS

3–1.
$$\dot{r} = -\frac{1}{y^2}\theta\frac{dy}{d\theta} = -h\frac{dy}{d\theta}.$$

$$\ddot{r} = -h\dot{\theta}\frac{d^2y}{d\theta^2} = \frac{-h^2}{r^2}\frac{d^2y}{d\theta^2}.$$

Substituting in (3–30), we see that

$$\frac{d^2y}{d\theta^2} + y = \frac{MG}{h^2}.$$

Substitution of $y = B\cos(\theta - \theta_0) + MG/h^2$ satisfies the equation.

3–2. $m\ddot{r} = GmM_E/(R_E + H)^2$ at a height $H \ll R_E$.
If we take $M_E = \rho_E(4/3)\,\pi R_E^3$, where the symbols represent the earth's mass, density, and radius, we can estimate G from the measured acceleration, $G \sim g\,[\rho_E(4\pi/3)\,R_E]^{-1}$.

3–3. At the center of mass of the swarm, the centrifugal and gravitational forces are equal: $(r\dot{\theta})^2 = GM/r$. A particle, p, at the swarm's near surface, will experience a centrifugal acceleration away from M, smaller than that of the swarm's center by MGr'/r^3. It will also experience a stronger gravitational acceleration towards M, by

$$\frac{MG}{r^2}\left[-1 + \frac{r^2}{(r - r')^2} \right].$$

For disruption to occur, these accelerations must be stronger than mG/r^2. Expanding this inequality gives

$$\frac{3M}{r^3} > \frac{m}{r'^3}. \tag{3–87}$$

This solution assumes no rotation of the swarm.

3–6. Let m be the lunar mass and M the terrestrial mass. The distance R of the earth from the center of mass is then given by

$$RM = (D - R)\,m$$

The apparent displacement of Mars is $2R/L$, where L is the distance to Mars. Hence $2R = 1.7 \times 10^{-4}\,L$. $R = 4.8 \times 10^3$ km and with $M = 6.0 \times 10^{27}$ obtained in Problem 3.2 we can now evaluate m as $\sim 7.4 \times 10^{25}$ g.

3–7. Call V the velocity the meteor has at grazing incidence, that is, when it hits the earth tangentially. Then this velocity is perpendicular to the radius vector R: We can therefore write conservation of angular momentum as

$$sv_0 = RV$$

Conservation of energy per unit meteor mass is

$$\frac{v^2}{2} = \frac{V^2}{2} - \frac{MG}{R}$$

Eliminating V from the equation one obtains the expression

$$s = \left(R^2 + \frac{2MGR}{v^2} \right)^{1/2}$$

Clearly all meteors with impact parameter less than s also can hit the earth. This gives rise to the desired expression.

3–8. The number of meteors hitting the earth per second is given by the density of meteors in space, times the volume of the cylinder of radius impact parameter s swept up in unit time:

$$\pi s^2 \cdot v_0 \cdot \rho$$

s is given in Problem 3–7.

3–9. Assume circular motion. The mass of the galaxy M acting on a star at its periphery is then given by the relation between kinetic and potential energy per unit mass of the star:

$$\frac{V^2}{2} = \frac{MG}{2R}$$

3–10. This problem is somewhat complex. Initially the linear momentum of each star about the center of mass is mv. If the star M explodes and leaves a remnant of mass $M/10$, the two remaining stars will move with a momentum $0.9mv$ relative to the initial center of mass. This gives rise to a kinetic energy of translation of the center of mass, and additional kinetic energy of rotation of these stars about their new center of mass. The new potential binding energy is only one tenth the initial binding energy; but if $m \gg M/10$, much of this decreased potential energy just goes into translational motion of the system. In that case the two surviving stars remain bound and $V \sim 0.9 \, mv/(m + M/10)$. If $M/10$ is still large, the decrease in potential energy permits m to escape, and much of V then represents a true velocity of separation of the surviving stars.

3–11. The centrifugal force > gravity: $r\omega^2 > mG/r^2$; $\omega > \sqrt{mG/r^3}$.

3–12. Suppose an observer falls toward the sun in a space ship. Light rays passing by the sun enter the window of his cabin. The equivalence principle states that he should see the light moving in a straight line. But since he is falling toward the sun, this means that the rays must actually be following a parabolic path relative to a stationary observer.

3–13. For a body rotating as a solid with constant angular velocity, ω,

$$\frac{dv}{dr} = \frac{V}{r} = \omega .$$

But by (3–44)

$$v(r) = \left(\frac{M(r) G}{r} \right)^{1/2} .$$

Here $M(r)$ is the mass enclosed by the circle of radius r:

$$M(r) = \frac{\omega^2 r^3}{G}, \qquad \frac{dM(r)}{dr} = \frac{3\omega^2 r^2}{G}$$

4

Random Processes

4:1 RANDOM EVENTS

If a bottle of ether is opened at one end of a room, we can soon smell the vapors at the other end. But the ether molecules have not traversed the room in a straight line, nor in a single bound. They have undergone myriad collisions with air molecules, bouncing first one way, then another in a random walk that takes some molecules back into the bottle from which they came, others through a crack in the door, and others yet into the vicinity of an observer's nose where they can be inhaled to give the sensation of smell.

In general, molecules diffuse through their surroundings by means of two processes: (i) individual collision with other atoms and molecules and (ii) turbulent and convective bulk motions that involve the transport of entire pockets of gas. These, too, are the mechanisms that act to mix the constituents of stellar and planetary atmospheres. Both processes give rise to random motions that can best be statistically described.

In an entirely different context, think of a broadband amplifier whose input terminals are not connected to any signal source. On displaying the output on an oscilloscope, we would find that the trace contains nothing but spikes, some large, others smaller, looking much like blades of grass on a dense lawn. An exact description of this pattern would be laborious; but a statistical summary in terms of mean height and mean spacing of spikes can be provided with ease and may in many situations present all the information actually needed.

The spikes are the noise inherent in any electrical measurement. If we are to detect, say, a radio-astronomical signal fed into the amplifier, we must be able to distinguish the signal from the noise. That can only be done if the statistics of the noise are properly understood.

Again, consider a third situation, a star embedded in a dense cloud of gas. Light emitted at the surface of the star has to penetrate through the cloud if it is to reach clear surroundings and travel on through space. An individual photon may be absorbed, re-emitted, absorbed again, and re-emitted many times in succession. The direction in which the photon is emitted may bear no relation at all to the direction in which it was traveling just before absorption. The photon may then travel about the cloud in short, randomly directed steps until it eventually reaches the edge of the cloud and escapes. This *random walk* can be described statistically. We can estimate the total distance covered by the photon before final escape and, at any given time in its travel, we can predict the approximate distance of the photon from the star.

These three physically distinct situations can all be treated from a single mathematical point of view. In its simplest form each problem can be reduced to a random walk. We picture a man taking a sequence of steps. He may choose to take a step forward, or a step backward; but, for simplicity, we will assume that his step size remains constant. If the direction of each step is randomly determined, say by toss of a coin, the man will execute a random walk. The toss of the coin might tell him that his first step should be backward, the next forward, the next forward again, backward, backward, forward, and so on. After 10 steps, how far will the man have moved from his initial position? How far will he be after 312 steps or after 10,000,000? We cannot give an exact answer, but we can readily evaluate the probability of terminating at any given distance from the starting point.

4:2 RANDOM WALK

Consider a starting position at some zero point. We toss a coin that tells the man to move forward or backward. He ends up at either the $+1$ or the -1 position (Fig. 4.1). If he ends up in the $+1$ position, the next toss of the coin will take him to the $+2$ or the 0 position, depending on whether the toss tells him to move forward or back. Similarly from the -1 position he could move to 0 or -2.

There exist two possible ways of arriving back at the zero position, and only one possible way of getting to the -2 or to the $+2$ position. Since all of these sequences are equally probable, there is a probability of $\frac{1}{4}$ that the man ends up in the $+2$ position, a probability of $\frac{1}{4}$ that he ends up at -2, and a probability of $\frac{1}{2}$ that he ends up in the zero position after two steps. The zero position is more probable because there are two distinct ways of reaching this position, while there is only one way to get to the $+2$ or -2 positions when only two steps are allowed.

Let us denote by $p(m, n)$ the number of ways of ending up at a distance of m steps from the starting point, if the man executes a total of n steps. We will call m the deviation from the starting position. We will call $p(m, n)$ the *relative probability* of

4:2

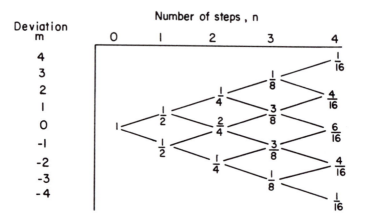

Fig. 4.1   Probability $P(m, n)$ of terminating at position m after n steps.

terminating at distance m. The *absolute probability* $P(m, n)$ of terminating at position m, after n steps is displayed in Figure 4.1 and is

$$P(m, n) = \frac{p(m, n)}{\sum_{k} p(k, n)} = \frac{\text{Number of paths leading to position } m}{\text{Sum of all distinct paths leading to any position, } k} \quad (4\text{--}1)$$

The numerators $p(m, n)$ of the fractions in Figure 4.1 have a binomial distribution; they are the same numbers that appear as coefficients in the expansion

$$\left[\frac{1}{x} + x\right]^{n} = x^{n} + nx^{n-2} + \frac{n(n-1)}{2!} x^{n-4} + \frac{n! \, x^{n-2r}}{(n-r)! \, r!} \cdots + \frac{1}{x^{n}} \quad (4\text{--}2)$$

Knowing this, we can easily evaluate the sum of coefficients in the series, $\sum_{k} p(k, n)$. It is the sum of the coefficients in the binomial expansion and can be obtained by setting $x = 1$ on the right side of equation 4–2.

Substituting $x = 1$ on the left side of (4–2) shows that the sum of terms must have the value 2^{n}:

$$\sum_{k=-n}^{n} p(k, n) = 2^{n} \quad (4\text{--}3)$$

and

$$P(m, n) = \frac{p(m, n)}{2^{n}} \quad (4\text{--}4)$$

4:2

We note also that if the exponent of a given term in equation 4–2 represents the deviation m, in Fig. 4.1, then the coefficient of that term represents the relative probability $p(m, n)$. In that sense we can rewrite (4–2) as

$$\left(\frac{1}{x} + x\right)^n = \sum_{k=-n}^{n} p(k, n)\, x^k \qquad (4\text{–}5)$$

Every second term of this series has a coefficient zero. We now wish to determine the mean deviation from the zero position after a random walk of n steps. By this we mean the sum of distances reached in any of the 2^n possible paths that we could take, all divided by 2^n. Since there are $p(k, n)$ ways of reaching the distance k, the numerator of this expression is $\sum_k kp(k, n)$ and we see that the *mean deviation*, $\langle k \rangle$, is

$$\langle k \rangle \equiv 2^{-n} \sum_{k=-n}^{n} kp(k, n) = \frac{\text{sum of all possible terminal distances after } n \text{ steps}}{\text{number of all possible paths using } n \text{ steps}}$$

$$(4\text{–}6)$$

We notice from Fig. 4.1 and from the binomial distribution (4–2) that the relative probability $p(k, n)$ of having a deviation k equals the relative probability of having deviation $-k$: $p(-k, n) = p(k, n)$. Since the summation in (4–6) is carried out over values from $-n$ to n, there will be an exact cancellation of pairs involving $k = m$ and $-m$, and the only uncancelled term is the one having $k = 0$. This shows that the value of $\langle k \rangle$ must be zero also. The mean deviation from the starting position must be zero, no matter how many steps we take.

This does not mean that the absolute value of the deviation is zero. Far from it. But there are equally many ways of ending up at a positive as at a negative distance and the average position is right at the starting point itself.

This much is evident from symmetry. However, we usually need to know something about the actual distance reached after n steps. For example, we want to know the actual distance from a star that a photon has traveled after n absorptions and re-emissions in a surrounding cloud. A useful measure of such distances is the *root mean square deviation* Δ

$$\Delta \equiv \langle k^2 \rangle^{1/2} = \left[\frac{\sum\limits_{k=-n}^{n} k^2 p(k, n)}{\sum\limits_{k=-n}^{n} p(k, n)}\right]^{1/2} = \left[\frac{\text{sum of (distances)}^2}{\text{sum of all possible paths}}\right]^{1/2} \qquad (4\text{–}7)$$

4:2

This is obtained by first taking the mean of the deviation squared $\langle k^2 \rangle$, and then taking the root of this mean value to obtain a deviation in terms of a number of unit length steps. If we did not take the square root, the quantity obtained would have to be measured in units of $(\text{step})^2$; this is an area, rather than a length or distance. To evaluate the sum

$$\sum_{k=-n}^{n} k^2 p(k, n) \tag{4-8}$$

we can employ a simple technique. We substitute the quantity $x = e^y$ in equation 4–5 and differentiate twice in succession with respect to y. In the limit of small y values, we then obtain

$$\sum_{k=-n}^{n} k^2 p(k, n) = \frac{d^2}{dy^2} \sum_{k=-n}^{n} p(k, n) e^{ky} = \lim_{y \to 0} \frac{d^2}{dy^2} (e^{-y} + e^y)^n$$

$$= n(n-1)(e^{-y} + e^y)^{n-2}(e^y - e^{-y})^2 + n(e^{-y} + e^y)^n]_{y=0} = n2^n \tag{4-9}$$

In summary, we can write

$$\sum_{k=-n}^{n} k^2 p(k, n) = n2^n \tag{4-10}$$

Equations 4–3 and 4–10 can now be substituted into (4–7) to obtain a root mean square deviation

$$\Delta = n^{1/2} \tag{4-11}$$

After n steps of unit length the absolute value of the distance from the starting position is therefore approximately $n^{1/2}$ units.

The following four problems widen the applications of the random walk concept.

PROBLEM 4–1. For a one-dimensional random walk, involving steps of unequal length, prove that the mean position after a given number of steps is zero, the starting position.

Note that for a finite number of differing step lengths, this walk can be reduced to a succession of random walks, each walk having only one step length.

PROBLEM 4–2. Prove that the root mean square deviation for a walk involving the sum of different numbers n_i of steps of length λ_i, is

$$\Delta = N^{1/2} \lambda_{\text{rms}} \tag{4-12}$$

where $N = \sum_i n_i$ and λ_{rms} is the root mean square value of the step length

$$\lambda_{rms} = \left[\frac{\sum_i n_i \lambda_i^2}{N}\right]^{1/2} \tag{4-13}$$

PROBLEM 4–3. Show that the root mean square deviation in a three-dimensional walk with step lengths L_0 is $s^{1/2}L_0$ after s steps. To show this, take the three Cartesian components of the ith step (see Fig. 4.2) as

$$L_0 \cos \theta_i, \qquad L_0 \sin \theta_i \cos \phi_i, \qquad L_0 \sin \theta_i \sin \phi_i \tag{4-14}$$

Fig. 4.2 Polar coordinate system used in describing the three-dimensional random walk.

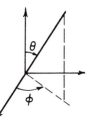

The mean square deviations along the three coordinates are, respectively,

$$\Delta_z^2 = \sum_{i=1}^{s} L_0^2 \cos^2 \theta_i \qquad \Delta_x^2 = \sum_{i=1}^{s} L_0^2 \sin^2 \theta_i \cos^2 \phi_i$$

$$\Delta_y^2 = \sum_{i=1}^{s} L_0^2 \sin^2 \theta_i \sin^2 \phi_i \tag{4-15}$$

These components can be added by the Pythagorean theorem to give the overall mean square deviation as

$$\Delta^2 = sL_0^2 \tag{4-16}$$

PROBLEM 4–4. A hot star is surrounded by a cloud of hydrogen that is partly ionized, partly neutral. Radiation emitted by the star at the wavelength of the Lyman-α line can be absorbed and re-emitted by the neutral atoms. Let the mean path traveled by a photon, between emission and absorption have length L. Let the radius of the cloud be R. About how many absorption and re-emission processes are needed before the photon finally escapes from the cloud? We make use of this result in section 9:6.

The random walk concept provides an essential basis for all radiative transfer computations. We will tackle such problems later, in discussing the means by which energy can be transported from the center of a star, where it is initially released, to the surface layers, and then through the star's atmosphere out into space. In the general theory of radiative transfer, the *opacity* of the material is inversely proportional to the step length we assumed for the random walk above. The added complication that arises in most practical problems is that the mean energy per photon becomes less and less as energy is transported outward from the center of a star. Energy initially found in hard gamma rays eventually leaves the stellar surface as visible and infrared radiation. One gamma photon released in a nuclear reaction at the center of the star provides enough energy for about a million photons emitted at the stellar surface. The walk from the center of a star, therefore, involves not a single photon alone, but also all its many descendants.

4:3 DISTRIBUTION FUNCTIONS, PROBABILITIES, AND MEAN VALUES

In section 4:2 we calculated the mean deviation and root mean square deviation after a number of steps in a random walk. Often we are interested in computing mean values for functions of the deviation; and for distributions other than binomial distributions there is also a procedure for obtaining such values.

Suppose a variable x can take on a set of discrete values x_i. Let the absolute probability of finding the value x_i in any given measurement be $P(x_i)$. If we pick a function $F(x)$ that depends only on the variable x, we can then compute the mean value that we would obtain for $F(x)$ if we were to make a large number of measurements. This mean is obtained by multiplying $F(x_i)$ by the probability $P(x_i)$ that the variable x_i will be encountered in any given measurement; summation over all i values then yields the mean value $\langle F(x) \rangle$

$$\langle F(x) \rangle = \sum_i P(x_i) F(x_i) \tag{4-17}$$

Sometimes the absolute probability is not immediately available but the relative probability $p(x_i)$ is known. We then have the choice of computing $P(x_i)$ as in equation 4–1, or else we can proceed directly to write

$$\langle F(x) \rangle = \frac{\sum_i p(x_i) F(x_i)}{\sum_i p(x_i)} \tag{4-18}$$

where the denominator gives the normalization that is always needed when relative probabilities are used.

4:3

If x can take on a continuum of values within a certain range, the integral expressions corresponding to equations 4–17 and 4–18 are

$$\langle F(x) \rangle = \int P(x) F(x)\, dx = \frac{\int p(x) F(x)\, dx}{\int p(x)\, dx} \tag{4–19}$$

where the integrals are taken over the range of the variable for which a mean value $\langle F(x) \rangle$ is of interest. In some situations this range is $-\infty < x < \infty$.

We note that the expressions 4–6 and 4–7 already have the general form required by equations 4–17 to 4–19. Basically, in equation 4–6 the function $F(x)$ is just x itself, while in (4–7) it is x^2. We have merely substituted a new symbol x, for the values previously denoted by the position symbol k.

4:4 PROJECTED LENGTH OF RANDOMLY ORIENTED RODS

Let a system be viewed along a direction defining the axis of polar coordinates (θ, ϕ) (Fig. 4.3). A rod of length L has some arbitrary orientation θ with respect to the axis, and its projected length transverse to the line of sight is $L \sin \theta$, independent of ϕ, $0 \leq \phi < 2\pi$.

Fig. 4.3 Polar coordinate system for discussion of projected lengths.

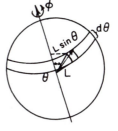

We wish to determine the mean value of the observed length, the average being taken over all possible orientations of the rod: The probability of finding the rod with an orientation that lies within an increment $d\theta$ at angle θ is proportional to the area that the strip $d\theta$ defines on the surface of a sphere of unit radius. The normalized probability $P(\theta)$ is

$$P(\theta)\, d\theta = \frac{1}{2\pi} \int p(\theta, \phi)\, d\theta\, d\phi = \sin \theta\, d\theta \tag{4–20}$$

We see that this is a properly normalized probability since

$$\int_0^{\pi/2} P(\theta)\, d\theta = -\cos \theta \Big|_0^{\pi/2} = 1 \tag{4–21}$$

4:4

that is, the probability of finding the rod with *some* orientation between 0 and $\pi/2$ is unity.* The probability of finding the rod with projected length $L \sin \theta$ is therefore $\sin \theta$, and the mean value of the projected length averaged over all position angles is

$$\frac{\int_0^{\pi/2} P(\theta)\, L \sin \theta\, d\theta}{\int_0^{\pi/2} P(\theta)\, d\theta} = \int_0^{\pi/2} L \sin^2 \theta\, d\theta = \frac{\pi}{4} L \qquad (4\text{-}22)$$

Here, the integral in the numerator is a summation over the lengths obtained over all orientations, and the integral in the denominator assures an average value by dividing the numerator by the whole range of probabilities. This division is not strictly necessary because we already have normalized correctly. However, had we, for example, wished to find the mean projected lengths only for those rods having inclinations to the polar axis in the range $0 < \theta \leq \pi/4$, the limits of integration both in the numerator and denominator would be 0 and $\pi/4$, and the integral in the denominator would no longer be trivial. Reversing the problem, we can ask for the actual value of a length S when only the random projected lengths can be observed to have mean value D. Then

$$S = \frac{4\langle D \rangle}{\pi} \qquad (4\text{-}23)$$

by simple inversion of the argument developed in (4–22). We can ask a slightly different question, "Given a particular observed value of D, what is the mean of all the values S could have?" To answer this, we average $D/\sin \theta$ over the interval $0 \leq \theta \leq \pi/2$, for a fixed value of ϕ since the orientation ϕ is implicitly the direction along which D has been measured. This average has an infinite value since $(\sin \theta)^{-1}$ becomes large as θ approaches zero. The value of $\langle 1/S \rangle$ however is finite.

Similarly we can use our approach to decide whether elliptical galaxies are prolate—cigar-shaped—or oblate—disk-shaped. To make such an analysis, we do have to assume that all elliptical galaxies have roughly the same shape; so that according to this view, the globular galaxies would just be ordinary ellipticals viewed along their symmetry axis.

PROBLEM 4-5. When a series of double galaxies is observed, the total mass of each pair can be statistically determined by measuring the projected separation between the galaxies and the projected radial component of their motions about

---

* The limits of integration are $0 \leq \phi < 2\pi$, $0 \leq \theta \leq \pi/2$, since a rod with orientation (θ, ϕ) is equivalent to one with orientation $(-\theta, \phi + \pi)$.

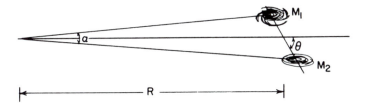

Fig. 4.4 Diagram to illustrate estimation of the total mass found in binary galaxies.

each other. If R is the distance to the pair as determined by their mean red shift, and α is the angular separation, then we can obtain the projected separation d_p. The difference between the two red shifts gives the projected orbital velocity component v_p. Assuming that the galaxies move in circular orbits about each other, and that one of the galaxies is much more massive, show that the mass of the pair is statistically given by

$$M_{\text{pair}} = \frac{\langle v^2 \rangle}{G \langle 1/r \rangle} = \frac{3\pi}{2} \frac{\langle v_p^2 \rangle}{G \langle 1/d_p \rangle} \qquad (4\text{--}24)$$

It may first be useful to show that $\langle v^2 \rangle = 3 \langle v_p^2 \rangle$ and that $\langle 1/r \rangle = (2/\pi) \langle 1/d_p \rangle$. Note that $\langle r \rangle \neq \langle 1/r \rangle^{-1}$. Actually the projection angle in this case is not independent for r and v, and alternate forms of (4–24) can be presented which take this correlation into account.

When we talk about clusters of galaxies, the same considerations apply, because the virial theorem (3–83) again sets the mean potential energy equal to twice the (negative of the) kinetic energy; and the mass of the entire cluster is then substituted on the left side of equation 4–24. The right side gives the mean squared velocities of the cluster galaxies and their mean reciprocal distances from the cluster center. As discussed in 3:15, when the cluster mass is estimated in this way, it always turns out to be some 10 times greater than the sum of the masses of the individual galaxies determined as in Problem 3–9. We will return to this puzzle in Chapter 10; but even there we will not be able to resolve the difficulty.

Salpeter and Bahcall (Sa69a) used (4–24) to obtain an upper limit on the mass of the quasar B264 which appears to lie in a cluster of galaxies. Their upper limit was $5 \times 10^{13}\ M_\odot$. Their estimate essentially said that the quasar must be less massive than the total mass of the cluster. Since nothing at all had been previously known about quasar masses, even this very high upper limit was interesting.

4:4

4:5 THE MOTION OF MOLECULES

An assembly of molecules surrounding an interstellar dust grain exerts pressure on the grain's surface. This pressure arises because the molecules are moving randomly and sometimes collide with the dust. A molecule initially moving toward the grain is deflected at the grain's surface and recedes following the collision. Since the particle's velocity is changed, its momentum $\mathbf{p}$ also is altered. For a brief interval the surface, therefore, exerts a force on the molecule, because, by definition, a force is required to produce the change of momentum. This follows from Newton's law, equation 3–4, which can be rewritten as

$$\mathbf{F} = m\ddot{\mathbf{r}} = \dot{\mathbf{p}} \tag{3–4}$$

If the grain exerts a force on a molecule during a given time interval τ, the molecule too must be reacting on the grain in that time. The sum of all the forces exerted by all the individual molecules impinging on unit grain area at any given time then constitutes the pressure—or force per unit area—acting on the dust.

To calculate the pressure we must first decide how many molecules hit a grain per unit time. Figure 4.5 shows a spherical polar coordinate system by means of which we can label the direction from which the particles initially approach. That direction is given by angles (θ, ϕ). If there are $n(\theta, \phi, v)$ molecules per unit volume coming from an increment of solid angle $d\Omega = \sin\theta \, d\theta \, d\phi$ about the direction (θ, ϕ) with a speed v to $v + dv$, then the number of particles incident on unit surface area in unit time is

$$\int\int\int v \cos\theta \, n(\theta, \phi, v) \sin\theta \, d\theta \, d\phi \, dv \tag{4–25}$$

The factor $\cos\theta$ has to be included because the volume of an inclined cylinder that contains all the incident particles is the product of the base area and the height (Figure 4.6).

Expression 4–25 is proportional to v since particles with larger speeds can reach the impact area from greater distances in any given time interval.

If we assume that each molecule is reflected specularly—as from a mirror— then the angle of incidence is equal to the angle of reflection from the surface, and the total change in momentum for a reflected particle is

$$\Delta p = -2p \cos\theta \tag{4–26}$$

Only the momentum component normal to the surface changes in such a reflection and this gives rise to the factor $\cos\theta$. We can now compute the pressure

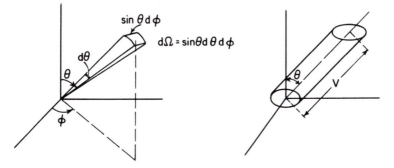

Fig. 4.5 Spherical polar coordinates for computing of pressure.

Fig. 4.6 Inclined cylindrical element containing all molecules striking the surface from direction θ, with speed v, in unit time interval.

that is just (the negative of) the total change of momentum suffered by all molecules incident on unit area in unit time.

$$P = \int_0^{2\pi} d\phi \int_0^{\infty} dv \int_0^{\pi/2} d\theta (2p \cos \theta) \, v \cos \theta \, n(\theta, \phi, v) \sin \theta \qquad (4\text{--}27)$$

In an *isotropic gas* the number of molecules arriving from unit solid angle is independent of θ and ϕ and we can write

$$n(\theta, \phi, v) \, dv = \frac{n(v)}{4\pi} \, dv \qquad (4\text{--}28)$$

Here $n(v)$ is the number density of molecules with speeds in the range v to $v + dv$ and the factor $1/4\pi$ is a normalization constant that arises because 4π steradians are needed to describe all possible approach directions.

Expression 4–28 allows us to separate out a velocity-dependent part of the integrand in (4–27). It is independent of the direction coordinates θ and ϕ. If $v \ll c$—where c is the speed of light—then $p = mv$, where m is the mass of a molecule. We can then write

$$\int_0^{\infty} n(v) \, v^2 \, dv = n \langle v^2 \rangle \qquad (4\text{--}29)$$

where n is the number density of particles per unit volume regardless of speed and direction, and $\langle v^2 \rangle$ is the mean squared value of the velocity. Equation 4–29 simply is a definition of the mean squared velocity.

4:5

The other part of the integral in (4–27) now can be written as

$$\frac{1}{2\pi} \int_0^{2\pi} \int_0^{\pi/2} \cos^2 \theta \sin \theta \, d\theta \, d\phi = \frac{1}{3} = \langle \cos^2\theta \rangle \tag{4–30}$$

This integral defines the mean value of $\cos^2 \theta$ averaged over a hemisphere $0 \leq \theta \leq \pi/2$. This is the hemisphere from which all particles striking the wall must approach. Because of symmetry about $\theta = \pi/2$, the mean squared value of the cosine function actually is 1/3 even if we integrate over all possible directions, rather than just one hemisphere.

Substituting equations 4–29 and 4–30 into 4–27 we can rewrite the expression for pressure as

$$P = \frac{nm\langle v^2 \rangle}{3} \tag{4–31}$$

Writing the product of the pressure P with the volume V that encloses N particles of the assembly, we then have the expression

$$PV = \frac{Nm\langle v^2 \rangle}{3} = N\Theta \tag{4–32}$$

where $N = nV$ and $\Theta = m\langle v^2 \rangle/3$.

PROBLEM 4–6. The random velocity of galaxies is thought to amount to $v \sim 100 \, \text{km sec}^{-1}$. Their number density is $n \sim 10^{-1} \, \text{Mpc}^{-3}$. Typical galaxies have a mass of 3×10^{44} g. What is the cosmic pressure due to galaxies? This pressure will be considered in Chapter 10, where the cosmic pressure needs to be known to determine the dynamics of the universe in its expansion or contraction.

PROBLEM 4–7. The number density of stars in the sun's vicinity is $n \sim 10^{-57} \, \text{cm}^{-3}$. The sun's velocity relative to these stars is $v \sim 2 \times 10^6 \, \text{cm sec}^{-1}$ and we can take the cross section for collision with another star to be $\sigma \sim 5 \times 10^{22}$ cm^2. In the Jeans theory of the birth of the solar system, such an encounter is considered responsible for the formation of the planets. How probable is it that the sun would have formed planets in $P = 5 \times 10^9$ y? How many planetary systems would we expect altogether in 'the Galaxy if there are 10^{11} stars and if the sun is representative?

4:5

4:6 IDEAL GAS LAW

Tenuous gases obey a simple law at temperatures far above the temperature of condensation. This law relates the temperature of a gas to its pressure and density. Since it becomes exact only at extremely high temperatures and low densities, it represents an idealization that a real gas can only approach; and we speak of the *ideal gas law*. In practice, deviations from ideal behavior are small for a large variety of gases in many different situations, and the law is therefore very useful.

To understand this law, we must first know what is to be meant by *temperature*. We can easily "feel" whether a body is hot or cold; but it is not simple to describe this feeling in terms of a measurable physical quantity. One way of prescribing temperatures is in terms of a device, for example, an ordinary mercury bulb thermometer. When the thermometer is dipped into a bowl of water that feels hot, the mercury expands out of the bulb and rises in the capillary tube. When the thermometer is placed into a cold bowl of water the mercury contracts. We can attach an arbitrary scale to the capillary portion of the thermometer and take readings to obtain the temperature in terms of the location of the mercury meniscus in the capillary. To show just how arbitrary such a scale may be, we need only recall that there are at least five different temperature scales in common use in the Western world.

Choosing a given mercury thermometer as a standard, we can make observations of the behavior of gases and eventually arrive at a relation between the density, pressure, and temperature of a given gas. This relation is called an *equation of state*. It has the functional form

$$F(T, P, \rho) = 0 \tag{4-33}$$

The density is sometimes expressed in terms of its reciprocal, the volume per unit mass, or more often in terms of the *molar volume*, or volume per mole of gas. The *mole* is a quantity of matter represented by $\mathcal{N} = 6.02 \times 10^{23}$ molecules. $\mathcal{N}$ is called *Avogadro's number*. Avogadro's number is the number of atoms of the carbon isotope C^{12} weighing exactly 12 grams—one gram-atomic-weight of C^{12}.

Writing the molar volume as $\mathscr{V}$, we obtain the ideal gas law as

$$P\mathscr{V} = RT \tag{4-34}$$

where R is a constant called the *gas constant*. At constant pressure the volume of a given amount of gas increases linearly with temperature. At fixed volume the pressure rises linearly with temperature. Some gases, notably helium, behave very nearly like an ideal gas and can, therefore, be used to define a *gas thermometer* temperature scale. The important point to realize is that, in any event, temperature has to be defined operationally in terms of a convenient device.

We note the similarity between equations 4–32 and 4–34. When N in equation 4–32 is chosen to be Avogadro's number $\mathcal{N}$, we find that

$$\frac{RT}{\mathcal{N}} = \Theta = \frac{m\langle v^2 \rangle}{3} = mc_s^2 \qquad (4\text{–}35)$$

Here we have introduced a new symbol c_s, standing for the *speed of sound* in an ideal gas, mainly because it is interesting to note that sound and pressure waves are propagated at a speed equal to the root mean square velocity component of the gas molecules along the direction of propagation. In solids and liquids incompressibility—stiffness—dominates over molecular motion in propagating pressure, and the speed of sound is appreciably higher than the molecular speeds (Mo68).

We can define a new constant $k = R/\mathcal{N}$, called Boltzmann's constant. Equation 4–35 then becomes

$$\frac{3}{2}kT = \frac{m\langle v^2 \rangle}{2} \qquad (4\text{–}36)$$

The right side of equation 4–36 is the mean kinetic energy per particle in the assembly, and the temperature is therefore nothing other than an index of the mean kinetic energy. In a hot gas the molecules move at high velocity; in a cooler gas they move more slowly. The Boltzmann constant k has to be experimentally determined by direct or indirect measurement of the kinetic energy of molecules in a gas at a given temperature: $k = 1.38 \times 10^{-16}$ erg $°K^{-1}$.

Equation 4–32 can now be rewritten as

$$P\mathcal{V} = \mathcal{N}kT \qquad \text{or} \qquad P = nkT \qquad (4\text{–}37)$$

This is straightforward as long as we deal with one particular kind of gas or one given type of molecule. But what happens if the gas consists of a mixture of different molecules? The kinetic theory developed thus far predicts that the total pressure should still be determined by the total number density of molecules as prescribed by equation 4–37. If there are j different kinds of molecules present in thermal equilibrium, each with number density n_i, the complete relation would read

$$P = \sum_{i=1}^{j} P_i = \sum_{i=1}^{j} n_i kT = nkT \qquad (4\text{–}38)$$

where P_i is the *partial pressure* exerted by molecules of type i alone. Equation 4–38 expresses *Dalton's law* of partial pressures: The total pressure of an ideal gas is the sum of the partial pressures of the various constituents.

PROBLEM 4–8. Interstellar atomic hydrogen is often found in neutral, H I clouds whose temperature is 100°K. What is the root mean squared velocity at which the hydrogen atoms travel? If the number density $n = 1$ cm^{-3}, what is the pressure in interstellar space?

PROBLEM 4–9. These clouds also contain dust grains that might characteristically have diameters 5×10^{-5} cm and unit density. Treating the dust as though it were an ideal gas, what would be the random velocity of dust grains?

PROBLEM 4–10. If the gas had systematic velocity v relative to the dust grains, how much momentum would be transferred to each dust grain per unit time, and what is the acceleration? Assume that the gas density $n = 1$ cm^{-3}, $v = 10^6$ cm sec^{-1}, and that the gas atoms stick to the grain in each collision.

PROBLEM 4–11. What would be the rate of mass gain for this grain? How soon would its mass increase by 1%?

PROBLEM 4–12. In an ionized hydrogen (H II) region, protons and electrons are dissociated. If the temperature of this interstellar gas is 10^{4}°K, calculate electron and proton velocities.

4:7 RADIATION KINETICS

Electromagnetic radiation is transmitted in the form of photons—discrete quanta having momentum p and energy $\mathscr{E}$. The experimentally determined relationship between the spectral frequency v—color of the radiation—and the energy and momentum is

$$p = \frac{hv}{c} \tag{4-39}$$

$$\mathscr{E} = hv \tag{4-40}$$

where h is Planck's constant, c is the speed of light, and v is the spectral frequency.

We can substitute expression 4–39 into the pressure equation 4–27, replacing v by c, and neglecting the integration over velocity since all photons have the same speed c. Expression 4–27 then reads

$$P(v)\, dv = \int_0^{2\pi} d\phi \int_0^{\pi/2} d\theta\, \frac{2hv}{c} \cos\theta\, c \cos\theta\, n(\theta, \phi, v) \sin\theta\, dv \tag{4-41}$$

The two factors c cancel, and hv can be replaced by $\mathscr{E}$. For an isotropic radiation field, $n(\theta, \phi, v) = n(v)/4\pi$, and use of equation 4–30 leads to

$$P(v) = \frac{n(v)\,\varepsilon}{3} = \frac{hvn(v)}{3} \tag{4–42}$$

If there are quanta of j different spectral frequencies present, expression 4–42 becomes

$$P = \frac{U}{3} \tag{4–43}$$

where U is the total energy density summed over all spectral frequencies:

$$U = \sum_{i=1}^{j} n_i h v_i \tag{4–44}$$

PROBLEM 4–13. The radiation energy density in the universe is of the order of $6 \times 10^{-13}\,\text{erg cm}^{-3}$, if primarily a $3°\text{K}$ radiation field (4:13) is considered to exist on a cosmic scale. What is the pressure due to this field and how does it compare to the galactic pressure of Problem 4–6?

PROBLEM 4–14. The radiation energy incident from the sun on unit area per unit time is $1.37 \times 10^6\,\text{erg cm}^{-2}\,\text{sec}^{-1}$, at the earth. This quantity is called the *solar constant*. Find the radiative repulsive force on a 10^{-2} cm diameter black (totally absorbing) grain, at the distance of the earth.

PROBLEM 4–15. A 10^{-4} cm radius grain absorbs $\frac{1}{3}$ of solar radiation incident on its surface and scatters the remainder isotropically. Calculate the ratio of gravitational attraction to radiative repulsion from the sun, assuming that the grain has density $6\,\text{g cm}^{-3}$. Show that this ratio is constant as a function of distance from the sun.

PROBLEM 4–16. If the repulsive force of radiation for a grain is $\frac{1}{3}$ the attraction to the sun due to gravitation, we can define an "effective" gravitational constant $G_{\text{eff}} = \frac{2}{3}G$, where G is the gravitational constant. This will characterize the motion of the grain. What is the orbital period of such a grain moving along the earth's orbit? How does its orbital velocity compare to that of the earth?

4:8 ISOTHERMAL DISTRIBUTIONS

We say that a gas is *isothermal* if its temperature is the same throughout the volume it occupies. Consider an isothermal spherically symmetric gas con-

4:8

figuration in space. There is a decreasing gas density and pressure at increasing central distance r. The pressure change dP between positions r (Fig. 4.7) and $r + dr$ is given by the gravitational force acting on matter between r and $r + dr$:

$$dP = - dr\, \rho(r) \nabla \mathbb{V}(r) \tag{4–45}$$

Here $\rho(r) = n(r)\, m$ and $\mathbb{V}(r)$ is the gravitational potential due to mass enclosed by the sphere r. For an ideal gas (see equation 4–38) $P/\rho = kT/m$. Dividing this expression into equation 4–45 we have

$$\frac{dP}{P} = - \frac{m}{kT} \nabla \mathbb{V}(r)\, dr$$

which integrates to

$$P = P_0 e^{-m \mathbb{V}(r)/kT} \tag{4–46}$$

Reapplying the ideal gas law, we can also obtain the forms

$$n = n_0 e^{-m\mathbb{V}(r)/kT} \qquad \text{or} \qquad \rho = \rho_0 e^{-m\mathbb{V}(r)/kT} \tag{4–47}$$

The exponential term appearing in equations 4–46 and 4–47 is called the *Boltzmann factor*. It plays an important role throughout the theory of statistical thermodynamics and, as we will see in section 4:21, gives a useful starting point for describing the distributions of stars in globular clusters, and molecules in protostars.

Fig. 4.7 Pressure-distance relation for a spherically symmetric configuration.

4:9 ATMOSPHERIC DENSITY

Using equation 4–47, we can readily find the density distribution in the atmosphere of a star, planet, or satellite. In what follows we will keep referring to the parent body as a planet, but the theory holds equally well for a star, moon, or any other massive body.

The gravitational potential at any location in the atmosphere is given by

$$\mathbb{V}(r) = - \frac{MG}{r} \tag{4–48}$$

where r is the distance measured from the center of the planet and M is its mass. Expression 4-48 also assumes that the atmosphere is tenuous so that M can be assumed to be constant and independent of r. Let R be the planet's radius, and consider a point at height x above the surface. The difference between the potential at height x and at the surface is

$$\mathbb{V}(R + x) - \mathbb{V}(R) = -\frac{MG}{R + x} + \frac{MG}{R} = \frac{MGx}{R^2}, \quad x \ll R \qquad (4\text{-}49)$$

Equation 4-47 then becomes

$$n = n_0 e^{-(mMG/kTR^2)x} = n_0 e^{-mgx/kT} \qquad (4\text{-}50)$$

where n_0 now represents the density at the surface and $MG/R^2 \equiv g$ is the *surface gravity* of the planet. It is clear that the atmospheric density decreases exponentially with height. Moreover, we can define a *scale height*

$$\Delta = \frac{kTR^2}{mMG} = \frac{kT}{mg} \qquad (4\text{-}51)$$

We see that the density at height $x + \Delta$ is reduced by a factor e below the value at height x. It is worthwhile noting that the scale height is small for low temperature gases composed of heavy molecules—m large—and for dense parent bodies—large M, small R.

PROBLEM 4-17. Show that an atmosphere consisting of a combination of gases has a variety of scale heights, one for each gas component present. Show that the total pressure is

$$P = \sum_i P_i = \sum_i P_{i0} e^{-(m_i gx/kT)} \qquad (4\text{-}52)$$

and that the total density is

$$\rho = \sum_i n_i m_i = \sum_i n_{i0} m_i e^{-(m_i gx/kT)} \qquad (4\text{-}53)$$

where the subscript 0 denotes a value at the base of the atmosphere. Assume that there is no convection in the atmosphere. (Convection normally requires bulk motion of entire volumes of gas and gives rise to winds that do not allow complete separation of different gaseous constituents. The concept of scale height then cannot be applied.

The earth's atmosphere exhibits some of these features. At the low densities found in the upper atmosphere, there is some separation of gases with different scale heights. Helium, for example, appears in appreciable concentrations only at high altitudes. In the lower atmosphere three features complicate any analysis. There are winds, temperature gradients, and atmospheric water vapor. The vapor is near the condensing point and a local atmospheric temperature drop can give rise to condensation and a decrease in pressure. This gives rise to winds. More important, the lower atmosphere is not isothermal and does not behave in the simple way described here.

PROBLEM 4-18. The mass of the atmosphere is negligible compared to the mass of the earth. If the gravitational attraction at the surface of the earth is 980 dynes g^{-1}, calculate the scale height of the atmosphere's main constituent, molecular nitrogen N$_2$.

4:10 PARTICLE ENERGY DISTRIBUTION IN AN ATMOSPHERE

The exponential decline of particle density with height is an important clue to the velocity distribution of particles. We note that molecules at a height x_1, having an upward directed velocity component $v_x = (2gh)^{1/2}$, have enough energy to reach a height $x_1 + h$. Whether a given molecule with this instantaneous velocity actually reaches height $x_1 + h$ cannot be predicted. The molecule might collide with another one, and lose most of its energy. However, as long as thermal equilibrium exists, and the gas temperature remains stable, we can be sure that, for every molecule that loses energy through a collision, there will be a *restituting collision* at some nearby point in which some other molecule gains a similar amount of energy. This concept, sometimes referred to as *detailed balancing*, allows us to neglect the effect of collisions in the remainder of our argument.

Since the temperature is the same at all levels of an isothermal atmosphere, the velocity distribution must also be the same everywhere, and only the number of particles changes with altitude. The ratio of the particle densities at heights $x_1 + h$ and x_1, (see equation 4–53) is $\exp(-mgh/kT)$. Since the particles encountered at height $x_1 + h$ have all come up from the lower height x_1, to which they will eventually return—fall back—we can be certain that the fraction of particles passing through a plane at height x_1 and having speeds greater than $v_h = (2gh)^{1/2}$ is going to be precisely that fraction of particles having enough energy to reach altitudes above $x_1 + h$. We can therefore express the two-way flux of particles with vertical velocity v_x greater than v_h as

$$\frac{N(v_x > v_h)}{N(v_x > 0)} = \left[\int_h^\infty e^{-mgx/kT} \, dx \Big/ \int_0^\infty e^{-mgx/kT} \, dx \right] = e^{-mgh/kT} = e^{-mv_h^2/2kT} \quad (4\text{–}54)$$

4:10

Note that N is not a density; it is a number crossing unit area in unit time interval. Note also that collisions will isotropize the velocity distribution. Hence, we consider a velocity distribution $f(v)$ that is normalized by the integral

$$\iiint_{-\infty}^{\infty} f(v_x, v_y, v_z)\, dv_x\, dv_y\, dv_z = 1 \tag{4-55}$$

As a trial solution for the function f, we can use an exponential v_x dependence, like that given by equation 4–54. The isotropy requirement then demands a similar dependence on v_y and v_z, and equation 4–55 gives the full function as

$$f(v_x, v_y, v_z) = \left(\frac{m}{2\pi k T}\right)^{3/2} e^{-(m/2kT)(v_x^2 + v_y^2 + v_z^2)} \tag{4-56}$$

where the coefficient is a normalization factor required by (4–55). This function is separable in the variables v_x, v_y, and v_z. To test whether it also obeys equation 4–54 we note that

$$\frac{N(v_x > v_h)}{N(v_x > 0)} = \frac{\displaystyle\int_{v_h}^{\infty} v_x e^{-(m/2kT)v_x^2}\, dv_x}{\displaystyle\int_0^{\infty} v_x e^{-(m/2kT)v_x^2}\, dv_x} = e^{-mv_h^2/2kT} \tag{4-57}$$

The quantity v_x in the integrand plays the same role here as in equation 4–27. It takes into account that the higher velocity particles can reach a given surface from a larger distance, and from a larger volume, in unit time. We can write the distribution (4–56) in terms of the speed

$$v = (v_x^2 + v_y^2 + v_z^2)^{1/2} \tag{4-58}$$

We then obtain

$$f(v) = \left(\frac{m}{2\pi k T}\right)^{3/2} e^{-mv^2/2kT} \tag{4-59}$$

PROBLEM 4-19. Satisfy yourself that the normalization condition for $f(v)$ is

$$4\pi \int_0^{\infty} f(v)\, v^2\, dv = 1 \tag{4-60}$$

Show also that, in terms of momentum, the distribution function is

$$f(p) = \frac{1}{(2\pi m k T)^{3/2}} e^{-p^2/2mkT} \tag{4-61}$$

and

$$\int_0^\infty 4\pi f(p)\, p^2\, dp = 1 \qquad (4\text{-}62)$$

Note that equations 4–56, 4–59, and 4–61 all are independent of the gravitational potential initially postulated. The equations derived here therefore have much wider applicability than just to the gravitational problem. We will discuss this further in section 4:15.

The velocity and momentum distribution functions (4–59) and (4–61) are called Maxwell-Boltzmann distributions, after L. Boltzmann and J. C. Maxwell who were two of the founders of classical kinetic theory. The momentum distribution is plotted in Fig. 4.8. These distribution functions have extremely wide application.

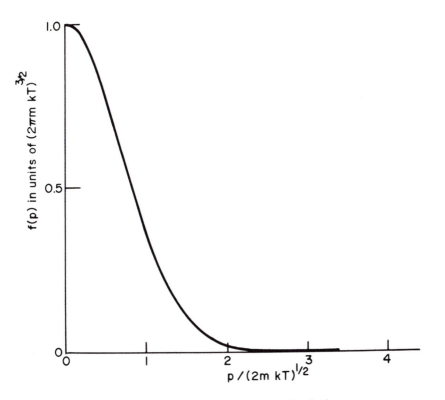

Fig. 4.8 Maxwell-Boltzmann momentum distribution.

PROBLEM 4–20. If the moon had an atmosphere consisting of gases at $300°K$, calculate the mass of the lightest gas molecules for which $3kT/2 < MmG/R$. m is the mass of the molecule, M and R are the mass and radius of the moon, respectively, 7.3×10^{25} g and 1.7×10^8 cm. Note that the quantity on the left is related to the escape velocity at the moon. What is this velocity? Actually heavier molecules than those with mass m, calculated above, can escape from the moon, because (a) in a Maxwell-Boltzmann distribution, gases have many molecules with speeds larger than the mean speed and (b) because the sunward side of the moon reaches temperatures near $400°K$ during the two weeks it is exposed to the sun.

Despite their great usefulness, Maxwell–Boltzmann statistics cannot be applied under certain conditions, such as those encountered at high densities in the centers of stars. Neither do they apply to radiation emitted by stars. There, we need to consider quantum effects that have no classical basis. The next few sections describe these effects.

4:11 PHASE SPACE

The quantum effects that lead to deviations from classical statistical behavior always involve particles that are identical to each other. We might deal with electrons that have almost identical positions and momenta, and spin; or we might have photons with identical frequency, position, direction of propagation, and polarization.

For the case of electrons an important restriction comes into play. The *Pauli exclusion principle* forbids any two electrons from having identical properties. *Neutrons, protons, neutrinos,* and, in fact, all particles with odd half-integral *spin* $(\frac{1}{2}, \frac{3}{2}, ...)$ also obey this principle. *Photons* and *pions*, on the other hand, have integral or zero spin, and any number of these particles can have identical momenta, positions, and spins. The first group of particles—those that obey the injunction of the Pauli principle—are called *Fermi–Dirac particles* or *Fermions*; the others are called *Bosons* and their behavior is governed by the *Bose–Einstein statistics*.

Thus far we have not stated what we mean by "identical." Clearly we could always imagine an infinitesimal difference in the momenta of two particles, or in their positions. Should such particles still be termed identical, or should they not? The question is essentially answered by *Heisenberg's uncertainty principle*, which denies the possibility of distinguishing two particles if the difference in the momentum δp, multiplied by the difference in position δr, is less than h. This restriction derives from the uncertainty in the simultaneous measurement of momentum and position components for any given particle

$$\Delta p_x \Delta x \sim \hbar, \quad \Delta x \equiv \langle (x - \langle x \rangle)^2 \rangle^{1/2} = \langle x^2 - \langle x \rangle^2 \rangle^{1/2}$$

$$\Delta p_x = \langle p_x^2 - \langle p_x \rangle^2 \rangle^{1/2} \tag{4-63}$$

where $\hbar = h/2\pi$, and h is *Planck's constant*. If two particles are to be distinguishable, then $\delta p_x \delta x$ should be somewhat greater than $h = 6.626 \times 10^{-27}$ erg sec.

Quantum mechanically one can show that two particles are to be considered identical if their momenta and positions are identical within values

$$\delta p_x \delta x = h, \quad \delta p_y \delta y = h, \quad \delta p_z \delta z = h \tag{4-64}$$

and if their spins are identical.

In this description each particle is characterized by a position (x, y, z, p_x, p_y, p_z) in a six-dimensional *phase space*. It occupies a six-dimensional *phase cell* whose volume is (Figs. 4.9 and 4.10):

$$\delta x \delta y \delta z \delta p_x \delta p_y \delta p_z = h^3 \tag{4-64a}$$

Particles within one phase cell are identical—physically indistinguishable—while those outside can be distinguished. Since δx is the dimension of the phase cell, it must be at least twice as large as Δx, the root mean square deviation from the central position. The same relation holds between δp_x and Δp_x. That is why the right side of equation 4–63 involves $\hbar$ while equation 4–64 contains the larger value, h (Fig. 4.9).

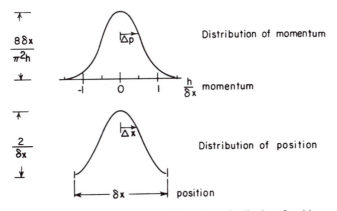

Fig. 4.9 Relation between phase cell dimensions, distribution of positions and momenta, and uncertainties in these variables. Only the simplest of a large family of distribution functions corresponding to different energies are shown.

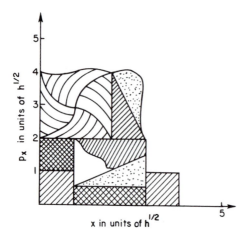

Fig. 4.10   Phase space is a six-dimensional hypothetical space, having three momentum and three spatial dimensions. Projected onto the $p_x - x$ plane, individual cells always present an area h. Although their shapes may be quite arbitrary, as shown, it is often useful to think of them as square or rectangular since that makes computations simpler. In section 4:14 (Fig. 4.13) we will see how an initially rectangular cell becomes distorted.

We can now ask how many electrons could be fitted into a box with volume V? The answer depends on how high a particle momentum we wish to consider. If momenta up to a maximum value p_m are permitted, the available volume in phase space is $2 \cdot 4\pi/3 \cdot p_m^3 \cdot V$. The factor 2 accounts for two distinct spin polarizations, since electrons whose spins differ, always can be distinguished and therefore must belong to different phase cells. Hence the number of available phase cells is $8\pi/3 \cdot p_m^3 \cdot V/h^3$, and that also is the maximum number of electrons that could occupy the box.

In general, the number of phase cells with momenta in a range p to $p + dp$ is

$$Z(p)\, dp = 2V \frac{4\pi p^2\, dp}{h^3} \qquad (4\text{--}65)$$

At the center of a star, ionized matter is sometimes packed so closely that all the lowest electron states are filled. Further contraction of the star can then force the electrons to assume much higher momenta than the value $(3kTm)^{1/2}$ normally found in tenuous gases. Such a closely packed gas of Fermions is said to be *degenerate*. We will study this form of matter in section 4:14 and in Chapter 8, where very dense cores of stars are discussed.

Sometimes we may prefer to talk about *frequency space* instead of *momentum space*. Defining the particle frequency v, by $v \equiv pc/h$, we obtain the number of phase cells with frequencies between v and $v + dv$ as

$$Z(v)\, dv = 2\left[\frac{4\pi v^2\, dv}{c^3} \right] V \qquad (4\text{--}65a)$$

4:11

4:12 ANGULAR DIAMETERS OF STARS

The fact that two photons sometimes occupy the same phase cell allows us to measure the angular diameter of stars. The idea is this: Two photon counters are placed a distance D apart, transverse to the direction of the star. If D is small enough, we have the possibility that one photon from a cell will hit one detector, while the other photon hits the other detector, the simultaneous arrival being detected by a coincidence counter. Let the diameter of the star be d and its distance R. The angle it subtends is $\theta = d/R$. The photon pair impinging on either detector has a distribution in momentum, along the direction of D, amounting to $\Delta p_D = p\theta = (h v/c)\,\theta$ where v is the frequency of radiation to which the detector is sensitive. But the nonzero value of Δp_D makes it necessary that D itself be small so that photons reaching either detector may be in the same phase cell. That is, it is necessary that

$$D\Delta p_D \lesssim h$$

$$\text{or} \qquad \frac{Dhv}{c}\,\theta \lesssim h \qquad\qquad (4\text{--}66)$$

$$\text{or} \qquad D\theta \lesssim \lambda, \qquad \lambda = c/v$$

λ is the *wavelength* of the radiation. By increasing D a decreasing coincidence rate is observed, and for values of D at which coincidences no longer occur the angular diameter is $\theta \lesssim \lambda/D$. The stellar *angular* diameter (Figure 4.11) is

$$\theta \sim d/R \sim \lambda/D \qquad\qquad (4\text{--}67)$$

in such observations. This technique was first discovered by R. Hanbury Brown and R. Q. Twiss (Ha54). A second method of measuring angular diameters of stars makes use of the Michelson stellar interferometer that essentially depends on the same type of phenomenon in that the normal diffraction peak width again is given by uncertainty principle considerations.

Fig. 4.11   The   Hanbury-Brown-Twiss   interferometer.

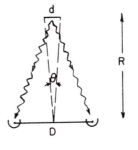

4:13 THE SPECTRUM OF LIGHT INSIDE AND OUTSIDE A HOT BODY

Any opaque body is permeated by a radiation bath. Atoms, molecules, or ions and electrons are continually absorbing and re-emitting quanta of light. From time to time a photon approaches the edge of the body and escapes. This diffusion of photons from the interior of the hot body out to its boundary, and the subsequent escape into empty space is an important process in stars. Energy generated at the center of the star slowly diffuses outward and escapes. The escaping radiation gives the star its luminous appearance.

To understand this phenomenon in some detail, we want to describe the radiation to be expected inside a body. The spectrum of the radiation is going to be different for different temperatures, and we would like to deduce that spectrum and obtain a complete description of the radiation bath.

We first consider a photon gas embedded in material at temperature T. The radiation is in thermal equilibrium with the material if there is ample opportunity for the photons to interact with the atoms through scattering or absorption and re-emission. Two factors have to be considered:

(a) Photons are Bose–Einstein particles and can aggregate in single phase cells.

(b) If the frequency of the photons aggregating in a phase cell is v, and if there are n photons in the cell, we can consider the assembly of photons in this phase cell to be in an energy state with energy $(n + \frac{1}{2})hv$. We sometimes speak of a *quantum oscillator* in the nth state when we describe this phenomenon. Even when a phase cell is completely empty, in the ground state, a residual ground state energy of $hv/2$ is present. This energy normally cannot be observed, because it does not participate in the absorption and emission processes. [However, there exist (Fr46) processes involving the surfaces of deformable bodies in which this ground state energy can give rise to surface tension effects.]

We can compute the probability of finding a quantum oscillator in the nth excited state. The relative probability of that state is given by the Boltzmann factor $e^{-(n+1/2)hv/kT}$. The absolute probability is given by dividing the relative probability by the sum of all the relative probabilities:

$$P(v, T) = \frac{e^{-(n+1/2)hv/kT}}{\sum_n e^{-(n+1/2)hv/kT}} = \frac{e^{-(nhv/kT)}}{\sum_n e^{-nhv/kT}} \tag{4–68}$$

In these terms we can give the average energy $\langle \mathscr{E} \rangle$, per phase cell, of all phase cells corresponding to frequency v. We sum the energies of all the oscillators

4:13

and divide by the total number of oscillators. Choosing $x \equiv h\nu/kT$, we obtain

$$\langle \mathscr{E} \rangle = \sum_n \left(n + \frac{1}{2} \right) h\nu e^{-nh\nu/kT} \left[\sum_n e^{-nh\nu/kT} \right]^{-1}$$

$$= \frac{kT(xe^{-x} + 2xe^{-2x} + 3xe^{-3x} + ...)}{1 + e^{-x} + e^{-2x} + e^{-3x} + ...} + \frac{h\nu}{2} \qquad (4\text{-}69)$$

This allows us to evaluate the denominator of equation 4–69 as $(1 - e^{-x})^{-1}$. To evaluate the numerator, we can make use of the same binomial expansion formula twice in succession.

$$kT\{x(e^{-x} + e^{-2x} + e^{-3x} + ...)$$
$$+ x(e^{-2x} + e^{-3x} + ...)$$
$$+ x(e^{-3x} + ...)$$
$$+ \dots\dots\dots\dots\dots\dots\dots\dots\dots\dots\dots)\}$$

$$= kT\left\{ \frac{xe^{-x}}{1 - e^{-x}} + \frac{xe^{-2x}}{1 - e^{-x}} + \frac{xe^{-3x}}{1 - e^{-x}} + ... \right\}$$

$$= kT \frac{xe^{-x}}{(1 - e^{-x})^2}$$

In these terms

$$\langle \mathscr{E} \rangle = \frac{kTxe^{-x}}{1 - e^{-x}} + \frac{h\nu}{2} = \frac{kTx}{(e^x - 1)} + \frac{h\nu}{2} = \frac{h\nu}{(e^{h\nu/kT} - 1)} + \frac{h\nu}{2} \qquad (4\text{-}70)$$

Knowing the number of phase cells per unit volume, $8\pi\nu^2 \, d\nu/c^3$, and the mean energy per phase cell, we can write the energy density of photons as a function of frequency and temperature. This is the *blackbody radiation* spectrum:

$$\rho(\nu, T) \, d\nu = \frac{8\pi\nu^2 \, d\nu}{c^3} \left(\frac{h\nu}{e^{h\nu/kT} - 1} + \frac{h\nu}{2} \right)$$

$$n(\nu, T) \, d\nu = \frac{8\pi\nu^2}{c^3} \left(\frac{1}{e^{h\nu/kT} - 1} \right) d\nu \qquad (4\text{-}71)$$

As previously stated, the term $h\nu/2$ is not observable in terms of photon absorption or emission. We will therefore neglect it from now on, and concentrate on the remainder of the expression, which can give rise to observable astronomical signals. Integrating equation 4–71 over all frequencies from zero to infinity, we can obtain the total energy density. The total photon density is similarly obtained:

$$\rho(T) = \frac{8}{15}\frac{\pi^5}{c^3}\frac{k^4}{h^3}T^4 = aT^4 = U = 7.56 \times 10^{-15}T^4 \text{ erg cm}^{-3}$$

$$n(T) = \frac{8\pi}{c^3}\int_0^\infty \frac{v^2\, dv}{e^{hv/kT} - 1} = 16\pi\left(\frac{kT}{hc}\right)^3 \zeta(3) \approx 20T^3 \text{ photons cm}^{-3}$$

(4–72)

Here, $\zeta(n)$ is the Riemann zeta function, $\sum\limits_{k=1}^{\infty} k^{-n}$, and $\zeta(3) = 1.20206$.

PROBLEM 4-21. Note that all this is strictly true only if the index of refraction, n, in the medium is $n = 1$. For general values of n, show that

$$\rho(T) = n^3 a T^4$$

This is more generally the case inside a star. Show also what happens if the index of refraction is frequency dependent—which it always is.

Equation 4–72 is a well-known definite integral. We note that the second term in parentheses in equation 4–71 would give rise to an infinite zero point energy as written there. The coefficient of the T^4 term is sometimes abbreviated (see equation 4–72) by the symbol a. We can also define another useful constant $\sigma = ac/4$, *the Stefan–Boltzmann constant*. This constant allows us to write the energy emitted per unit area of a hot blackbody in unit time, as

$$W = \sigma T^4 \tag{4–73}$$

To see this, we can think of photons that escape from the surface as being representative of the density of photons immediately within the surface of the body. Only those photons can be considered with velocities directed outward through the surface. So only one half of the photons come into consideration. These photons have an average velocity component normal to the surface equal to $c\langle\cos\theta\rangle$ where θ is the angle of emission with respect to the direction normal to the surface. We therefore have to evaluate $\langle\cos\theta\rangle$, averaged over all possible angles. This is

$$\langle\cos\theta\rangle = \frac{1}{2\pi}\int_0^{2\pi}\int_0^{\pi/2}\cos\theta\sin\theta\, d\theta\, d\phi = \frac{\cos^2\theta}{2}\Big|_0^{\pi/2} = \frac{1}{2}$$

(4–74)

$$\therefore c\langle\cos\theta\rangle = \frac{c}{2}$$

But since only half the photons are outward directed, the total flux is $1/2 \cdot c/2 \cdot aT^4 = acT^4/4$, as previously stated.

The spectrum of most stars is closely approximated by a blackbody spectrum, with individual spectral emission and absorption lines superposed. To the extent that the blackbody approximation holds, it is possible to ascertain the temperature of the star's photosphere where most of the light is emitted. Using two different wide band filters, say the *B* and *V* filters often used in observations, we can determine the ratio of intensities in these spectral ranges. This ratio can be uniquely related to the temperature. The temperature derived in this way is called the *color temperature*. A useful formula is (A163):

$$T_c = \frac{7300}{(B - V) + 0.73} \tag{4–75}$$

PROBLEM 4-22. Using the effective wavelengths given in Table A.1, compare the ratio of blue and visual intensities predicted, respectively, by equations 4–71 and 4–75 for a star at temperature $6000°K$ (spectral class G) and one at $10,000°K$ (spectral class A).

Another means of defining temperature involves the luminosity of the star. Since the total power emitted per unit area is a function of temperature alone, we can calculate an *effective temperature* T_e for the star if both its luminosity and surface area can be determined:

$$L = \sigma T_e^4 4\pi R^2 \tag{4–76}$$

If the distance of the star is known from observations of the kind described in Chapter 2, the stellar radius can be obtained using the Michelson or Hanbury Brown–Twiss interferometers discussed in section 4:12. From (4–76) it is readily seen that

$$\log \frac{L}{L_\odot} = 4 \log \frac{T_e}{T_{e\odot}} + 2 \log \frac{R}{R_\odot} \tag{4–77}$$

where $T_{e\odot} \sim 5800°K$ and $R_\odot = 6.96 \times 10^{10}$ cm, are the solar values. $T_{e\odot}$ is uncertain by as much as $\sim 50°K$. When the Hertzsprung–Russell diagram is plotted in terms of the logarithm of luminosity and effective temperature, as in Fig. 1.5, stars with identical radii lie on lines of constant slope, as required by equation 4–77.

It is worth mentioning two typical astrophysical situations in which temperature is a useful concept.

(a) Temperatures in the Solar System

The temperature of a black interplanetary object is determined by the energy equilibrium equation

$$\frac{L_\odot}{4\pi R^2} \pi r^2 = \sigma T^4 4\pi r^2$$

where $L_\odot$ is the solar luminosity, R is the distance from the sun, and r is the radius of the object. If the mean efficiency for absorption (in the visible) is ε_a and the mean efficiency of reradiation (at infrared wavelengths) is ε_r, we have

$$T = \left(\frac{\varepsilon_a}{\varepsilon_r} \frac{L_\odot}{16\pi\sigma R^2} \right)^{1/4} \tag{4-78}$$

We note that:

(i) At the earth's distance

$$T \sim \left(\frac{\varepsilon_a}{\varepsilon_r} \right)^{1/4} \left(\frac{4 \times 10^{33}}{16\pi(5.7 \times 10^{-5})2.3 \times 10^{26}} \right)^{1/4} \sim 282 \left(\frac{\varepsilon_a}{\varepsilon_r} \right)^{1/4} \,{}^\circ K$$

(ii) A gray body ($\varepsilon_a = \varepsilon_r$) has the same temperature as a black one.

(iii) For increasing distance from the sun $T \propto R^{-1/2}$.

(iv) If the thermal conductivity of the body is small and its rotation slow, as for the moon, the subsolar point assumes a temperature

$$T \sim \left(\frac{\varepsilon_a}{\varepsilon_r} \frac{L_\odot}{R^2 4\pi\sigma} \right)^{1/4} \tag{4-79}$$

which is $(4)^{1/4} \sim 1.4$ higher than the temperature of an equivalent rotating body.

(b) Radio Astronomical Temperatures

Some characteristics of radio astronomical measurements can be understood in terms of temperatures. At very low frequencies, $v \ll kT/h$—often called the Rayleigh–Jeans limit—the energy density in a source can be written (equation 4–71) as

$$\rho(v) = \frac{8\pi k T v^2}{c^3} = \frac{8\pi k T}{c\lambda^2} \tag{4-80}$$

where $\lambda \equiv c/v$ is the wavelength. The flux emanating per solid angle and area normal to the surface is called the *intensity* $I(v)$, at frequency v. This is the *surface brightness* of the source, and

$$I(v) = \frac{c\rho(v)}{4\pi} = \frac{2v^2 kT}{c^2} = \frac{2kT}{\lambda^2} \tag{4-81}$$

If a flux $I(v)$ is measured in an observation then, regardless of whether the source is thermal, we can pretend that a temperature parameter can be assigned to the observation. This is called the *brightness temperature* T_b and is defined at frequency v as

$$T_b(v) \equiv \frac{I(v)\, c^2}{2kv^2} = \frac{I(v)}{2k}\lambda^2 \tag{4-82}$$

T_b then is the temperature of an ideal blackbody whose radiant energy in the particular energy range v to $v + dv$ is the same as that of the observed source. A related concept is that of *antenna temperature*—which has nothing to do with the temperature that the antenna actually assumes under ambient climatic conditions. To examine this concept we must first consider some practical properties of antennas. In general, an antenna absorbs differing amounts of power depending on the direction of the source. If we draw a *directional diagram* of an antenna, it usually has the shape of Figure 4.12. The response $A(\theta, \phi)$ of the antenna is called its *effective area*. The power absorbed is

$$P \equiv \frac{1}{2} \int A(\theta, \phi)\, I(v, \theta, \phi)\, dv\, d\Omega$$

and for a small source,

$$P(v, \theta, \phi)\, dv = \frac{1}{2} F(v)\, A(\theta, \phi)\, dv, \qquad F(v) = \int I(v)\, d\Omega \tag{4-83}$$

Fig. 4.12 Directional diagram of an antenna, showing a main lobe and a set of sidelobes. The angle θ is the beamwidth (see text).

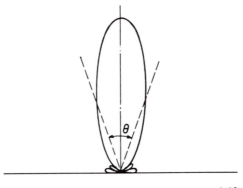

Here $F(v)$ is the flux density at the antenna, and the factor $\frac{1}{2}$ comes about because the antenna only accepts one component of polarization. Now if A is independent of the angle ϕ, and a diagram like Fig. 4.12 is drawn, $A(\theta)$ normally has a very large value in one particular direction, $\theta = 0$, and the large lobe around this direction is called the *main lobe*. The smaller lobes in the diagram are called side lobes. *Back lobes* can also occur. A well-designed radio telescope has a narrow main lobe for greatest positional accuracy, and minimized sidelobes to minimize the confusion produced by sources outside the desired field of view (Sh60).

We can define a mean value of the effective area of the antenna taken over all directions as

$$\langle A \rangle \equiv \frac{1}{4\pi} \int A(\theta, \phi) \, d\Omega \tag{4-84}$$

and the *gain* of the antenna is the dimensionless quantity

$$G(\theta, \phi) = \frac{A(\theta, \phi)}{\langle A \rangle} \tag{4-85}$$

which gives the ratio of the effective area in a given direction to the mean effective area. The function G has a maximum value in the direction $\theta = \phi = 0$ in a properly designed instrument. The *beamwidth* is the angle θ between points in the directional diagram at which $A = A(\theta, \phi)/2$.

In these terms, we can now return to the concept of *antenna temperature* T_a. If a source has directional and spectral brightness $I(v, \theta, \phi)$, then a radio telescope with effective area $A(\theta, \phi)$ receives an amount of power

$$P(v) = \frac{1}{2} \int A(\theta, \phi) \, I(v, \theta, \phi) \, dv \, d\Omega \tag{4-86}$$

On the other hand, we could replace the antenna by a resistor at temperature T connected to the receiver. Such a resistor can be shown experimentally and theoretically to produce *thermal noise* power in an amount

$$P = kT\Delta v \tag{4-87}$$

where Δv is the receiver bandwidth. We can therefore define an antenna temperature T_a, so that

$$T_a = \frac{1}{k\Delta v} \cdot \frac{1}{2} \int A(\theta, \phi) \, I(v, \theta, \phi) \, dv \, d\Omega \tag{4-88}$$

This equation is useful for practical reasons. It is relatively easy to compare the power received from a celestial source to that received from a resistor switched to the receiver input in place of the antenna. The noise in (4–87) is sometimes called *Johnson noise* or *Nyquist noise*. Johnson (Jo28) and Nyquist (Ny28), respectively, supplied the experimental data and theoretical explanation leading to (4–87).

4:14 BOLTZMANN EQUATION AND LIOUVILLE'S THEOREM

Define a function $f(\mathbf{r}, \mathbf{p}, t)$, the density of particles in phase space, so that the number of particles in volume element $d\mathbf{r}$ at position $\mathbf{r}$, having momenta that lie in some momentum-space volume $d\mathbf{p}$, is $f(\mathbf{r}, \mathbf{p}, t) \, d\mathbf{r} \, d\mathbf{p}$. We ask how the function f evolves with time. Since each particle in the assembly can be described in terms of three momentum and three spatial coordinates, the general form of the equation reads

$$\frac{\partial f}{\partial t} + \sum_i \frac{\partial f}{\partial \mathbf{r}_i} \frac{d\mathbf{r}_i}{dt} + \sum_i \frac{\partial f}{\partial \mathbf{p}_i} \frac{d\mathbf{p}_i}{dt} = \frac{df}{dt}\bigg|_{\text{collisions}} \qquad (4\text{–}89)$$

The left side of this equation gives the time rate of change of particles in the volume element $d\mathbf{r} \, d\mathbf{p}$ as a function of the coordinates $\mathbf{r}_i$, $\mathbf{p}_i$, $i = 1, 2, 3, \ldots, n$, for an n particle assembly. As the particles move, the surface enclosing them in phase space becomes distorted and the expression gives the rate of change of density through this distortion and through any other effects. The right side gives the loss or gain of particles through collisions. Equation 4–89 is called the *Boltzmann equation*.

To see how the evolution proceeds in the case of a collisionless situation, in which the right side of equation 4–89 is zero, we draw a simple two-dimensional picture. In Figure 4.13 we have an assembly of particles initially confined between positions $\mathbf{r}_1$ and $\mathbf{r}_2$ and between momentum values $\mathbf{p}_a$ and $\mathbf{p}_b$. Some time later, the momentum values are unchanged, but the particles have moved so that the higher momentum particles are now at positions between $\mathbf{r}_1'$ and $\mathbf{r}_2'$ while the lower momentum particles are at positions between $\mathbf{r}_1''$ and $\mathbf{r}_2''$. However, since the base and height of the enclosing area has not changed, the number density of particles per unit area has remained constant.

A similar argument holds when forces are applied to the particles. In that case the momenta of particles are not constant and the parallelepiped in Fig. 4.13 will also be displaced in a vertical direction. However, a similar argument can then be applied to show that the area covered by the particles still remains constant and the density of particles in this two-dimensional situation is unchanged. This is particularly easy to see if the force is the same on all particles. In that case, $d\mathbf{p}/dt$ is uniform and the difference in values $\mathbf{p}_a - \mathbf{p}_b$ is maintained constant. When different forces are exerted on differing constitutents of the gas, the area occupied

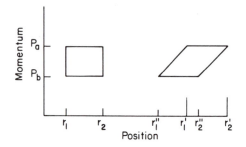

by each constitutent remains constant. These results also hold when there are gradients in the force fields.

A further extension of the argument can be applied to the full six-dimensional situation. Unless there are some means for creating or destroying particles in the assembly (such as through collisions or through, say, particle-antiparticle pair formation) the density of particles will be constant along the trajectory in the six-dimensional space.

This is the sense of *Liouville's theorem*: The six-dimensional space density of particles in an assembly remains unchanged unless collisions occur:

$$df/dt = 0 \qquad\qquad (4-90)$$

This theorem has interesting applications to cosmic ray particles. Many of these particles are so energetic that they must be able to escape from the magnetic fields resident in our Galaxy. Their density in space outside the Galaxy must therefore be the same as the density that we measure in the vicinity of the earth, provided only that they have had enough time since their creation to traverse distances comparable to those of remote clusters of galaxies. This follows at once if we consider an assembly of energetic particles passing by the earth. These particles move on throughout the Galaxy, guided by magnetic field lines (see section 6:6 for further discussion of this topic). Eventually they escape the Galaxy, because the Galaxy's magnetic field is not strong enough to keep them trapped. As these particles go out into intergalactic space their density in phase space must still be constant. If the arrival rate of cosmic ray particles at the earth could be shown to be constant in time this would mean that the spatial density of cosmic rays in extragalactic space is the same as that measured at the earth. This argument need not be true for low energy particles if these particles can remain bottled up in local magnetic fields within our Galaxy. However, if these magnetic fields allow a small leakage of low energy particles into extragalactic space, Liouville's theorem again demands that the density of the particles eventually become uniform throughout these accessible spaces. A local probe of cosmic ray particle intensities can, for this reason, potentially be useful in giving information on particle densities throughout the universe. On the other hand, we should not be overly optimistic. Proposals that supernovae or pulsars in our Galaxy may be responsible for most of the cosmic ray particles we see, suggest that local measurements at the earth may not have a direct bearing on the density of particles in extragalactic space, because there may not have been enough time to produce a homogeneous distribution of cosmic ray particles in extragalactic regions.

Another interesting application of Liouville's theorem concerns the use of optical telescopes in concentrating light beams onto small detectors. In many applications we could obtain very high instrumental sensitivity if light from some cosmic source could be concentrated onto the smallest possible detector. Let the solid angle subtended by the astronomical object be Ω and the telescope area be A. Then Liouville's theorem states that the smallest detector area onto which the light could be focused is

$$a = \frac{A\Omega}{4\pi} \tag{4-91}$$

and that is only possible if light can be made to impinge on the detector from all sides. Usually we are able to make light fall onto the detector only from some smaller solid angle $\Omega' < 4\pi$ so that the minimum area of the detector becomes

$$a = A\Omega/\Omega' \tag{4-92}$$

A violation of this situation would also imply a violation of the second law of thermodynamics which states that heat cannot freely flow from a cold to a hot object. For the density of photons in phase space would then be greater at the detector than at the source. That would imply that the radiation temperature at the source was lower than at the detector and that radiation was actually flowing from a cooler to a hotter object.

Finally we should still mention the problem discussed in section 3:16, where a swarm of particles moves through a gravitational field. There we were concerned with tidal disruption of globular clusters, but noted that while the clusters became extended along a direction pointing toward the Galactic center, the gravitational forces also tended to produce a compression lateral to that direction. This compression produces some additional transverse velocities so that the overall evolutionary pattern becomes quite complex. Liouville's theorem, however, gives us at least one solid guide toward understanding the overall development. It tells us that whatever detailed dynamical arguments we apply— such as those of section 3:16—the results must always agree at least with Liouville's requirement of a constant phase space density.

4:14

4:15 FERMI-DIRAC STATISTICS

In a *Fermi–Dirac assembly* a phase cell can contain only one particle or none. For any given assembly there exists a *Fermi energy*, $\mathscr{E}_F$, up to which all states are filled at zero temperature. At $T > 0$ excitation to a higher state of energy $\mathscr{E}$ can take place. The relative probabilities of being at energy $\mathscr{E}$ and αkT are, respectively,

$$e^{-(\mathscr{E} - \alpha kT)/kT} \quad \text{and } 1 \tag{4-93}$$

The relative probability of occupancy of a state of energy $\mathscr{E}$, in an assembly at temperature T, therefore, is

$$\frac{e^{\alpha - \mathscr{E}/kT}}{1 + e^{\alpha - (\mathscr{E}/kT)}} = \frac{1}{1 + e^{(\mathscr{E}/kT) - \alpha}} \tag{4-94}$$

Here we have not specified the energy αkT, but we can see that at very low temperatures, $T \sim 0$, αkT must approach $\mathscr{E}_F$ because the Fermi function

$$F(\mathscr{E}) = [1 + e^{(\mathscr{E} - \mathscr{E}_F)/kT}]^{-1} \tag{4-95}$$

has the form shown in Fig. 4.14.

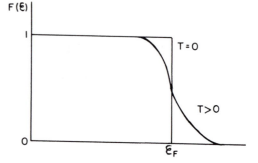

Fig. 4.14   The Fermi function $F(\mathscr{E})$.

We define the Fermi energy $\mathscr{E}_F$ as that energy for which $F(\mathscr{E}) = \frac{1}{2}$. Note that for $T = 0$, the exponent in (4–95) has a large absolute value, whenever $\mathscr{E} - \mathscr{E}_F \neq 0$, so that

$$F(\mathscr{E}) = 1 \quad \text{for} \quad \mathscr{E} < \mathscr{E}_F$$

$$F(\mathscr{E}) = 0 \quad \text{for} \quad \mathscr{E} > \mathscr{E}_F \tag{4-96}$$

This gives rise to the step function in Figure 4.14. Whenever all the available energy levels are filled—which means whenever $T = 0$—we say that the gas of

4:15

Fermions is completely *degenerate*. When $T > 0$, the step is seen to roll off more gently. The product of this probability $F(\mathscr{E})$ and $\mathscr{E}$ gives a mean value for the energy contained in all phase cells corresponding to an energy $\mathscr{E}$. Filled as well as empty cells have to be considered to obtain this value.

$$\text{mean value} = \frac{\mathscr{E}}{1 + e^{(\mathscr{E}/kT) - \alpha}} \tag{4-97}$$

We know that the number of states in the momentum range

$$p \quad \text{to} \quad p + dp \quad \text{is} \quad Z(p)\,dp = \frac{8\pi p^2 V\,dp}{h^3}$$

But

$$\mathscr{E} = \frac{p^2}{2m}, \qquad d\mathscr{E} = \frac{p}{m}\,dp$$

$$\therefore \; Z(\mathscr{E})\,d\mathscr{E} = \frac{8\pi}{h^3}\,V\sqrt{2m\mathscr{E}}\;m\,d\mathscr{E} = \frac{4\pi V}{h^3}(2m)^{3/2}\,\mathscr{E}^{1/2}\,d\mathscr{E} \tag{4-98}$$

The overall mean energy of the particles integrated over all values $\mathscr{E}$ is therefore

$$\langle \mathscr{E} \rangle = \frac{\displaystyle\int_0^\infty Z(\mathscr{E})\,F(\mathscr{E})\,\mathscr{E}\,d\mathscr{E}}{\displaystyle\int_0^\infty Z(\mathscr{E})\,F(\mathscr{E})\,d\mathscr{E}} \tag{4-99}$$

Again, setting αkT equal to the Fermi energy, $\mathscr{E}_F$, for an assembly of particles at temperature $T = 0$, we obtain

$$\langle \mathscr{E} \rangle_{T=0} = \int_0^{\mathscr{E}_F} \mathscr{E}^{3/2}\,d\mathscr{E} \left[\int_0^{\mathscr{E}_F} \mathscr{E}^{1/2}\,d\mathscr{E} \right]^{-1}$$

$$\langle \mathscr{E} \rangle_{T=0} = \frac{3}{5}\mathscr{E}_F \tag{4-100}$$

One can show that for $T > 0$, $\mathscr{E}_F < \mathscr{E}_{F_0}$. The Fermi energy drops slightly. For $\mathscr{E} - \mathscr{E}_F \gg kT$, that is, in the limit of large particle energies, we have

$$F(\mathscr{E}) \sim e^{-(\mathscr{E} - \mathscr{E}_F)/kT}$$

which approaches a Boltzmann distribution for very energetic Fermions.

4:15

At the center of stars degenerate conditions often exist. This is true mainly for electrons since at a given energy $\mathscr{E}$, $p = \sqrt{2\mathscr{E}m}$ is less for electrons than for protons, by a factor $\sqrt{m_p/m_e}$. The lower energy electron states therefore become fully occupied—degenerate—much more readily than proton states.

PROBLEM 4-23. Suppose the universe is filled with completely degenerate neutrinos up to an energy Φ_ν at a neutrino temperature $T_\nu = 0$. Show that ρ_ν, the mass density (energy density divided by c^2) of neutrinos is

$$\rho_\nu = \frac{\pi \Phi_\nu^4}{h^3 c^5} \tag{4-101}$$

Note that neutrinos exist in only one spin state, not two (Wa67).

To see why electrons and protons, which are actually Fermions, appear to have the characteristics of Maxwell–Boltzmann particles in many astrophysical situations we note the following. We can derive the velocity distribution for classical particles in a way similar to the derivation of the Fermi–Dirac distribution. Assume that particles can occupy arbitrary positions in momentum and configuration space. This is equivalent to saying that the phase cells are infinitessimally small. We can obtain such a system by pretending that Planck's constant goes to zero as a limit: $h \to 0$. This makes $\mathscr{E}_F = 0$, since arbitrarily many particles can have zero and near-zero energies, and the probabilities in (4–93) become $e^{-\mathscr{E}/kT}$ and 1. We now write the number of particles in the assembly, having momenta near p:

$$n(p)\, dp \propto \frac{8\pi p^2 V}{h^3}\, dp\, e^{-p^2/2mkT} \tag{4-102}$$

Integrating over all p values,

$$n = C \int_0^\infty \frac{8\pi V}{h^3} p^2 e^{-p^2/2mkT}\, dp \tag{4-103}$$

where C is the proportionality constant. This is an error function integral whose value is the total number of particles

$$n = C \frac{8\pi V}{h^3} \left(\frac{1}{4} \sqrt{\pi(2mkT)^3} \right)$$

Hence

$$C = \frac{nh^3}{2V(2\pi mkT)^{3/2}} \tag{4-104}$$

so that

$$n(p) = \frac{4\pi n p^2 e^{-p^2/2mkT}}{(2\pi mkT)^{3/2}}$$

a result obtained earlier.

The Maxwell–Boltzmann statistics apply in all problems dealing with the motion of particles in the atmospheres of stars and planets, with nondegenerate matter in the interior of stars, and with situations involving gas and the random motion of dust grains in interplanetary and interstellar space. These statistics also apply in some problems of stellar dynamics in which the stars can be thought of as members of an interacting assembly. Galaxies moving within a cluster are also believed to obey the M–B statistics. The formulas developed in the next sections therefore have wide applications in astrophysics.

4:16 THE SAHA EQUATION

At high temperatures, we often find multiply ionized states of any given atom in thermal equilibrium with each other. Consider two populations of particles labeled r and $r + 1$, respectively, representing an atomic species A in states of ionization r and $r + 1$. We can think of a reaction $A_r \rightleftharpoons A_{r+1} + e^-$ which might be driven to the right by ionizing photons and collisions, and to the left by the recombination of electrons with ions. Quite generally, the number densities of populations are then related by an expression similar in form to equation 4–50, but now written as $n_{r+1}/n_r = [g'_{r+1}/g_r]\exp - \mathscr{E}/kT$, where g'_{r+1} and g_r are the degeneracies—the statistical weights—in the upper and lower ionization states and $\mathscr{E}$ is the total energy difference between upper and lower states. That energy includes the kinetic energy $p^2/2m_e$ given to the electron if an energy $\mathscr{E}$ were imparted to the particle in ionization state r. So far, however, we have not considered that each state of ionization also comprises several states of excitation. Calling the excitation of the particle in the rth state i, and that in the $[r + 1]^{\text{st}}$ state j, we write

$$\frac{n_{r+1,j}}{n_{r,i}} = \frac{g_{r+1,j}}{g_{r,j}}\exp - [\chi_r + \mathscr{E}_{r+1,j} - \mathscr{E}_{r,i} + p^2/2m_e]/kT, \qquad (4\text{--}105)$$

where χ_r is the energy required to ionize the atom from the lowest excitation level in the rth state of ionization to the corresponding level in the $(r + 1)^{\text{st}}$ state, m_e is the electron mass, and $\mathscr{E}_{r+1,j}$ and $\mathscr{E}_{r,i}$, respectively, are the excitation energies within the corresponding ionization states.

Now, $g'_{r+1,j}$ consists of the product of two degeneracies, $g_{r+1,j}$ and g'_e, where $g_{r+1,j}$ is simply the degeneracy of the ionized particle in state $[r + 1, j]$, and g_e is the electron degeneracy given by the number of states available for the liberated electron to enter upon ionization with kinetic energy $p^2/2m_e$. This electron degeneracy is (see (4–65)) just $g'_e = Z(p)\,dp/n_e V = 4\pi g_e p^2\,dp/n_e h^3$, corresponding to the number of phase space states available in volume V and momentum range p to $p + dp$ to a single electron when the electron density is n_e. Here $g_e = 2$ is the electron spin degeneracy.

Recalling equations 4–103 and 4–104 and integrating equation 4–105 over all values of p, we obtain

$$\frac{n_{r+1,j}n_e}{n_{r,i}} = \frac{g_{r+1,j}g_e}{g_{r,i}} \frac{[2\pi mkT]^{3/2}}{h^3} \exp - \left[\frac{\chi_r + \mathscr{E}_{r+1,j} - \mathscr{E}_{r,i}}{kT} \right] \qquad (4\text{–}106)$$

We will need this equation to discuss reactions inside stars.

PROBLEM 4-24. In the solar corona, collisional excitation of atoms predominates over other processes of excitation. Among the identified spectral lines are those of CaXIII (12 times ionized calcium) and CaXV (14 times ionized). The ionization potentials of these ions are 655 and 814 ev, respectively. The lines from CaXIII are considerably stronger than those of CaXV. This fact alone can tell us very roughly what the temperature of the corona is. What is it?

4:17 MEAN VALUES

Once the energy, frequency, or momentum distribution of particles in an assembly are known, mean values of various functions of these parameters can be computed. For particles obeying Maxwell–Boltzmann statistics, the mean value of a function $F(p)$ is

$$\langle F(p) \rangle = \frac{\displaystyle\int_0^\infty Z(p)\,F(p)\,e^{-\,p^2/2mkT}\,dp}{\displaystyle\int_0^\infty Z(p)\,e^{-\,p^2/2mkT}\,dp} \qquad (4\text{–}107)$$

This equation has exactly the form of equation 4–19. The integrand in the denominator is the probability of finding a particle with momentum p

PROBLEM 4-25. Two frequently used quantities are $\langle p \rangle$ and $\langle p^2 \rangle$. Show that

$$\langle p \rangle = \frac{\displaystyle\int_0^\infty p^3 e^{-p^2/2mkT}\,dp}{\displaystyle\int_0^\infty p^2 e^{-p^2/2mkT}\,dp} = \frac{\frac{1}{2}(2mkT)^2\,\Gamma(2)}{\frac{1}{2}(2mkT)^{3/2}\,\Gamma(\frac{3}{2})} = \sqrt{\frac{8mkT}{\pi}} \qquad (4\text{-}108)$$

Note that this is the mean magnitude of the momentum. The mean momentum $\langle \mathbf{p} \rangle$ is zero because momenta along different directions cancel. Show also that

$$\langle p^2 \rangle = \frac{\displaystyle\int_0^\infty p^4 e^{-p^2/2mkT}\,dp}{\displaystyle\int_0^\infty p^2 e^{-p^2/2mkT}\,dp} = 3mkT \qquad (4\text{-}109)$$

PROBLEM 4-26. In section 6:16 we will make use of the quantity $\langle 1/v \rangle$. Show that

$$\left\langle \frac{1}{v} \right\rangle = \sqrt{\frac{2m}{\pi kT}} \qquad (4\text{-}110)$$

PROBLEM 4-27. In astronomical spectroscopy we can only measure velocities of atoms along a line of sight when observing the shape of a spectral line. To determine the temperature of a gas, whose mean squared random velocity $\langle v_r^2 \rangle$ along the line of sight is known, we therefore have to know how $\langle v_r^2 \rangle$ and T are related. For a Maxwell–Boltzmann distribution show that

$$\langle v_r^2 \rangle = kT/m$$

These integrals all have the form

$$\int_0^\infty x^{n-1} e^{-ax^2}\,dx = \frac{\Gamma(n/2)}{2a^{n/2}} \qquad (4\text{-}111)$$

where $\Gamma(n) = (n-1)!$ and $\Gamma(\frac{1}{2}) = \sqrt{\pi}$.

The analogous integrals required for computing mean values for energies or momenta for Fermions or Bosons involve the Fermi-Dirac or Bose–Einstein distribution functions.

4:17

4:18 THE FIRST LAW OF THERMODYNAMICS

The first law of thermodynamics expresses the conservation of energy. If a gas is heated, the supplied energy can act in one of two ways. It can raise the gas temperature, or it can perform work by expanding the gas against an externally applied pressure. Symbolically we write*

$$dQ = dU + Pd\mathscr{V} \tag{4-112}$$

where all quantities are normalized to one mole of matter and where the left-hand side gives the amount of heat dQ supplied to the system; dU is the change in internal energy and $Pd\mathscr{V}$ is the work performed. The nature of this last term is easily understood if we recall that *work* is involved in any displacement D against a force F. If the change in volume $d\mathscr{V}$ involves, say, the displacement of a piston of area A, then the force involved is $F = PA$ and the distance the piston moves is $D = d\mathscr{V}/A$

The *internal energy* U of the gas is the sum of the kinetic energy of translation, as the molecules shoot around; the kinetic and potential energy involved in the vibrations of atoms within a molecule; the energy of excited electronic states; and the kinetic energy of molecular rotation.

Q is the *heat content* of the system. The heat Q that must be supplied to give rise to a one degree change in temperature is called the *heat capacity* of the system. Clearly the heat capacity depends on the amount of work that is done. If no work is involved—which means that the system is kept at constant volume—all the heat goes into increasing the internal energy and

$$c_v = \left[\frac{dQ}{dT} \right]_{\mathscr{V}} = \frac{dU}{dT} \tag{4-113}$$

The subscript $\mathscr{V}$ denotes constant volume.

Sometimes we need to know the heat capacity under constant pressure conditions. For an ideal gas, this relation is quite simple. In differential form, the ideal gas law (4–34) reads

$$Pd\mathscr{V} + \mathscr{V}dP = RdT \tag{4-114}$$

so that the first law becomes

$$dQ = \left(\frac{dU}{dT} + R \right) dT - \mathscr{V}dP \tag{4-115}$$

---

* dQ is not an *exact differential*. This means that the change of heat, dQ depends on how the change is attained. For example, it can depend on whether we first raise the internal energy by dU, and then do work $Pd\mathscr{V}$, or vice versa.

For constant pressure

$$c_p = \frac{dU}{dT} + R \qquad (4-116)$$

For an ideal gas we therefore have the important relation

$$c_p - c_v = R = \mathcal{N}k \qquad (4-117)$$

This follows from (4–113) and (4–116). $\mathcal{N}$ is Avogadro's number and the heat capacities are figured for one mole of gas.

We have already stated that the internal energy involves the translation, vibration, electronic excitation, and rotational energy of the molecules. We can ask ourselves how these energies are distributed in a typical molecule. We know that the probability of exciting any classical particle to an energy $\mathscr{E}$ is proportional to the Boltzmann factor $e^{-\mathscr{E}/kT}$. This is true whether $\mathscr{E}$ is a vibrational, electronic, rotational, or translational energy. For an assembly of classical particles, then, the mean internal energy per molecule depends only on the number of ways that energy can be excited, that is, the number of *degrees of freedom* multiplied by $kT/2$. This factor $kT/2$ is consistent with our previous finding: That the total translational energy, including three degrees of freedom, is $(\frac{3}{2})\mathcal{N}kT$ per mole. Each translational degree of freedom, therefore, has energy $\frac{1}{2}kT$ and each other available degree of freedom in thermal equilibrium will also be excited to this mean energy. This is called the *equipartition principle*.

PROBLEM 4-28. Show that an interstellar grain in thermal equilibrium with gas at $T \sim 100°$K rotates rapidly. If its radius is $a \sim 10^{-5}$ cm, and the density is $\rho \sim 1$, show that the angular velocity is $\omega \sim 10^{5.5}$ rad sec^{-1}.

The equipartition principle is a part of classical physics. It does not quite agree with observations; the actual values can be explained more easily by quantum mechanical arguments. The difference between classical and quantum theory hinges to a large extent on what is meant by "available" degrees of freedom. The electronically excited states of atoms and molecules normally are not populated at low temperatures. Hence, at temperatures of the order of several hundred degrees Kelvin, no contribution to the heat capacity is made by such states. Even the vibrational states make a relatively small contribution to the heat capacity because vibrational energies usually are large compared to rotational energies. Aside from the translational contribution, it is therefore the low energy rotational states that make a major contribution to the internal energy and the specific heat at constant volume.

The rotational position of a diatomic molecule can be given in terms of two coordinates θ and ϕ. It therefore has two degrees of freedom. A polyatomic molecule having three or more atoms in any configuration excepting a linear one requires three coordinates for a complete description and therefore has three degrees of freedom. A diatomic or linear molecule makes a contribution of kT to the heat capacity and a nonlinear molecule contributes $3kT/2$. Even these relatively simple rules hold only at low temperatures. At higher temperatures the situation is complicated because a quantum mechanical weighting function has to be introduced to take into account the number of identical (degenerate) states that a particle can have at higher rotational energies.

We will be interested in the heat capacity of interstellar gases where temperatures are low and many of the above mentioned difficulties do not arise. Let us define the ratio of heat capacities at constant pressure and volume as $c_p/c_v \equiv \gamma$. Then by (4–117) we have

$$\gamma = \frac{c_v + \mathcal{N}k}{c_v} \qquad (4\text{--}118)$$

For monatomic gases we simply deal with the translational internal energy and $c_v = 3k\,\mathcal{N}/2$; $\gamma = \frac{5}{3}$. For diatomic molecules two rotational degrees of freedom are available in addition to the three translational degrees of freedom, so that $c_v = 5k\,\mathcal{N}/2$ and $\gamma = \frac{7}{5}$.

4:19 ISOTHERMAL AND ADIABATIC PROCESSES

The contraction of a cool interstellar gas cloud or, equally well, the expansion of a hot ionized gas cloud can proceed in a variety of ways. Some cosmic processes involving the dynamics of gases can occur quite slowly at constant temperature. These are called *isothermal processes*. The internal energy does not change and the heat put into the system equals work done by it. Another type of process that describes many fast evolving systems, is the *adiabatic process* in which there is neither heat flow into the gas nor heat flowing out, $dQ = 0$.

$$dQ = c_v\,dT + Pd\mathcal{V} = 0 \qquad (4\text{--}119)$$

For an ideal gas

$$c_v\,dT + \frac{RT}{\mathcal{V}}\,d\mathcal{V} = 0 = c_v\,\frac{dT}{T} + (c_p - c_v)\frac{d\mathcal{V}}{\mathcal{V}} \qquad (4\text{--}120)$$

Integrating, we have

$$\log T + (\gamma - 1)\log \mathcal{V} = \text{constant} \qquad (4\text{--}121)$$

or

$$T\mathcal{V}^{\gamma - 1} = \text{constant} \qquad (4\text{--}122)$$

$$P\mathcal{V}^{\gamma} = \text{constant} \qquad (4\text{--}123)$$

and

$$P^{(1-\gamma)}T^{\gamma} = \text{constant} \qquad (4\text{--}124)$$

These are the adiabatic relations for an ideal gas. They govern the behavior, for example, of interstellar gases suddenly compressed by a shock front heading out from a newly formed 0-star or from an exploding supernova. We will study these phenomena in Chapter 9.

For electromagnetic radiation the internal energy per unit volume is

$$U = aT^4 \qquad (4\text{--}125)$$

This is just the energy density. The pressure therefore has one-third this value (4–43), and for volume V we can describe an adiabatic process by

$$dQ = dU + P\,dV = 4aT^3 V\,dT + \tfrac{4}{3}aT^4\,dV = 3V\,dP + 4P\,dV = 0 \quad (4\text{--}126)$$

$$P \propto V^{-4/3}, \qquad \gamma = \tfrac{4}{3} \qquad (4\text{--}127)$$

Since $c_p = \infty$ in this case, γ is defined by the form of (4–123), and not by c_p/c_v. In the interstellar medium where, to a good approximation, we deal only with monatomic or ionized particles—particles having no internal degrees of freedom—and with radiation, γ varies from $\tfrac{4}{3}$ to $\tfrac{5}{3}$ depending on whether gas particles or radiation dominate the pressure.

4:20 FORMATION OF CONDENSATIONS AND THE STABILITY OF THE INTERSTELLAR MEDIUM

We think that the stars were formed from gases that originally permeated the whole galaxy, and that galaxies were formed from a medium that initially was more or less uniformly distributed throughout the universe.

There is strong evidence that star formation is going on at the present time. Many stars are in a stage that can only persist for a few million years because the stellar luminosity—energy output—is so great that these stars soon would deplete their available energy and evolve into objects with entirely different appearances. These bright stars are generally found in the vicinity of unusually high dust and gas concentrations and the belief is that the stars were formed from this dense gas.

We now ask how an interstellar gas cloud could collapse to form a star? To answer this question we can study the stability of gases under various conditions. The results indicate that the stability is dependent on the ratio of heat capacities γ, and that collapse of a gas cloud might be explained in these terms. Whether such views have any bearing on the real course of events in star formation is not yet understood.

Consider an assembly of molecules. Their kinetic energy $\mathbb{T}$ per mole is

$$\mathbb{T} = \tfrac{3}{2}(c_p - c_v)\, T \tag{4-128}$$

or

$$\mathbb{T} = \tfrac{3}{2}(\gamma - 1)\, c_v T \tag{4-129}$$

The internal energy is

$$U = c_v T \tag{4-130}$$

Hence

$$\mathbb{T} = \tfrac{3}{2}(\gamma - 1)\, U \tag{4-131}$$

By the virial theorem (3–83) we then have

$$3(\gamma - 1)\, U + \mathbb{V} = 0 \tag{4-132}$$

as long as inverse square law forces predominate among particles. This means that the equation holds true both when gravitational forces are important and where charged particle interactions dominate the behavior on a small scale (see section 4:21 below). In some situations it can even hold when light pressure from surrounding stars acts on particles of the assembly.

If the total energy per mole is

$$\mathscr{E} = U + \mathbb{V} \tag{4-133}$$

we have from equation 4–132 that

$$\mathscr{E} = -(3\gamma - 4)\, U = \frac{3\gamma - 4}{3(\gamma - 1)}\, \mathbb{V} \tag{4-134}$$

Three results are apparent (Ch39)*:

(a) If $\gamma = \tfrac{4}{3}$, $\mathscr{E}$ is always zero independent of the configuration. Expansion and contraction are possible and the configuration is unstable. This case corresponds to a photon gas (4–127). In its early stages, a planetary nebula has radiation dominated pressure acting to produce its expansion (Ka68). It should therefore be only marginally stable.

(b) For $\gamma = 1$, $\mathbb{V}$ always is zero for any $\mathscr{E}$ value and again no stable configuration exists.

(c) For $\gamma > \tfrac{4}{3}$, equation 4–134 shows that $\mathscr{E}$ always is negative and the system is bound. If the system contracts and the potential energy changes by $\Delta\mathbb{V}$ then, we see that

$$\Delta\mathscr{E} = +\frac{(3\gamma - 4)}{3(\gamma - 1)}\,\Delta\mathbb{V} = -(3\gamma - 4)\,\Delta U \tag{4-135}$$

4:20

An amount of energy $-\Delta\mathcal{E}$ is lost by radiation

$$-\Delta\mathcal{E} = -\frac{3\gamma - 4}{3(\gamma - 1)}\Delta\mathbb{V} \qquad (4\text{--}136)$$

while the internal energy increases by

$$\Delta U = -\frac{1}{3(\gamma - 1)}\Delta\mathbb{V} \qquad (4\text{--}137)$$

through a rise in temperature.

As the protostar contracts to form a star it therefore becomes hotter and hotter.

Two comments are necessary:

(a) When theories of the kind developed here are applied on a cosmic scale, say, to formation of galaxies or clusters of galaxies, we run into difficulties in defining the potential $\mathbb{V}$. The zero level of the potential can no longer be defined using Newtonian theory alone, and some more comprehensive approach such as that of general relativity should be used. This complicates the treatment of the problem considerably, and no properly worked out theory exists.

(b) In practice, star formation probably takes place in strong interaction with the surrounding medium. This is indicated by the formation of stars in regions where other stars have just formed and where pressures on the surrounding medium are setting in. There are theories that describe the formation of stars through the compression of cool gas clouds either by surrounding hot ionized regions, or through the compression by starlight emitted from nearby hot stars. The stability of an isolated medium, as treated above, may therefore not be strictly relevant to the discussion. Nevertheless, it is clear that very stable gas configurations will resist applied compressive forces, while intrinsically unstable gases will readily respond to such pressures. As already indicated in section 1:4, magnetic fields, rotational motion about the Galactic center, and rotation about the protostar center, endows protostellar matter with stability against collapse. Random gas motions and radiation energies therefore are not the sole contributors to U in equations 4–131 to 4–134.

4:21 IONIZED GASES AND ASSEMBLIES OF STARS

The behavior of large clusters of stars or galaxies can be described statistically much as we describe the behavior of gases. There are many striking similarities particularly in the physics of ionized gases (plasma) and aggregates of stars. These similarities come about because Newton's gravitational attraction can be written in a form similar to Coulomb's electrostatic force:

Newton's force   Coulomb's force

$$\frac{(iG^{1/2}m_1)(iG^{1/2}m_2)}{r^2} \qquad \frac{Q_1Q_2}{r^2} \tag{4-138}$$

Here the gravitational analogue to electrostatic charge is the product of mass, the square root of the gravitational constant, and the imaginary number, i. The correspondence can be extended to include fields, potentials, potential energies, and other physical parameters. The primary difference between gravitational processes and electrostatic interactions lies in the fact that electric charges can be both positive and negative while the sign of the gravitational analogue to charge is always the same—mass is always positive.

We will first derive some properties of assemblies of gravitationally interacting particles and then make a comparison to plasma behavior. If we take a spherical distribution of particles—a set of stars or galaxies—the force acting on unit mass at distance r from the center is

$$F(r) = -\frac{1}{r^2}\int_0^r 4\pi Gr^2\rho(r)\,dr \tag{4-139}$$

This means that

$$\frac{d}{dr}r^2F(r) = -4\pi Gr^2\rho(r) \tag{4-140}$$

and setting the force equal to a potential gradient

$$F(r) = -\nabla\mathbb{V}(r) = -\frac{d}{dr}\mathbb{V}(r) \tag{4-141}$$

we have

$$\frac{1}{r^2}\frac{d}{dr}r^2\frac{d}{dr}\mathbb{V}(r) = +4\pi G\rho(r) \tag{4-142}$$

which is *Poisson's equation*. Substituting from equation 4–47 we have the *Poisson–Boltzmann equation* for a gas at temperature T:

$$\frac{1}{r^2}\frac{d}{dr}r^2\frac{d\mathbb{V}}{dr} = 4\pi\rho_0 Ge^{-[m\mathbb{V}(r)/kT]} \tag{4-143}$$

It is worthwhile noting that the potential appearing in the exponent on the right of this equation really was obtained through the integration of $\nabla\mathbb{V}(r)$ in

4:21

equation 4–46. The behavior of an assembly of stars or galaxies would, therefore, be no different if some constant potential, present throughout the universe, were added to $\mathbb{V}(r)$. This is essentially the point that was already raised in section 4:20 in connection with the stability of uniform distributions of gas.

PROBLEM 4–29. Show that the substitutions

$$\frac{m\mathbb{V}(r)}{kT} \equiv \psi \quad \text{and} \quad r \equiv \left(\frac{kT}{4\pi\rho_0 mG}\right)^{1/2} \xi \qquad (4\text{–}144)$$

turn equation 4–143 into

$$\frac{1}{\xi^2} \frac{d}{d\xi}\left(\xi^2 \frac{d\psi}{d\xi}\right) = e^{-\psi} \qquad (4\text{–}145)$$

We now have to decide on the boundary conditions that have to be imposed on this differential equation. At the center of the cluster there are no forces and the first derivative of $\mathbb{V}$ or ψ must be zero. Since the potential can have an arbitrary additive constant, we can choose the potential to be zero at the center. In terms of the new variables these two conditions are

$$\psi = 0 \quad \text{and} \quad \frac{d\psi}{d\xi} = 0 \quad \text{at} \quad \xi = 0 \qquad (4\text{–}146)$$

Taken together with equation 4–145 they lead to a solution that has no closed form (Ch 43).

PROBLEM 4–30. Show that in the limit of very small and of large ξ values the respective solutions are (Ch 39)

$$\psi \sim \tfrac{1}{6}\xi^2 - \tfrac{1}{120}\xi^4 + \tfrac{1}{1890}\xi^6 + \cdots \quad , \quad \xi \ll 1 \qquad (4\text{–}147)$$

$$\psi \sim \log\left(\frac{\xi^2}{2}\right), \quad \xi^2 > 2 \qquad (4\text{–}148)$$

This can be verified by substitution in equations 4–145 and 4–146. From this, the radial density and mass distributions can be found (Ch 43). The density distribution is plotted in Fig. 4.15.

One difficulty with this plot and with the asymptotic solution (4–148) is that the density $\rho_0 e^{-\psi}$ is proportional to ξ^{-2}. This causes the total mass integrated to large distances to become infinite. We therefore need a cut-off mechanism that will restrict the radius of a cluster of stars to a finite value. We already men-

4:21

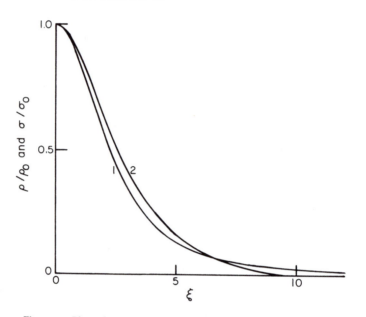

Fig. 4.15 Plot of density (1) and areal density (2) against radial distance from the center of a distribution. For a cluster, σ/σ_0 represents the star density drop with radial distance, as measured directly on a photographic plate. ρ_0 and σ_0 represent values at the center. $\rho/\rho_0 = \exp(-\psi)$, $\xi = (4\pi\rho_0 mG/kT)^{1/2} r$. (From *Principles of Stellar Dynamics* by S. Chandrasekhar. Reprinted through permission of the publisher, Dover Publications, Inc. New York.)

tioned the interaction with the galactic nucleus as representing one such mechanism (section 3:16).

It is interesting to compare these results to those obtained for a plasma in which both positive and negatively charged particles are present. The derivation of the Poisson–Boltzmann equation was in no way based on particle charges. It concerned itself only with an inverse square law force and a uniform mass distribution that could equally well have been a charge distribution. Using the density for an assembly of dissimilar particles (see equation 4–53), the Poisson–Boltzmann equation can be written as

$$\frac{1}{r^2} \frac{d}{dr} r^2 \frac{d\mathbb{V}}{dr} = -4\pi \sum_i n_{i0} q_i e^{-q_i \mathbb{V}/kT} \qquad (4\text{--}149)$$

for plasma. Here q_i is the charge of particles of type i. If we restrict ourselves to large interparticle distances—a condition that holds in intergalactic, interstellar, and interplanetary space—then

$$q_i \mathbb{V} \ll kT \tag{4–150}$$

and we can use the Taylor expansion

$$e^{-q_i \mathbb{V}/kT} = 1 - \frac{q_i \mathbb{V}}{kT} + \frac{1}{2}\left(\frac{q_i \mathbb{V}}{kT}\right)^2 + \cdots \tag{4–151}$$

Neglecting quadratic and higher terms, the charge density on the right of (4–149) then becomes

$$\rho = \sum_i n_{i0} q_i - \frac{\mathbb{V}}{kT}\sum_i n_{i0} q_i^2 \tag{4–152}$$

The first term vanishes because of charge neutrality for the bulk of the plasma. Note that this term does not vanish in the gravitational case. There it is dominant. The second term can be written as

$$\rho = -\frac{\mathbb{V}e^2}{kT}\sum_i n_{i0} Z_i^2 \qquad \text{where} \qquad q_i = eZ_i \tag{4–153}$$

Substituting in (4–149) we have

$$\frac{1}{r^2}\frac{d}{dr}\left(r^2 \frac{d\mathbb{V}}{dr}\right) = \frac{4\pi e^2 \mathbb{V}}{kT}\sum_i n_{i0} Z_i^2 \tag{4–154}$$

or

$$\frac{d}{dr}\left(r^2 \frac{d\mathbb{V}}{dr}\right) = r^2 \mathbb{V}\left(\frac{1}{L^2}\right) \tag{4–155}$$

where

$$\frac{1}{L^2} = \frac{4\pi e^2}{kT}\sum_i n_{i0} Z_i^2 \tag{4–156}$$

L has the dimension of a length. It is called the *Debye shielding length* and is a distance over which a charged particle embedded in a plasma can exert an appreciable electrostatic field. Beyond that distance its electrostatic influence rapidly diminishes. For fully ionized hydrogen, $L = 6.90(T/n)^{1/2}$ cm.

One reason why the shielding length is of interest in astrophysics is because it points out the impossibility of maintaining an electric field over any large scale. A field cannot be influential over distances much larger than L. For interstellar gas clouds with $n_{i0} = 10^{-3}$ cm^{-3}, $Z_i = 1$, and $T = 100°$K, L turns out to be about 20 meters. This is completely negligible compared to typical interstellar distances ~ 1 pc.

Electrostatic forces may be important in large-scale processes, but only when they appear in conjunction with large-scale magnetic fields, that can prevent the flow of charged particles along the electric field lines and therefore prevent the charge separation required for electrostatic shielding. The behavior of plasmas in the presence of magnetic fields is treated in the theory of magnetohydrodynamic processes (Co57), (Sp62). We will briefly consider these processes in Chapter 6.

ANSWERS TO SELECTED PROBLEMS

4–2. Suppose we take n_i steps of length λ_i. The mean square deviation then is $n_i \lambda_i^2$; and a similar result holds for all step sizes. Hence the final mean square deviation is $\sum_i n_i \lambda_i^2 = N \langle \lambda^2 \rangle$.

4–4. For escape, the deviation has to be $\sim n^{1/2}$ steps of length L. Hence $n \sim R^2/L^2$.

4–5. For a given value of $R\alpha$, the value of $1/r$ averaged over all θ values is

$$\langle 1/r \rangle = \langle \sin \theta \rangle / R\alpha = 2/\pi R\alpha .$$

If we also average over different values of $1/R\alpha$ we obtain

$$\left\langle \frac{1}{r} \right\rangle = \frac{2}{\pi} \left\langle \frac{1}{R\alpha} \right\rangle$$

and

$$\text{total mass} = \frac{\langle v^2 \rangle}{G \langle 1/r \rangle} = \frac{3 \langle v_p^2 \rangle \dfrac{\pi}{2}}{\left\langle \dfrac{1}{R\alpha} \right\rangle}$$

4–6. $\dfrac{nm \langle v^2 \rangle}{3} \sim 3 \times 10^{-17}$ dynes cm^{-2}.

4–7. The collision probability per star pair in unit time is

$$\text{Probability} = nv\sigma \text{ sec}^{-1}.$$

In P sec the probability is $nv\sigma P$ per star pair or about 1.5×10^{-11} that the sun would have formed a planetary system in the time available. If there are 10^{11} stars altogether, 1.5 pairs, or 3 solar systems would have formed in this way in 5×10^9 y. The sun, Vega, HL Tau, and β Pic all have ambient dust clouds. These potentially could be remnants of protoplanetary clouds, or for HL Tau, possibly a protoplanetary cloud itself (see Chapter 1). These system are unlikely to have had a common origin. Their number density in the vicinity of the sun already appears too great to be accounted for by the Jeans hypothesis.

4–8. $P \sim 10^{-14}$ dynes cm^{-2}.

4–9. $v \sim (3kT/M)^{1/2} \sim 0.8$ cm sec^{-1} for $M = 4\pi\rho r^3/3 \sim 7 \times 10^{-14}$ g.

4–10. The momentum transfer rate is $\pi r^2 n v^2 m = dp/dt = 3.1 \times 10^{-21}$ g cm sec^{-1}. The mass of the grains is given in Problem 4–9 and, hence, $dv/dt = 4.8 \times 10^{-8}$ cm sec^{-2}. This gives the initial acceleration. However, as the grain gains velocity and mass, the acceleration decreases toward a value of zero, which is reached when the grain reaches velocity v.

4–11. The mass gain $dM/dt = \pi r^2 n v m = 3 \times 10^{-27}$ g sec^{-1}. At this rate the grain would gain 1 % of its mass in 2×10^{11} sec.

4–13. $P \sim 2 \times 10^{-13}$ dyn cm^{-2}.

4–14. If n is the number of photons passing through unit area per unit time, the pressure is

$$P = \frac{\int n(v)\, hv\, dv}{c}.$$

that is, the energy incident on unit area per unit time, divided by the speed of light. The force on the grain is its area, multiplied by P.

$$P = \frac{1.37 \times 10^6}{3 \times 10^{10}} = 4.6 \times 10^{-5} \text{ dynes cm}^{-2}$$

$$F = 3.6 \times 10^{-9} \text{ dynes}$$

4–15. For isotropic scattering one averages the function $(1 - \cos\theta)$ over all solid angle increments to get the mean momentum transfer. Thus

$$P = \mathscr{E}/3c + \frac{2\mathscr{E}/3c}{4\pi} \iint (1 - \cos\theta) \sin\theta\, d\phi\, d\theta$$

with θ chosen as zero in the forward scattering direction, and $\mathscr{E}$ the energy incident on the grain per unit time. Thus $P = \mathscr{E}/c$, and the grain has a force of 1.44×10^{-12} dynes acting on it at the earth's distance from the sun. The gravitational force is mMG/R^2; here m is the particle mass, M is the solar mass, and R is the distance of the sun from the earth:

$$m = \rho 4\pi s^3/3 = 2.5 \times 10^{-11} \text{ g}$$

the gravitational force is 1.5×10^{-11} dynes and the radiative repulsion is about 10 % of the gravitational attraction.

Finally, the ratio of gravitational to radiative force remains constant because the angle subtended by the grain, as seen from the sun, diminishes as R^2—in the same way as the gravitational force.

4–16. From the equation for the period of a grain in an elliptic orbit

$$\tau = 2\pi a^{3/2}/(MG_{\text{eff}})^{1/2}$$

We note that the period depends on the square root of the effective gravitational constant. The period of the grain will be $(\tfrac{3}{2})^{1/2}$ years, that is, 1.22 y. The orbital velocity will be $v_E/1.22$, where v_E is the earth's orbital velocity; that speed will be roughly 24 km/sec. The collision velocity of the earth with such a grain would therefore be ~ 6 km sec^{-1}.

4–18. The thinness of the atmospheric layer implies that g is constant throughout. Hence the scale height h is determined by the equation $ghm/kT = 1$ and

$$h \sim 10^6 \text{ cm.}$$

4–20. $m \sim 2 \times 10^{-24}$ g

and $v \sim 2.3$ km sec^{-1}.

4–23. $Z(p)\, dp = \dfrac{4\pi p^2\, dp}{h^3} \cdot V,$ $p = \dfrac{\mathscr{E}}{c},$ $dp = \dfrac{d\mathscr{E}}{c}$

$$\therefore\ Z(\mathscr{E})\, d\mathscr{E} = \dfrac{4\pi \mathscr{E}^2\, d\mathscr{E}}{h^3 c^3}\, V.$$

The total energy density is

$$\dfrac{\mathscr{E}}{V} = \int_0^{\Phi_v} \dfrac{\mathscr{E}Z(\mathscr{E})\, d\mathscr{E}}{V} = \dfrac{\pi \Phi_v^4}{h^3 c^3}$$

and the mass density $\rho_v = \dfrac{\mathscr{E}}{c^2 V} = \dfrac{\pi \Phi_v^4}{h^3 c^5}$

4–24. $kT \sim 655$ ev $\sim 10^{-9}$ erg

$$T \sim \dfrac{10^{-9}\text{ erg}}{1.4 \times 10^{-16}\text{ erg/}^\circ} \sim 7 \times 10^6\,^\circ\text{K}$$

We reason that $kT \sim$ excitation energy; the higher ionized state gives rise to a weak line because T is not sufficiently high to lead to frequent ionization to this level. That is, for the higher ionized state $kT <$ ionization energy. For the lower ionized states kT is probably more comparable to the ionization energy. Actually T is $\sim 1.5 \times 10^6\,^\circ$K in the corona.

5

Photons and Fast Particles

5:1 THE RELATIVITY PRINCIPLE

When we discussed Newton's laws of motion, we were careful to note that they only held under restricted conditions. All motions had to be described with respect to *inertial frames of reference*—frames at rest or moved at constant velocity with respect to the mean motion of ambient galaxies.

Under these restricted conditions not only Newton's laws but all other laws of physics are obeyed. This general statement—first formulated by Einstein—is called the *principle of relativity*. It implies that an observer cannot determine the absolute motion of his inertial frame of reference—only its motion relative to some other frame. The principle also has many other important consequences which, taken together, form the basis of the theory of *special relativity* (Ei05a). As already mentioned in section 3:8, Einstein also widened the concept of the inertial frame beyond Newton's scope of a frame moving at constant velocity with respect to fixed stars. He showed that we can include coordinate frames fixed in any freely falling, non-rotating bodies. Such local inertial frames may accelerate with respect to frames that are far from any massive objects but are fully equivalent to them as far as the principle of relativity is concerned. Finally, Einstein postulated that the speed of light is the same in all reference frames, whether they move or are stationary. This actually is a consequence of the relativity principle. If this speed were not the same, an observer could determine whether he was at rest or moving.

We should note that the relativity principle is founded on observations. It could not have been predicted from logic alone.

In recent centuries, long before Einstein's birth, there has always been an awareness that some sort of relativity principle might exist. In Galileo's time, when the speed of light was believed to be infinite, the statement of the principle was almost exactly the same as Einstein's. At that time the velocity of light was believed to be infinite as measured in any reference frame. The instantaneous transmission of signals and messages over large distances seemed possible. Since any velocity added to an infinite velocity still gave infinite speed, it was clear that no matter how an observer moved he would always see light traveling at the same, infinite, speed; and similarly all other laws of physics seemed identical for Newton's inertial frames.

Then in 1666 Roemer discovered that the speed of light is finite, though large. This tended to detract from the *Galilean relativity principle* because it seemed that an observer moving into the direction of a light source would see the light wave moving faster than an observer moving away from the source. But at the end of the 19th century Michelson and others discovered that the speed of light was identical in all the moving reference frames they were able to check. Independently Einstein postulated a new principle of relativity similar to Galileo's except that the speed of light now was finite and equal in all reference frames. To some extent this concept had already been present in Maxwell's theory of electromagnetism, but the required constancy of the speed of light was considered a weakness of the theory, not a strength.

As we will see, Einstein's relativity principle also led him to conclude that no physical object could travel at a speed in excess of light, $c = 2.998 \times 10^{10}\,\text{cm sec}^{-1}$. The concept of an infinite velocity simply had no correspondence in physical moving objects.

The theory of relativity has the task of formulating the laws of physics in such a way that physical processes can be accurately described in any moving coordinate system. This study conveniently divides into two parts. The first theory is more restricted. It deals with physical processes as viewed from inertial reference frames and specifically excludes any consideration of gravity. This is *special relativity*. The second, more general theory incorporates not only special relativity but also the study of gravitational fields and arbitrarily accelerated motions. It is therefore called the *general theory of relativity*. We will not discuss this theory here, partly because it is complicated, and partly because in its present form its application to astrophysics is not entirely established. There may exist rival theories of similar scope that would give greater insight into astrophysical process. One such theory has been advocated by Dicke (Di67a).

5:2 RELATIVISTIC TERMINOLOGY

Suppose a physical process occurs in a system at rest with respect to some inertial reference frame K'. An observer in some other inertial frame K views this process. If K and K' are moving at large volocities V, relative to each other, the observer will see the processes distorted both in space and in time. But the special theory of relativity will allow him to reconstruct the proper scale of events as they occur in system K. This is a very useful property of relativity theory. We will find many applications of it in astrophysics where high velocities are often encountered. The special theory, however, goes beyond this limited function of reconstructing clear pictures from apparently distorted observations. It gives new insight into the interrelation between time and space, and between momentum, energy, and mass; it justifies the impossibility of exceeding the speed of light, and yields many other new results.

To make full use of this theory, we will need to take a few preparatory steps. We must define new concepts and formulate them in mathematical terms.

(a) To the extent that it is valid, the special theory abolishes an absolute standard for a state of rest. It states that there is no way of defining zero speed in an absolute way. Bodies may be at rest—but only *relative* to some other body or frame of reference.

We know that this statement need not be quite true: A preferred natural state of rest does exist for any locality in the universe. It is the state of rest relative to the mean motion of ambient galaxies. This fact was not known at the time relativity theory was established. It tends to weaken the statement we formulated above, and allows us to state only that an absolute standard of rest is inconsequential to special relativity. The theory draws no distinction between absolute rest and constant velocity.

(b) In relativity we will talk about *events* that have to be described both by a place and a time of occurrence. We need four coordinates to define an event—three space coordinates and a time coordinate.

Correspondingly there exists a hypothetical four-dimensional space having spatial and time coordinates. In this space, events are represented by *world points* (x, y, z, t). Any physical process can be described as a sequence of events and can be represented as a grouping or continuum of world points in the four-dimensional space-time representation. Each physical particle can be represented by a *world line* in this four-dimensional plot.

(c) Two distinct events labeled a and b are separated by an *interval*, s_{ab}, of length

$$s_{ab}^2 = -\left[(x_a - x_b)^2 + (y_a - y_b)^2 + (z_a - z_b)^2 - c^2(t_a - t_b)^2\right] \quad (5\text{--}1)$$

This suggests that we could define a new coordinate $\tau = ict$, where i is the imaginary number, to obtain (5–1) in the form

$$s_{ab}^2 = -\left[(x_a - x_b)^2 + (y_a - y_b)^2 + (z_a - z_b)^2 + (\tau_a - \tau_b)^2\right] \quad (5\text{--}2)$$

5:2

This form brings out a symmetry between time and space coordinates. Equation 5–2 is just the Pythagorean expression for the separation of two points in a four-dimensional flat space. Such a space is also called a *Euclidean space*, and the particular four-dimensional space described in (5–1) is known as a *Minkowski space* (Mi08). Equation 5–2 helps to point out some of the properties of space and time coordinates. The time coordinate in the formulation (5–2) is an imaginary quantity, while the spatial coordinates are real. Unfortunately, the substitution $\tau = ict$ is not very useful. Special relativity, in its full form, deals with quantities that are best described in tensor notation. But that notation cannot be properly used if time is taken to be an imaginary quantity. Rather, as we will see, x, y, z, and ct should be considered to be components of a four vector in a space that is said to have *signature* $(+ + + -)$, meaning that the Pythagorean expression for the square of the interval between events is the sum of the squares of the spatial components of the separation, with the square of the time increment subtracted.

(d) We can formulate equation 5–1 in differential form

$$ds^2 = -(dx^2 + dy^2 + dz^2 - c^2dt^2) \qquad (5\text{--}3)$$

where ds is called the *line element*.

(e) The interval between two events is said to be *timelike* if $s_{ab}^2 > 0$ and *spacelike* if $s_{ab}^2 < 0$. When s_{ab}^2 just equals zero, we see from either equation 5–1 or 5–3, that

$$v_x^2 + v_y^2 + v_z^2 \equiv v^2 = c^2 \qquad (5\text{--}4)$$

The surface containing all intervals $s_{ab} = 0$, or line elements $ds = 0$ is called the *light cone*. It contains all trajectories going through a point (x, y, z, t) with the speed of light. A two-dimensional projection of this cone is shown in Fig. 5.1. This means that we choose coordinates $y = z = 0$ and the projection of the surface

$$x^2 + y^2 + z^2 = c^2t^2 \qquad (5\text{--}5)$$

now becomes $x = \pm ct$ with slope

$$\frac{dt}{dx} = \pm \frac{1}{c} \qquad (5\text{--}6)$$

Consider an observer placed at the origin of the coordinate system. All lines representing physical particles must indicate velocities $v < c$ and, therefore, are contained in that part of the light cone containing the t-axis. The lower half of the diagram represents the past. The upper half contains all world points lying in the future. The two parts of the diagram containing the x-axis are absolutely inaccessible in the sense that velocities greater than the speed of light would be required to reach them.

It is interesting that the concept of absolute past and future depends on the

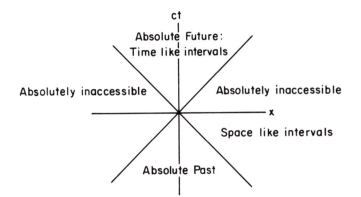

Fig. 5.1 World diagram to show relation between different kinds of events.

fact that the speed of light cannot be exceeded. If it could, we would be able to travel to a sufficiently distant point and "catch up" with light that had been emitted, say, in the supernova of 1054 A.D. With a sufficiently good telescope we could then "look back" and see the star just prior to explosion. The event could thus be brought into our "present," but it would still be inaccessible to us in the sense that we would not be able to influence the event in any way. This problem is looked at more thoroughly in section 5:12.

(f) The time read on a clock moving with the reference frame of an observer is called the *proper time* for that frame; and the length of an object measured in that frame is called the *proper length*.

5:3 RELATIVE MOTION

Consider two inertial frames of reference K and K', whose axes x, y, z and x', y', z' are parallel (Figure 5.2). Relative to K, K' moves with velocity V along the x-axis. An event has coordinates (x, y, z, t) as measured by an observer at rest in system K, and coordinates (x', y', z', t') as measured by an observer at rest in K'.

At some time $t = t' = 0$, let the origins of the two reference frames coincide. The subsequent motion will not affect the identity of the y and z components: $y' = y$ and $z' = z$; but t and x will be related to t' and x' through a more complicated set of relations, the *Lorentz transformations*, which read (Lo04):

$$x = \frac{x' + Vt'}{\sqrt{1 - V^2/c^2}}, \quad y = y', \quad z = z', \quad t = \frac{t' + V(x'/c^2)}{\sqrt{1 - V^2/c^2}} \qquad (5-7)$$

5:3

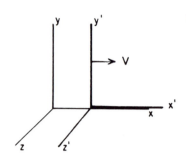

Fig. 5.2   Notation for moving coordinate frames.

or

$$x_1 = (x'_1 + \beta x'_4)\gamma(V), \quad x_2 = x'_2, \quad x_3 = x'_3, \quad x_4 = (x'_4 + \beta x'_1)\gamma(V) \quad (5\text{-}7a)$$

if we set $x = x_1$, $y = x_2$, $z = x_3$, $ct = x_4$, $\beta = V/c$, and $\gamma(V) = (1 - \beta^2)^{-1}$. The second formulation shows the symmetry between space and time coordinates. These equations can be derived and follow directly from the principle of relativity and the constancy of the speed of light. Here we will only show that the equations are consistent with some of the predictions of the principle.

For example, since the speed of light is the same in systems K and K', we would expect that a light wave emitted at $t = t' = 0$—that is, when the origins of the coordinate systems coincide—would propagate spherically in both systems.

PROBLEM 5-1.   Equation 5–5 describes the propagation of a spherical wavefront in the coordinate system K. Show that according to (5–7), the corresponding equation describing the propagation of the wavefront in K' would be

$$x'^2 + y'^2 + z'^2 = c^2 t'^2$$

so that this wave appears spherical too.

Another consequence of the relativity principle is that formulas expressing x', y', z', and t' in terms of x, y, z, and t can be easily obtained by changing V to $-V$.

PROBLEM 5-2.   Show that this procedure is valid by actually solving equations 5–7 for x', y', z', and t'.

We also want to examine whether the speed of light will always appear to be c, viewed from any reference frame. We can answer this question by discussing how velocities transform according to equations 5–7. Let us write the expressions

5:3

in differential form:

$$dx = (dx' + V\,dt')\,\gamma(V), \quad dy = dy', \quad dz = dz', \quad dt = \left(dt' + \frac{V}{c^2}\,dx'\right)\gamma(V)$$

$$(5\text{–}8)$$

where

$$\gamma(V) \equiv \frac{1}{\sqrt{1 - V^2/c^2}}$$

This allows us to write the derivatives

$$v_x = \frac{dx}{dt} = \frac{dx' + V\,dt'}{dt' + \dfrac{V}{c^2}\,dx'} = \frac{v'_x + V}{1 + v'_x\dfrac{V}{c^2}}$$

$$v_y = \frac{dy}{dt} = \frac{v'_y}{\left(1 + v'_x\dfrac{V}{c^2}\right)\gamma(V)} \tag{5–9}$$

$$v_z = \frac{dz}{dt} = \frac{v'_z}{\left(1 + v'_x\dfrac{V}{c^2}\right)\gamma(V)}$$

These equations prescribe the *composition* (addition) *of velocities*. If $v'_z = v'_y = 0$, and we write $v'_x = v'$, then equations 5–9 show that $v_y = v_z = 0$ and $v_x = v$ where

$$v = \frac{v' + V}{1 + \dfrac{v'V}{c^2}} \tag{5–10}$$

We interpret this equation in the following way: When all motions are along the x-axis, a velocity measured as having a value v' in reference frame K', will appear to have velocity v in a frame K. v, v' and V are related by equation 5–10. V is the velocity of K' relative to K (Fig. 5.1).

Three cases are of interest

(a) If $v' = V = c$, then substitution shows that $v = c$.

(b) If $v' < c$, $V = c$ or if $v' = c$, $V < c$ then $v = c$. This also can be shown by substitution in equation 5–10. It means that the speed of light is constant and has a value c in all inertial frames of reference.

(c) Finally, if $v' < c$ and $V < c$, then $v < c$.

PROBLEM 5–3. Show that the result (c) is always true by writing $v' = (1 - \delta)\,c$, $V = (1 - \Delta)\,c$ where $0 < \delta, \Delta < 1$.

PROBLEM 5–4. If the speed of light is infinite, Galilean relativity results. Give the transformations equivalent to (5–7) to (5–9) and obtain the law of composition of velocities. These expressions should be consistent with Newtonian physics.

Expression 5–10 is interesting because it also shows that a particle traveling at a speed less than the speed of light can never be accelerated to a speed equaling c. To see that, suppose that the particle initially was moving with velocity V. It is now given an extra velocity v' that also is less than c. From (c), above, we see that the resultant velocity is always less than c. We can keep adding small increments to the particle's velocity, but to no avail. It will always remain at a speed less than the speed of light. This situation characterizes the highly energetic cosmic ray particles. They move very close to the speed of light. When accelerated, they move a little faster, but never faster than c.

The Lorentz transformation leaves the interval s between two events invariant, but this is done at the expense of changes in the apparent time and spatial separations of events. The time separation is affected in the following way: If a clock is at rest at position $x = 0$ in K, then the proper time for K is given by t and the time measured by an observer O' at rest in the K' frame is

$$t' = \left(t - V\frac{x}{c^2} \right) \gamma(V) \Bigg|_{x=0} = t \, \gamma(V) \tag{5–11}$$

Actually, we are not interested in an absolute time, only in time intervals $\Delta t = t_1 - t_2$ and $\Delta t' = t_1' - t_2'$. Hence the equations 5–11 reduce to

$$\Delta t' = \frac{\Delta t}{\sqrt{1 - V^2/c^2}} \equiv \Delta t \, \gamma(V) \tag{5–12}$$

A time interval measured as Δt in the K frame would appear longer, $\Delta t' > \Delta t$, in the K' frame. To the observer O', the clock appears to be going slower. We speak of a *time dilation* in moving reference frames. We note that the choice of $x = 0$ was not necessary. The relation between Δt and $\Delta t'$ is independent of the choice of position, x.

In Problem 5–9 we will see that this time dilation can prolong the decay time of fast-moving, unstable, cosmic ray particles by many orders of magnitude. The time dilation is a dominant effect for the decay of such particles.

We can similarly derive the change in spatial separation between simultaneously observed events. If the positions of two points at rest in the K system are x_a and x_b as measured by an observer O at rest in K, the *proper length* of a line joining the two points is $\Delta x = x_b - x_a$. O', the observer at rest in K' measures the separation of the two points at some given time t'. We use the equations

$$x_a = (x'_a + Vt')\gamma(V) \qquad x_b = (x'_b + Vt')\gamma(V) \qquad (5\text{--}13)$$

where t' is the same in both expressions since O' sees both points simultaneously. The spatial separation observed from the K' frame, then, is

$$\Delta x' = x'_a - x'_b = (x_a - x_b)\sqrt{1 - \frac{V^2}{c^2}} = \frac{\Delta x}{\gamma(V)} \qquad (5\text{--}14)$$

Since the square root term is always less than unity, this means that the length measured by O' is shorter than the proper length. We call this the *Lorentz contraction* (Lo04).

The Lorentz contraction is found only along the direction of motion, while the transverse dimensions y and z remain unaffected according to equations 5–7. This could at first sight lead us to believe that a moving sphere should appear flattened into an oblate ellipsoid, and that a cube would appear distorted in some way dependent on its orientation with respect to the moving axes.

This view was held for more than half a century after the discovery of the special relativistic transformations by Lorentz and Einstein. But in 1959 Terrell (Te59) suggested that a sphere should always appear spherical, a cube cubical, and so on. The Lorentz transformations, while producing some distortions, primarily act to change the apparent orientation of the object by effectively rotating it.

To see how this comes about, suppose that a cube is moving with velocity V along the x direction. This motion is relative to an observer who looks at the cube in a direction transverse to its motion.

Fig. 5.3 The sides of a rapidly moving cube.

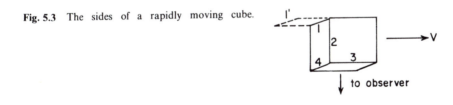

to observer

We will be interested in the apparent length of the edges 1, 2, and 3. Let the length of each edge be L, as measured by an observer at rest with respect to the cube and let edge 1 be perpendicular both to the direction of motion and the direction of the observer (Figure 5.3).

When the observer sees both edges 1 and 4 simultaneously—as he would when taking a photographic picture—he does not observe photons that were simultaneously emitted at these two edges. The light reaching him from edge 1 was emitted a time L/c earlier than light arriving from edge 4. But at that earlier time, edge 1 occupied position 1'. A photograph will therefore show a view of the cube with the far edge occupying position 1' and the near edge occupying position 4. The projected length of side 2 is the projected distance between 1' and 4, namely Lv/c.

This factor does not enter in discussing the length of edge 4, since the ends of these edges simultaneously emit those light rays that later are simultaneously observed. Side 4 is perpendicular to the direction of motion and its length is left unchanged by the Lorentz transformation; the Lorentz transformations also leave sides 1 and 2 unchanged. But side 3 is shortened by a factor $\sqrt{1 - (V^2/c^2)}$ (see equation 5–14).

A photograph will show sides 1, 2, and 3 having lengths L, LV/c, and $L/\gamma(V)$, respectively. If we define an angle ϕ by $V/c = \sin \phi$, then it is easy to see that these sides have apparent lengths, L, $L \sin \phi$, and $L \cos \phi$. The cube appears rotated by an angle ϕ.

While this is true for a small distant cube at its point of nearest approach, there are added distortions if the same cube is seen, say, earlier in its trajectory: Light arriving from the nearest edge is then emitted later—when the cube is closer—than light from edge 4, the trailing edge. The near edge therefore appears disproportionately long. In general, the cube appears both distorted and rotated (Ma72a)*.

5:4 FOUR-VECTORS

Let us now turn to the interrelationship between the world diagrams of two observers O and O' moving with inertial frames K and K'. As in Fig. 5.2 we will take K' to be moving in the direction of K's positive x-axis with velocity V. The origin of coordinates will then have components $y = y' = 0$ and $z = z' = 0$ at all times.

Let us also choose the origins of K and K' to coincide at some time $t = t' = 0$. This means that $x = x' = 0$ at that time. As seen by O, the origin of K' then has the world line t', shown in Fig. 5.4. The line passes through the origin and has a slope

$$\frac{c\,dt}{dx} = \frac{c}{V} \qquad (5\text{–}15)$$

That t' actually is the time axis for O' follows from the first and last equations of (5–7), for $x' = 0$. Again, if we set $t' = 0$ in these two equations, we see that the slope of the x' axis in O's world diagram must be $cdt/dx = V/c$. The angle ψ between the ct and ct' therefore equals the angle between the x and x' axes:

$$\psi = \tan^{-1} \frac{V}{c} \qquad (5\text{–}16)$$

Fig. 5.4 Minkowski diagram showing characteristics of a moving inertial coordinate system K' as seen by another inertial observer.

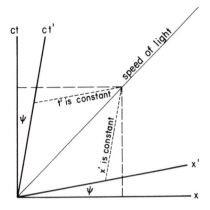

Different observers O and O′ have different spaces described by a fixed proper time. Events that appear simultaneous in one frame no longer appear simultaneous in others. This concept is foreign to Galilean relativity, where the speed of light is infinite and where the slope V/c would be zero. In Galilean relativity, the x and x', and the t and t' axes would therefore coincide. The light cone bisects both the spaces K and K' in this diagram, often called the *Minkowski diagram* (Ro68)*.

A vector in the four-dimensional spaces K and K' appears equally long to both observers. If the vector joins events $(0,0,0,0)$ and (x_1,y_1,z_1,t_1) as seen by O, it will join $(0,0,0,0)$ and (x'_1,y'_1,z'_1,t'_1) as seen by O′. But since we can always choose the x direction to coincide with the direction of motion, we can again set $y = y'$, $z = z'$, so that the lengths squared of the two vectors become

$$L^2 = -\{x^2 + y^2 + z^2 - c^2t^2\}$$

$$= -\{[(x' + Vt')^2 - (ct' + Vx'/c)^2]\gamma^2(V) + y'^2 + z'^2\} \tag{5-17}$$

$$= -\{x'^2 + y'^2 + z'^2 - c^2t'^2\} = L'^2$$

$$L = L' \tag{5-18}$$

The vector therefore has the same length, judged by either observer. Such a vector with components x, y, z, ct is called a *four-vector*. Four-vectors play a particularly important role in special relativity; first, because the theory's natural setting is a *four-space*; and, second, because the length of the vectors is *invariant* with respect to coordinate transformations: This means that one observer measures exactly the same vector magnitude as any other. But since relativity postulates that the laws of physics are invariant in all inertial frames, these invariant lengths assume a special significance in the formulation of the laws of physics.

We note that the length L specified here corresponds to the interval s defined in equation 5–1. The interval therefore is an invariant. If two events 1 and 2 occur in one and the same place for an observer O', we see that $s_{12}^2 = c^2 t_{12}^2 - l_{12}^2 = c^2 t_{12}'^2 > 0$. The square of the interval, s_{12}^2, is positive since the elapsed time t_{12}' is a real quantity. l_{12} is the spatial separation in O's frame. We see that if an interval between events is timelike, there exists a frame in which the events occur in the same place. If the interval is spacelike we can similarly show that a frame exists in which the two events are simultaneous.

The general transformation of a four-vector with components A_1, A_2, A_3, and A_4 reads

$$A_1 = [A_1' + \beta A_4'] \gamma(V), \qquad A_2 = A_2', \qquad A_3 = A_3'$$

$$A_4 = [A_4' + \beta A_1'] \gamma(V) \tag{5-19}$$

We will find for example, that the *four-momentum* with components $(p_x, p_y, p_z, \mathscr{E}/c)$ transforms as a four-vector. So also does a four-vector (A_x, A_y, A_z, ϕ) having the electromagnetic vector and scalar potentials as components (section 6:13). There are many other such four-vectors comprising useful physical quantities and, in fact, all physically significant quantities must—according to special relativity—mold themselves to conform to this four-dimensional point of view.

5:5 ABERRATION OF LIGHT

Next we will want to use the Lorentz transformations to see how the measurement of angles depends on the relative motion of an observer. We will find here that the measurement of an angle—or rather the sine or cosine of an angle—does not at all involve the measurements of two lengths. Rather it requires the simultaneous measurement of two velocities. This comes about because a distant observer must make his angular measurements using light signals received from an object, and the law of composition of velocities determines the angles these light rays subtend at the observer.

Suppose a particle has a velocity vector that lies in the xy plane. The velocity v has components $v_x = v \cos \theta$ and $v_y = v \sin \theta$ along the x and y axes of the reference frame K. Viewed by an observer at rest in the frame K', the velocity components are $v_y' = v' \sin \theta'$ and $v_x' = v' \cos \theta'$. The velocity transformation equations then allow us to write

$$\tan \theta = \frac{v_y}{v_x} = \frac{v' \sin \theta'}{[v' \cos \theta' + V] \gamma(V)} \tag{5-20}$$

When we deal with a light ray, $v' = c = v$; and the angles subtended by the light ray transform as

$$\tan \theta = \frac{\sin \theta'}{[V/c + \cos \theta'] \gamma(V)} \tag{5-21}$$

for, v_x and v_y alone lead to

$$\sin \theta = \frac{\sin \theta'}{\gamma(V) [1 + \beta \cos \theta']}, \qquad \cos \theta = \frac{\cos \theta' + \beta}{1 + \beta \cos \theta'} \tag{5-22}$$

When $V \ll c$ and the terms in β^2 are negligible, the sine equation becomes

$$[\sin \theta + \beta \sin \theta \cos \theta'] = \sin \theta' \tag{5-23}$$

And the *aberration angle* $\Delta\theta = \theta' - \theta$ is

$$\Delta\theta \sim \beta \sin \theta', \qquad \text{for } \Delta\theta \ll 1, \quad \beta \ll 1 \tag{5-24}$$

We note that since the light travels in a direction opposite to that in which the telescope moves, $\sin \theta'$ has a negative value, that is, $\theta' < \theta$, as shown in Fig. 5.5. This angle is of great practical importance in observational astronomy. For, if a star's position (direction in the sky) is measured at different times of year it will be found to undergo a small annual motion.

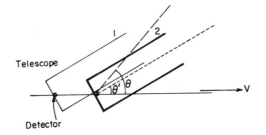

Fig. 5.5 Aberration of light. A telescope (stationary in some coordinate frame K') moves with velocity V, relative to a star. When the starlight enters, the telescope is in position 1. By the time the light has traveled the length of the instrument, the telescope has moved to position 2. All this time the telescope is pointed into direction θ', with respect to the x'-axis. But an observer whose telescope was at rest with respect to the star's reference frame K would have to point it at angle θ relative to x. Aberration is also found in Newtonian physics for a finite speed of light c. This is already indicated by the magnitude of the effect, which is of order V/c, while purely relativistic corrections always are of order V^2/c^2 and, therefore, are smaller. However, a relativistic correction for the Newtonian aberration is needed to obtain the correct value of the aberration angle.

PROBLEM 5-5. The orbital velocity of the earth about the sun is 30 km/sec, which means that the velocity of the earth changes by 60 km/sec over a six-month interval. Taking $V \sim 60$ km/sec show that

$$\Delta\theta \sim \frac{60}{3 \times 10^5} \sim 40'' \text{ of arc}$$

This is an easily measured angle in precision astronomy.

The aberration of light was first noticed by James Bradley in 1728. This represented the first verification of the earth's annual motion about the sun and the first conclusive proof that Copernicus had been right in the hypothesis he had advanced nearly two centuries earlier, in 1543.

5:6 MOMENTUM, MASS, AND ENERGY

The velocities we have talked about thus far are three-dimensional velocities. In relativity, however, the proper form to use is a four-vector because, as emphasized in section 5:4, four-vectors have *invariant magnitude*. We define a four-velocity with components

$$u_1 = \frac{dx}{ds}, \quad u_2 = \frac{dy}{ds}, \quad u_3 = \frac{dz}{ds}, \quad u_4 = c\frac{dt}{ds} \tag{5--25}$$

Since

$$ds = \sqrt{c^2 dt^2 - dx^2 - dy^2 - dz^2} = \frac{cdt}{\gamma(v)} \tag{5--26}$$

the equations 5–26 become

$$u_1 = \frac{dx}{dt} \frac{1}{c\sqrt{1 - \frac{v^2}{c^2}}} = \frac{v_x \gamma(v)}{c} \tag{5--27}$$

$$u_2 = v_y \gamma(v)/c \quad u_3 = v_z \gamma(v)/c \quad u_4 = \gamma(v)$$

or

$$u_i = [\gamma(v) dx_i/dt] c^{-1} \tag{5--28}$$

If we use the notation introduced in section 5:3, we see that u is a dimensionless quantity that does not have the units of velocity: cm sec^{-1}. We can also obtain the square of the magnitude of u:

$$u_1^2 + u_2^2 + u_3^2 - u_4^2 = -1 \tag{5--29}$$

This is an invariance property, which we will need below.

Frequently we are interested in a particle's *momentum*. This can be written in the form

$$\mathbf{p} = m_0 \mathbf{v} \gamma(v) \qquad (5\text{–}30)$$

and involves three components that correspond to the quantities

$$p_1 = m_0 c u_1, \qquad p_2 = m_0 c u_2, \qquad p_3 = m_0 c u_3 \qquad (5\text{–}31)$$

m_0 is the particle's mass measured at rest—the *rest mass*. The momentum therefore accounts for the first three components of a four-vector whose fourth component has the form $m_0 c u_4$. In relativity the fourth component is associated with the particle energy $\mathscr{E}$ in the following way

$$p_4 = \frac{\mathscr{E}}{c} = m_0 c u_4 \qquad (5\text{–}32)$$

The complete relativistic momentum four-vector then has components

$$(p_x, p_y, p_z, \mathscr{E}/c) \qquad (5\text{–}33)$$

It is clear that in the limit $v \ll c$, the first three components give the classical momentum $\mathbf{p} = m_0 \mathbf{v}$. However, the energy takes on a new form. Written explicitly, the energy equation is

$$\mathscr{E} = \frac{m_0 c^2}{\sqrt{1 - \dfrac{v^2}{c^2}}} = m_0 c^2 \gamma(v) \qquad (5\text{–}34)$$

At zero velocity this reduces to

$$\mathscr{E} = m_0 c^2 \qquad (5\text{–}35)$$

an expression that states that mass and energy are equivalent (Ei05b)*. It is precisely this equivalence that allows stars to radiate because the nuclear reactions that ultimately give rise to stellar radiation always involve a mass loss with the simultaneous liberation of energy in the form of photons or neutrinos. As the star radiates it conserves mass-energy by becoming less massive.

For small velocities equation 5–34 can be approximated by the expansion

$$\mathscr{E} = m_0 c^2 + \tfrac{1}{2} m_0 v^2 + \dots \qquad (5\text{–}36)$$

where the second term represents kinetic energy. The next higher term would be of order $m_0 v^4/c^2$. In mechanical or chemical processes m_0 remains essentially constant and we normally see changes only in the $mv^2/2$ term. This is why that term has classically been so important even though it is far smaller than the energy contained in a particle's mass.

Equation 5–34 shows that $\mathscr{E} \to \infty$ as $v \to c$, which means that an infinite amount of work would be required to accelerate a particle to the speed of light. As with all special relativistic effects, this statement is valid in inertial frames but need not be true for others. There need therefore be no conflict with the observations that distant galaxies travel at nearly the speed of light and that some may pass across the cosmic horizon when their speed, relative to our galaxy, exceeds the speed of light. Since these distant galaxies are at rest in reference frames that are accelerated relative to ours, special relativity need not hold, and we can make no general statements about speed limitations unless we talk in terms of a less specialized theory, such as the general theory of relativity.

Two important relations should still be stated. First, equations 5–30 and 5–34 show that

$$\mathbf{p} = \frac{\mathscr{E}}{c^2}\mathbf{v} \tag{5–37}$$

Second, writing the four-vector components as

$$p_i = m_0 c u_i \tag{5–38}$$

we obtain the square of the magnitude of the four-momentum vector

$$-(p_1^2 + p_2^2 + p_3^3 - p_4^2) = -p^2 + \frac{\mathscr{E}^2}{c^2} = m_0^2 c^2 \tag{5–39}$$

which again is invariant. Equations 5–39 can be rewritten as

$$\mathscr{E}^2 = p^2 c^2 + m_0^2 c^4 \tag{5–40}$$

Since photons have zero rest mass, and velocity c, (5–37) and (5–40) become

$$p = \frac{\mathscr{E}}{c} \tag{5–41}$$

The relations (5–37) and (5–40) are of particular importance in cosmic ray physics, where particle energies may be as high as $\sim 10^{20}$ ev. The rest mass of a proton with this energy is only 931 Mev, so that the total energy of a cosmic ray particle may be $\sim 10^{11}$ times its rest mass energy. This feature allows a cosmic ray primary, incident on the top layers of the earth's atmosphere, to undergo collisions that produce millions of shower particles whose total rest mass exceeds that of the primary proton by many orders of magnitude. The classical concept of conservation of mass is violated here, but the more encompassing principle of conservation of mass-energy permits this kind of process to take place.

5:6

5:7 THE DOPPLER EFFECT

Since energy is the fourth component of a four-vector $(\mathbf{p}, \mathscr{E}/c)$ it transforms as (see equations 5–19 and 5–33)

$$\mathscr{E} = \gamma(V)\left[\mathscr{E}' + Vp'_x\right] \tag{5–42}$$

when the relative motion is along the x-direction.
If we wish to see how photon energies will transform, we note for (5–41) that for a ray directed at an angle θ' with respect to the x'-axis

$$\mathscr{E} = \frac{\mathscr{E}' + (\mathscr{E}'V/c)\cos\theta'}{\sqrt{1 - \dfrac{V^2}{c^2}}} = \mathscr{E}'\left[1 + \beta\cos\theta'\right]\gamma(V) \tag{5–43}$$

The angle θ' is that shown in Fig. 5.5, but we have to recall that the direction of the photon's travel is opposite to the viewing direction. We know from quantum mechanics that the energy $\mathscr{E}$ is equal to the radiation frequency v, multiplied by Planck's constant, h—a universal constant. Using this in equation 5–43 gives,

$$v = v'(1 + \beta\cos\theta')\gamma(V) \tag{5–44}$$

which gives the Doppler shift in frequency for radiation emitted by a moving

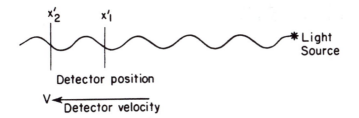

Fig. 5.6 The Doppler effect. The detector moves at velocity V, relative to the source. It starts measuring the frequency of radiation at time t'_1 and finishes at time t'_2. During this interval it is receding from the source, moving from position x'_1 to x'_2. A wave that would just have reached x'_1 by time t'_2 is therefore not counted, nor are any waves lying between x'_1 and x'_2, at time t'_2. The detector therefore senses a lower frequency v'. This explanation basically gives the first order Doppler shift proportional to V/c which is also present classically. The correct relativistic expression contains an additional factor $(\sqrt{1 - V^2/c^2})^{-1}$ as shown in equation (5–44).

5:7

source. In contrast to the classical situation, we see that there is a red shift even when the motion of the source is purely transverse ($\cos \theta = 0, \cos \theta' = -V/c$). This corresponds to a time dilation—a frequency decrease. When the source radiates in a direction opposite to its direction of motion, $\beta \cos \theta' < 0$, and $v < v'\gamma(V)$. When it radiates in the forward direction, $v > v'$.

In quasars we frequently see emission lines at some red shifted frequency v_0 and a series of absorption lines, for the same transition at higher frequencies $v_0 + \Delta_1, v_0 + \Delta_2, \ldots$ Here Δ is a shift corresponding to a few hundred or thousand km sec^{-1}. These are likely to be absorption lines due to cooler clouds along the line of sight and moving at different velocities. Some of these clouds may have been ejected from the quasar.

Line of sight components of stellar velocities within the Galaxy are generally determined by the Doppler shift observed in the spectra. Quasars have large red shifts that must be interpreted largely in terms of the cosmological red shift of galaxies at large distances, since the quasars often appear to be associated with galaxies of identical red shift.

5:8 POYNTING-ROBERTSON DRAG ON A GRAIN

Consider a grain of dust orbiting about the sun in interplanetary space. It absorbs sunlight, and re-emits this energy isotropically. We can view this two-step process from two differing viewpoints.

(a) Seen from the sun, the particle absorbs light coming radially from the sun and re-emits it isotropically in its own rest frame. A re-emitted photon carries off angular momentum proportional (i) to its equivalent mass hv/c^2, (ii) to the velocity of the grain $R\dot\theta$, and (iii) to the grain's distance from the sun R. Considering only terms linear in V/c, and neglecting any higher terms, we see that the grain loses orbital angular momentum L about the sun at a rate

$$dL = \frac{hv}{c^2} \dot\theta R^2 \qquad (5\text{–}45)$$

$$\frac{1}{L} dL = \frac{hv}{mc^2} \qquad (5\text{–}46)$$

for each photon whose energy is absorbed and re-emitted, or isotropically scattered in the grain's rest-frame. m is the grain's mass.

(b) Seen from the grain, radiation from the sun comes in at an aberrated angle θ' from the direction of motion, instead of at $\theta' = 270°$ (see equation 5–22).

$$\cos \theta = \frac{\cos \theta' + \dfrac{V}{c}}{1 + \dfrac{V}{c}\cos \theta'} = 0, \qquad \cos \theta' = -\frac{V}{c} \tag{5-47}$$

Here V is $\dot{\theta}R$, the grain's orbital velocity. Hence the photon imparts an angular momentum $pR \cos \theta' = -(h\nu/c^2)R^2\dot{\theta}$ to the grain.

For a grain with cross section σ_g

$$\frac{dL}{dt} = -\frac{L_\odot}{4\pi R^2}\frac{\sigma_g}{mc^2}L \tag{5-48}$$

where $L_\odot$ is the solar luminosity.

Either way the grain velocity decreases on just absorbing the light. From the first viewpoint, because the grain has gained mass which it then loses on re-emission; from the second, because of the momentum transfer.

PROBLEM 5-6. A grain having $m \sim 10^{-11}$ g, $\sigma_g \sim 10^{-8}$ cm^2 circles the sun at 1 AU. Calculate the length of time needed for it to fall into the sun, that is, to reach the solar surface, assuming that the motion is approximated by circular orbits throughout the whole time.

PROBLEM 5-7. Suppose one part in 10^8 of the sun's luminosity is absorbed, or scattered isotropically by grains circling the sun. What is the total mass of such matter falling into the sun each second?

5:9 MOTION THROUGH THE COSMIC MICROWAVE BACKGROUND RADIATION

We can derive the apparent angular distribution of light emitted isotropically in the reference frame of a moving object. Let the object be at rest in the K' system. Then the intensity $I(\theta')$ has the same value I', for all directions θ' (Figure 5.7). The energy radiated per unit time into an annular solid angle $2\pi \sin \theta' d\theta'$ is $2\pi I' \sin \theta' d\theta'$. In the K reference frame the intensity distribution is $I(\theta)$ and we would like to find the relation between $I(\theta)$ and I'.

The relativity principle requires that a body in thermal equilibrium in one inertial frame of reference also be in thermal equilibrium in all others. A blackbody radiator will therefore appear black in all inertial frames.

$$I(\nu) = \frac{2h\nu^3}{c^2}\left[\frac{1}{e^{h\nu/kT}-1}\right] \tag{5-49}$$

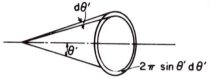

Fig. 5.7 Distribution of radiation, viewed in spherical polar coordinates.

For this to be true, we see that the ratios $I(v)/v^3$ and v/T both must be invariant under a Lorentz transformation. The total intensity seen from an arbitrary direction θ and integrated over all frequencies then is $\int I(v)\,dv$. This is proportional to v^4 and leads to

$$\frac{I(\theta)}{I'} = [(1 + \beta \cos \theta')\gamma(V)]^4, \tag{5-50}$$

which is the fourth power of the Doppler shift (5–44).

We see that an isotropically radiating, fast-moving body appears to radiate the bulk of its energy in the forward direction, ($\beta \cos \theta' \sim 1$), and only a small amount in the backward direction, ($\beta \cos \theta' \sim 1$). From the relation $d(\sin \theta) = \cos \theta\, d\theta$ and equation (5–22), we can obtain the expression $d\theta = d\theta'/[\gamma(V)(1 + \beta \cos \theta')]$. This allows us to write

$$2\pi \int I(\theta)\sin \theta\, d\theta = 2\pi I' \int [(1 + \beta \cos \theta')\gamma(V)]^2 \sin \theta'\, d\theta' = 4\pi I', \tag{5-51}$$

which is important. It means that the total power radiated by a source is the same for any set of observers in inertial frames. We will make use of this fact in section 6:19 to compute the total power emitted by a relativistic electron spiraling in a magnetic field.

Current observations indicate that the universe is bathed by an isotropic bath of microwave radiation. It is interesting that the presence of such a radiation field should allow us to determine an absolute rest frame on the basis of a local measurement. Such a frame would in no way violate the validity of special relativity which, as stated earlier, does not distinguish between different inertial frames. Rather, the establishment of an absolute rest frame would emphasize the fact that special relativity is really only meant to deal with small-scale phenomena and that phenomena on larger scales allow us to determine a preferred frame of reference in which cosmic processes look isotropic.

At each location in an isotropic universe—we will be discussing such cosmic models in section 10:5—there must exist one frame of reference from which the universe looks the same in all directions. This isotropy of course need hold only on a large scale, beyond the scale of clumping of the nearer galaxies. In the immediate vicinity of any observer, anisotropies on the scale of clusters of galaxies would always persist. The interesting feature of the radiation bath is that we should be able to determine the velocity of the earth relative to any coordinate system in which the cosmic flux appears isotropic. The earth might be expected to move at a speed of the order of 300 km sec^{-1} with respect to such a coordinate system, because the sun's motion about the center of the Galaxy has a velocity of the order of 250 km sec^{-1} and, in addition, galaxies are known to have apparently random velocities of the order of 100 km sec^{-1} relative to one another. These velocities would sum, vectorially, to something of the order of 300 km sec^{-1}.

If the isotropic microwave component of the cosmic flux has a blackbody spectrum (4–71) (Pe65)*; then the Doppler shift will transform the observed radiation to higher frequencies in the direction into which the earth is moving. This effect would make the factor v^3 in equation 5–49 systematically larger without changing the spectral shape of the curve that is given by the expression in brackets. An observer would therefore decide that he was seeing a blackbody spectrum at a temperature that was increased in the same amount as the Doppler increase in frequency, since the ratio v/T would then remain unchanged.

An observer moving relative to the cosmic radiation bath will then see a hotter blackbody flux in the direction into which he is moving than in the direction from which he came; and in fact the temperature of the observed flux will increase slowly, as a function of angle, starting from the trailing direction and reaching a maximum in the direction of motion. At each angle with respect to this direction of motion, the observer will see a blackbody spectrum but with a temperature dependent on the angle, as in (5–44). The difference between blackbody flux in the forward and backward direction should be most apparent near the peak of the spectrum.

Such observations have by now been carried out, and some of the results are shown in Figure 12.3. The Local Group of galaxies appears to be moving through the microwave background at ~600 km·sec^{-1}, falling toward the center of the Virgo cluster. Note that if there existed large-scale motions of galaxies and clusters relative to the microwave background, the Local Group's velocity with respect to these sources could considerably differ from its velocity through the background radiation.

PROBLEM 5-8. The Lorentz contraction is an important effect for extreme relativistic cosmic ray particles. For a proton with energy 10^{20} ev, for example, the disk is of the Galaxy would appear extremely thin. If the width of the disk is of the order of 100 pc in the frame of an observer at rest in the Galaxy, show that this width would appear to be $\sim 3 \times 10^9$ cm, comparable to the earth's circumference, to an observer moving with the cosmic ray proton.

PROBLEM 5-9. The time dilation factor similarly is important at cosmic ray energies. Consider the decay time of a neutron that has an energy comparable to the 10^{20} ev energies observed for protons. How far could such a neutron move across the Galaxy before it beta decayed? In the rest frame of the neutron this decay time is of the order of 10^3 sec; but in the framework of an observer at rest in the Galaxy it would be much longer. Show that the neutron could more than traverse the Galaxy.

PROBLEM 5–10 If an energetic cosmic gamma ray has sufficiently high energy it can collide with a low energy photon and give rise to an electron-positron pair. Because of symmetry considerations, this electron-positron pair has to be moving at a speed equal to the center of momentum of the two photons. The pair formation energy is of the order of 1 Mev. The energy of a typical $3°K$ cosmic background photon is of the order of 10^{-3} ev. What is the energy of the lowest energy gamma photon that can collide with a background photon to produce an electron-positron pair? Show that in the frame within which the pair is produced at rest, energy conservation gives

$$\frac{h v_1 \left(1 - \dfrac{v}{c} \right)}{\sqrt{1 - \dfrac{v^2}{c^2}}} + \frac{h v_2 \left(1 + \dfrac{v}{c} \right)}{\sqrt{1 - \dfrac{v^2}{c^2}}} = 2 m_0 c^2. \tag{5–52}$$

and momentum conservation requires the two terms on the left to be equal. These two requirements give

$$(h v_1)(h v_2) = (m_0 c^2)^2 \tag{5–53}$$

$$h v_1 \sim 10^{15}/4 \text{ ev}$$

Interestingly, the cross section for this process is sufficiently high so that no gamma rays with energies in excess of $h v_1$ can reach us if they originate beyond 10 kpc—that is, beyond the distance of the Galactic center.

5:10 PARTICLES AT HIGH ENERGIES

Cosmic rays are extremely energetic photons, nuclei, or subatomic particles that traverse the universe. Occasionally such a particle or photon impinges on the earth's atmosphere, or collides with an ordinary interstellar atom. What happens in such interactions?

We have no experimental data on particles whose energies are as high as, say, 10^{20} ev, because our laboratories can only accelerate particles to energies of the order of 3×10^{12} ev. However, the relativity principle permits us some insight even into such interactions. For example, we ask ourselves, how 10^{20} ev protons would interact with low energy photons in interstellar or intergalactic space. Such 3°K blackbody photons have a frequency $v \sim 3 \times 10^{11}$ Hz.

As seen by the proton, these millimeter-wavelength photons appear to be highly energetic gamma rays. This follows because $\gamma(v)$ must be $\sim 10^{11}$ for the proton, since its rest mass is only 9.31×10^8 ev; and by the same token the proton sees the photon Doppler shifted by a factor of 10^{11} (5–34 and 5–44). In the cosmic ray proton's rest frame, the photon appears to have a frequency of $\sim 3 \times 10^{22}$ Hz. That corresponds to a gamma photon with energy ~ 100 Mev, and the proton acts as though it were being bombarded by 100 Mev photons.

This simplifies the problem appreciably. One hundred Mev photons can be produced in the laboratory; and in fact we find that the main effect of photon-proton collisions at this energy is the production of π-mesons, through the interactions

$$\mathscr{P} + \gamma \rightarrow \mathscr{P} + \pi^\circ \tag{5–54}$$

$$\mathscr{P} + \gamma \rightarrow \mathscr{N} + \pi^+ \tag{5–55}$$

In the first reaction the proton-photon collision produces a neutral pion π° and a proton having a changed energy. The second reaction produces a neutron and a positively charged pion.

The cross sections for these interactions can be measured in the laboratory, and the results are then immediately applicable to our initial query. It turns out that the cross sections are so large that the highest energy cosmic ray protons can probably not traverse intergalactic space over distances $\gtrsim 20$ Mpc even if the only extragalactic photons are those found in the 3°K microwave photon flux (Gr66, St68).

For some time this presented a puzzle, because quasi-stellar objects appeared to be the most likely source of energetic cosmic ray protons and quasars appear to be far more distant than 20 Mpc. However, the suggestion that pulsars or expansion shells exploding out of a supernova (see section 6:6) may generate sufficiently energetic protons (Gu69) has removed some of these difficulties, since the cosmic ray protons can now be considered to be a more local phenomenon.

Similar relativistic arguments are useful in a variety of other problems involving cosmic ray particles. In Chapter 6, we will discuss the Compton and inverse Compton effects involving collisions of photons and electrons, and we will also consider several other relativistic effects. Frequently a physical problem can be considerably simplified if we choose to view the situation from a favorable inertial frame. The relativity principle shows us how to do that, and gives us many new insights into the symmetries of physical processes.

5:11 HIGH ENERGY COLLISIONS

Consider the elastic collision of a low energy particle with a similar particle initially at rest. If we view this interaction in the resting particle's frame, and both particles have mass m, then the center of mass will move with velocity $v/2$ as shown in Fig. 5.8.

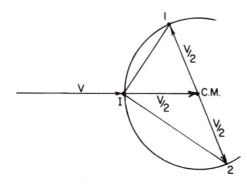

Fig. 5.8 Illustration of elastic collisions for identical particles.

For an initial approach velocity v of the moving particle, conservation of momentum will require that the two particles have velocities $v/2$ relative to the center of mass—after the collision, as well as before. For any time after the collision, a circle can be drawn through the impact point I, and the particle positions 1 and 2 that define a diameter on the circle. This means that the particles always subtend a right angle at the impact point.

So far the treatment has been nonrelativistic. In the relativistic case, the center of mass still lies on a line joining 1 and 2. Effectively, particles 1 and 2 are scattered away from the center of mass in opposing directions. Seen from a rest frame, however, they will appear to be scattered predominantly into the forward direction. This is precisely the same concentration into the forward direction that we saw for the rapidly moving light source that emits radiation isotropically in its own rest frame (5–50).

When a cosmic ray proton collides with the nucleus of a freely moving interstellar atom or with an atom that forms part of an interstellar grain, a fraction of the nucleus can be torn out. This may just be a proton or a neutron, or it could be a more massive fragment, say, a He^3 nucleus. Such *knock-on* particles always come off predominantly in the forward direction, close to the direction along which the primary proton was moving.

Similarly, when a primary arrives at the top of the earth's atmosphere, after its long trek through space, it collides with an atmospheric atom's nucleus, giving rise to energetic secondary fragments, mesons, baryons, and their decay products. These decay into mesons, gammas, electron-positron pairs, or neutrinos, or they may collide with other atoms until a whole shower of particles rains down. Such a *cosmic ray air shower* consisting of electrons, gamma rays, mesons, and other particles, even if initiated at an altitude of 10 or more kilometers, often arrives at ground level confined to a patch no more than some hundred meters in diameter. The forward concentration is so strong that the showers are well confined even though they sometimes consist of as many as 10^9 particles.

This close confinement allows us to deduce the total energy originally carried by the primary; we need only sample the energy incident on a number of rather small detector areas. In fact, much of our information about high energy cosmic ray primaries has come from just such studies made with arrays of cosmic ray shower detectors. These arrays usually sample an area not much more than a few hundred meters in diameter although an array as big as 3.6 km was used by an MIT group. The total energy in the shower can be determined from these samplings, and the time of arrival at each detector shows the direction from which the primary came (Ro64a). Fig. 5.9 shows some of the constituents of cosmic ray air showers.

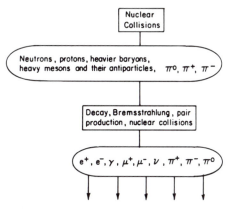

Fig. 5.9 Constituents of a cosmic ray air shower. The primary particle, here shown as a proton, collides with the nucleus of an atmospheric atom, producing a number of secondary particles that suffer nuclear collision, decay, pair production, or *bremsstrahlung*–a process in which a charged particle is slowed down by the emission of a gamma photon. A large succession of such events takes place. By the time the shower arrives at the surface of the earth, most of the charged particles we observe are electrons, positrons, and muons. Although most of the primary nuclei are protons, several percent can be alpha particles (helium nuclei) and about one percent are heavier nuclei. Electrons and positrons also can be primary particles. The air showers are a prime example of the conversion of energy into rest mass. On occasion, the energy of a single primary is sufficient to produce 10^9 shower particles.

5:12 FASTER-THAN-LIGHT PARTICLES

When Einstein first discovered the special relativistic concept he clearly stated that matter could not move at speeds greater than the speed of light. He argued that the relation between rest mass and energy (5–3) already implied that an infinite amount of energy was needed to accelerate matter to the speed of light and that particles with nonzero rest mass could therefore never quite reach even the velocity of light let alone higher velocities.

In recent years, this question has been re-opened by a number of researchers. They have argued that it certainly is not possible to actually reach the speed of light by continuous acceleration, but that this alone does not rule out the existence of faster than light matter created by some other means. They have called particles moving with speeds greater than *c tachyons*, and have examined the possible properties of such entities.

The basic argument in favor of even examining the possibility of tachyon existence is the formal similarity of the Lorentz transformations for velocities greater than and less than the speed of light, and the fact that the transformations taken by themselves say nothing that would rule out tachyon existence.

The similarity of course does not imply that particles and tachyons behave in precisely the same manner. If we look at equation 5–34, we note that a particle moving with speed $V > c$ has an imaginary quantity in the denominator. If the mass of the tachyon is real its energy would therefore be imaginary. In practice, one chooses the mass of the tachyon to be imaginary mainly on the grounds that this is not observationally ruled out. Perhaps this is a somewhat negative approach, but if this assumption is not made, it is more difficult to make headway to the point where predictions on the probable outcome of experiments become possible.

Choosing the mass to be an imaginary number makes the energy $\mathscr{E}$ real. The momentum too is then real as shown by (5–37).

We now combine equations 5–34 and 5–40:

$$\mathscr{E}^2 = m_0^2 c^4 \gamma^2 (V) = p^2 c^2 + m_0^2 c^4 \tag{5–56}$$

As V becomes large, $\mathscr{E}$ is seen to become small. In the limit of infinite velocity the energy becomes zero, but the momentum stays finite, asymptotically approaching the value $|m_0 c|$.

Thus far we have only departed from more standard concepts in accepting the possibility of imaginary mass values.

Now, however, we come to an argument that has caused considerable conceptual difficulty in the past few years. This argument considers sequences of events if tachyons can be used as carriers of information. Consider the world lines of two observers O and O', shown in Fig. 5.10. Let O emit a first tachyon toward O'. The slope of this trajectory is less than that for a photon, in the world diagram. The tachyon enters the "absolutely inaccessible" domain that was shown in Fig. 5.1.

Seen from O''s vantage point, tachyon 1 arrives from above the x'-axis, apparently moving backward in time. This is bad enough; but let us go a little further.

Fig. 5.10   World diagram for tachyon transmission.

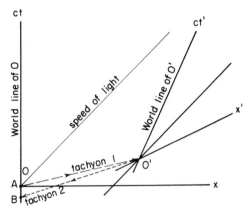

O′ can also send out tachyons and, in principle, the tachyon 2 he transmits towards O can be a faster moving tachyon than tachyon 1.

We can now set up the following paradox (Th69): Observer O waits until he sees no event at all during a long period of time. He then sends out a tachyon 1 at time A. O′ has previous instructions to send out a fast tachyon immediately upon receiving one from O. He does this now, sending out tachyon 2 that arrives at O at time B, during the time interval when O was sure no event had taken place. Cause and effect are thrown into complete confusion. The situation is not helped much even if a tachyon absorbed moving in a negative time direction is really equivalent to a tachyon emitted in the positive time direction.

Should we then disregard tachyons altogether? This seems an unnecessarily harsh edict: The causal argument we have just applied is strongly influenced by traditional special relativity and may be misleadingly restrictive. For example, the concept of simultaneity used in the argument, depends on the idea that no messages propagate faster than the speed of light. We noted in section 5:4 that the Minkowksi diagram would look quite different if the speed of transmission of information was higher than c: The angles ψ in Fig. 5.4 would become smaller, and in the limit of infinite velocity—the Galilean limit—we found $\psi = 0$. For tachyons then, we would expect smaller angles ψ; the angle between the ct'- and x'-axes in Fig. 5.10 should approach a right angle, and the causal paradox should disappear, because as the angle between the tachyon trajectory and the x'-axis increases, it becomes impossible to transmit signals into another observer's past. Arguments of this kind still leave the existence or inadmissibility of tachyons an open question. As we will see in Chapters 10 and 11, the existence of tachyons would have important consequences in cosmology and in rapid communication across large distances within our galaxy or across the Universe.

5:12

Preliminary experiments to search for tachyons have been carried out (A168). Thus far none have been detected, but perhaps some day they will be.

5:13 STRONG GRAVITATIONAL FIELDS

As already stated, the introduction of gravitational fields requires a theory more general than the special theory of relativity which restricts itself to inertial frames. For general problems involving gravitation, the general theory of relativity (Ei16) or similar gravitational theories (Di67) have to be used. However, some simple gravitational results can be obtained without such theories if we remember that the set of inertial frames also includes freely falling frames of such small size that the gravitational field can be considered to be locally uniform. We will consider two such local inertial frames in a centrally symmetric gravitational potential Φ.

Consider an observer O' at distance r from the central mass distribution Fig. 5.11. We would like to know the form which the line element ds^2 would take in his frame of reference. We will assume that he was at rest initially, and that he has only just started falling toward the central mass, quite recently. His velocity is therefore still essentially zero, but he is accelerating toward the center. Alternately, we can suppose that O' was initially moving outward from the star but at a speed less than the escape velocity. He only had enough kinetic energy to reach r. Here his velocity reached zero, and he is just beginning his journey back into the center. We see him when his velocity is zero.

Since O' is freely falling, his line element will seem to him to have the form (5–3). In spherical polar coordinates this is

$$ds^2 = c^2 dt'^2 - r'^2(\sin^2\theta' d\phi'^2 + d\theta'^2) - dr'^2 \qquad (5\text{--}57)$$

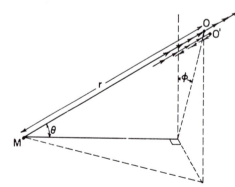

Fig. 5.11 Freely falling observer near a mass M.

We ask ourselves, however, what ds^2 would seem to be, seen by an observer O far enough away from the mass distribution so that Φ essentially is zero or negligibly small. Φ as used here will be the negative of $\mathbb{V}$ in (3–54).

We could of course assume that O gets all his information about O′'s system from light signals. But that is not necessary. The physical relationship between O and O′ is independent of how the observational information is conveyed.

Let us therefore suppose that O has taken a trip to find out for himself. We can suppose that he was near the central mass distribution, that he is now on his way out, and that he is in unpowered motion, freely falling radially outward, with just exactly enough energy to escape to infinity.

O goes through the radial distance r, close to O′, just as O′ passes through zero velocity and begins his infall. Since both observers are in inertial frames, the Lorentz transformations can be used to determine what O′'s line element would seem like to O. Once again, the components perpendicular to the direction of relative motion should be identical, and

$$r^2(d\theta^2 + \sin^2\theta\, d\phi^2) = r'^2(d\theta'^2 + \sin^2\theta'\, d\phi^2) \tag{5–58}$$

The radial components, however, will appear changed because of the relative motion. If the gravitational potential is weak, the velocity of O relative to O′ is immediately obtained from the fact that O just barely has enough kinetic energy to go to infinity, so that, equating kinetic and potential energy, for unit mass

$$\tfrac{1}{2} V^2 = \Phi \tag{5–59}$$

Hence equations 5–12 and 5–14 would now read

$$\Delta t = \frac{\Delta t'}{\sqrt{1 - (2\Phi/c^2)}} \qquad \Delta x = \Delta x'\sqrt{1 - (2\Phi/c^2)} \tag{5–60}$$

This also is the correct form for strong potentials Φ, where the classical concept of kinetic energy no longer has a clear meaning. We can therefore write that the line element (5–57) has the form

$$ds^2 = (c^2 - 2\Phi)\, dt^2 - r^2(\sin^2\theta\, d\phi^2 + d\theta^2) - \frac{dr^2}{1 - (2\Phi/c^2)} \tag{5–61}$$

as seen by O. This represents a translation of the clock rate and scale length in O′'s frame as seen from O's coordinate system. O notes this down, and is able to convey these impressions when he reaches infinity. He has been traveling in an inertial frame all this time, and his results are therefore not suspect.

The line element, or *metric*, depicted in (5–61) is called the *Schwarzschild line element.* When the potential is generated by a mass M, we can rewrite (5–61) as

$$ds^2 = \left(c^2 - \frac{2MG}{r} \right) dt^2 - \left(1 - \frac{2MG}{rc^2} \right)^{-1} dr^2 - r^2 (\sin^2 \theta \, d\phi^2 + d\theta^2)$$

(5–62)

We see that something odd must take place at the *Schwarzschild radius*

$$r_s = \frac{2MG}{c^2}$$

(5–63)

Here, according to (5–60) the clocks would appear to run infinitely slowly. A message emitted at some time, t_0, would not arrive at larger radial distances until an infinite time later. In fact signals emitted at $r < r_s$ never come out. A massive object that was completely enclosed in r_s could therefore not radiate out into the rest of the universe and would appear invisible. Such objects have been called *black holes*. They are only detectable through the gravitational and electromagnetic field they set up, but not through emitted radiation (Ru71a). A star could be orbiting about a black hole companion; its orbital motion about an apparently dark region in space is a sign that a black hole might be there.

For an ordinary stellar mass, collapsed through a neutron star state, $r_s \sim 3 \times 10^5$ cm. This is not much smaller than the radius of the neutron star itself. For an object as massive as a galaxy $r_s \sim 10^{16}$ cm.

We will mention black holes again later. However, for the moment, it is still worth discussing two matters.

First, there cannot be a preponderant number of black holes within the Galaxy, because we can account for roughly 60% of the Galactic mass in terms of ordinary stars and interstellar matter observed by visual and radio observers. Only some 40% of the mass appears unobserved, and some of this is likely to be in the form of nonluminous, cool stars, interstellar molecular hydrogen, and so on. We cannot rule out, however, that something of the order of one-third the Galactic mass might exist in the form of black holes.

Secondly, space travelers must be careful. Once they enter a black hole they can never return. The interior of such an object is as isolated from us as a separate universe.

ANSWERS TO SELECTED PROBLEMS

5–3. By (5–10)

$$v = \frac{c(2 - \delta - \Delta)}{(2 - \delta - \Delta + \Delta\delta)} < c.$$

5:13

For opposing velocities we obtain an expression of form

$$-c < v = \frac{(\Delta - \delta)c}{\delta + \Delta - \delta\Delta} = \frac{(\Delta^2 - \delta^2)c}{\Delta^2 + \delta^2 + \delta\Delta(2 - \delta + \Delta)} < c.$$

5–4. $x = x' + Vt', y = y', z = z', t = t'$, and since $dx = dx' + V\,dt', dt = dt',$
$v_x = v'_x + V, v_y = v'_y,$ and $v_z = v'_z$.

5–6. $\dfrac{dL}{dt} = -\dfrac{L}{R^2} \cdot \left(\dfrac{L_\odot}{4\pi}\right) \cdot \dfrac{\sigma_g}{(mc^2)}.$

Since $L = mvR$ and $v^2/R = GM_\odot/R^2$,

$$\frac{dL}{L} = \frac{dR}{2R}.$$

$$t = \int_{R_\odot}^{1AU} R\,dR \left\{\frac{2\pi}{L_\odot} \cdot \frac{mc^2}{\sigma_g}\right\} = 5 \times 10^3 \text{ y.}$$

5–7.   From Problem 5–6, for each grain i

$$m_i c^2 = t_i \frac{\sigma_g}{\pi(R_I^2 - R_\odot^2)} L_\odot$$

where R_I is the grain's initial position.
For all grains $M_{TOT}c^2 \approx t_{TOT} \cdot (10^{-8} L_\odot)$.

∴ Mass/sec falling into sun is $\dfrac{(10^{-8})(4 \times 10^{33} \text{ erg sec}^{-1})}{9 \times 10^{20} \text{ cm}^2 \text{ sec}^{-2}}$

$$= 4.5 \times 10^4 \text{ g.}$$

5–8. $\mathscr{E} = \gamma(V) m_0 c^2$; since $\mathscr{E} = 10^{20}$ ev, $\gamma(V) = 10^{11}$ and
$\Delta x' = \Delta x/\gamma(V) = 3 \times 10^{20}/10^{11} \sim 3 \times 10^9$ cm.

5–9. $\Delta t' = \gamma(V)\Delta t \sim 10^{14}$ sec. At $v \sim c$, the distance traveled is
3×10^{24} cm = 1 Mpc.

6

Electromagnetic Processes in Space

6:1 COULOMB'S LAW AND DIELECTRIC DISPLACEMENT

In earlier chapters we noted the similarities between Coulomb's law for the attraction of charged particles and Newton's law for the attraction of masses. Both are inverse square law forces. Coulomb's law states that the attraction between two charges q_1 and q_2 is

$$\mathbf{F} = \left(\frac{q_1 q_2}{r^3} \right) \mathbf{r} \tag{6-1}$$

The charges q can be either positive or negative. In the general case where a large number of separate charges exert a force on a given charge q, this force is the vector sum of a whole series of terms of the form of equation 6–1.

$$\mathbf{F} = q \sum_i \left[\frac{q_i}{r_i^3} \right] \mathbf{r}_i \tag{6-2}$$

and we can define an *electric field* $\mathbf{E}$

$$\mathbf{E} = \frac{\mathbf{F}}{q} \tag{6-3}$$

which can be considered the seat of the force. All this assumes that the charges q_i and q are at rest in a vacuum. If the charges q_i are moving, the charge q will experience an additional force that is magnetic in character, and if the charges are not in vacuum but in a polarizable dielectric material, the material will adjust itself to cancel out some of the force. The actual force acting on q then becomes less than that given in equation 6–2.

To specify this situation completely, we therefore define one more vector quantity, the *dielectric displacement* **D**, which is independent of the properties of the material in which the charges are imbedded. **D** is strictly a geometric quantity and specifies the field that would be obtained if all charges were in a vacuum. In the presence of a uniform dielectric, equation 6–2 becomes

$$\mathbf{F} = \frac{q}{\varepsilon} \sum_i \frac{q_i}{r_i^3} \mathbf{r}_i \tag{6–4}$$

Equation 6–3 still holds true since it, in fact, is the definition of electric fields; but the dielectric displacement now becomes

$$\mathbf{D} = \varepsilon \mathbf{E} \tag{6–5}$$

which is seen to be independent of ε, and dependent only on the position and magnitudes of the charges, that is, on the quantities q_i and $\mathbf{r}_i$. ε is the *dielectric constant* which, for most real materials, can be taken to be independent of **E** for field strengths below a critical value.

If we draw a spherical surface around a charge q_1, the displacement produced by this charge at the surface obeys the relation

$$\mathbf{D} = \left(\frac{q_1}{r^3} \right) \mathbf{r} = \left(\frac{4\pi q_1}{4\pi r^3} \right) \mathbf{r} \tag{6–6}$$

so that

$$\mathbf{D} \cdot \mathbf{n} = \left(\frac{4\pi q_1}{rA} \right) \mathbf{r} \cdot \mathbf{n} \tag{6–7}$$

where **n** is the normal to the surface at point **r** and A is the total area of the enclosing surface. The dots denote a *scalar product*. If a large number of charges are involved, or if the charge distribution becomes continuous, a more general form of expression (6–7) is applicable:

$$\int \mathbf{D} \cdot d\mathbf{s} = \int 4\pi \rho \, dV = \int \nabla \cdot \mathbf{D} \, dV \tag{6–8}$$

where the last equality is obtained from *Gauss's theorem* on vector integration which states that for an arbitrary vector **X**

$$\int \mathbf{X} \cdot d\mathbf{s} = \int \nabla \cdot \mathbf{X} \, dV \tag{6–9}$$

where the integral on the left is a *surface integral* and $d\mathbf{s}$ is an element of the surface over which the integration takes place. $\nabla \cdot$ is the *divergence operator*.

One may wonder why we emphasize the relation between **D** and **E** in such detail when we have set out to discuss electromagnetic processes in space. We might expect that the emptiness of the cosmos would assure that **D** and **E** are always identical. In actual fact this is not quite true, and much of our knowledge of the contents of interstellar space depends on small differences between **E** and **D**. We define one more quantity that will be useful later. It is the *polarization field* **P** which is a measure of the difference between the displacement and electric fields:

$$\mathbf{P} = \frac{[\mathbf{D} - \mathbf{E}]}{4\pi} = \frac{(\varepsilon - 1)\mathbf{E}}{4\pi} \tag{6-10}$$

$4\pi\mathbf{P}$ is the field set up through the rearrangement of charges in the polarizable material. It tends to oppose the externally applied field, reducing its value from **D** to **E**. The factor 4π introduced here is a matter of convention and has the following significance: At a plane boundary, with charge density σ per unit area, **D** just equals $4\pi\sigma$. The polarization field, instead, will depend on σ', the induced charge density per unit area. Now, **P** is the electric dipole moment per unit volume. If this volume contains n dipoles having charge q and separation d, $\mathbf{P} = nqd$. The charge density σ' then is nqd also, because we can visualize a cube of unit volume, made up of d^{-1} dipole layers each of thickness d and containing nd dipoles. This makes P numerically equal to σ'—no factor of 4π occurs.

Thus far we have acted as though static fields perhaps were important on a scale of cosmic dimensions. In general, this is probably not true, because in a near vacuum electric charges generally can quickly rearrange themselves into a configuration where all electric fields are neutralized, that is, into a charge neutralized configuration where any small volume element basically contains the same number of positive and negative charges. The dimensions of such volumes are given by the Debye shielding length discussed previously in Chapter 4. There is one exception to this general rule, and it is important. We will show in the next section that electric charges generally are tied to magnetic field lines in space; and if an electric field is applied perpendicular to the direction of a cosmic magnetic field, the charges cannot flow across the magnetic field lines to neutralize the electric field. In such a situation large-scale electric fields may persist.

6:2 COSMIC MAGNETIC FIELDS

An electric charge q traveling through a cosmic magnetic field experiences a force **F** called the *Lorentz force*:

$$\mathbf{F} = \frac{q\mathbf{v} \wedge \mathbf{B}}{c} \tag{6-11}$$

6:2

where **v** is the velocity of the charge, **B** is the *magnetic field* and c is the speed of light. The *cross product*, in equation 6–11 shows that the force, and hence the acceleration experienced by the charge, is perpendicular to both the velocity and the direction of the magnetic field. The charge therefore spirals (see Fig. 6.1)

Fig. 6.1 Diagram to illustrate spiral motion in a magnetic field.

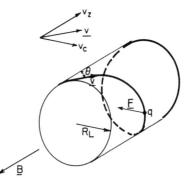

along the magnetic field lines without changing energy. (To do work on the particle one would require a force that has some component along the direction of motion). In a constant magnetic field, the particle describes a helical motion with constant pitch. The velocity component v_z along the direction of the field, is a constant of the motion, and the circular velocity v_c about the field lines then defines a *pitch angle* θ so that

$$\tan \theta = \frac{v_c}{v_z} \qquad (6-12)$$

The *gyroradius* or *Larmor radius*, R_L, of this motion is easily obtained by setting the magnetic force equal to the centrifugal force acting on the particle. If the particle has transverse momentum p_c and *gyrofrequency* $\omega_c = v_c/R_L$, the force has magnitude

$$\dot{p} = \frac{p_c v_c}{R_L} = \frac{qBv_c}{c}, \quad R_L = \frac{p_c c}{qB} \quad \text{and} \quad \omega_c = \frac{v_c}{R_L} = \frac{qBv_c}{p_c c} \qquad (6-13)$$

The gyrofrequency is sometimes also called the *cyclotron frequency.*

PROBLEM 6–1. Show that the Larmor radius of a proton, moving at 10 km sec^{-1} through a field of 10^{-6} gauss is small compared to interstellar and even interplanetary distances.

Because the Larmor radius is small compared to the expected dimensions of interstellar and interplanetary fields, we have a situation in which charged particles, moving with thermal velocities characteristic of cosmic gases, are effectively tied to the magnetic field lines. They can move along the field lines but cannot cross them any appreciable distance. We say that the particles are "frozen" to the field and the motion of such particle-field combinations is called *frozen-in flow*. *Magnetohydrodynamics* is the subject that deals with problems arising from such flow (Co57).

We notice that the only way a particle can escape from being frozen to the lines of force is through an encounter with another particle. Each particle then assumes a completely new orbit. If such collisions are sufficiently frequent, the particles can diffuse across magnetic fields.

Inasmuch as cosmic magnetic fields have their origins in the organized motion of particles, the frozen-in flow not only is due to the presence of a magnetic field, but also maintains the field which causes it. This self-consistent motion of charges is not an obvious result, but magnetohydrodynamics shows that it is real. The collisional processes just mentioned, therefore, conspire not only to prevent freezing-in, but also as a consequence tend to destroy the magnetic fields that are maintained by the frozen in flow. For this reason, frozen-in fields cannot be maintained in dense gases where collisions are frequent. These collisions of charges with surrounding particles provide an electrical resistance that dissipates particle motions and the energy resident in the magnetic field. Frozen-in flow therefore has a short life in a dissipative medium (Sp62).

Magnetohydrodynamics also tells us that the presence of a force, such as a gravitational or electrostatic force, acting normal to the magnetic field, can produce a drifting motion in which charges move in directions perpendicular both to the applied force and to the magnetic field direction. Such particle drifts occur in the *Van Allen belts* of charged particles that constitute part of the earth's *magnetosphere*. These drifts, however, do not directly act to dissipate cosmic magnetic fields, unless the drifting particles suffer collisions.

6:3 OHM'S LAW AND DISSIPATION

A current generally consists of two types of terms. The first of these expresses the actual flow of charge in response to an applied electric field. The second term corresponds to a virtual current representing a change in the applied field. This change gives rise to a magnetic field (see section 6:5) just as a moving charge would. It is genuinely important. One writes

$$\mathbf{j} = \rho\mathbf{v} + \frac{1}{4\pi} \frac{\partial \mathbf{D}}{\partial t} \qquad (6\text{--}14)$$

6:3

The value of the velocity **v** in this equation is determined by two competing effects. The applied electric field seeks to continuously accelerate the charge, while distant collisions with other electric charges continually seek to slow the particle down. The resistivity of the medium is a measure of this slowing down. Its reciprocal is the *conductivity* σ. In terms of **E** and σ, equation 6–14 can be written as

$$\mathbf{j} = \sigma\mathbf{E} + \frac{1}{4\pi}\frac{\partial\mathbf{D}}{\partial t} \tag{6–15}$$

In general, the conductivity depends on the density of the gas, its temperature, the state of ionization, and, hence, also on the chemical composition. Distant collisions provide the main process that slows down the motion of charged particles in space, and they determine the value of σ. These collisions were discussed in sections 3:13 and 3:14 and will be fully treated in 6:16.

6:4 MAGNETIC ACCELERATION OF PARTICLES

One of Faraday's contributions to electromagnetism was his discovery that a time-varying magnetic field gives rise to electric currents in a conducting medium that encircles the field. The plane in which this current flows is perpendicular to the direction of the time-varying field component. In integral form *Faraday's law* is expressed as

$$\frac{1}{c}\frac{\partial}{\partial t}\int\mathbf{B}\cdot d\mathbf{s} = -\oint\mathbf{E}\cdot d\mathbf{l} \tag{6–16}$$

where the integral on the left is a surface integral over the area enclosed by the loop through which the current is flowing (Fig. 6.2a). The integral on the right is a line integral taken over that loop and the current observed by Faraday has been replaced by the electric field that gives rise to it in accordance with equation 6–15. We note now that, if any region of interstellar space should suddenly be subjected to a rising magnetic field, electric charges would experience an effective electrical field **E** proportional to the time rate of change of **B**. In the laboratory this effect is used to elevate charges to very high energies. The first device that successfully accomplished this acceleration is the betatron constructed by D. W. Kerst in 1940. Since that time, it has been suggested by various astrophysicists that the betatron process might also be active in interstellar space in accelerating charged particles to the very high energies observed in cosmic rays.

A rapid rise in magnetic field strength could be produced by the compression of a cosmic cloud in a direction perpendicular to its magnetic field. Such a compression can occur in the collision of interstellar clouds, either with one another, or with high velocity gases ejected from exploding supernovae. This process may

6:4

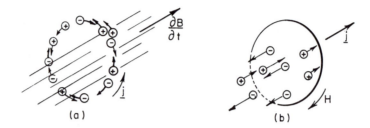

Fig. 6.2 Illustrations for Faraday's and Ampère's Laws. (*a*) Faraday's law states that the current in a conducting loop, and the associated electric field are determined by the rate of change in the number of magnetic *lines of force* enclosed by the loop (see equations 6–15 and 6–16). The number of lines of force crossing unit area is proportional to the magnetic field, *B*.

(*b*) Ampère's law states that the magnetic field, integrated along a loop enclosing a current, is determined by the total current crossing the enclosed area (see equation 6–17).

produce low energy cosmic rays, sometimes called *suprathermal particles*. It is not powerful enough to produce extremely energetic particles. More effective mechanisms are discussed in section 6:6 below.

PROBLEM 6–2. Suppose the magnetic field in a region of space increases from 10^{-6} to 10^{-5} gauss over a period of 10^7 y. To what energy would electrons and protons be accelerated, if they moved perpendicular to the field and suffered no collisions? How does the final energy depend on the initial energy? To do this, it is useful first to derive the energy-field relationship $d\mathscr{E}/\mathscr{E} = dB/B$ which follows from (6–13) and (6–16).

6:5 AMPÈRE'S LAW AND THE RELATION BETWEEN COSMIC CURRENTS AND MAGNETIC FIELDS

In section 6:2 we had noted that cosmic magnetic fields exist by virtue of the gyrating electric charges that are frozen to the field. This idea is more precisely expressed by *Ampère's law* which states that a current produces an encircling magnetic field (Fig. 6.2*b*):

$$\frac{4\pi}{c} \int \mathbf{j} \cdot d\mathbf{s} = \oint \mathbf{H} \cdot d\mathbf{l} \qquad (6-17)$$

Here again, the left side of the equation is a surface integral taken over the entire surface encircled by the magnetic field in the line integral on the right.

We believe that cosmic magnetic clouds are configurations in which equation 6–17 is obeyed in every locale. The shapes of the magnetic fields and currents are therefore likely to be quite complicated. One can think of initial configurations called *"force-free" magnetic fields* in which the magnetic fields and the flow of charges are so arranged that no forces result to destroy the configuration. These force-free configurations may well represent the structure of cosmic magnetic fields. Equation 6–11 tells us that a "force-free" structure must have $\mathbf{j} \wedge \mathbf{B} = 0$ everywhere.

6:6 MAGNETIC MIRRORS, MAGNETIC BOTTLES, AND COSMIC RAY PARTICLES

In section 6:4 we described the acceleration of charged particles by a betatron process. Another scheme for magnetically accelerating cosmic ray particles was suggested by Fermi. In the Fermi mechanism the cosmic ray particles are thought to travel between cosmic gas clouds. Each cloud has an embedded magnetic field. When a particle approaches the cloud and enters its field perpendicular to the field direction, it is turned back by virtue of the magnetic force given by equation 6–11. For, after traveling in a semicircle, the particle once again finds itself at the edge of the cloud and headed into the direction from which it came. As shown in Fig. 6.3—and as explained below—a similar reflection occurs for particles approaching along the lines of force.

If the particle impinges on a cloud that is receding from it, the particle's momentum after encounter is smaller than before. If the particle impinges on an approaching cloud, its final momentum is greater than before the collision. In general the probability for collision is greater for an approaching than for a

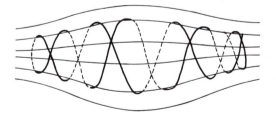

Fig. 6.3 Charged particle trajectories in a magnetic bottle. Light lines denote magnetic lines of force. A high density of field lines indicates a strong magnetic field.

receding cloud. (This corresponds to normal experience. On a highway we pass more cars going in the opposite direction than going along the direction in which we are traveling.)

Statistically, therefore, particles will derive increased momentum from encounters with clouds and can be accelerated to high energies. The process is similar to the acceleration of a ping-pong ball between two slowly approaching paddles. After the ball has made many bounces off each paddle, it is going far faster than either of these reflecting surfaces.

A number of variants on the Fermi process have been suggested. One of these involves a particle's successive impacts on an expanding supernova ejection shell. After each impact the accelerated particle returns—deflected back toward the shell by an external magnetic field.

The cosmic ray particle eventually must suffer some destructive collision due to one of several competing processes, such as inelastic impact on another particle, loss from the Galaxy and, hence, loss from contact with the accelerating clouds, and so on. Because of this, the number of accelerating reflections between destructive processes is limited. To reach truly high energies, therefore, cosmic ray particles must be injected into the accelerating fields with rather high initial energies. It is possible that sufficiently energetic particles are produced in supernova explosions. It is also possible that quasars are capable of producing the high energy cosmic rays in processes such as those described above, and that the highest energy cosmic rays reach our galaxy from these distant objects.

In any case, the Fermi mechanism is no longer thought to be the sole process producing cosmic rays. First, we know that the most energetic protons, those having energies of order 10^{20} ev, have a gyroradius comparable to the galactic radius. These particles could therefore not be retained in the Galaxy for any length of time, certainly not long enough to go through the final stages of acceleration to the observed energies. Such particles must therefore be extragalactic or else produced in very intensely magnetized regions within the Milky Way.

Second, we know that the heavy nuclei, which form an abundant part of the cosmic ray flux, would suffer destructive collisions during the long stay in interstellar space required by the rather slow Fermi acceleration mechanism. Yet we find iron nuclei to be abundant at least up to energies of the order of 10^{12} to 10^{13} ev. Above these energies we have little information about the chemical abundance. Again, we are driven toward a mechanism that could energize these particles rapidly.

Currently, pulsars and perhaps rapidly rotating white dwarfs are considered possible sources of cosmic rays. Further observations may lead to a clarification of these hypotheses. We can also hope that the study of solar flares, which are responsible for the solar cosmic ray component, will give us a better understanding of at least one mechanism for accelerating these energetic particles.

Generally, a charged particle moves along the lines of force of a magnetic field, spiraling as it goes. The pitch angle is given by equation 6–12. If the particle encounters a region of the magnetic field where the lines are more compressed, it experiences an increase in the field strength and by Faraday's law (6–16) its circular velocity v_c increases. However, since the field itself is not doing any work on the charge, the increase in v_c must be bought at the expense of kinetic energy initially resident in the longitudinal motion, that is, at the expense of a reduction in v_z. When the particle has advanced into the intense magnetic field to such a depth that all its kinetic energy is spent in circular motion, the pitch angle θ becomes $\pi/2$; the particle is reflected and spirals back out of the intense field.

As the particle first spirals into the intensifying magnetic field, its angular momentum about the axis of symmetry of the motion is conserved. Hence the *magnetic moment* **M**:

$$M = \frac{\mathbf{j} \wedge \mathbf{r}}{2c}, \quad \mathbf{j} = q\mathbf{v} \tag{6–18}$$

also is conserved. Substituting the Larmor radius for r (equation 6–13) we find

$$M = \frac{v_c p_c}{2B} \tag{6–19}$$

along the direction of the magnetic field. From this it follows that the transverse kinetic energy is directly proportional to the field B. If a particle has an initial pitch angle θ in a field B, it can therefore penetrate the field until it reaches a region where the field is B_0 and $\sin \theta = 1$:

$$B_0 = \frac{B}{\sin^2 \theta} \tag{6–20}$$

Here it is reflected and spirals back out of the intense magnetic field.

A *magnetic bottle* consists of two such *magnetic mirrors* between which a particle is reflected going back and forth without possibility of escape. The Fermi mechanism ping-pong acceleration could involve a (shrinking) magnetic bottle in which the two magnetic mirrors approach.

We sometimes characterize cosmic ray particles by a *magnetic rigidity* BR_L which equals pc/q for motion strictly perpendicular to the field (equation 6–13). The rigidity has dimensions of energy per charge.

PROBLEM 6–3. Consider an interstellar cloud moving with velocity V. It acts as a magnetic mirror so that a particle suffers a change in speed $\Delta V = 2V$, added to its own initial velocity, in any reflection off the cloud. With two approaching clouds, a succession of collisions can occur. Using the law of composition of velocities, compute how many collisions a proton with initial energy $\mathscr{E}$ would need in order to double its energy. Let $V = 7$ km sec^{-1}, typical of interstellar cloud velocities, let the distance between approaching clouds (magnetic mirrors) be of the order of 10^{17} cm, and let $\mathscr{E} = 10^{10}$ ev. How long would it take to double the particle's energy? Is this time appreciably different for protons and electrons?

PROBLEM 6-4.   Pulsars are now believed to be the source of at least some cosmic ray particles. The particles are thought to be accelerated by magnetic lines of force that co-rotate with the central neutron star. Suppose that the field velocity is simply ωr, where ω is the star's angular velocity and r is the radial distance from the star. Moreover, consider the particles to be dragged along, frozen to the magnetic field lines. What is the energy of the particles, then, as a function of radial distance, if special relativistic physics is approximately valid in this problem? Beyond what radial distance can the particles and magnetic field not co-rotate?

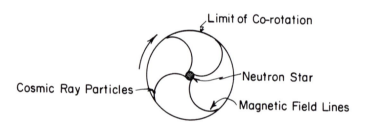

Fig. 6.4   Cosmic ray acceleration near a neutron star.

6:7 MAXWELL'S EQUATIONS

Four equations of electromagnetism allow us to derive all classical electromagnetic effects. They are

$$\nabla \cdot \mathbf{D} = 4\pi\rho \text{ (see equation 6-8)} \tag{6-21}$$

$$\nabla \wedge \mathbf{E} = -\frac{1}{c}\frac{\partial \mathbf{B}}{\partial t} \text{ (equivalent to equation 6-16)} \tag{6-22}$$

$$\nabla \wedge \mathbf{H} = \frac{4\pi}{c}\mathbf{j} \text{ (equivalent to equation 6-17)} \tag{6-23}$$

and finally

$$\nabla \cdot \mathbf{B} = 0 \tag{6-24}$$

This last equation states that there exist no *magnetic monopoles* (charges) analogous to electric charges. Only magnetic dipoles and higher multipole configurations occur in nature. Compare equation 6–24 to 6–21 in this respect.

Despite this statement, a search for magnetic monopoles has gone on ever since Dirac (Di31) pointed out that quantization of the electron's charge could be understood if a few or even only one such magnetic monopole existed in nature. Thus far no such *Dirac monopoles* have been found.

The four Maxwell equations generally must be supplemented by four auxiliary expressions.

$$\mathbf{D} = \varepsilon\mathbf{E} \quad \text{(equation 6–5)}$$

$$\mathbf{B} = \mu\mathbf{H} \tag{6–25}$$

$$\mathbf{j} = \sigma\mathbf{E} + \frac{1}{4\pi}\frac{\partial\mathbf{D}}{\partial t} \quad \text{(equation 6–15)}$$

$$\nabla\cdot\mathbf{j} = 0 \tag{6–26}$$

Equation 6–25 expresses a relation between the magnetic vectors $\mathbf{B}$ and $\mathbf{H}$ which is similar to that between $\mathbf{D}$ and $\mathbf{E}$. Equation 6–26 states that currents in the sense defined by equation 6–15 are continuous, having no sources or sinks. The *magnetic permeability* μ can have values greater than or less than unity, depending on whether the medium is *paramagnetic or diamagnetic*. In most cosmic gases $\mu = 1$, for all practical purposes, but (see section 9:8) paramagnetic grains in interstellar space may be responsible for an observed slight polarization of starlight.

6:8 THE WAVE EQUATION

From equations 6–22 and 6–23 and from the relations (6–15) and (6–25), we can obtain the expression

$$\nabla \wedge (\nabla \wedge \mathbf{E}) = -\frac{1}{c}\frac{\partial}{\partial t}\nabla \wedge \mathbf{B} \tag{6–27}$$

$$= \frac{-4\pi\mu}{c^2}\frac{\partial}{\partial t}\left(\sigma\mathbf{E} + \frac{\varepsilon}{4\pi}\frac{\partial\mathbf{E}}{\partial t}\right)$$

provided the dielectric constant ε and permeability μ do not vary with time, and μ is scalar. Actually both μ and ε are tensor quantities, but they frequently act like scalars. Let us use the identity

$$\nabla \wedge (\nabla \wedge \mathbf{E}) = \nabla(\nabla \cdot \mathbf{E}) - \nabla^2\mathbf{E} \tag{6–28}$$

6:8

and consider only regions in which the space charge is neutral; then $\nabla \cdot \mathbf{E} = 0$ and

$$\nabla^2 \mathbf{E} = \frac{\mu\varepsilon}{c^2} \frac{\partial^2 \mathbf{E}}{\partial t^2} + \frac{4\pi\mu\sigma}{c^2} \frac{\partial \mathbf{E}}{\partial t} \qquad (6\text{–}29)$$

In a nonconducting medium $\sigma = 0$ so that

$$\nabla^2 \mathbf{E} - \frac{\mu\varepsilon}{c^2} \frac{\partial^2 \mathbf{E}}{\partial t^2} = 0 \qquad (6\text{–}30)$$

which is the equation for waves propagating with speed

$$V = \frac{c}{\sqrt{\mu\varepsilon}} \qquad (6\text{–}31)$$

PROBLEM 6–5.   Derive a similar expression for the magnetic field:

$$\nabla^2 \mathbf{H} - \frac{\mu\varepsilon}{c^2} \frac{\partial^2 \mathbf{H}}{\partial t} = 0 \qquad (6\text{–}32)$$

paying particular attention to the limitations imposed on ε, μ, and σ in arriving at this result.

The operator ∇^2 is called the *Laplacian*, sometimes written as Δ. If ε and $\mu = 1$, we can define another operator $\Box = \nabla^2 - (1/c^2)(\partial^2/\partial t^2)$, called the *d'Alembertian*.

We note, in passing, that there is no conflict between our previous suggestion (section 6:3) that the conductivity in space is large, while simultaneously setting $\sigma = 0$ in equations 6–30 and 6–32. The conductivity is a frequency dependent quantity. At optical and even at radio frequencies σ usually is very low. Certainly, at optical frequencies the wavelength of the electromagnetic wave is short compared to the distance between charges, and the wave effectively propagates through a vacuum. At the longer radio wavelengths a transition occurs: Charges in the medium can respond to the electric and magnetic fields of a propagated wave, and σ becomes finite. When the second term on the right of equation 6–29 dominates, the expression assumes the form of a diffusion equation and the wave is *damped*.

We note that the propagated waves are *transverse* (Fig. 6.5). If the direction of propagation for a plane wave is the x direction, symmetry dictates that all partial derivatives with respect to y and z are zero. The divergence relations give

$$\frac{\partial E_x}{\partial x} = 0 \quad \text{and} \quad \frac{\partial H_x}{\partial x} = 0 \qquad (6\text{–}33)$$

and the curl equations, (6–22) and (6–23), give

$$\frac{\partial E_y}{\partial x} = -\frac{1}{c}\frac{\partial H_z}{\partial t}, \qquad \frac{\partial H_y}{\partial x} = \frac{1}{c}\frac{\partial E_z}{\partial t}$$

$$\frac{\partial E_z}{\partial x} = \frac{1}{c}\frac{\partial H_y}{\partial t}, \qquad \frac{\partial H_z}{\partial x} = -\frac{1}{c}\frac{\partial E_y}{\partial t} \tag{6–34}$$

$$0 = \frac{\partial H_x}{\partial t}, \qquad 0 = \frac{\partial E_x}{\partial t}$$

If **n** is the unit vector along the direction of propagation, we see that (6–33) and (6–34) are satisfied by an expression of the form

$$\mathbf{H} = \mathbf{n} \wedge \mathbf{E} \tag{6–35}$$

so that the **E** and **H** fields always are perpendicular and the solution of the *wave equation* (6–30) has the form

$$f_i = A\cos(2\pi v t - kx), \qquad i = y, z \tag{6–36}$$

where $k = \sqrt{\mu\varepsilon}\,(\omega/c)$ and $\omega = 2\pi v$. v is called the *frequency* and ω the *angular frequency* of the wave. k is the *wave number*—the number of waves per unit length along the direction of propagation. f_i represents electric and magnetic field components.

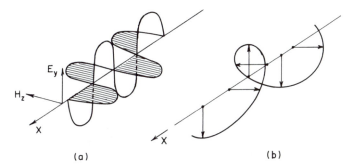

Fig. 6.5 Electromagnetic waves. (a) Wave propagating along the x direction with electric field plane polarized along the y direction.

(b) Circularly polarized wave propagating along the x direction. For simplicity, only the electric field direction is shown here. The direction of the **E** vector rotates about the x-axis. The sense of rotation shown here is said to be left-handed circularly polarized (LHP). Any electromagnetic wave can be constructed from a suitable superposition of left and right-handed circularly polarized waves. Plane polarized waves, for example, are obtained by superposing LHP and RHP waves of the same amplitude. Their relative phase determines the plane of the **E** vector (see Fig. 6.7).

6:8

6:9 PHASE AND GROUP VELOCITY

We write the equations for propagation of two waves f^- and f^+ that have angular frequencies $\omega - \Delta\omega$ and $\omega + \Delta\omega$, respectively:

$$f^- = A \cos \left[(\omega - \Delta\omega) t - (k - \Delta k) x \right]$$
$$f^+ = A \cos \left[(\omega + \Delta\omega) t - (k + \Delta k) x \right]$$

The superposition of these waves gives

$$f = f^- + f^+ = A\{\cos \left[(\omega - \Delta\omega) t - (k - \Delta k) x \right]$$
$$+ \cos \left[(\omega + \Delta\omega) t - (k + \Delta k) x \right]\} \qquad (6\text{--}37)$$
$$f = 2A \cos (\omega t - kx) \cos \left[(\Delta\omega) t - (\Delta k) x \right]$$

This means that there is a *carrier wave* frequency represented by $\cos (\omega t - kx)$ that is *amplitude modulated* by a wave $\cos (t\Delta\omega - x\Delta k)$. The carrier wave velocity is called the *phase velocity* (6–31), (6–36):

$$V = \frac{\omega}{k} \qquad (6\text{--}38)$$

while the velocity of the modulation is called the *group velocity*:

$$U = \frac{\partial\omega}{\partial k} \qquad (6\text{--}39)$$

We will note later that U is the physically more interesting quantity. It represents the speed at which *information* can be conveyed or energy transported. As long as the medium is purely dispersive, that is, $\omega = \omega(k)$, there is no difficulty in defining U. But if the conductivity σ becomes appreciable, one has absorption, the quantity A becomes complex, and U no longer has a clear physical meaning. For long wavelength cosmic radio waves, this kind of absorption prevents transit through the earth's ionosphere. These long waves must then be observed from rockets or satellites. At even longer wavelengths the interstellar medium absorbs, and such waves are not transmitted at all. We will return to this problem in section 6:11.

6:10 ENERGY DENSITY, PRESSURE, AND THE POYNTING VECTOR

The scalar product of equation 6–22 with **H**, subtracted from the product of (6–23) with **E** is

$$\frac{1}{c} \mathbf{H} \cdot \frac{\partial \mathbf{B}}{\partial t} + \frac{\mathbf{E}}{c} \cdot \frac{\partial \mathbf{D}}{\partial t} + \frac{4\pi\sigma\mathbf{E} \cdot \mathbf{E}}{c} = - (H \cdot \nabla \wedge E) + (E \cdot \nabla \wedge H) \qquad (6\text{--}40)$$

Using the vector identity

$$\nabla \cdot (\mathbf{A} \wedge \mathbf{B}) = \mathbf{B} \cdot \nabla \wedge \mathbf{A} - \mathbf{A} \cdot \nabla \wedge \mathbf{B} \qquad (6\text{--}41)$$

we find that

$$\frac{1}{8\pi} \frac{\partial}{\partial t} (\varepsilon E^2 + \mu H^2) = -\sigma E^2 - \nabla \cdot \mathbf{S} \qquad (6\text{--}42)$$

where

$$\mathbf{S} = (c/4\pi) \mathbf{E} \wedge \mathbf{H}. \qquad (6\text{--}43)$$

$\mathbf{S}$ is called the *Poynting vector*. If we apply Gauss's theorem (6–9) relating volume and surface integrals, (6–42) can be written as

$$\frac{\partial}{\partial t} \int \frac{\varepsilon E^2 + \mu H^2}{8\pi} \, dV = - \int \sigma E^2 dV - \oint \mathbf{S} \cdot d\mathbf{s} \qquad (6\text{--}44)$$

Here the first term on the right is equivalent to the rate of change of kinetic energy of moving charges. It involves the scalar product of the force on the particles and their velocity since σE represents a current, that is, the motion of charged particles:

$$\int \sigma E^2 dV \rightarrow \sum e\mathbf{v} \cdot \mathbf{E} = \sum \mathbf{v} \cdot \dot{\mathbf{p}} \qquad (6\text{--}45)$$

This is the time derivative of the kinetic energy summed over all particles. The other two terms in (6–44) represent the flow of electromagnetic energy. The term on the left of equation 6–44 is the rate of change of energy in the volume; $(\varepsilon E^2 + \mu H^2)/8\pi$ is the energy density of the fields. The second term on the right represents the flow of energy through the enclosing surface and S therefore is the *electro-magnetic flux density*. Equation 6–44 states that the rate of change of energy in a volume equals the rate of change of the kinetic energy of charges plus the rate at which energy is radiated away.

Previously we found that the pressure P due to randomly oriented electromagnetic waves is just 1/3 the numerical value of the energy density. In section 4:7 we determined this on kinetic grounds:

$$P = \frac{1}{3} \frac{1}{8\pi} (\varepsilon E^2 + \mu H^2) \qquad (6\text{--}46)$$

The case of static fields is similar except that a magnetic pressure can now exist without an accompanying electric pressure; the conductivity σ is high and a current σE maintains the magnetic field. There exists a kinetic pressure due to the flow of charges, and this will depend on σE. The situation is further complicated since the magnetic pressure actually is a tensor quantity that depends on the orientation of the fields. For a magnetic field there always exists a tension along the lines of force and an outward pressure perpendicular to the lines of force.

We can see this in the following way. The magnetic energy density in a cube of unit dimension is $\mu H^2/8\pi$. If the cube is compressed an amount dl along a direction parallel to the field lines, the field strength remains constant, but the volume decreases by dl. Since the energy density remains constant while the volume decreases, the total energy in the volume decreases by $(\mu H^2/8\pi)\, dl$, and this amount of energy is given off in the contraction. That means that the amount of work done to compress the cube is $-(\mu H^2/8\pi)\, dl$, and indicates that there is a pressure $-\mu H^2/8\pi$ along the field lines.

If the cube is compressed along a direction transverse to the field lines, the number of lines of force in the volume does not change, and a compression Δl increases the field strength to $H/(1 - \Delta l)$. The energy density now becomes $\sim (\mu H^2/8\pi)(1 + 2\Delta l)$, and because of the decrease in volume $(1 - \Delta l)$, the total energy change on compression is $\sim (\mu H^2/8\pi)\,\Delta l$. In this case an amount of work $(\mu H^2/8\pi)\,\Delta l$ must be done to compress the cube, and the pressure resisting compression is $\mu H^2/8\pi$. For a volume containing randomly directed bundles of field lines, the net effect of averaging over two transverse and one longitudinal direction is an overall outward pressure $P = (\mu H^2/8\pi)/3$.

This is the reason why difficulties arise in the problem of star formation in the presence of fields (section 4:20). It is relatively simple to see how matter can contract along the lines of force in that situation, but it is more difficult to understand how condensation takes place perpendicular to the direction of the field because the gases are frozen to the field lines and the pressure of the magnetic field attempts to resist any contraction. To see how severe this problem is, we note that the transverse pressure is $H^2/8\pi$.

PROBLEM 6–6. The transverse pressure of a static magnetic field is $P_s = H^2/8\pi$; the magnetic part of the radiation pressure (6–46) is $P_r = \frac{1}{3}H^2/8\pi$. What is the significance of the factor $\frac{1}{3}$?

Initially a typical field strength might be 10^{-6} gauss, so that $P_{\text{initial}} \sim 10^{-13}$ dyn cm^{-2}. As a protostar contracts from $\sim 10^{18}$ cm down to 10^{11} cm, conservation of the number of field lines requires that $H \propto r^{-2}$, so that $H^2 \propto r^{-4}$ and we would end up with a protostar having 10^8 gauss magnetic fields and 10^{15} dyn cm^{-2} magnetic pressures. The gravitational forces are far too weak to produce such high fields. We conclude that somehow we are looking at the problem in the wrong way. Stars manage to form despite these difficulties.

6:11 PROPAGATION OF WAVES THROUGH A TENUOUS IONIZED MEDIUM

Consider an ionized medium without electric or magnetic fields. Let this medium be tenuous, so that collisions between ions and electrons are rare. Then, for small departures from equilibrium, electric fields resident in the electromagnetic wave accelerate the electrons in the medium relative to the more massive ions:

$$m\ddot{\mathbf{r}} = e\mathbf{E}(\mathbf{r}, t) \tag{6-47}$$

Here e and m are the charge and mass of the electron, and $\mathbf{E}$ is the field associated with the wave. Let the wave have the form

$$\mathbf{E}(\mathbf{r}, t) = E_0(\mathbf{r})\, e^{i\omega t} \qquad \text{(Real part)} \tag{6-48}$$

where only the real part will be considered. The displacement of the electron from its equilibrium position then is

$$\mathbf{r} = -\frac{e}{m\omega^2}\mathbf{E} \tag{6-49}$$

This satisfies both equations 6–47 and 6–48. The displacement of the electrons effectively sets up a large number of dipoles which, as discussed in section 6:1, give rise to a polarization field $\mathbf{P}$. If n is the number density of electrons, the polarization field is to be expressed as the sum of the individual dipole fields produced by the passing wave

$$\mathbf{P} = ne\mathbf{r} = -\frac{ne^2}{m\omega^2}\mathbf{E} \tag{6-50}$$

The definition of the polarization field, equation 6–10, then tells us that the dielectric constant of the medium must be

$$\varepsilon = 1 - \frac{4\pi ne^2}{m\omega^2} \tag{6-51}$$

Since the propagation phase velocity is inversely proportional to the refractive index at frequency ω, $n_\omega = \varepsilon^{1/2}$ (we can set $\mu = 1$ in all problems dealing with cosmic wave propagation), the *phase* velocity in a plasma will be greater than the speed of light! But no information and no energy is transmitted at this velocity. Therefore no violation of special relativity is involved. The more significant *group velocity* is always less than c.

If a wave propagates along the x-direction through the cosmic medium, the transverse $\mathbf{E}$ and $\mathbf{B}$ field components have the form (6–36):

$$f = f_0 \cos(kx \pm \omega t) \tag{6-52}$$

and

$$\omega^2 = \frac{k^2 c^2}{\varepsilon} = \frac{k^2 c^2}{1 - (4\pi ne^2/m\omega^2)} \tag{6-53}$$

6:11

where equation 6–51 has been invoked with $\mu = 1$. This can be written as

$$\omega^2 = k^2 c^2 + \frac{4\pi n e^2}{m} \equiv k^2 c^2 + \omega_p^2 \tag{6-54}$$

where

$$\omega_p \equiv \left(\frac{4\pi n e^2}{m}\right)^{1/2} \sim 5.6 \times 10^4 n^{1/2} \text{ radians sec}^{-1} \tag{6-55}$$

is called the *plasma frequency*. It is related to the Debye length L (see equation 4–156) by $(mL^2/kT)^{1/2} = \omega_p^{-1}$. ω_p^{-1} is the time for an electron to cross a Debye length at a velocity $(kT/m)^{1/2}$.

If $\omega < \omega_p$, k becomes imaginary and the wave will not propagate through the medium.

In radio astronomy—as mentioned in section 6:9—observations at low frequencies cannot be carried out from below the ionosphere. Radio waves cannot be transmitted at frequencies below the ionospheric plasma frequency. That frequency varies since the electron density is not uniform. Typically, however, the cut-off is at frequencies of a few megahertz.

When $\omega > \omega_p$, propagation can take place. The group velocity of the wave is

$$U = \frac{d\omega}{dk} = \frac{c}{\sqrt{1 + \omega_p^2/c^2 k^2}} \tag{6-56}$$

The velocity of propagation therefore is frequency dependent. A situation in which this phenomenon is important concerns the propagation of pulses emitted by a *pulsar* (He68b)*. If the emitted pulse contains a range of frequency components, the arrival time of these frequencies at the earth will be delayed more at lower frequencies.

We can write (6–56) as

$$U = \frac{c}{\sqrt{1 + \omega_p^2/(\omega^2 - \omega_p^2)}} \tag{6-57}$$

The arrival time of a pulse that has traveled a distance D is D/U and the frequency dependence of the arrival time is

$$\frac{d(D/U)}{d\omega} \sim -\frac{D}{c}\frac{\omega_p^2}{\omega^3} \quad \text{for} \quad \omega \gg \omega_p \tag{6-58}$$

Observations of pulsars show that the pulse arrival time is frequency dependent and the observed delays in arrival time actually take the form (6–58). We therefore conclude that the time delay takes place in a medium whose plasma frequency is less than the radiation frequency. Equation 6–55 therefore puts an upper limit on the number density of electrons in the dispersing medium. More important, however, is the conclusion that the frequency dependence of the time delay is directly proportional to Dn, the total number of electrons per unit cross-sectional area along the line of sight to the emitting object. This useful relation follows directly from (6–55) and (6–58). The integrated electron number density along the line of sight, S, is called the *dispersion measure* $\mathscr{D}$:

$$\mathscr{D} \equiv \int_0^D n(s)\,ds = D\langle n \rangle \tag{6–59}$$

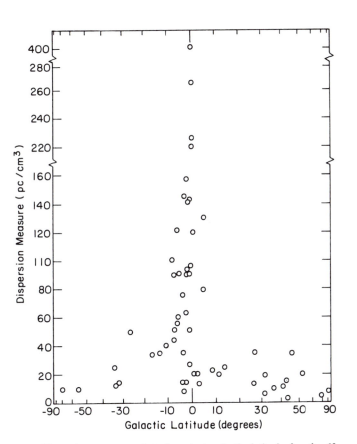

Fig. 6.6 Dispersion measure plotted against galactic latitude for the 63 pulsars known in early 1972 (Te72).

If the mean number density of electrons in the interstellar medium is known, the dispersion relation (6–58) can give us the pulsar distance. Conversely, if D is known from other sources, a mean value of n along the line of sight is obtained. The mean value, estimated in this way, would include a contribution due to any electrons surrounding the emitting region and part of the emitting object, as well as true interstellar electrons. The dispersion measure within the pulsar would not be distance dependent, while the dispersion due to the interstellar medium would. On this basis we distinguish the two contributions; and we find that the pulsar itself contributes negligibly. The dispersion measure along lines of sight leading to sources at known distances gives a mean electron density of about $\langle n \rangle = 0.03$ cm^{-3}. This value varies from source to source, depending on the number of bright, hot, ionizing stars along the line of sight. Using a mean density of electrons in interstellar space, amounting to ~ 0.03 cm^{-3}, we find that the distribution of the nearer pulsars fits the distances of the nearer spiral arms in the Galaxy. (Da69) These pulsars also show a tendency to cluster close to the Galactic plane (Fig. 6.6).

Throughout this section we have assumed that the collision frequency ν_c between ions and electrons is low. However, when ν_c becomes high, energy losses through dissipation no longer can be neglected. This problem is treated in section 6:16.

6:12 FARADAY ROTATION

Information about electron number densities in the cosmic medium can also be obtained from the *Faraday rotation* of a wave's plane of polarization. To understand this effect, consider an electron moving in a plane perpendicular to the direction of a magnetic field, **B**. It will be deflected by a force (6–11)

$$\mathbf{F} = \frac{e\mathbf{v} \wedge \mathbf{B}}{c}$$

If the electron is also under the influence of an electromagnetic wave, it will experience a force due to the wave's E field. Finally the gyrations under the combined influence of these fields must be balanced by an outward directed centrifugal force. The relation between these three forces is given by

$$e\mathbf{E} \pm \frac{eB\omega\mathbf{r}}{c} = -m\omega^2\mathbf{r} \qquad (6\text{–}60)$$

where **E** is the component of the field vector perpendicular to the magnetic field, and the second term on the left has a negative sign when the electron rotates counterclockwise viewed along the direction of the **B** field. This is the motion induced by an electromagnetic wave with a right-handed circular polarization

RHP propagating parallel to **B**. A left-handed circular polarization LHP gives rise to a force $+eB\omega\mathbf{r}/c$ directed along the direction of displacement from the electron's equilibrium position. Note, however, that the value of e is negative for an electron. Solving for **r** gives

$$\mathbf{r} = -\frac{e}{m}\left(\frac{1}{\omega^2 \pm \dfrac{eB\omega}{mc}}\right)\mathbf{E} \tag{6-61}$$

The dielectric polarization, as in (6–50) becomes $\mathbf{P} = ne\mathbf{r}$, giving rise to a dielectric constant

$$\varepsilon = 1 - \frac{4\pi ne^2}{m\omega(\omega \pm \omega_c)}, \qquad \omega_c \equiv \frac{eB}{mc} \tag{6-62}$$

Here ω_c is the gyro- or cyclotron frequency (6–13). Since the index of refraction $\varepsilon^{1/2}$ is not the same, it follows that the left- and right-handed polarized radiation will travel at different velocities through an ionized medium in a longitudinal magnetic field.

If a wave initially is plane polarized with a given direction of polarization, the polarization angle can be expressed as a superposition of two circularly polarized waves of given phase, say θ_0, and equal amplitude. As the waves propagate the phase relationship will change, because one wave lags behind the other, and the direction of polarization therefore rotates. Sometimes the **E** vectors will be in phase; at other times they will be out of phase.

Fig. 6.7 Addition of circularly polarized waves, to give plane polarized radiation. The phase at time $t = 0$ is θ_0.

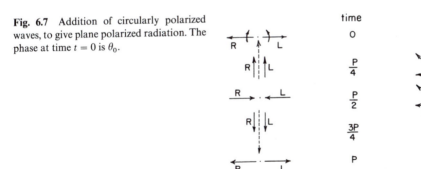

Figure 6.7 shows two sets of superposed, opposite circularly polarized waves. The one on the left has $\theta_0 = 180°$. The one on the right has $\theta_0 = 90°$. The **E** vectors and their sums are shown at different times during the waves' period P. The sum of the vectors is indicated by the dashed line. We can see that the direction of the plane polarized wave is given by an angle equaling half the phase lag.

However, the initial direction of one of the **E** vectors must also be specified. In Fig. 6.7, for example, we took the left-handed polarized **E** vector to point to the right at $t = 0$.

Turning now to the velocity of propagation and refractive index, we find the difference Δn between indices n_L and n_R to be

$$n_L^2 - n_R^2 = \varepsilon_L - \varepsilon_R = 2n_\omega \Delta n \qquad (6\text{–}63)$$

We write

$$n_\omega \sim 1 - \frac{4\pi n e^2}{2m\omega^2} \qquad (6\text{–}64)$$

where (6–64) is obtained from equation 6–51 provided $n_\omega - 1 \ll 1$. Substituting the dielectric constants from (6–62) into (6–63) we obtain

$$\Delta n = \frac{\dfrac{4\pi n e^2 (2\omega_c)}{m\omega(\omega^2 - \omega_c^2)}}{2\left(1 - \dfrac{2\pi n e^2}{m\omega^2}\right)} \qquad (6\text{–}65)$$

If $\omega \gg \omega_c$ and $ne^2/m\omega^2 \ll 1$,

$$\Delta n = \frac{4\pi n e^2 \omega_c}{m\omega^3} \qquad (6\text{–}66)$$

This distance lag per unit time is $c\,\Delta n/n_\omega^2 \sim c\,\Delta n$. The phase lag of the LHP relative to the RHP wave therefore becomes $\omega\,\Delta n$, and the plane of polarization rotates through half this angle in unit time:

$$\Delta\theta \sim \frac{\omega\,\Delta n}{2} \qquad (6\text{–}67)$$

The difference in velocity of propagation, and hence the rate at which the polarization vector rotates, is therefore proportional to the number density n and to B. For a given velocity difference, the phase rotates at a rate inversely proportional to the wavelength λ, because the distance one wave has to lag behind the other becomes greater for larger wavelengths. On the other hand, the velocity difference between the waves is proportional to ω^{-3}, according to (6–66), and therefore is proportional to λ^3. Hence the angle $\theta(D)$ through which the plane of polarization is rotated over distance D is proportional to λ^2. In observing distant radio sources emitting polarized radiation, we can determine the angle θ as a function of wavelength. This gives a value for the product of electron density n and the magnetic field component along the viewing direction (provided the path length is known). More correctly, since the rotation depends on the presence of both a properly oriented magnetic field and the local particle density at the field's position, the rotation actually gives a value of the product of particle density and magnetic field, integrated along the line of sight.

If, as is sometimes supposed, the particles and fields actually do not occupy the same positions in space, but are physically separated from one another, then the Faraday rotation only produces a lower limit to the field strength and particle density. Nonetheless, since we know relatively little about the interstellar medium, even this much information is of current astrophysical interest.

In the case of pulsars, the dispersion measure tells us the mean number density of electrons along the line of sight (6:11). The Faraday rotation can then be used to estimate the mean component of the magnetic field strength along the line of sight. This procedure has been followed in obtaining the local Galactic magnetic field (see Fig. 9.9). Since the field direction changes along this path, only a statistical estimate of actual field strength is obtained in this way.

PROBLEM 6-7. Suppose the field strength is B everywhere, that it varies randomly in direction from region to region, but that its direction is constant over any region of length L. If the source distance is NL, and the electron number density is n, show by a random walk procedure that

$$\theta \sim \sqrt{NL} \left(\frac{2\pi n e^3 B}{m^2 c^2 \omega^2} \right)$$

For simplicity assume that B always points directly toward or away from the observer.

6:13 LIGHT EMISSION BY SLOWLY MOVING CHARGES

When an electric charge is set into accelerated motion, it can emit radiation. If this motion is induced by an incident electromagnetic wave, we may find that the charge—or group of charged particles—absorbs or scatters the radiant energy. To see this, consider the current associated with the accelerated charge. This current will induce a magnetic field at some distance from the position of the charges, but the magnetic field strength variations will normally be somewhat out of phase with the variations in the current. This is due to the time delay involved in transmitting the information about the current strength from one position to another. That information can only be transmitted at the speed of light. For the moment we will regard the charges and currents as sources of electric and magnetic fields. If we use the Maxwell equations 6–22 and 6–23 for empty space where $\mathbf{E} = \mathbf{D}$ and $\mathbf{H} = \mathbf{B}$, and we now write $\mathbf{j}_c$ to symbolize the conduction current $\sigma \mathbf{E}$,

$$\nabla \wedge \mathbf{H} = \frac{4\pi}{c} \mathbf{j}_c + \frac{1}{c} \frac{\partial \mathbf{E}}{\partial t} \tag{6-68}$$

$$\nabla \wedge \mathbf{E} = -\frac{1}{c} \frac{\partial \mathbf{H}}{\partial t} \tag{6-69}$$

Now consider a *vector potential* $\mathbf{A}$ as giving rise to the magnetic field, while a *scalar potential* ϕ, together with $\mathbf{A}$, gives rise to the electric field; then we can write

$$\mathbf{H} = \nabla \wedge \mathbf{A} \tag{6-70}$$

and

$$\mathbf{E} = -\nabla\phi - \frac{1}{c} \frac{\partial \mathbf{A}}{\partial t} \tag{6-71}$$

which are consistent with the Maxwell equations, above. Separable equations, each depending on only one of these potentials, can then be obtained provided that

$$\nabla \cdot \mathbf{A} + \frac{1}{c} \frac{\partial \phi}{\partial t} = 0 \tag{6-72}$$

holds. Equation 6–72 is called the Lorentz condition.

PROBLEM 6-8.   Check the validity of this statement by direct substitution of equations 6–70, 6–71, and 6–72 into the Maxwell equations. In this way obtain

$$\nabla^2 \mathbf{A} - \frac{1}{c^2} \frac{\partial^2 \mathbf{A}}{\partial t^2} = -\frac{4\pi}{c} \mathbf{j}_c \tag{6-73}$$

and

$$\nabla^2 \phi - \frac{1}{c^2} \frac{\partial^2 \phi}{\partial t^2} = -4\pi\rho \tag{6-74}$$

In empty space, equations 6–73 and 6–74 have the right side equal to zero; the right side is nonzero only at the actual location of charges and currents. Furthermore, in a static case where the time derivative vanishes, ϕ obeys the Poisson equation (4–142) which we used earlier in discussing plasmas. When we solve the Poisson equation, the potential is expressed in terms of an integral over the volume distributed charges, divided by the distance of the charges from the point at which the potential is evaluated. In view of this, we write the potential as

$$\phi(\mathbf{R}_0, t) = \frac{1}{R_0} \int \rho\left(t - \frac{R_0}{c} + \frac{\mathbf{r} \cdot \mathbf{n}}{c}\right) dV \tag{6-75}$$

where R_0 is the distance from the *center of charge*, and $\mathbf{r} \cdot \mathbf{n}$ is the projected distance of a point in the charge distribution, measured from the center of charge, along the direction joining the point $\mathbf{R}$ to the center of charge (see Fig. 6.8). $\mathbf{n}$ is a unit vector along that direction. Equation 6–75 tells us that the potential at any given time is determined by the charge distribution at a time $R/c = (R_0 - \mathbf{n} \cdot \mathbf{r})/c$ earlier. The similarity between equations 6–73 and 6–74 suggests that we can also write

$$\mathbf{A}(\mathbf{R}_0, t) = \frac{1}{cR_0} \int \mathbf{j}_c\left(t - \frac{R_0}{c} + \frac{\mathbf{r} \cdot \mathbf{n}}{c}\right) dV \tag{6-76}$$

We note that a plane wave in vacuum obeys the relation (6–35)

$$\mathbf{H} = \mathbf{n} \wedge \mathbf{E} \tag{6-35}$$

and that (6–71) therefore leads to

$$\mathbf{H} = \frac{1}{c} \dot{\mathbf{A}} \wedge \mathbf{n} \tag{6-77}$$

as long as the magnetic field strength is measured at a large distance from the charge distribution so that $\nabla\phi$ can be neglected.

It is now a simple matter to determine the energy radiated away by the moving charges. The Poynting vector is immediately obtained from equations 6–35 and 6–43.

$$\mathbf{S} = \frac{c}{4\pi} H^2 \mathbf{n} \tag{6-78}$$

Fig. 6.8 Diagram to illustrate radiation by a dipole, see equations 6–75 to 6–85.

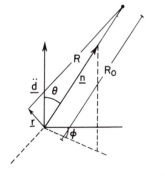

For the simple case of radiation by a *dipole*—that is, two slightly separated dissimilar charges—we can take the integral over the current distribution in (6–76) to be just the rate of change of the dipole moment

$$\mathbf{A} = \frac{1}{cR_0} \dot{\mathbf{d}} \qquad (6\text{–}79)$$

where

$$\dot{\mathbf{d}} = \frac{d}{dt} \sum e\mathbf{r} \qquad (6\text{–}80)$$

Here $\mathbf{d} = \sum e\mathbf{r}$ is the dipole moment of the charge distribution; the time derivative refers to a time $t' = t - (R_0/c)$, and the dimension of the dipole must be small compared to the radiated wavelength λ; for then

$$\frac{\mathbf{r} \cdot \mathbf{n}}{c} \ll \frac{\lambda}{c} = P \qquad (6\text{–}81)$$

and neglect of the term $\mathbf{r} \cdot \mathbf{n}/c$ in equation (6–76) involves a neglect only of a time increment small compared to the *period of oscillation P*. We now see from (6–77) and (6–35) that the field strengths at a distance R_0 from the dipole are

$$\mathbf{H} = \frac{1}{c^2 R_0} \ddot{\mathbf{d}} \wedge \mathbf{n} \qquad (6\text{–}82)$$

$$\mathbf{E} = \frac{1}{c^2 R_0} (\ddot{\mathbf{d}} \wedge \mathbf{n}) \wedge \mathbf{n} \qquad (6\text{–}83)$$

and the *intensity* of radiation dI radiated into a solid angle $d\Omega$ is given by the Poynting vector integrated over that angle:

$$dI = \frac{1}{4\pi c^3} (\ddot{\mathbf{d}} \wedge \mathbf{n})^2 \, d\Omega \qquad (6\text{–}84)$$

Integrating over all angles $d\Omega = \sin \theta \, d\theta \, d\phi$

$$I = \int\int \frac{\ddot{d}^2}{4\pi c^3} \sin^3\theta \, d\theta \, d\phi \qquad (6\text{–}85)$$

$$= \frac{2}{3c^3} \ddot{d}^2 \qquad (6\text{–}86)$$

For two opposite charges e and $-e$, separated by a distance $\mathbf{r}$, the dipole moment is

$$\mathbf{d} = e\mathbf{r} \qquad (6\text{–}87)$$

and the *total radiated energy* per second is

$$I = \frac{2e^2\ddot{r}^2}{3c^3} \tag{6-88}$$

PROBLEM 6-9. A magnetic dipole can be considered to consist of two fictitious magnetic charges Q and $-Q$ separated by a distance $\mathbf{a}$. The magnetic dipole moment would then be $\mathbf{M} = Q\mathbf{a}$. (a) Show that the magnetic field along the axis of this configuration is $H = 2aQ/r^3$. (b) At the surface of a pulsar $H \sim 10^{12}$ gauss, $r \sim 10^6$ cm, $\omega \sim 10^2$. By analogy to equation 6-88 show, for $\mathbf{a}$ perpendicular to the axis of the star's rotation, that the intensity of radiation is (Pa68)

$$I = \frac{2}{3c^3} \ddot{\mathbf{M}}^2 \sim 10^{36} \text{ erg sec}^{-1} \tag{6-89}$$

We should still note that a system of charged particles all of which have the same charge-to-mass ratio cannot radiate as a dipole. The center of charge and the center of mass coincide for such a system; and if the center of mass $\sum m\mathbf{r}$ remains stationary, the derivatives $\ddot{d}$ all vanish:

$$\ddot{\mathbf{d}} = \sum e\ddot{\mathbf{r}} = \sum \frac{e}{m} m\ddot{\mathbf{r}} = 0 \tag{6-90}$$

For such an assembly of charges we can still obtain *electric quadrupole radiation*, or radiation generated by higher electric or magnetic *multipole processes*. These processes depend on the inclusion of terms in $\mathbf{r} \cdot \mathbf{n}/c$ that we had previously neglected. The current j_c is now expressed as an expansion in $\mathbf{r} \cdot \mathbf{n}/c$:

$$j_c\left(t' + \frac{\mathbf{r} \cdot \mathbf{n}}{c}\right) = j_c(t') + \frac{\partial}{\partial t}\left(\frac{\mathbf{r} \cdot \mathbf{n}}{c}\right)j_c(t') + \cdots \tag{6-91}$$

where again $t' = t - R_0/c$. If we only retain the first two terms of the expansion and sum over all charges, equation 6-76 yields

$$\mathbf{A} = \frac{\sum e\mathbf{v}}{cR_0} + \frac{1}{c^2 R_0}\frac{\partial}{\partial t}\sum e\mathbf{v}(\mathbf{r} \cdot \mathbf{n}) \tag{6-92}$$

where the first term again is produced only by a time varying dipole moment, and we now understand that $\mathbf{v}$ and $\mathbf{r}$ values are measured for time t', although all primes have been dropped for ease in writing. One can show (see La51) that this leads to

$$\mathbf{A} = \frac{\dot{\mathbf{d}}}{cR_0} + \frac{1}{6c^2 R_0}\frac{\partial^2}{\partial t^2}\mathbf{D} - \frac{1}{cR_0}(\dot{\mathbf{M}} \wedge \mathbf{n}) \tag{6-93}$$

where

$$\mathbf{M} = \frac{1}{2c} \sum e\mathbf{v} \wedge \mathbf{r} \quad \text{and} \quad \mathbf{D} = \sum e\left(3\mathbf{r}(\mathbf{n} \cdot \mathbf{r}) - \mathbf{n}r^2\right) \quad (6\text{--}94)$$

are the *magnetic dipole moment* and *electric quadrupole moment*, respectively. Note that the magnetic dipole term also vanishes when the *charge-to-mass ratio* is the same for all particles. This comes about because angular momentum is proportional to $\mathbf{M}$ and conservation of angular momentum implies $\dot{\mathbf{M}} = 0$. The second term on the right of equation 6–93 is called the *electric quadrupole term*.

The higher multipole terms are small compared to dipole radiation terms, since they effectively involve an expansion in v/c and, as (6–81) shows, this is a small quantity when the dimensions of the system are small compared to the wavelength.

The considerations presented here in classical terms also apply in the *quantum theory of radiation*. Instead of talking about the intensity of radiation given off by a moving system of charges, we then talk about the probability for emission of radiation. Again, in those situations in which the charge-to-mass ratio does not vanish, the emission probability is normally much higher for electric dipole radiation than for multipole radiation. In those systems having the same e/m ratio for all constituent particles, electric dipole radiation is "*forbidden*" by the quantum mechanical *selection rules*. For example, we now recognize large masses of interstellar or intergalactic molecular hydrogen H_2 in the Universe. But the presence of this gas for many years could not be established by observations of its infrared spectrum, primarily because the symmetry of the hydrogen molecule forces us to look for lines that are emitted or absorbed only through the very weak electric quadrupole process. (The ultraviolet transitions break this symmetry and were, in fact, observed earlier (Ca70a).

This incidentally brings up one last important point: that *emission* of radiation is just the reverse of *absorption* and the probability for absorption in an atomic system is identical to the probability for *induced emission* (see section 7:10). Induced emission is the process in which an atom or molecule emits radiation in response to stimulation by a light wave that has exactly the same frequency as the wave that the atom can emit. We then find that the stimulated and stimulating radiation have exactly the same characteristics, that is, the photons all belong to the same phase cell. This induced emission is different from the quantum mechanical *spontaneous emission* which has no direct analogue in terms of absorption. The spontaneous emission corresponds to emission by an unperburbed atom or molecule, giving off radiation on its own without any external influence.

It is also interesting that the general approach to radiative processes presented here is relevant to *gravitational radiation*. As discussed in earlier sections, both gravitational forces and electrostatic forces drop as the square of the distance separating the masses or charges. This allows us to use a formalism somewhat similar to electromagnetic theory in dealing with gravitational radiation. One immediate consequence of such considerations is a statement about the strength of the expected gravitational radiation. Since the ratio of inertial to gravitational mass is constant for all matter, gravitational dipole radiation is not permitted. The much weaker quadrupole radiation is the first allowed multipole emission process. Furthermore, the magnitude of the expected radiation at any given multipole level will also be considerably smaller, simply because the ratio of gravitational mass to inertial mass is much smaller than the ratio of electric charge to mass. The ratio of intensities can therefore be expected to differ by factors of order $e^2/m^2G \sim 10^{42}$ if the electron charge-to-mass ratio is used. It is clear then that gravitational radiation can only be expected for large masses and also when the accelerations involved are very large. Such situations would require very compact massive systems to exist in the Universe. We are currently searching to see whether such objects exist. Ordinary binary stars are not massive or compact enough to yield measurable amounts of gravitational radiation, but gravitational radiation can considerably affect the orbits of some compact binaries and their emission can be deduced from the orbital evolution of compact binary pulsars.

PROBLEM 6–10. Using equation 6–77 together with expression 6–93 for the quadrupole moment, to obtain a Poynting vector of the form (6–78), show that the radiation intensity for quadrupole radiation is proportional to $(\dddot{D})^2$ and c^{-5}. The dots indicate the third derivative with respect to time.

The actual intensity for gravitational quadrupole radiation (La51) is

$$I = \frac{G}{45c^5}\,\dddot{D}^2 \tag{6–95}$$

where D is a tensor having the form of (6–94) but with mass replacing the electric charge e. The quadrupole moment is proportional to the ellipticity $\xi \sim \frac{1}{2}[1 - a_{min}^2/a_{max}^2]$ and for an ellipsoid rotating with the mass symmetry axis perpendicular to the axis of rotation:

$$I \sim \frac{GM^2a^4\xi^2\omega^6}{c^5} \tag{6–96}$$

(Ch70). Here we assume that $\xi \ll 1$.

For a pulsar with $a \sim 10^6$ cm, $M \sim 10^{33}$ g, $\xi \sim 10^{-5}$, and $\omega \sim 10^2$ we see that the intensity of gravitational radiation is 3×10^{32} erg sec^{-1}. This is smaller than the magnetic dipole radiation; but very early in the pulsar's career, when it spins with a period of the order of one millisecond, the ω^6 dependence and a possibly increased ℓ value allows the gravitational radiation to equal or dominate the magnetic dipole radiation.

In addition to pulsars, supernovae, quasars and galactic nuclei may also be sources of gravitational radiation.

6:14 LIGHT SCATTERING BY UNBOUND CHARGES

When a plane polarized electromagnetic wave moving along the z-direction is incident on a charged particle having mass m and charge e, the particle is subjected to an electric field of form

$$\mathbf{E} = \mathbf{E}_0 \cos{(\mathbf{k} \cdot \mathbf{r} - \omega t + \alpha)} \tag{6-97}$$

If the field is weak enough so that the velocity imparted to the charge is always small—$v \ll c$—then the force $e\mathbf{E}$ is always large compared to the force $e\mathbf{v} \wedge \mathbf{H}/c$ acting on the particle. This is evident from (6–35). The acceleration experienced by the particle is given by

$$m\ddot{\mathbf{r}} = e\mathbf{E} \tag{6-98}$$

and the dipole moment produced by the displacement of the charge, $\mathbf{d} = e\mathbf{r}$, has a second time-derivative

$$\ddot{\mathbf{d}} = \frac{e^2}{m} \mathbf{E} \tag{6-99}$$

We now see that equation 6–84 predicts a scattered light intensity per solid angle along direction $\mathbf{n}$:

$$dI = \frac{e^4}{4\pi m^2 c^3} (\mathbf{E} \wedge \mathbf{n})^2 \, d\Omega \tag{6-100}$$

We speak of a *differential scattering cross section*

$$d\sigma(\theta, \phi) = \frac{dI(\theta, \phi)}{S} = \left[\frac{e^2}{mc^2} \right]^2 \sin^2\theta \, d\Omega \tag{6-101}$$

where θ is the angle between the scattering direction $\mathbf{n}$ and the electric field $\mathbf{E}$ of the incident wave. S is given by (6–43). We note:

(a) That the frequency of the radiation is not changed by scattering.
(b) That the angular distribution of scattered light is not dependent on the frequency.

6:14

Fig. 6.9 Direction of incident and scattered waves (see equations 6–101 and 6–105).

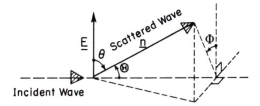

Incident Wave

(c) That the total cross section is not frequency dependent. The total cross section is obtained by integrating $d\sigma(\theta, \phi)$ over all angles θ, ϕ:

$$\sigma = \int_0^\pi \int_0^{2\pi} \sigma(\theta, \phi) \sin\theta \, d\phi \, d\theta$$

For electrons

$$\sigma_e = \frac{8\pi}{3} \left(\frac{e^2}{mc^2} \right)^2 = 6.65 \times 10^{-25} \text{ cm}^2 \tag{6-102}$$

This is called the *Thomson scattering* cross section.

(d) The differential scattering cross section is symmetrical in θ about $\theta = \pi/2$.

(e) σ is a factor of $(m_p/m_e)^2 \sim 10^6$ times less for protons than for electrons.

Let us still consider the influence that the polarization of the **E** vector has on the actual angular distribution of radiation. If the initial wave incident on the particle is unpolarized, we obtain a scattering cross section, independent of Φ, but dependent on the polar angle Θ included between the directions of the incident and scattered waves (Fig. 6.9). In this context we note that the angle θ has to be considered a function of the angles Φ and Θ, and that

$$\cos\theta = \sin\Theta \cos\Phi \tag{6-103}$$

For a given angle Θ

$$\therefore \langle \sin^2\theta \rangle = 1 - \sin^2\Theta \langle \cos^2\Phi \rangle = 1 - \frac{\sin^2\Theta}{2} = \frac{1}{2}(1 + \cos^2\Theta) \tag{6-104}$$

where we have made use of the fact that $\langle \cos^2\Phi \rangle = 1/2$ when the average is taken over all angles Φ. For unpolarized radiation we can therefore write

$$d\sigma = \frac{1}{2} \left(\frac{e^2}{mc^2} \right)^2 (1 + \cos^2\Theta) \, d\Omega \tag{6-105}$$

This yields the important result that

(f) For unpolarized radiation the cross section has peak values in the forward and backward directions, that is, most of the light is scattered along the direction in which the wave was moving initially—or backward into the direction from which the wave came.

PROBLEM 6–11.   Show that there is a force component

$$F(\Theta) = (1 - \cos \Theta) \, d\sigma \, \frac{\mathbf{S}}{c}$$

acting on the scattering charge along the direction of propagation. Show that when this is averaged over all values Θ, one obtains a total force $\mathbf{F}$ along the direction of incidence

$$\mathbf{F} = \frac{2}{3} \left(\frac{e^2}{mc^2} \right)^2 E^2 = \frac{\sigma \mathbf{S}}{c} \tag{6–106}$$

In the vicinity of bright hot stars, this can be the dominant force acting on electrons. Much of the visible light reaching us from the solar corona also seems to be due to scattering by electrons. However, the zodiacal glow, that is, the diffuse scattered sunlight in the ecliptic plane, is due to radiation scattered off small solid grains circling the sun in the orbital plane of the planets. This glow extends into the corona and weakly contributes to its brightness.

We can now also consider scattering by a *harmonically bound charge* that would normally oscillate at a natural frequency ω_0. The electric field attempts to force the oscillator to vibrate at a frequency ω instead. The equation of motion for this *forced oscillation* is

$$\ddot{\mathbf{r}} + \omega_0^2 \mathbf{r} = \frac{e\mathbf{E}}{m} \tag{6–107}$$

If $\mathbf{E} = 0$, we obtain oscillation at frequency ω_0. If $\mathbf{E}$ has the form (6–97), equation 6–107 has the solution

$$\mathbf{r} = \frac{e\mathbf{E}}{m} \frac{1}{(\omega_0^2 - \omega^2)} \tag{6–108}$$

and

$$\ddot{\mathbf{d}} = \frac{e^2}{m} \mathbf{E} \left(\frac{1}{1 - (\omega_0^2/\omega^2)} \right) \tag{6–109}$$

It is clear then (see equation 6–99) that the scattering cross section is

$$\sigma = \frac{\sigma_{\text{Thomson}}}{(1 - \omega_0^2/\omega^2)^2} \tag{6–110}$$

6:14

When $\omega \gg \omega_0$, the electron acts as though it were free and we again have $\sigma = \sigma_{\text{Thomson}}$. If $\omega_0 \gg \omega$ we obtain

$$\sigma = \frac{8\pi}{3} \frac{e^4}{m^2 c^4} \frac{\omega^4}{\omega_0^4} \tag{6-111}$$

called the *Rayleigh scattering cross section*. Rayleigh scattering is responsible for the scattering of visible light in the daytime sky. The electrons are strongly bound to their parent molecules so that ω_0 is large compared to the frequency of visible light ω. For red light $(\omega/\omega_0)^4$ is smaller than for blue light by a factor close to $(2)^4 = 16$. Hence blue light is scattered most strongly. Red light therefore passes more easily straight through the atmosphere without deflection, while blue light is scattered out of a straight path, and the sky appears blue when we look away from the sun.

There is one more case of scattering that is interesting in astronomy. This is the scattering by fine dust grains. For spherical grains with refractive index n, the cross section for scattering can be shown to be

$$\sigma = 24\pi^3 \left[\frac{n^2 - 1}{n^2 + 2} \right]^2 \frac{V^2}{\lambda^4} \tag{6-112}$$

if the radius a of the sphere is much smaller than the wavelength λ. V is the volume $(4\pi/3)\,a^3$. We note that the factor λ^{-4} is reminiscent of Rayleigh scattering and, of course, the two types of scattering are related. The differential cross section has exactly the same angular dependence as found for Thomson or Rayleigh scattering (see 6–105). This kind of scattering may be approximately characteristic of the zodiacal (interplanetary) grains mentioned above, and of the interstellar grains discussed in the next section.

6:15 EXTINCTION BY INTERSTELLAR GRAINS

Interstellar grains absorb and scatter radiation so that starlight does not reach an observer directly. We talk about *extinction* by the grains (Gr68). The term extinction refers to the fractional amount of light prevented from reaching us. It is a useful concept when we do not know how much of the radiation is scattered and how much is absorbed. The scattered radiation sometimes can be observed in *reflection nebulae*—clouds of dust grains illuminated by a bright star. These clouds show spectra remarkably similar to those of the illuminating star; and it therefore appears that the scattered portion of the radiation is very much like scattering off snow. The particles are basically white or gray in this sense. On the other hand, when we see starlight that has passed through a cloud, we find an amount of extinction that to first approximation is inversely proportional to

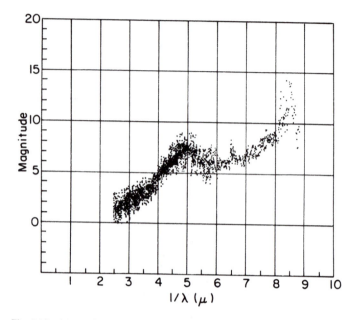

Fig. 6.10 Interstellar extinction curve showing magnitudes of extinction as a function of reciprocal wavelength. The data were obtained from observations of the stars ζ-Persei and ε-Persei, and have been normalized to an extinction difference $E(B - V)$ of one magnitude and $V \sim 0$. The curve would therefore roughly characterize extinction over a path length of order ~ 1 kpc through the galactic plane. (After Stecher St69.)

the wavelength, λ. The data are shown in Fig. 6.10. It may be that λ^{-4} scattering is the predominant scattering process; the size of grains may be roughly comparable to the wavelength of the radiation, and the size distribution may be such that the apparent scattering cross section integrated over all different particle sizes gives an overall mean cross section proportional to λ^{-1}. These considerations all are very uncertain.

From current observations it is not clear what kind of material the grains represent. Large organic molecules or graphite particles are a plausible constituent, and there is evidence for ice and silicates. Some recent infrared spectral observations indicate a weak absorption band at 3μ wavelength that corresponds to the position at which ice crystals would be expected to absorb. Quite possibly some grains consist of an ice mantle deposited on a graphite core: The graphite could condense into grains in the dense atmosphere of a *carbon star*; water would then be frozen out on these grains in the outer, cooler portions of the star's atmosphere, as the grains streamed outward into interstellar space. There is a strong feeling that interstellar grains have to be formed in some relatively dense regions of the universe, such as in the atmospheres of stars, or in dense clouds, because the general interstellar medium is so tenuous that collisions among molecules are too rare to permit growth of material into micron-sized grains. This is discussed further in section 9:4.

One more factor might be mentioned. Thus far we have only talked about spherical grains that are purely dielectric. It is, however, also possible for grains to have a metallic character; they can then absorb and emit radiation—they do not merely scatter. For metallic grains the dielectric constant has an imaginary component and we talk about a *complex refractive index m*. In these terms, for grains small compared to the wavelength, absorption dominates over scattering, and the extinction is given (Gr68) by

$$E = 6\pi N \left(\frac{1 - m^2}{m^2 + 2} \right) \frac{V}{\lambda} \quad (\text{IP}) \qquad (6\text{--}113)$$

where E is the total light extinguished for unit incident energy, and the symbol (IP) means that the imaginary part of the expression in parentheses should be used.

For particles with dimensions comparable to the wavelength of the extinguished radiation, the expressions become quite complicated even for spherical particles; and for nonspherical grains the theory of extinction is extremely laborious.

We might be tempted to attribute interstellar extinction to metallic absorption alone, because then the $1/\lambda$ relation might be directly obtained. However, matters are not that simple. The refractive indices m and n are wavelength dependent, and that wavelength dependence is determined by the chemical makeup of the grains. With all these free parameters, a $1/\lambda$ dependence can therefore be fitted with relative ease—and this is corroborated by the number of theoretical models that astronomers have produced to describe interstellar grains. As in other parts of astronomy, a wealth of models reflects a large degree of uncertainty.

Radiation scattered by any of the three processes mentioned in section 6:14 should lead to a polarization that can be shown to have the Θ-dependence

$$P = \frac{\sin^2 \Theta}{1 + \cos^2 \Theta} \qquad (6\text{--}114)$$

In addition, polarization, both in absorption and in emission, can be produced by systematically oriented elongated or disk-shaped grains. The light that reaches us from distant stars located close to the plane of the Galaxy shows polarization believed due to this process. How small grains could be aligned to give consistently

6:15

polarized radiation is discussed in section 9:8. The interaction of grains with radiation is a complicated subject. Detailed discussions can be found in references (Gr68) and (vdHu57).

6:16 ABSORPTION AND EMISSION OF RADIATION BY A PLASMA

In section 6:11, we treated the propagation of radiation through a tenuous ionized medium. Sometimes, however, the plasma encountered in astrophysical situations is dense in that collisions between ions and electrons have a relatively high frequency of occurrence, v_c. In this sense, a medium may be tenuous for high frequency waves, but dense at lower frequencies. This is a quite common situation. Because the nature of the transmission is so different in these two cases, the spectrum of radiation received from a source also may be quite different at high and low frequencies. A relatively abrupt change in spectrum taken together with other data then allows us to determine the collision frequency and, hence, as we will show, also the density of the medium. Radioastronomy therefore provides an extremely useful technique for measuring the density of interstellar ionized gases.

To show how all this comes about, we consider an ionized medium and define the *collision frequency* v_c as the frequency with which an electron successively becomes deflected through a total angle of 90°, usually through a series of small collisions with ions. We choose an angle of 90°, because this is the deflection a particle has to suffer to give up all the directed momentum it had at some previous period, that is, to lose all sense of the direction into which it was accelerated by some previously applied force. We only consider collisions with ions, because we will be interested in the dissipation of energy through collisions. When an electron collides with another electron, the motion is that of a symmetric dipole and no energy is radiated away in such a process. Electron-electron collisions can therefore be neglected.

We now consider a situation in which an electron is accelerated by an applied electric field **E**—in this case the electric field component of an electromagnetic wave. As the particle reaches appreciable velocity induced by the field, it suffers a collision and gives up all its directed momentum. This means that there is an acceleration by the electromagnetic wave traveling through the medium and deceleration through collisions. The net effect is to impose forces on the electron so that

$$m\ddot{\mathbf{r}} = e\mathbf{E}(r, t) - m\dot{\mathbf{r}}v_c \qquad (6\text{--}115)$$

Here m is the reduced electron mass. The second term on the right shows a momentum loss equal to the instantaneous momentum $m\dot{\mathbf{r}}$ of the electron every time there is a collision, or v_c times in unit time interval. This is just what we stated formally in defining v_c; and our problem will be to calculate the actual value v_c has. However, before we do that, we can proceed to solve equation 6–115 and obtain the transmission properties of the plasma in different frequency ranges relative to v_c.

We have already noted that some of the momentum conferred on the electrons by the electromagnetic wave is lost in collisions. This means that the energy transferred from the wave to the particles becomes dissipated, and since the energy of the electromagnetic wave depends on E^2, the square of the wave amplitude, we can expect to find that E will decrease as the wave propagates through the medium. We will therefore make use of a function $E(r, t)$ of the form

$$E(r, t) = E_0 e^{-Kx/2} \cos \omega t \qquad (6-116)$$

in equation 6–115. Here K is the *absorption coefficient*. The factor 2 in the exponent of this *damping term* is provided so that the energy in the wave, rather than its amplitude, may decay be a factor of $1/e$ in distance $x = 1/K$. Note that the absorption coefficient always has units (length)$^{-1}$.

Because the rate of energy loss from the wave will be determined by the total number of collisions per unit volume, we rewrite equation 6–115 as

$$nm\ddot{r} + nmv_c\dot{r} = neE_0 e^{-Kx/2} \cos \omega t \qquad (6-117)$$

where n is the electron density.

If we use a complex field E, instead of a field with real values in equation 6–117, the solution of the differential equation becomes much simpler. However, in order to remember that only the real parts of the equation have physical significance, we add the annotation (RP) in this case:

$$nm\ddot{r} + nmv_c\dot{r} = neE_0 e^{-(Kx/2)+i\omega t} = neE \qquad \text{(RP)} \qquad (6-118)$$

PROBLEM 6–12. By substitution, show that a particular solution of (6–118) is

$$r = -\left(\frac{eE_0}{m\omega}\right) e^{(i\omega t - Kx/2)} \left[\frac{iv_c + \omega}{v_c^2 + \omega^2}\right] \qquad (6-119)$$

The current due to the n particles per unit volume can now be written as

$$j = ne\dot{r} = \frac{ne^2}{m} \left[\frac{v_c - i\omega}{v_c^2 + \omega^2}\right] E \qquad \text{(RP)} \qquad (6-120)$$

As in equation 6–50, $ne r$ is an induced polarization field, and the imaginary term in the brackets on the right of (6–120) is the induced polarization current

$$\frac{d\mathbf{P}}{dt} = i\omega\mathbf{P} = i\omega\frac{\varepsilon - 1}{4\pi}\mathbf{E} \quad \text{(RP)} \tag{6–121}$$

Here the imaginary number i enters as a consequence of the assumed field in (6–118), and (6–121) then is a direct consequence of the definition (6–10). The real term in the brackets of (6–120) is just the current $\sigma\mathbf{E}$ due to the flow of charge. We note two features of equation (6–120). The term proportional to v_c on the right represents the dissipation of energy, and will therefore be directly related to the absorption coefficient K. The second term, proportional to $i\omega$, depends on the dielectric constant in the medium, and hence will yield the phase velocity $c\varepsilon^{-1/2}$ of the wave through the medium. Formally written:

$$\mathbf{j} = \left(\sigma + i\omega\frac{\varepsilon - 1}{4\pi}\right)\mathbf{E} \quad \text{(RP)} \tag{6–122}$$

with

$$\varepsilon = 1 - \frac{4\pi e^2 n}{m(\omega^2 + v_c^2)} \quad \text{and} \quad \sigma = \frac{e^2 n v_c}{m(\omega^2 + v_c^2)} \tag{6–123}$$

If we write the imaginary and *complex dielectric constants* as

$$\varepsilon_i = -\frac{i4\pi\sigma}{\omega} \quad \text{and} \quad \varepsilon_c = \varepsilon + \varepsilon_i \tag{6–124}$$

then equation 6–122 can be written in a form characteristic of a pure dielectric. In fact, all of Maxwell's equations take on this form. This can be seen directly by noting that $\mathbf{j}$ appears only in equations 6–15 and 6–23 in the set of Maxwell's differential equations. For a complex field, as it appears in (6–118), a propagating wave will have a form (see equation 6–36)

$$\mathbf{E} = \mathbf{E}_0 \exp i\left[\omega t \pm \frac{\omega\varepsilon_c^{1/2}x}{c}\right] \quad \text{(RP)} \tag{6–125}$$

so that (6–118) will hold if

$$\frac{K}{2} = \frac{i\omega}{c}\varepsilon_c^{1/2} \quad \text{(RP)} \tag{6–126}$$

We can always write ε_c in the form

$$\varepsilon_c = (N + iQ)^2 \tag{6–127}$$

where N and Q are real quantities as long as we choose

$$\varepsilon_i = 2NQi = -\left(\frac{4\pi\sigma}{\omega}\right)i \quad \text{and} \quad \varepsilon = N^2 - Q^2 \tag{6–128}$$

We are therefore interested in the quantity

$$\frac{K}{2} = -\frac{\omega Q}{c} = \frac{4\pi\sigma}{2Nc} \tag{6-129}$$

In practice $\omega \gg v_c$, ω_p, in all radio-astronomical situations, so that (6–123) and (6–128) give

$$|\varepsilon| \gg \frac{4\pi\sigma}{\omega} \quad \text{and} \quad N \sim \varepsilon^{1/2} \tag{6-130}$$

With this same approximation we then also obtain

$$K = \frac{4\pi(e^2 n/m\omega^2)\, v_c/c}{\sqrt{1 - 4\pi e^2 n/m\omega^2}} = \frac{v_c(\omega_p^2/\omega^2)/c}{\sqrt{1 - \omega_p^2/\omega^2}} \tag{6-131}$$

where ω_p is the plasma frequency (6–55).

We still need to calculate the collision frequency v_c, but that should be simple because most of the work has already been done. Equation 3–72 gives the force acting on a particle of reduced mass μ deflected in the superposition of inverse square law fields produced by a density of n scattering centers per unit volume. To avoid confusion with the symbol μ used here for magnetic permeability, we will continue to use the symbol m for the electron's reduced mass. In equation (6–115) we had defined the drag force $m\dot{r}v_c$ and we now set this equal to the right side of (3–67), noting that since $\dot{r}$ is the velocity before collision, it plays the same role as v_0 in (3–67):

$$mv_0 v_c = m2\pi n v_0^2 \int_{s_{\min}}^{s_{\max}} s(1 - \cos\Theta)\, ds \tag{6-132}$$

For small deflections Θ

$$1 - \cos\Theta \approx 2\tan^2\frac{\Theta}{2} = 2\left[\frac{Ze^2}{v_0^2 sm}\right]^2 \tag{6-133}$$

The second half of this inequality is based on an analogy with equation 3–69 but with Coulomb forces replacing gravitational forces. Z is the typical charge on an ion. We therefore have

$$v_c = 4\pi n \left\langle\frac{1}{v_0}\right\rangle \frac{1}{\langle v_0^2 \rangle} \frac{Z^2 e^4}{m^2} \int_{s_{\min}}^{s_{\max}} s^{-1}\, ds$$

$$\tag{6-134}$$

$$= \frac{4\sqrt{2\pi}}{3} n \frac{Z^2 e^4}{\sqrt{m}\,(kT)^{3/2}} \ln\frac{s_{\max}}{s_{\min}}$$

6:16

where expressions 4–109 and 4–110 have been used. From (6–131) and (6–134)

$$K(\omega) = \frac{32\pi^{3/2}e^6 n^2 Z^2}{3\sqrt{2}\,c\omega^2 (kTm)^{3/2}} \ln \frac{s_{max}}{s_{min}} \tag{6-135}$$

Here the assumption of very many, very weak deflections has been made so that the minimum impact parameter s'_{min} must be large enough to give a potential energy small compared to the kinetic energy. Specifically, for $Z = 1$,

$$s'_{min} \gg \frac{2e^2}{mv_0^2} \tag{6-136}$$

In the interstellar medium, s rarely will be less than 10^{-2} cm, while $2e^2 = 5 \times 10^{-19}$; and typically mv^2 is 10^{-12} erg in ionized regions. This shows that the condition (6–136) usually is well satisfied except in very rare chance collisions with a small impact parameter.

A second lower bound is given by the *de Broglie wavelength* of the electron $\lambda_e = h/mv$. At closer distances than this, the electron no longer behaves like a point charge, and we can use a lower limit $s''_{min} > \lambda_e/2$. There also are two upper bounds we can set. First, we want a collision to appear instantaneous, that is, the time $1/\omega \gg s'_{max}/v_0$; the time during which the electric field changes is long compared to the time in which the electron suffers a collision, or goes through its minimum approach. The second upper limit is $s''_{max} = L$, the Debye length given by equation 4–156. Shielding by nearer particles screens out the effects of charges at distances greater than L. Using the limits s'_{max} and s'_{min} for the ionized interstellar matter, we can then write

$$\frac{s'_{max}}{s'_{min}} = \frac{mv_0^2}{2e^2} \frac{v_0}{\omega} \sim \frac{(2kT)^{3/2}}{2e^2 \omega m^{1/2}}$$

The full expression for ionized hydrogen, $Z = 1$, reads

$$K(\omega) = \frac{32\pi^{3/2}}{3\sqrt{2}} \frac{e^6 n^2}{c(mkT)^{3/2}\,\omega^2} \ln \left[\frac{1.32(kT)^{3/2}}{e^2 m^{1/2}\omega} \right]$$

$$K(\nu) = \frac{8}{3\sqrt{2\pi}} \frac{e^6 n^2}{c(mkT)^{3/2}\,\nu^2} \ln \left[\frac{1.32(kT)^{3/2}}{2\pi e^2 m^{1/2}\nu} \right] \tag{6-137}$$

6:17 RADIATION FROM THERMAL RADIO SOURCES

If we now look at the results obtained in the previous section, we note that we have an absorption coefficient $K(\nu)$ that tells us the amount of absorption obtained per unit length of travel through an ionized medium. An electromagnetic wave traveling a distance $D = \int dx$ through the medium will encounter an *optical depth*

$$\tau(v) = \int K(v)\,dx \qquad (6\text{-}138)$$

If the temperature throughout the region is constant, only the density will vary with position x and we find that

$$\tau(v) = F(T, v)\int n^2\,dx \sim 8.2 \times 10^{-2}T^{-1.35}v^{-2.1}\int n^2\,dx \qquad (6\text{-}139)$$

(La74) where the function F is $K(v)/n^2$ (see equation 6–137) and the integral

$$\mathscr{E}_m = \int n^2\,dx = \langle n^2 \rangle D \qquad (6\text{-}140)$$

is called the *emission measure*; it is a measure of the amount of absorption and emission expected along D. It is customary in radio-astronomy to express the electron concentration n in terms of cm^{-3} and D, the path length covered, in parsecs. The emission measure then has units cm^{-6} pc.

The emission measure is just a measure that tells how frequently atomic particles approach each other closely along a line of sight through a given region. For this reason, such quantities as the number of atomic recombinations giving rise to a given emission line are also proportional to $\mathscr{E}_m$. Usually the recombination line strength R_1 for any given line v_1 also is a known function of the temperature, so that

$$R(v_1) = F_1(T)\,\mathscr{E}_m \qquad (6\text{-}141)$$

Hence, if we measure both the recombination line strength—possibly in the visible part of the spectrum—and also the radio thermal emission, both the emission measure and the temperature of the region can be determined. For this to be true, radio measurements are best taken at frequencies for which the region is optically thin, so that self-absorption of radiation by the cloud need not be considered.

An interesting feature of the self-absorption by an optically thin cloud is that the brightness should be independent of the frequency v. This comes about because the absorption $K(v)$ is inversely dependent on v^2—if we neglect the weak frequency dependence of the logarithmic term in (6–137). At the same time, the energy density of radio waves corresponding to a blackbody at gas temperature, T, would be

$$\rho(v) \sim \frac{8\pi k T v^2}{c^3} \qquad hv \ll kT \qquad (4\text{-}80)$$

at very long wavelengths. The product of optical depth or effective emissivity for the gaseous region, and blackbody intensity $I(v) = \rho(v)\,c/4$, is therefore frequency independent as long as the region is optically thin. At low frequencies, where $\tau(v) \gtrsim 1$, this behavior ceases to be true. The effective emissivity then remains close to unity, and the only frequency dependent term is $I(v)$. This is the Rayleigh–Jeans limit, where a thermal source exhibits a spectrum proportional to v^2.

For the flat part of the spectrum the product $S(v) = \tau(v)\,I(v)$ is proportional to $T^{-1/2}\mathscr{E}_m$; and this latter product can be immediately determined from a measurement of the surface brightness anywhere in this frequency range. In the steep part of the spectrum, where the region is opaque, the measured surface brightness at frequency v depends only on T (see equation 4–81). These two sets of observations, taken together, provide data both on the temperature and on the emission measure $\mathscr{E}_m$. Figure 6.11 shows spectra for some very compact ionized hydrogen regions and these show the expected form very clearly.

On this log-log plot the low frequency spectrum has the expected slope of 2. At high frequencies it is flat.

NGC 7027 is a planetary nebula for which Mezger (Me68)—see the data of Fig. 6.11—finds an emission measure of 5.4×10^7 cm^{-6} pc and a temperature of

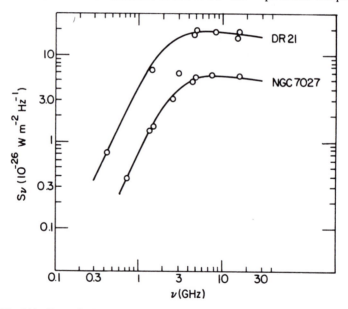

Fig. 6.11 Data obtained by a number of observers on the compact H<small>II</small> region DR21 and the planetary nebula NGC 7027 (see text). After P.G. Mezger (Me68). (From *Interstellar Ionized Hydrogen*, Y. Terzian, Ed., 1968, W.A. Benjamin, Inc. Reading, Massachusetts.)

$\sim 1.1 \times 10^{4\circ}$K. If the object is assumed to have a depth along the line of sight similar to the observed diameter, then an actual density can be computed. Mezger gives the value $n \sim 2.3 \times 10^4$ cm^{-3} for this object. From the density and the total volume, we can also obtain the nebular mass, which in this case is roughly $0.25 \, M_\odot$, or 5×10^{32} g. The observed diameter of NGC 7027 is about 0.1 pc.

Young compact HII regions, which may be found in the plane of the Galaxy in the vicinity of bright young stars, tend to have somewhat lower temperatures, about the same densities, but sometimes much greater masses—up to several stellar masses. These are believed to possibly represent the remains of clouds from which massive stars were formed. When massive protostars light up as they approach the main sequence, they emit an intense ultraviolet flux that heats and ionizes the gas. Such HII regions will be discussed in Chapter 9.

6:18 SYNCHROTRON RADIATION

When a charged particle moves at relativistic velocity across a magnetic field, it describes a spiral motion. The axis of this spiral lies along the direction of the magnetic field and the acceleration experienced by the particle is along directions perpendicular to the field lines. As the particle moves, the direction of the acceleration vector continually changes.

We first consider the motion of a relativistically moving particle orbiting in a plane perpendicular to a magnetic field. This constitutes no restriction in generality, because a constant velocity component along the magnetic field lines would leave the radiation rate unaffected.

If we recall that a force corresponds to a rate of change of momentum, we can use (6-11) to calculate the rate at which the particle is deflected. Consider the direction of motion in Fig. 6.12 to be the x direction, and let the radial direction be the y direction. In a time Δt_0, the momentum change, which is along the y direction, will amount to

$$\Delta p_y = \frac{evB}{c} \Delta t_0 \tag{6-142}$$

Since the initial relativistic momentum p_x is

$$p_x = \frac{m_0 v}{\sqrt{1 - v^2/c^2}} \tag{5-30}$$

we can see that the angular deflection during time interval Δt_0 is,

$$\delta = \frac{\Delta p_y}{p_x} = \frac{eB}{m_0 c} \sqrt{1 - \frac{v^2}{c^2}} \, \Delta t_0 = \frac{eB \, \Delta t_0}{m_0 c \gamma(v)} \tag{6-143}$$

6:18

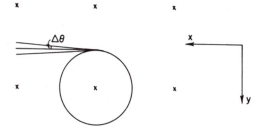

Fig. 6.12 Relativistic charged particle orbiting in a magnetic field. The direction of the field lines is into the paper. Because the acceleration of the particle has no z-component, radiation reaching an observer along the x-direction will be linearly polarized along the y-direction—perpendicular to the magnetic field.

m_0 is the particle's rest mass. From this it follows that the time Δt_1, required for the particle to orbit one radian $[\delta = 1]$ is

$$\Delta t_1 = \frac{m_0 c}{eB} \gamma(v) \tag{6-144}$$

The gyrofrequency given in equation 6–13 is the reciprocal of Δt_1; we see that, if we use (5–30) to substitute $m_0 \gamma(v)$ for p_c/v_c in (6–13).

Having obtained the gyrofrequency of the particle, we might think that the problem is completely solved and that the particle will simply radiate energy at that frequency. However, that is not so. The spectrum radiated by the moving charge actually lies at frequencies often many orders of magnitude higher than ω_c. The reason for this is directly related to the strong concentration of emitted radiation into a narrow beam of angular half-width (obtainable from equation 5–50):

$$\Delta \theta \sim \sqrt{1 - \frac{v^2}{c^2}} = \gamma(v)^{-1} \tag{6-145}$$

about the forward direction of motion. Because of this, an observer is not properly oriented to receive radiation emitted by the particle except during a very short-time interval

$$\Delta t_2 = 2\Delta \theta \, \Delta t_1 = \frac{2m_0 c}{eB} \tag{6-146}$$

of each orbit.

But the radiation emitted during interval Δt actually arrives at the observer over an even smaller time span, because radiation emitted by the particle at the beginning of the interval Δt has a longer distance to travel to the observer than radiation emitted at the end of the interval when the particle is nearer to him. If the particle travels a distance of length L during interval Δt, radiation emitted at the end of the interval will only arrive at a time

$$\Delta t \sim -\left(\frac{L}{c} - \frac{L}{v}\right) \tag{6-147}$$

later than radiation emitted at the beginning of the interval. Since

$$L \sim v \, \Delta t_2 \tag{6-148}$$

we obtain

$$\Delta t \sim \left(1 - \frac{v}{c}\right)\Delta t_2 \sim \frac{m_0 c}{eB}\left(1 - \frac{v^2}{c^2}\right) \tag{6-149}$$

because for highly relativistic particles

$$\left(1 - \frac{v}{c}\right) \sim \frac{1}{2}\left(1 + \frac{v}{c}\right)\left(1 - \frac{v}{c}\right) = \frac{1}{2}\left(1 - \frac{v^2}{c^2}\right) \tag{6-150}$$

The radiation frequency corresponding to the reciprocal of this time interval is

$$\omega_m \sim \frac{1}{\Delta t} \sim \frac{eB}{m_0 c}\left(1 - \frac{v^2}{c^2}\right)^{-1} = \gamma^2(v)\,\omega_c = \frac{eB}{m_0 c}\left(\frac{\mathscr{E}}{m_0 c^2}\right)^2 \tag{6-151}$$

where $\mathscr{E}$ is the total energy of the particle, $\mathscr{E} \gg m_0 c^2$ and ω_c is the radiation frequency of a non-relativistic particle moving in a magnetic field. We expect to see radiation at frequencies of this order of magnitude from relativistic particles moving in a magnetic field. Since $(1 - v^2/c^2)$ is a very small number, it is clear that ω_m is many orders of magnitude greater than the gyrofrequency,

$$\omega_m \gg \omega_c = \frac{eB}{m_0 c} \tag{6-152}$$

Let us summarize what we have done:

(1) First, we computed the orbital frequency of a particle moving in a magnetic field.

(2) Next, we calculated the time in the observer's frame during which the particle was capable of emitting radiation into his direction.

(3) Finally, we computed the length of time elapsing between the arrival of the first and last portion of the electromagnetic wave train at the position of the observer. This elapsed time was very small compared to the period of the particle's gyration in the magnetic field and the corresponding frequency $\omega_m \sim 1/\Delta t$ was found to be $(\mathscr{E}/m_0 c^2)^2$ higher than the nonrelativistic gyrofrequency for this field.

6:18

6:19 THE SYNCHROTRON RADIATION SPECTRUM

The actually expected synchrotron radiation spectrum obtained when the above sketched calculations are done rigorously for monoenergetic electrons is a set of extremely finely spaced lines at high harmonics of the gyrofrequency. The peak of the spectral distribution occurs at a frequency $\omega = 0.5\ \omega_m$. The shape of the spectral function $p(\omega/\omega_m)$ is shown in Fig. 6.13. The maximum value of p is $p(0.5) = 0.10$. Details of the theory are discussed in references (Gi64)* and (Sh60)*.

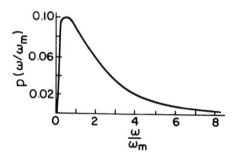

Fig. 6.13   Envelope of the narrowly spaced lines comprising the synchrotron spectrum of a particle whose frequency $\omega_m = (eB/m_0 c)(\mathscr{E}/m_0 c^2)^2$. In an actual situation the particle energies vary to some extent and, hence, the finely spaced lines are never seen. One observes a continuum of the shape of the envelope (after I.S. Shklovsky, Sh60).

One can show that the energy actually radiated by a particle of energy $\mathscr{E}$ per unit time into unit frequency interval $d\nu$ is

$$P(\nu, \mathscr{E}) = 2\pi\ P(\omega, \mathscr{E}) = \frac{16e^3 B}{m_0 c^2}\ p(\omega/\omega_m) \qquad (6\text{–}153)$$

In the limit of very high and very low frequencies the function $p(\omega/\omega_m)$ has asymptotic values

$$p(\omega/\omega_m) = 0.256 \left(\frac{\omega}{\omega_m} \right)^{1/3}, \qquad \omega \ll \omega_m$$

$$p(\omega/\omega_m) = \frac{1}{16} \left(\frac{\pi\omega}{\omega_m} \right)^{1/2} \exp\left(-\frac{2\omega}{3\omega_m} \right), \qquad \omega \gg \omega_m \qquad (6\text{–}154)$$

PROBLEM 6–13.   In section (5:9) we saw that the power radiated by a body is independent of an observer's rest frame, as long as both the source and observer move in inertial frames of reference. Make use of this fact to obtain the total power radiated in the form of synchrotron radiation, by computing the emission of the spiraling charge as viewed from an inertial frame moving with the charge's instantaneous velocity. For a charge whose total energy is $\mathscr{E}$, this total power is

$$P(\mathscr{E}) = \frac{2}{3} \frac{e^4 B^2}{m_0^2 c^3} \left(\frac{\mathscr{E}}{m_0 c^2} \right)^2 \tag{6-155}$$

for motion perpendicular to the magnetic field. For an electron

$$P(\mathscr{E}) = 1.58 \times 10^{-15} B^2 \left(\frac{\mathscr{E}}{m_0 c^2} \right)^2 \text{ erg sec}^{-1}$$

$$= 2.48 \times 10^{-2} \left(\frac{B^2}{8\pi} \right) \left(\frac{\mathscr{E}}{m_0 c^2} \right)^2 \text{ ev sec}^{-1}$$

Verify that these expressions are at least approximately consistent with expressions 6–151, 6–153, and 6–154 by noting that $P(\omega, \mathscr{E})\,\omega_m$ roughly corresponds to $P(\mathscr{E})$, and that equation 6–153 with a numerical integration under the curve in Fig. 6.13 gives the same result.

PROBLEM 6–14. One astronomical object in which synchrotron radiation is important is the Crab Nebula. Take the magnetic field strength in some of the Crab's bright filaments to be of order 10^{-4} gauss, and show that a classically moving electron would radiate at a frequency of about 300 Hz, independent of the energy. On the other hand, if the energy becomes 10^9 ev, some 2×10^3 times the rest-mass energy, the peak radiation will occur at about 600 MHz and if the energy becomes 10^{12} ev, the radiation peaks in the visible part of the spectrum at 6×10^{14} Hz.

It is clear that the exact form of the observed spectrum will depend on the energy spectrum of the radiating particles as well as on the function $P(v, \mathscr{E})$. If we integrate the radiation coming from different distances r along the observed line of sight out to some distance R, the resulting spectral intensity at frequency v will be

$$I_v \, dv = \int_0^{\mathscr{E}_{max}} \int_0^R P(v, \mathscr{E})\, n(\mathscr{E}, r)\, dr \, d\mathscr{E} \, dv \tag{6-156}$$

where $n(\mathscr{E}, r)$ is the number density of particles of energy $\mathscr{E}$ at distance r.

In the frequently encountered situation where $n(\mathscr{E}) \propto \mathscr{E}^{-\gamma}$, that is, where the electrons have an exponential spectrum with constant exponent $-\gamma$, the intensity obeys the proportionality relation $I_v \propto v^{-\alpha}$ where $\alpha = (\gamma - 1)/2$. To show this relationship, we note from Fig. 6.13 and from (6–153) that $P(v, \mathscr{E})$ is equal to $16e^3 B/m_0 c^2$ multiplied by an amplitude 0.1 and 2π, since the bandwidth is $\Delta v = \Delta\omega/2\pi \sim 3\omega_m/2\pi$.

Let us suppose now that every electron deposits its total radiated power at frequency ω_m. Equation 6–151 shows that $\mathscr{E} \propto \omega_m^{1/2}$ so that $\mathscr{E}^{-\gamma} \propto \omega_m^{-\gamma/2}$, $\Delta\mathscr{E} \propto \Delta\omega/\omega_m^{1/2}$, and the total radiated power in (6–156) obeys the proportionality relation

$$I(v)\,\Delta v \propto \omega_m \mathscr{E}^{-\gamma} \Delta\mathscr{E} \propto \omega_m^{(1-\gamma)/2}\,\Delta\omega \qquad (6\text{–}157)$$

for a constant spectrum along the path of integration. Hence

$$I(v) \propto v^{-\alpha}, \qquad \alpha = \frac{(\gamma-1)}{2} \qquad (6\text{–}158)$$

α is called the *spectral index* of the source. To obtain this relationship between electron energy and electromagnetic radiation spectra, the source must be optically thin. The optically thick (self-absorbing) sources are discussed below. For a wide variety of nonthermal cosmic sources $0.2 \lesssim \alpha \lesssim 1.2$. For extragalactic objects indices up to $\alpha = 2$ occur. In the frequency range below a few gigahertz, many quasars have $\alpha < 0.5$; but they often contain optically thick components. Most radio galaxies have $\alpha > 0.5$ (Co72). Spectra of some extragalactic sources appear in Fig. 9.13.

For the Galaxy, confirmation of these general concepts is good. We observe a Galactic cosmic ray electron spectrum with $\gamma \sim 2.6$. This is measured at the earth's position, but the electrons have reached us from great distances. Radio waves also show an overall Galactic spectrum with index $\alpha \sim 0.8$—in agreement with (6–158).

Equation 6–158 is of great importance in astrophysics because it permits us to estimate the relativistic electron energy spectrum by looking at the synchrotron radio emission from a distant region. The total intensity of the radio waves is, however, not only a function of the total number of electrons along a line of sight, it also depends on the magnetic field strength in the region where the relativistic electrons radiate. Proton synchrotron radiation may also be important (Re68b).

PROBLEM 6–15. Show that $I(v)$ is proportional to $B^{(\gamma+1)/2}$. For a randomly oriented field, B^2 takes on the mean value of the component of the (magnetic field)^2 perpendicular to the line of sight. Hence

$$I(v) \propto B^{(\gamma+1)/2}v^{-(\gamma-1)/2}$$

$$\propto B^{\alpha+1}v^{-\alpha} \qquad (6\text{–}159)$$

To conclude this discussion, we should still state that synchrotron emission, like any other emission process, also has a corresponding absorption process. Some strong extragalactic radio sources are believed to generate their radiation by means of synchrotron emission. Yet the spectrum of these sources appears black in just those regions where synchrotron radiation would have the highest emissivity. This is interpreted (see section 7:10) as meaning that the sources are opaque to their own radiation. Figure 9.13 shows that the flux for many nonthermal sources is high at low frequencies. On the other hand, at these low frequencies, the flux cannot exceed the flux of a blackbody

$$I(v)\, d\Omega = \frac{2kT}{c^2} v^2\, d\Omega \tag{4--81}$$

whose temperature T is determined by the electron energy, $kT \sim \mathscr{E}$. Now, equation 6--151 gives the relation between $\mathscr{E}$, the magnetic field in the source B, and the emitted frequency $v \sim \omega_{max}/2\pi$. Substituting for kT in (4--81), we then have

$$I(v)\, d\Omega = \left(\frac{8\pi v^5 m_0^3 c}{eB} \right)^{1/2} d\Omega$$

which expresses the magnetic field strength in the source in terms of the observed flux at frequency v, and the angular size of the source. The low frequency spectrum then is no longer a blackbody spectrum since the energy of electrons decreases at lower radiated frequencies. Effectively the temperature of the electrons, $\mathscr{E}/k$, is frequency dependent. The required data needed to compute the source magnetic field can be gathered with a radio interferometer (Ke71). The peaked spectrum of synchrotron self-absorption characterizes the radio source 3C147 (Fig. 9.13).

6:20 THE COMPTON EFFECT AND INVERSE COMPTON EFFECT

When a high energy photon impinges on a charged particle, it tends to transfer momentum to it, giving it an impulse with a component along the photon's initial direction of propagation. This is an effect that we had neglected in dealing with the low energy Thomson scattering process. Although we talk about *Compton scattering* when we discuss the interaction of highly energetic electromagnetic radiation with charged particles and *Thomson scattering* when lower energies are involved, we must understand that the basic process is exactly the same, and that we are only talking about differences in the mathematical approach convenient for analyzing the most important physical effects in different energy ranges.

Corresponding to the Compton effect, there is an exactly parallel situation, the *inverse Compton effect*, in which a highly energetic particle transfers momentum to a low energy photon and endows it with a large momentum and energy. These processes are exactly alike except that the coordinate frame from which they are viewed differs. To an observer at rest with respect to the high energy particle,

6:20

the inverse Compton effect will appear to be an ordinary Compton scattering process. To him it would appear that a highly energetic photon was being scattered by a stationary charged particle.

Because of this similarity we will only derive the expressions needed for the Compton effect, and then discuss the inverse effect in terms of a coordinate transformation. We will set down four equations governing the interaction of a photon with a particle. We note here that the effect is more conveniently described in terms of photons than in terms of electromagnetic waves, but again this is only a matter of convenience and does not reflect a physical difference in the radiation involved. The considerations we have to take into account are:

(i) Conservation of mass-energy, given by

$$m_0 c^2 + h\nu = \mathscr{E} + h\nu' \tag{6-160}$$

where ν and ν' are the radiation frequency before and after the collision, m_0 is the rest mass, and $\mathscr{E}$ the relativistic mass energy of the recoil particle (Fig. 6.14).

(ii) The relation of $\mathscr{E}$ to m_0 is (5–34):

$$\mathscr{E} = m_0 c^2 \left(1 - \frac{v^2}{c^2} \right)^{-1/2} \equiv m_0 \gamma(v)\, c^2 \tag{6-161}$$

(iii) Conservation of momentum along the direction of the incoming photon yields

$$\frac{h\nu}{c} = \frac{h\nu'}{c} \cos\theta + m_0\gamma(v)\, v \cos\phi \tag{6-162}$$

(iv) The corresponding expression for the transverse momentum is

$$0 = \frac{h\nu'}{c} \sin\theta - m_0\gamma(v)\, v \sin\phi \tag{6-163}$$

We now have four equations in four unknowns, v, v', θ, and ϕ.

PROBLEM 6–16.   Show that these four equations can be solved to give the expression

$$\frac{c}{h}\left(\frac{1}{\nu'} - \frac{1}{\nu} \right) = \frac{1 - \cos\theta}{m_0 c} \tag{6-164}$$

By taking the wavelength of radiation to be $\lambda = c/\nu$, $\lambda' = c/\nu'$, we obtain

$$\lambda' - \lambda = 2\lambda_c \sin^2\frac{\theta}{2} \tag{6-165}$$

Fig. 6.14 Compton scattering.

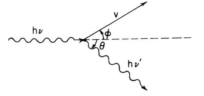

where

$$\lambda_c \equiv \frac{h}{m_0 c} \tag{6-166}$$

is called the *Compton wavelength* of the particle. For an electron $\lambda_c = 2.4 \times 10^{-2}$ Å or 2.4×10^{-10} cm. We note that for visible light, the change in wavelength amounts to only ~ 0.05 Å in 5000 Å, a nearly negligible effect. This is why momentum transfer could be neglected in Thomson scattering. However, in the X-ray region, say, at wavelengths of 0.5 Å, we encounter 10% effects; and at higher energies very large shifts can be expected, $(\lambda' - \lambda)/\lambda \gg 1$.

The cross section for Compton scattering must be computed quantum mechanically and turns out to be dependent on the energy of the incoming photon. The expression for this cross section (see Fig. 6.15), known as the *Klein-Nishina formula*, is

$$\sigma_c = 2\pi r_e^2 \left\{ \frac{1 + \alpha}{\alpha^2} \left[\frac{2(1 + \alpha)}{1 + 2\alpha} - \frac{1}{\alpha} \ln(1 + 2\alpha) \right] + \frac{1}{2\alpha} \ln(1 + 2\alpha) - \frac{1 + 3\alpha}{(1 + 2\alpha)^2} \right\} \tag{6-167}$$

where r_e is the *classical electron radius* and α is the ratio of photon to electron energy. For an electron

$$r_e \equiv \frac{e^2}{m_0 c^2} = 2.82 \times 10^{-13} \text{ cm}, \qquad \alpha = \frac{h\nu}{m_0 c^2} \tag{6-168}$$

In the extreme energy limit this is approximated by

$$\sigma_L = \sigma_e \left\{ 1 - 2\alpha + \frac{26}{5} \alpha^2 + \ldots \right\}, \qquad \alpha \ll 1, \qquad \text{low energies} \tag{6-169}$$

$$\sigma_H = \frac{3}{8} \sigma_e \frac{1}{\alpha} \left(\ln 2\alpha + \frac{1}{2} \right), \qquad \alpha \gg 1, \qquad \text{high energies} \tag{6-170}$$

6:20

where

$$\sigma_e = \frac{8\pi}{3} r_e^2 \qquad\qquad (6\text{--}171)$$

is the Thomson scattering cross section.

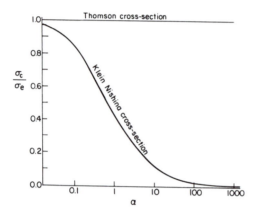

Fig. 6.15 Comparison of Compton and Thomson scattering cross sections as a function of $\alpha = h\nu/m_0 c^2$ (after Jánossy Já50).

For a proton the cross section would be smaller—inversely proportional to the mass. This means that Compton scattering is primarily an electron scattering phenomenon. Scattering by atoms takes place as though each atom had Z free electrons and the atomic scattering cross section is just Z times greater than that for an individual electron. The atomic binding energy is small compared to the photon energies encountered in Compton scattering and the electrons can be regarded as essentially free.

Let us still turn to the inverse Compton effect. Here we have a highly relativistic electron colliding with a low energy photon and transferring momentum to convert it into a high energy photon. This process can be followed from the point of view of an observer moving with the electron. He will see the incoming radiation blue-shifted (see equation 5–44) to a wavelength

$$\lambda_D = \lambda \sqrt{\frac{c - v}{c + v}} \qquad\qquad (6\text{--}172)$$

Still in this frame of reference, the scattered wave will have wavelength (see equation 6–165)

$$\lambda' = 2\lambda_c \sin^2 \frac{\theta}{2} + \lambda_D \qquad\qquad (6\text{--}173)$$

since this is a simple Compton process to the observer initially at rest with respect to the electrons. For backscattered radiation $\sin^2\theta/2 = 1$.

Now, when this wave is once again viewed from the stationary reference system—rather than from the viewpoint of the fast electron—a back-scattered photon will be found to have a wavelength

$$\lambda_s \sim \lambda' \sqrt{\frac{c-v}{c+v}} \sim \lambda\left(\frac{c-v}{c+v}\right) + 2\lambda_c\left(\frac{c-v}{c+v}\right)^{1/2} \qquad (6\text{--}174)$$

This is the same transformation as (6–172); a stationary observer also sees back-scattered radiation blue-shifted. Note that we have considered only direct back-scattering and that this expression does not consider scattering at other angles. Quite generally, however, it is quite clear that the wavelength of the photon becomes appreciably shortened in the process and its energy is increased by factors of order

$$\frac{c+v}{c-v} \sim \frac{(1+v/c)^2}{(1-v^2/c^2)} \sim \frac{\mathscr{E}^2}{m_0^2 c^4} \qquad (6\text{--}175)$$

where $\mathscr{E}$ is the initial energy of the particle. As will be discussed in section 9:10, the total power radiated by an electron in inverse Compton scattering is closely related to the power radiated in the form of synchrotron radiation. The total synchrotron emission is proportional to the magnetic field energy density in space, $B^2/8\pi$. The total inverse Compton scattering power loss for electrons is proportional to the electromagnetic radiation energy density in space. The proportionality constant for these two processes is identical.

6:21 SYNCHROTRON EMISSION AND THE INVERSE COMPTON EFFECT ON A COSMIC SCALE

We find many indications that relativistic processes are responsible for much of the observed radiation reaching us from distant quasars. First we find that those spectral lines that can be observed correspond to gases at temperatures of the order of $10^{6\circ}$K. These lines are very broad and indicate the existence of bulk velocities of the order of 10^3 km/sec. There often is rapid brightening and a subsequent drop of flux, indicating that there are violent events of short duration taking place on a regular basis. Second, we add the facts that a strong radio continuum flux is observed and that the objects have a blue appearance in the visible. All this indicates violent outbursts on a supernova—or possibly far more massive—scale, in which relativistic particles are formed.

Some of the energy goes into producing bulk motion and into ionizing gases.

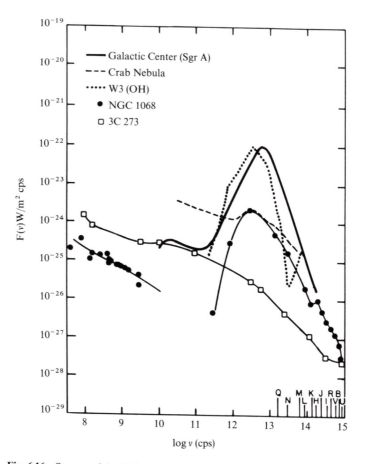

Fig. 6.16   Spectra of the Galactic center (SgrA); the Crab Nebula supernova remnant; the quasar 3C273; the Seyfert (active) galaxy NGC 1068; and the Galactic star-forming region W3(OH).

Other parts of the energy are to be found in relativistic particles that can give rise to radio and possibly infrared, visible, X-ray, and γ-radiation by means of synchrotron emission or the inverse Compton effect.

For the quasar component 3C273B, for example, we have a spectrum shown in Fig. 6.16. The bulk of the energy given off by the object lies in the far infrared at frequencies around 10^{13} Hz. There are a variety of theories proposing different origins for this flux. The most convincing ones are synchrotron radiation or the inverse Compton effect. The main difficulty with synchrotron emission lies in the fluctuations that are observed. A synchrotron radiating electron has a lifetime (see equation 6–155)

$$\tau \sim \frac{\mathscr{E}}{P(\mathscr{E})} = \frac{3m_0^4 c^7}{2\mathscr{E} e^4 B^2} \tag{6–176}$$

that for a 10^9 ev electron amounts to something of the order of 10^5 y for $B \sim 10^{-3}$ gauss, a field that would be consistent with the spectral curves of Fig. 6.16. But the observed visible fluctuations in radiation intensity are much shorter lived than a year. It is not clear whether the infrared emission also fluctuates; however, if it does, it would seem that the particles that emit radiation must be able to emit the bulk of their energy very rapidly, and the synchrotron emission or inverse Compton effect might do this well.

The inverse Compton effect has also been used to explain the presence of the Galactic gamma ray and X-ray flux and the isotropic X-ray flux that appears to be reaching us from outside the Galaxy. The idea here is that relativistic electrons can be produced readily enough by acceleration in magnetic fields, but there is no well-known way to produce highly energetic photons except by first producing energetic particles that then give up their energy in some exchange process. The inverse Compton effect is the most likely mechanism for this kind of energy conversion. The Galactic component may be due to relativistic electrons colliding with the visible flux emitted at the source of the electrons, say, in a supernova explosion. The extra-galactic flux might be due to the interaction of extra-galactic electrons with the 3°K cosmic blackbody radiation.

Interestingly a maximum brightness temperature of 10^{12}°K can be set on optically thick synchrotron emitting sources. At this brightness the inverse Compton scattering by the radiation emitted within the source quickly reduces the energy of the relativistic electrons and thus reduces the brightness temperature.

6:22 THE CHERENKOV EFFECT

We now come to a process that is primarily important for studying cosmic ray particles—the *Cherenkov effect*. This effect is not so important in the interaction of cosmic rays with other matter in the universe, as it is in the interaction of incoming particles with the earth's atmosphere. The Cherenkov effect causes these particles to decelerate radiatively and the light emitted can be used as a sensitive means for detecting the particles.

To see how the effect works, we consider a highly relativistic particle entering the earth's atmosphere. Because the particle is arriving from a region where the density has been very low and is entering a region of relatively high density, it finds that it has to make some adjustments. The presence of the electrically charged particle produces an impulse on atoms it approaches in the upper atmosphere and will cause the atoms to radiate. The impulse comes about because the particle is moving faster than the speed of propagation of radiation in this dense medium. The electric field due to the particle therefore appears to the atoms to be switched on very abruptly; a rapidly time-varying field arises at the position of the perturbed atom. This is just the condition required to cause the atom to radiate. The relativistic particle will continue to affect atoms along its path in this way, until it has slowed down to the local speed of propagation of light. At that time the electric field changes produced in the vicinity of atoms take on a less abrupt character and the radiative effects are diminished.

There are many parallels between Cherenkov radiation and hydrodynamic shocks. Just like a supersonic object that produces sonic booms and keeps on losing energy until it slows down to the local speed of sound, the cosmic ray particle also keeps losing energy, through the Cherenkov effect, until it slows down to the local speed of light in the medium through which it is traveling.

The radiation produced in this manner is emitted into a small forward angle of full width $\Delta\theta$ (see equation 5–50 and Fig. 6.17):

$$\Delta\theta \sim 2\sqrt{1 - \frac{v^2}{c^2}}$$

just as in synchrotron emission or in any other relativistic radiation effect. The time of arrival of the radiation is also dictated by considerations similar to those found in synchrotron radiation. If the layer through which the radiation passes, before it has become sufficiently slowed down, is of thickness d, the time elapsing between the arrival of the first and the last photons of a wave train at the observer is, in analogy to equation 6–147,

$$\Delta t_c \sim \left(\frac{d}{v} - \frac{d}{c}\right) \sim \frac{d}{v}\left(1 - \frac{v}{c}\right) \sim \frac{d}{2c}\left(1 - \frac{v^2}{c^2}\right) \qquad (6\text{–}177)$$

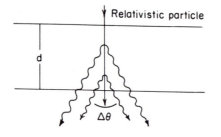

Relativistic particle

Fig. 6.17   Diagram to illustrate Cherenkov radiation.

so that the corresponding frequency is of the order of

$$\omega_c = \frac{1}{\Delta t_c} \sim \frac{2c}{d} \left(1 - \frac{v^2}{c^2}\right)^{-1} \sim \frac{2c}{d} \left(\frac{\mathscr{E}}{m_0 c^2}\right)^2 \tag{6-178}$$

If the distance traversed in the upper atmosphere is of the order of $d \sim 10^6$ cm = 10 km, and a proton of energy 3×10^{14} ev is considered, $\mathscr{E}/mc^2 \sim 3 \times 10^5$ and $\omega_c \sim 6 \times 10^{15}$ or $v_c \sim 10^{15}$ cps.

In many cases, an energetic primary produces a shower of secondary particles through collisions with atoms of the upper atmosphere. These secondaries also can give rise to Cherenkov radiation. The Cherenkov radiation spectrum does not normally peak at a frequency near ω_c. It depends more on atomic radiation properties of atmospheric gases and is relatively insensitive to the energy of the primary particle.

One interesting feature of Cherenkov detection is that it not only identifies the existence of cosmic ray particles, but also gives the direction of arrival with reasonable accuracy; the uncertainly $\Delta\theta$ in the direction from which the particles will appear to arrive is quite small. Gamma rays with energies $\gtrsim 10^{15}$ ev can be observed in this way, their direction of arrival lying along the direction of their celestial source. The detection process is indirect in this instance, and depends on the formation of very energetic secondary charged particles in the upper atmosphere. These, in turn, generate a visible light pulse through Cherenkov radiation.

ADDITIONAL PROBLEM 6-17. A rotating mass has energy $I\omega^2/2$, where I is the moment of inertia and ω is the angular frequency. Suppose that the rate of change of energy is proportional to the $n + 1$ power of ω:

$$\dot{\mathscr{E}} = K\omega^{n+1}$$

Show that

$$n = \frac{\ddot{\omega}\omega}{\dot{\omega}^2} \tag{6-179}$$

For the Crab nebula pulsars, we observe that $n \sim 2.5$. Does this more nearly match the result expected for magnetic dipole or for gravitational radiation?

ANSWERS TO SELECTED PROBLEMS

6-1. $R_L = \dfrac{p_c c}{qB} = \dfrac{m_p v_p c}{qB}$

$= 10^8$ cm,

$$1 \text{ AU} = 1.5 \times 10^{13} \text{ cm}$$

$$m_p = 1.6 \times 10^{-24} \text{ gm}$$

$$v_p = 10^6 \text{ cm/sec}$$

$$q = 4.8 \times 10^{-10} \text{ esu}$$

$$B = 10^{-6} \text{ gauss.}$$

6-2. For circular motion at the Larmor radius R_L:

$$\mathscr{E} = \frac{p_c v_c}{2} = \frac{q B v_c R_L}{2c} \tag{6-13}$$

$$\dot{\mathscr{E}} = -q\dot{B}\pi R_L^2 \frac{\omega}{2\pi} \tag{6-16}$$

Since the circular path takes the particle $\omega/2\pi$ full turns around the field.

$$\therefore \frac{d\mathscr{E}}{\mathscr{E}} = \frac{dB}{B}.$$

Hence the particles have a tenfold increase in energy.

6-3. $\Delta V = 2(7 \text{ km/sec}) \sim 14 \text{ km sec}^{-1}$ is the speed imparted to cosmic rays, measured in their rest frame. $\mathscr{E}_i = \gamma(v_i) m_0 c^2$, the initial energy ($\mathscr{E}_i = 10^{10}$ ev), where

$$\gamma(v_i) = \frac{1}{\sqrt{1 - v_i^2/c^2}}$$

and v_i is the initial velocity. v_f is the final velocity.

For protons say, $m_0 c^2 \sim 10^9$ ev, so that $\gamma(v_i) \sim 10$. If the energy is doubled $\gamma(v_e) \sim 20$, and

$$\left(\frac{v_i}{c}\right)^2 \cong (1 - 0.01) \Rightarrow \frac{v_i}{c} = 1 - 0.005$$

$$\left(\frac{v_f}{c}\right)^2 \cong (1 - 0.0025) \Rightarrow \frac{v_f}{c} = 1 - 0.00125$$

where v_f is the final velocity.

Hence $v_f - v_i = (.0037) c = 1.1 \times 10^3 \text{ km sec}^{-1}$.

By the law of composition of velocities, if v' is the velocity after one bounce:

$$v' \simeq (v_i + \Delta V)\left(1 - \frac{v_i \Delta V}{c^2}\right) \sim v_i - v_i^2 \frac{\Delta V}{c^2} + \Delta V - 0\left[\frac{(\Delta V)^2}{c^2}\right]$$

$$v' - v_i = \Delta V\left(1 - \frac{v_i^2}{c^2}\right) = \frac{\Delta V}{\gamma^2(v_i)}.$$

Now γ changes from 10 to 20. Using an average $\gamma^2 \sim 280$ always, $\Delta V/\gamma^2 \sim 14/280 \sim 0.05$ km sec^{-1}.

Hence the number of collisions is 2×10^4.

At a distance between bounces of 10^{17} cm, and a speed $\sim 3 \times 10^{10}$ cm sec^{-1}, the number of years to double the energy is $[10^{17}/(3 \times 10^7 \times 3 \times 10^{10})] 2 \times 10^4 \sim 2 \times 10^3$ y for protons. During this time the clouds approach and just about touch. Electrons have a higher γ value and are not energized nearly as much during this time.

6-4. $\mathscr{E} = m_0 c^2 \gamma(\omega r), \qquad \gamma(\omega r) = \dfrac{1}{\sqrt{1 - \omega^2 r^2/c^2}}.$

At $\omega r = c$, particles and field can no longer co-rotate.

6-7. Problem 4-3, gives the rms deviation of N steps of length L as $\sqrt{N} L$. In each step, the Faraday rotation angle is given by $\theta(L) = \frac{1}{2}(\omega/c) L \Delta n$. Substitution of (6-66) and the gyrofrequency from (6-62) gives the result.

6-9. If we imagine fictitious charges, by analogy with electric charges

$$\mathbf{H} = \frac{Q}{(r - a/2)^2} - \frac{Q}{(r + a/2)^2} = \frac{2Qa}{r^3} = \frac{2\mathbf{M}}{r^3}.$$

Hence **d** and **M** are analogous and by substitution we are led to the result (6-89).

6-10. $\mathbf{H} = \dfrac{\dot{\mathbf{A}} \wedge \mathbf{n}}{c} \qquad$ and $\qquad \mathbf{A} = \dfrac{1}{6c^2 R_0} \ddot{\mathbf{D}}$

$$\mathbf{H} = \frac{1}{6c^2 R_0} \frac{\dddot{\mathbf{D}} \times \mathbf{n}}{c}$$

$$\mathbf{S} = \frac{c}{4\pi} H^2 \mathbf{n} \propto \frac{(\dddot{\mathbf{D}})^2}{c^5}.$$

6-13. To show this, use the total radiated power (6-86):

$$\frac{2}{3c^3} \ddot{d}^2 = \frac{2e^2}{3c^3} \left(\frac{\dot{p}}{m_0}\right)^2. \qquad \text{(6-13) and (5-40) then give}$$

$$\dot{p} = \frac{eB}{m_0 c} p_c = \left(\frac{eB}{m_0 c}\right) \left(\frac{\mathscr{E}}{c}\right). \qquad \text{Substitute, to get the result (6-155).}$$

6-14. $B = 10^{-4}$ gauss

$$v_c = \frac{\omega_c}{2\pi} = \frac{eB}{2\pi m_0 c} = 300 \text{ Hz}$$

$$\omega_m = \gamma^2 \omega_c = 1200 \text{ MHz for } \gamma = 2 \times 10^3$$

Peak at $\omega = \omega_m/2 = 600$ MHz.

6-15. $I(v) \Delta v \propto \displaystyle\int_0^{\mathscr{E}} P(\mathscr{E}) n(\mathscr{E}) d\mathscr{E}$ (by 6–156) where

$P(\mathscr{E}) \propto \mathscr{E}^2 B^2$ by (6–155).

$\therefore I(v) \Delta v \propto B^2 \displaystyle\int_0^{\mathscr{E}} \mathscr{E}^2 \mathscr{E}^{-\gamma} d\mathscr{E} \propto K B^2 \mathscr{E}^{3-\gamma}.$

But $\mathscr{E} \propto \left[\dfrac{\omega_m}{B} \right]^{1/2}$ by (6–151)

$\therefore I(v) \Delta v \propto B^2 B^{(\gamma-3)/2} \omega_m^{(3-\gamma)/2}$

$I(v) \propto B^{(\gamma+1)/2} v^{(1-\gamma)/2}.$

6-16. Squaring (6–162) and (6–163) and adding gives

$-2h^2 vv' \cos\theta + h^2(v^2 + v'^2) = m_0^2 v^2 \gamma^2 c^2 = m_0^2 c^4 (\gamma^2 - 1).$

Squaring (6–160) gives $m_0^2 c^4 (\gamma^2 - 1) = h^2 (v^2 + v'^2 - 2vv') + 2h m_0 c^2 (v - v').$

Equating these two expressions we obtain $hvv'(1 - \cos\theta) = (v - v') m_0 c^2$ which is equivalent to (6–164).

6-17. $\dot{\mathscr{E}} = K\omega^{n+1} = I\omega\dot{\omega}$

$\therefore \dot{\omega} = \dfrac{K}{I} \omega^n, \qquad \ddot{\omega} = \dfrac{K}{I} n\omega^{n-1}\dot{\omega} \qquad \text{and} \qquad \ddot{\omega}\omega = n\dot{\omega}^2.$

Equations 6–88 and 6–96 show that $n + 1 = 4$ for the magnetic dipole, and $n + 1 = 6$ for the gravitational quadrupole radiation. The current data therefore are in closer agreement with a magnetic dipole mechanism. Since pulsars may emit predominantly gravitational radiation when they are first formed, observations leading to a value $n = 5$ could be obtained right after the formation of such an object. This would be interesting because the observations could be made in the radio domain, but would give evidence of gravitational radiation that might be too difficult to detect directly.

7

Quantum Processes in Astrophysics

7:1 ABSORPTION AND EMISSION OF RADIATION BY ATOMIC SYSTEMS

In Chapter 6, we considered a series of processes by means of which radiation could be absorbed or emitted by particles. But we restricted ourselves to situations in which the Maxwell field equations of classical electrodynamics could be applied. These equations break down on the scale of atomic systems. The electron bound to a positively charged nucleus does not lose energy because of its accelerated motion, although the classical theory of radiation predicts that it should. Instead, the ground state of hydrogen, or any atom, is stable for an indefinitely long period of time. Moreover, when energy is actually radiated away from one of the excited states, and the atomic ground energy state is finally reached, we always find that only discrete amounts of energy have been given off in each transition. Again, this is at variance with classical predictions.

Since the interpretation of astronomical observations depends on an understanding of transitions that occur between atomic levels, we will consider just how they take place, and what can be learned from them.

It is important in this connection that almost everything we know about stars or galaxies is learned through spectroscopic observations. Our ideas of the chemistry of the sun and of the chemical composition of other stars is based entirely on the interpretation of line strengths of different atoms, ions, or molecules. Our understanding of the temperature distribution in the solar corona is based on the strength of transitions observed for several highly ionized atoms, notably iron. Our picture of the distribution of magnetic fields across the surface of the sun is based entirely on the interpretation of the splitting of atomic lines by magnetic fields in the solar surface. What we know about the motion of gases and their temperature at different heights above the solar surface, again, is largely based on spectroscopic information. In this case, small shifts in line positions, and the shape and width of the lines yields much of the information we need. Some idea of the densities of atoms, ions, or electrons at different levels of the solar atmosphere can also be obtained from line width and shape.

Of course, for the sun, some of this information can be obtained by other means, because it is near, and can be clearly resolved. We therefore can determine velocities of gas clouds at the limb by direct observations and, eventually, we may expect to obtain radial velocity measurements from radar observations. Densities presumably can also be probed by radar and much data can be obtained by measuring the dispersion of cosmic radio waves passing near the sun. Thus direct visual observations and radio measurements based on classical theory do go a long way toward clarifying our understanding of motions and densities in the solar atmosphere. However, when it comes to more distant objects like emission line stars, or quasars, where detailed resoluion of the object does not appear possible, much new information can only be gained by an understanding of atomic processes observed using spectroscopic techniques. Of course, the dominance of quantum processes is not complete. For example, the knowledge we have about relativistic particles emitting synchroton radiation in quasars is based on classical theory only. Much of the thermal radio emission from interstellar clouds of plasma can also be understood classically (see section 6:16). However, almost everything else we know about these objects has some connection with the quantum theory of radiation.

In the next few sections we will describe how knowledge of quantum processes permits us to learn a great deal about the physical characteristics of astronomical objects. We will try to understand these processes in terms only of the elementary conditions that lie at the base of quantum theory. In general, this will only yield rough values of the parameters we need to know, but we will nevertheless be able to obtain a valid understanding of the role played by quantum processes in astronomy.

7:2 QUANTIZATION OF ATOMIC SYSTEMS

The classical theories of physics no longer apply on a scale comparable to the size of atoms, and many of our preconceptions have to be changed. However, a number of important features are shared by quantum and classical theory. Thus, in a closed system we find that:

(a) Mass-energy is always conserved.
(b) Momentum and angular momentum are always conserved.
(c) Electric charge is always conserved.

7:2

On the atomic scale, these conservation principles take on a form which deviates somewhat from classical formulations. However, when these differences are important, we can still be sure that:

(d) As the size of the atomic system grows, the features predicted by quantum theory approach those calculated on the basis of classical physics. This is called the *correspondence principle*.

In contrast to these similarities between classical and quantum behavior, there are three major differences:

(a′) *Action*, a quantity that has units of (energy × time) or (momentum × distance), is quantized. The unit of action is $\hbar$. By this we mean that in a bound atomic system action can only change by integral amounts of Planck's constant h divided by 2π; $h/2\pi \equiv \hbar$. This statement has many consequences, some of which will be described in this chapter.

(b′) Even if they existed—and they do not—states of an atomic system whose characteristic action differs by an amount less than $\hbar$, cannot be distinguished. This is *Heisenberg's uncertainty principle*.

(c′) Two particles having half-integral spin cannot have identical properties in the sense of having identical momentum, position, and spin direction. This is *Pauli's exclusion principle* (see section 4:11).

The three statements (a′), (b′), and (c′) are not axioms of quantum mechanics. Rather, they can be considered as useful rules that emerge from a more complicated theory of quantum mechanics that also makes quantitative predictions about the behavior of electrons, atoms, and nuclei.

The concept of action is not as familiar as the idea of *angular momentum*, which has the same units and is subject to quantization in the same way. We might therefore take a brief look at how angular momentum changes occur in atoms.

In any bound atomic system, a change of angular momentum along any given direction in which we choose to make a measurement, will always have a value $\hbar$. The direction of this angular momentum is important.* We shall therefore talk

---

* For, while the angular momentum along the measured direction can only change in steps whose size is $\hbar$, we have no such definite prescription for the changes which can simultaneously take place in the transverse angular momentum. The uncertainty principle precludes a simultaneous definitive measurement of the longitudinal and transverse angular momentum components. All that we can say is that there exist a number of *selection rules* that specify allowed changes of the *total angular momentum* of the system as will be shown in section 7:7. The rules state that the magnitude of the angular momentum squared J^2 changes by integral amounts of a basic step size $\hbar^2$. These integral amounts depend on the initial value of J characterizing the system, and on the multipole considerations mentioned in section 6:13.

7:2

about a *measured angular momentum component* with the understanding that we have a definite direction in mind whenever we make a measurement.

This angular momentum quantization can be understood in more basic terms. All the fundamental particles involved in building up atoms have definite measured *spin* values. For electrons, protons, and neutrons these values are $\pm \hbar/2$. A change from one spin orientation of an atomic electron, over to another orientation, therefore, amounts to a change of one unit of $\hbar$ in the measured angular momentum component. Such a change is readily brought about by the absorption or emission of a photon, because photons have spin angular momentum components $\pm \hbar$ along the direction of their motion. Because of the quantization of photon spin components, any change at all in angular momentum of an atomic system must have a component $\hbar$. For, all the different states of an atom can be reached from any other state through a succession of photon absorption or emission processes, or through a set of spin-flip transitions for the electrons or within the nucleus.

In any case, however, quantization is intrinsic to atoms, even without this argument concerning photons. We can therefore be sure that if an atomic system has a state of zero angular momentum, then all other states must have integral values of the angular momentum component. Similarly, if the lowest angular momentum state has a value $\hbar/2$, then all other states must have half integral values for their angular momentum components (see also section 7:7).

This is one way in which the statement (a'), above, provides insight into the structure of quantized systems. In addition, the principle (b') gives us some general quantitative information. Let us consider the simplest atom, hydrogen, in terms of this principle. The energy of the lowest state can then be estimated directly. For, the smallest possible size of an electrostatically bound atom must be related to the uncertainty in momentum through

$$p^2 r^2 \sim \langle \Delta p^2 \rangle \langle \Delta r^2 \rangle \sim \hbar^2 \tag{7-1}$$

Here we have taken the mean squared value of the radial momentum and the radial position as being equal to the uncertainty in these parameters. Through the virial theorem, applied to a system bound by inverse square law forces, we can write the energy of a state either as half the electrostatic potential energy for the interacting proton and electron, or as the negative of the system's kinetic energy (3–83). Thus the lowest energy state is

$$\mathscr{E}_1 = -\frac{Z e^2}{2 r} = -\frac{p^2}{2\mu} \sim -\frac{\hbar^2}{2\mu r^2} \tag{7-2}$$

where μ is the reduced mass of the electron. Here we have made use of equation 7–1 at the extreme right side of (7–2). By eliminating r from this equation, we

7:2

can immediately write

$$\mathscr{E}_1 = -\frac{Z^2 \mu e^4}{2\hbar^2} \qquad (7\text{-}3)$$

$$r = \frac{\hbar^2}{Z\mu e^2} \qquad (7\text{-}4)$$

This root-mean-square radius r is called the *Bohr radius* of the atom and $\mathscr{E}_1$ is the atom's *ground state energy*. For hydrogen $Z = 1$ and $\mathscr{E}_1 = -13.6$ ev, $r \sim 5.29 \times 10^{-9}$ cm. We have proceeded here on the assumption that the electrostatic potential confines the electron to a limited volume around the proton, and have derived a solution consistent with the uncertainty principle. Nothing has been assumed about possible orbits that the electron might describe about the proton; and, in fact, the very act of setting the mean squared value of position and momentum equal to the mean squared value of their uncertainties, implied that the electron is to be found in the whole volume, not just in a well-defined orbit having a narrow range of r or p values.

PROBLEM 7-1. If we wanted to distinguish successive states having differing radial positions and momenta, the product pr for these states would have to differ by $\hbar$; otherwise, they would not be distinguishable in Heisenberg's sense. Setting $p_n r_n = n\hbar$, show that

$$\mathscr{E}_n = -Z^2 \frac{\mu e^4}{2n^2} \hbar^{-2} \qquad (7\text{-}5)$$

If the nth radial state of the atom has a phase space volume proportional to $4\pi p_n^2 \, \Delta p_n$ and $4\pi r_n^2 \, \Delta r_n$, show that the number of possible states with principal quantum number n is proportional to n^2. We will find a great deal of use for this result.

In order to find the actual number of states corresponding to the quantum number n, we still have to invoke the Pauli exclusion principle, statement (c') above. We know that the state $n = 1$ corresponds to only one cell in phase space. Accordingly there can only be two states, one in which the nuclear spin and electron spin are parallel and the other in which they are antiparallel. Using the result of Problem 7-1, we then see that the nth radial state comprises $2n^2$ different substates, all having the same energy to the approximation considered here.

We see from this that just the most basic concepts (a'), (b'), and (c') suffice to tell a great deal about the structure of hydrogen and hydrogenlike atoms such as singly ionized helium, five times ionized carbon, or any other bare nucleus surrounded by only one electron.

We should, however, not pretend that all problems of atomic structure can be handled as simply. Equation 7–2, for example, makes use of Newtonian mechanics and electrostatic interactions alone. We have neglected all relativistic effects and all interactions of the spins of particles that constitute the atom. In dealing with such features, or with the interactions between particles and various types of fields, it is important to make use of the full mathematical structure provided by quantum mechanics. At the basis of any such structure, however, are the elementary principles (a) to (d) and (a') to (c'), and we shall make much use of them in the next few sections.

PROBLEM 7–2. We can show that the principles (a') to (c') also permit a determination of the size of the atomic nucleus. To see this consider the *nucleons*, that is, protons and neutrons to be bound to each other by a short-range attractive potential

$$\mathbb{V} = -\mathbb{V}_0 \quad \text{if} \quad r < r_0, \qquad \mathbb{V} = 0 \quad \text{if} \quad r \geq r_0 \qquad (7\text{–}6)$$

Using equation 7–1 show that

$$r \sim \frac{\hbar}{[2M(\mathbb{V}_0 - \mathscr{E}_b)]^{1/2}} \qquad (7\text{–}7)$$

Here M is the nucleon mass and $-\mathscr{E}_b$ is the binding energy per nucleon, whose value is roughly 6 Mev. If $\mathbb{V}_0 \sim 2\mathscr{E}_b$ show that a typical nuclear radius is of order 10^{-13} cm. This gives a characteristic interaction cross section of $\sim 10^{-26}$ cm^2 for nucleons. We will find this to be of interest in Chapter 8 where nuclear processes inside stars are described.

We note that this nuclear radius is quite insensitive to the dependence of the potential on distance. The depth of the potential well and the binding energy determine the size of the nucleus.

7:3 ATOMIC HYDROGEN AND HYDROGENLIKE SPECTRA

The considerations of the previous section permit us to discuss some of the main features of atomic hydrogen spectra observed in astronomy. The energy of the observed spectral lines simply represents the difference in energy of the atomic levels between which a transition occurs when a photon is absorbed or emitted.

To start with one of the simplest concepts, we notice that in (7–3) and (7–5) the energy depends on the reduced hydrogenic mass. It therefore has a somewhat different value for normal hydrogen, which has only a proton in its nucleus, and for deuterium which has a nucleus composed of a neutron and a proton. The extra neutron in deuterium makes the nucleus about twice as massive as in normal hydrogen so that the deuterium reduced mass μ_D has a value

$$\mu_D = \frac{m_e m_D}{m_e + m_D} \sim \frac{2 m_e m_P}{m_e + 2 m_P} \sim m_e \left(1 - \frac{m_e}{2 m_P} \right) \qquad (7\text{--}8)$$

while

$$\mu_P \sim m_e \left(1 - \frac{m_e}{m_P} \right) \qquad (7\text{--}9)$$

Subscripts e, D, and P, here represent electrons, deuterons, and protons. Corresponding to normal hydrogenic spectral lines, we would therefore expect to see deuterium lines whose energy was greater by about one part in $2 m_P / m_e \sim 3700$. In the visible part of the spectrum this corresponds to a line shift of the order of 1.5 Å toward shorter wavelengths. Although such a spectral shift could be easily determined, it is interesting that for many years, despite much searching, no deuterium was ever detected in any astronomical object, anywhere. In contrast, the terrestrial abundance of deuterium is readily measured and is roughly 2×10^{-4} by fraction of atoms. If such abundances existed elsewhere in the universe, deuterium should have been detected readily. However, in the interstellar medium, deuterium appears to be fractionated; it binds more readily to molecules than does ordinary hydrogen, and, therefore, appears underabundant, atomic D/H ratios being $\sim 2 \times 10^{-5}$ (Mu87). Most of the extant deuterium may be primordial, and as we will see in section 8:12, any deuterium that has ever been cycled into a star has been destroyed. If all interstellar matter in our galaxy had undergone such cycling at least once, planets formed early in the Galaxy's life and clouds containing gases that have never undergone nuclear processing in a star might be the only places where deuterium is now to be found.

The reduced mass just mentioned also helps to distinguish hydrogenic transitions from spectral lines of ionized helium. Singly ionized helium He II has one electron surrounding a nucleus with charge $Z = 2$. According to equation 7–5 the energy of any given state should therefore be just four times as great as the corresponding hydrogenic energy. This integral relation would sometimes lead to an exact identity of line energies for transitions involving *principal quantum numbers* n that were twice as great for helium as for hydrogen. The difference in reduced mass, however, shifts these lines sufficiently, so that ambiguities often can be avoided in astronomical observations. When the Doppler line shift for a moving source is not known, identification on the basis of one or two lines may, however, not be possible, and a search may have to be made for lines of other well-known atoms or for helium lines that are not common to the hydrogen spectrum, that is, transitions involving one level with an odd principal quantum number, and one level with an even value of n.

Although we have presented this similarity of spectra as though it were a matter of difficulty, it is in fact often a great help. After years of theoretical work that explains many fine details, we understand the hydrogen spectrum well. Whenever it becomes possible to relate properties of complicated atoms to specific similar properties of hydrogen, a whole body of theoretical knowledge therefore becomes available at once, and this often leads to a better understanding of the more complex system.

Until recent decades interest in hydrogenic spectra centered largely around transitions in which at least one of the states had a low principal quantum number, say, $n \lesssim 5$. Not much thought was given to very high lying states, and transitions to be found between such states were always thought to give rise only to very weak spectral lines. It therefore came as a surprise that transitions involving states with $n = 90, 104, 159, 166$, and many others in this same range were observable in radio astronomy (Hö65). Not only these, but also the correspondingly excited helium states could be identified and again distinguished on the basis of reduced mass differences. These lines have permitted the observation of ionized hydrogen regions over great distances in the Galaxy. Ionized regions that are not detectable in the photographic range now are readily accessible because radio waves are transmitted through the dust clouds that extinguish visible light from all but the nearest portions of the Galaxy. Often these regions had been previously known, because dense ionized plasmas emit readily measured thermal continuum radiation (section 6:16). However, the discovery of the line radiation allowed us to deduce the radial velocity of the region and its distance within the galaxy calculated on the basis of differential rotation models (3:12).

For completeness we should still present some of the terminology often used in discussions of the hydrogen spectrum. It is useful to know that transitions involving lower states $n = 1, 2, 3$, and 4, respectively, are members of the *Lyman, Balmer, Paschen,* and *Brackett* spectral series (See Fig. 7.1). The line with the longest wavelength in each of these spectral series is termed α; the second line is called β, and so on. Thus the transition $n = 4 \rightarrow n = 2$ gives rise to the Balmer-β line in emission. Members of the Balmer spectrum sometimes are written out as Hα, Hβ, and so on. Lyman spectral lines are written as Lα, Lβ, ... or Ly-α, Ly-β ....

In Problem 7–1, we had shown that there are $2n^2$ quantum states to be associated with the nth energy level of the hydrogen atom. We must still see how these states are distinguishable from each other.

7:3

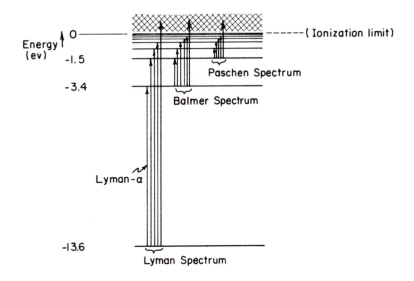

Fig. 7.1   Energy level diagram of atomic hydrogen.

The lowest or ground state actually consists of two distinct components, corresponding to the two differing orientations of the electron's spin relative to the nuclear spin direction. These two configurations have slightly differing energies and hence a transition from the higher to the lower state can occur spontaneously. In radio astronomy this transition has played a leading role. It occurs at 1420 MHz, a frequency corresponding to an energy difference of $\sim 6 \times 10^{-6}$ ev or less than one part in two million of the binding energy of the atom in its ground state ~ 13.6 ev. The distribution of hydrogen in our galaxy was first mapped by means of 1420 MHz observations. This was possible because, as already stated, radio waves are not absorbed by the dust that extinguishes visible light. It is interesting that we now have rather good maps showing the distribution of gas in the Galaxy, but no comparable map showing the distribution of stars. This comes about only because stars do not emit sufficiently great amounts of radiation in the radio part of the spectrum or in the far infrared.

The energy separation between the lowest level, in which the spins are opposed (total spin angular momentum quantum number $F = 0$) and the state with parallel spin orientation ($F = 1$), is called the *hyperfine splitting* of the ground state (Fig. 7.2). This splitting is present at all levels n, and assures a total multiplicity of $2n^2$ states at any given level.

When we take a look at the first excited state of the hydrogen atom $n = 2$, we encounter two types of sublevels whose energies happen to be close. First,

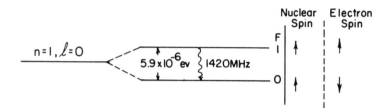

Fig. 7.2   Energy level diagram for the *hyperfine splitting* of the ground state of the hydrogen atom. The state in which the electron and proton spins are aligned has a slightly higher energy. The transition frequency 1420 MHz corresponds to a radio wavelength of 21 cm.

just as in the ground state, there are again two hyperfine states in which the electron has zero orbital angular momentum about the nucleus. A transition from this state to the ground state through the emission of a photon is forbidden, because the angular momentum of the atomic system would have to remain unchanged in such a transition; but that is not possible because the photon involved in that transition always carries off angular momentum. In tenuous ionized regions of interstellar space, the lifetime of atoms in such a state of $n = 2$ can therefore be very long. Eventually, the atom can revert to the ground state through the emission of two photons, rather than just one; but the lifetime against this two-photon decay is of the order of 0.12 sec (Nu84), in contrast to the usual 10^{-8} sec required for normal allowed transitions. Metastable helium atoms similarly have a two-photon decay time measured as $\sim 2 \times 10^{-2}$ sec (Va 70).

We might still ask whether the angular momentum criterion in such transitions could be satisfied if the electronic transition was accompanied by a spin flip from a parallel to an anti-parallel configuration. The coupling between the electron spin and the electromagnetic radiation, however, is low and does not suffice to make that transition as probable as the two-photon decay.

The second set of levels, within the state $n = 2$, all have an *orbital angular momentum quantum number* $l = 1$, with a corresponding (total angular momentum)^2 of $l(l + 1) \hbar^2 = 2\hbar^2$. The (angular momentum)^2 does not have the value $l^2 \hbar^2$, because, aside from a well-defined angular momentum component about one (arbitrarily) chosen axis, there always remains an uncertain angular momentum about two orthogonal axes, which adds an amount $l\hbar^2$ to the (angular momentum)^2 (see section 7:7). Corresponding to $l = 1$, there are three sublevels, each split into two further, hyperfine states. One of these sublevels has an angular momentum component $\hbar$ along some given direction, the second has a component 0, and the third has a component $-\hbar$ along that direction. These three components are labeled $m = 1, 0,$ and -1. The label m is called the *magnetic quantum number* because the states have differing energies when a magnetic field is applied to the atom. In the absence of a magnetic field, there is a splitting of the order of 10^{-5} ev

Fig. 7.3 Energy level diagram showing the fine structure of the $n = 2$ level of hydrogen. The labeling in the left-hand column has the following significance. The letters S and P denote the total orbital angular momenta 0 and 1, respectively. The right lower index gives the total angular momentum resulting from a vectorial addition of the electron and orbital angular momenta. The left upper index is the *multiplicity* $(2S + 1)$ of the term, where S now is the total electron spin. This two-fold meaning of S sometimes leads to confusion. As an example, the $^2P_{3/2}$ state has $l = 1$; the orbital and electron spins are parallel, giving a total spin 3/2; and, since the spin for a single electron has magnitude 1/2, the left superscript is 2.

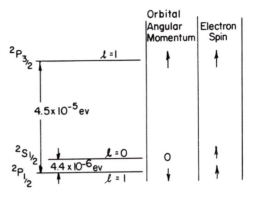

between some of these states. This is called the *fine structure* of the atom and is shown in Fig. 7.3.*

The excited state $n = 3$ again has a hyperfine split sublevel of zero angular momentum, $l = 0$. There are three such pairs of states with $l = 1$, and five pairs with $l = 2$, corresponding to magnetic quantum numbers $m = 2, 1, 0, -1$, and -2. Under normal conditions, these levels are *degenerate*, meaning that they have precisely the same energy. In an applied magnetic field, however, the energy of the states is shifted somewhat and the energy separation between states becomes proportional to the field strength H for low values of H. This splitting is called *Zeeman splitting*.

Zeeman splitting can be understood in the following way. The orbital angular momentum of the electron implies a loop current that has an associated magnetic dipole field. Depending on whether this dipole respectively is aligned along, perpendicular to, or opposed to the field, we have an atomic state with decreased, unaltered, or increased energy.

Quantitatively, the orbital angular momentum of the electron about the nucleus gives rise to a magnetic dipole moment with components along the field direction of

$$\mu_B m_i = \frac{e\hbar}{2mc} m_i, \qquad i = 0, \pm 1, \pm 2, \dots \qquad (7–10)$$

where μ_B is called the *Bohr magneton*. The energy of a state in a magnetic field,

---

* For atoms other than hydrogen, the labeling of states does not proceed in this particular way because the spins and orbital angular momenta of the electrons interact through their magnetic moments. However, the enumeration of the different quantum states still proceeds in terms of their distinguishing characteristics—that is, in terms of the Heisenberg and Pauli principles.

is then

$$\mathcal{E} = \mathcal{E}_0 + \mu_B H m_i = \mathcal{E}_0 + \hbar \omega_L m_i, \qquad \omega_L = \frac{eH}{2mc} \qquad (7\text{--}11)$$

The state with the smallest energy has its angular momentum antiparallel to the field direction, that is, the configuration in which the quantum number m_i has its lowest value. ω_L is the *Larmor frequency*. ω_L should be compared to the gyrofrequency (6–13) which is twice as large: $\omega_c = 2\omega_L$.

We note that the classical energy of a magnetic dipole in a magnetic field would be $\mathbf{M} \cdot \mathbf{H}$. But when this expression is introduced into equation 6–18, and we seek to find the energy of a magnetic dipole aligned with the field, we obtain

$$\mathcal{E} = \mathbf{M} \cdot \mathbf{H} = \frac{e(\mathbf{v} \wedge \mathbf{r}) \cdot \mathbf{H}}{2c} = \frac{eLH}{2mc} = \frac{\omega_c L}{2} \qquad (6\text{--}18a)$$

Here we have made use of equation 6–13 to see the classical energy dependence on ω_c. We can now see why ω_L is only half the gyrofrequency. On the other hand, if we make use of the Larmor frequency in (7–11) we preserve an analogy to photons in that, for $m_i = 1$, the magnetic energy becomes $\mathcal{E} - \mathcal{E}_0 = h\nu_L = \hbar\omega_L$.

Figure 7.4 shows the splitting in the energy levels corresponding to quantum numbers $l = 2$ and $l = 1$ and gives the spectral lines that arise from a transition between such states.

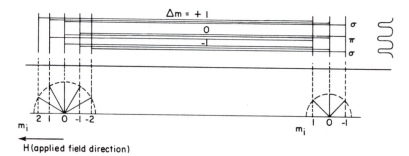

Fig. 7.4 Transitions between energy states shifted through the application of an external magnetic field. The figure shows both the orientation of the angular momentum components relative to the direction of the applied field, and the shifts in the transition energy. We note that the angular momentum components, along the field direction, always have values that are integral multiples of $\hbar$. The total length of the angular momentum vector is $[l(l + 1)]^{1/2}$, that is, nonintegral. σ and π denote the states of polarization of the emitted radiation (see text). A large horizontal distance between levels indicates a large transition energy between two states.

The Zeeman splitting provides us with useful information about the magnetic fields on the solar surface and in distant stars. In some strongly magnetic stars of spectral type A, fields higher than 30,000 gauss have been discovered. The general dipole field for the sun has a value of the order of a gauss, but the local field strength varies greatly. In sunspots, fields of 3×10^3 gauss are not unusual.

The determination of magnetic field strengths from the spectra alone would normally be very difficult, because the lines are very broad, and often overlap since the splitting is small. Fortunately, however, the lines marked σ, in Fig. 7.4 have a different polarization from the line marked π. The *magnetographic method* makes use of this polarization to separate out the different components through use of analyzers sensitive only to light of a given polarization. By carefully measuring the line centers of the variously polarized components, we can then obtain the energy splitting between states, even when the lines are strongly broadened through disturbing effects. For solar work, spectral lines of iron or chromium are often used. The energy splitting gives H directly through equation 7–11.

Interstellar magnetic fields have been measured in a similar way by means of radio observations. The principle of this technique is identical to that used in solar work.

Since a considerable number of theories depend on the existence of an interstellar magnetic field, the only direct way of measuring it is presented here in some detail.

In neutral hydrogen regions of interstellar space, the 21 cm line of atomic hydrogen can be used to determine the presence of magnetic fields. In such a field, the energy levels are split and three different transition lines are expected, corresponding to $\Delta m = 0, \pm 1$. Viewed along the direction of the magnetic field, only two lines appear, respectively, at frequencies $v = \omega/2\pi$ (see equation 7–11)

$$v = v_0 \pm \frac{eH}{4\pi mc}, \qquad v_0 = 1420 \text{ MHz} \tag{7–12}$$

These two components are circularly polarized in opposite senses (Fig. 7.5). They are called the σ-components (Fig. 7.4) and appear linearly polarized when viewed normal to the field direction; the direction of polarization is at right angles to the direction of the field.

There is also an unshifted component, the π-component, which appears at frequency v_0 when viewed normal to the direction of the field; it is linearly polarized with the direction of polarization parallel to the field. Viewed along the field lines, this unshifted component does not appear at all.

The observation of Zeeman splitting is made difficult by the rapid motion of the interstellar gas atoms. This produces a Doppler broadening of the lines (section 7:6). A 1 km sec^{-1} random motion gives a frequency shift of order

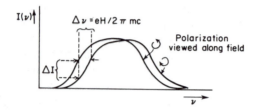

Fig. 7.5 Zeeman shift of the positions of the two circularly polarized components of the 1420 MHz hydrogen line viewed along the direction of the magnetic field.

$\delta v/v \sim 3 \times 10^{-6}$. At the 21-cm line frequency of 1420 MHz, this corresponds roughly to 4×10^3 Hz. In contrast, the frequency split Δv due to the magnetic field is $2.8 \times 10^6 H$ Hz between the two σ-components. This means that a field of order 10^{-5} gauss, only gives a split of $\Delta v \sim 30$ Hz.

Normally such a splitting would be all but impossible to observe in the presence of the overriding Doppler broadening. As already indicated, however, the saving feature in the situation is the difference in polarization. By working at the edge of the line, where the slope is steep, the difference in intensity ΔI of the two polarized components can be accentuated (Fig. 7.5). This technique has established the existence of fields as high as $\sim 2 \times 10^{-5}$ gauss, at least in some dense clouds.

7:4 SPECTRA OF IONIZED HYDROGEN

(a) Positive Ions

Hydrogen can become ionized through the absorption of a photon whose energy is 13.6 ev or higher. Once the minimum energy required to loosen the electron from the proton is reached, the excess energy can always be absorbed in the form of translational kinetic energy of the electron and proton. This feature is important in determining the appearance of very hot stars. We never observe ultraviolet photons just beyond the ionization limit—the *Lyman limit*. Photons emitted at these energies are immediately absorbed in the gas surrounding the star; and if there is not enough gas there, then the absorption will certainly take place in interstellar space, between the star and the earth.

We might think that the recombination of electrons with protons would then always regenerate the ultraviolet photons. However, this occurs only part of the time. Frequently the recombination leaves the atom in one of its excited states, with a subsequent cascade through lower excited states down to the ground state. In this process, a number of less energetic photons are created. If this type of energy degradation does not occur at the first recombination, it normally will take place on a later occasion. There will be ample opportunity, since the mean free path for ionizing radiation is very short—the probability of ionizing an atom is extremely large. At energies somewhat higher than the ionization limit, the absorption cross section is of order 10^{-17} cm². Hence, for typical interstellar densities of order 1 atom cm^{-3}, the mean free path for absorption is only of order 10^{17} cm, in contrast to standard interstellar distances of order 3×10^{18} cm or more. In a random walk, an ionizing photon would then have 10^{3} opportunities for ionization and recombination in crossing 3×10^{18} cm. The probability that an ionizing photon would penetrate the full 3×10^{18} cm without ever ionizing a single hydrogen atom is of order $e^{-30} \sim 10^{-13}$.

(b) Negative Ions

A hydrogen atom can become ionized not only through the loss of an electron, but also by gaining one to become the negative ion H⁻, the hydride ion.

$$H + e \rightarrow H^{-} + h\nu \qquad (7\text{--}13)$$

The structure of this ion is somewhat similar to that of the neutral helium atom in having two electrons bound to a nucleus. The second electron is only weakly bound, since the first electron is quite effective in screening out the nuclear charge. The binding energy is 0.75 ev and there is only one bound state. All transitions from or to this state therefore involve the continuum, that is, a neutral hydrogen atom and a free electron (Fig. 7.6).

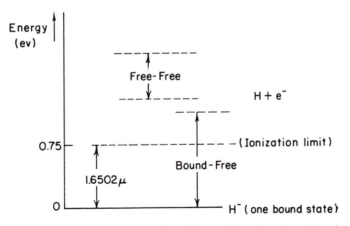

Fig. 7.6 Energy level diagram of the H⁻ ion. There is only one bound state with binding energy 0.75 ev. All radiative transitions must be either between the bound state and a state with a free electron, or else between two free states. (A wavelength of 1 μ is 10^{-4} cm.)

Because the 0.75 ev binding energy is so low, visible starlight, from cool stars like the sun, can be absorbed. This absorption continues out to 1.65μ wavelength, where the absorption due to bound-free transitions no longer can occur. However, at longer wavelengths, absorption can take place, since the H^- ion also can have free-free transitions—absorption in which energy is taken up by the translational energy of an unbound electron in the presence of a hydrogen atom. This is no small effect. The H^- ion plays an important role in the transport of energy through the solar atmosphere (Ch 58).

We should still explain how H^- even happens to exist in the atmosphere of cool stars: It survives, because metal atoms such as sodium, calcium and magnesium, which have low ionization potentials, can be easily ionized even by the light of cool stars. Some of the electrons generated in this way attach themselves to hydrogen atoms, to form the H^- ions.

Of course, the absorption due to H^- is always accompanied by subsequent reemission. It is interesting, therefore, that most of the light we receive from the sun is due to a continuum transition in which atoms and electrons recombine to form the hydride ion, H^-.

Many other elements of course also have ions that play an important role in astrophysics. Their physical properties are often similar to those exemplified by hydrogen.

Doubly ionized helium has physical properties very much the same as singly ionized hydrogen, except that the larger nuclear charge makes a difference in the transition energies involved. Singly ionized helium mimics many of the spectroscopic properties of the hydrogen atom, as already mentioned (7:3).

Molecules can of course be ionized too and molecular ions have characteristic spectra of their own. They are often observed in the ionized gaseous tails of comets, where one detects OH^+, CO^+, H_2O^+, H_3O^+, and others ions spectroscopically or by means of mass spectrographs aboard space probes sent into the comet's tail.

7:5 HYDROGEN MOLECULES

In general, molecules can have three types of quantized states. First, there is the possibility that atoms in the molecule vibrate relative to one another. In that case the *vibrational energies* are quantized. Second, there is the possibility of quantized rotation. This means that the angular momentum is quantized. Third, just as in atoms, there exist different quantized electronic states.

The binding energy between atoms is relatively weak. By this we mean that the energy required to separate two atoms that have formed a molecule is normally

7:5

Fig. 7.7 Vibrational levels in the hydrogen molecule H_2 ground electronic state. The lengths and positions of the lines along the abscissa indicate the range of separations d between nuclei in the molecule. The equilibrium separation is denoted by d_e. There are only 14 vibrational states below the energy continuum. Each of these states can be split into a number of substates corresponding to different amounts of angular momentum.

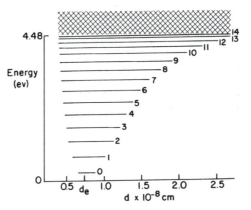

smaller than the energy required to ionize an atom. Only some large alkali atoms have ionizing energies lower than the highest molecular binding energies. Correspondingly also, the radiative transitions between excited vibrational states tend to occur at lower energies than those found for the transitions involving the lower electronic levels of most atoms. Characteristically, we tend to find that electronic transitions, that is, transitions between electronic excited states, in atoms or molecules, occur in the visible and ultraviolet part of the spectrum. Vibrational transitions occur in the near infrared part of the spectrum, roughly at wavelengths between 1 and 20 μ, and rotational transitions occur in the far infrared $\lambda \gtrsim 20\mu$ and microwave spectrum.

Of course this is just a rule of thumb and is not strictly obeyed. We already know that hydrogen atoms can have electronic transitions that reach way out to the very lowest energies associated with radio wavelengths. Those were discussed in section 7:3. It is true, however, that pure vibrational transitions do not often occur in the visible spectra of astrophysically important substances. And pure rotational spectra are not expected to occur at wavelengths shorter than a few microns.

For many purposes, the vibrations of two atoms relative to one another can be treated as harmonic oscillations. These vibrations are quantized. Figure 7.7 shows the energy levels for hydrogen H_2 in the lowest electronic state of the molecule. If the vibrations become too violent, the molecule dissociates into two separate atoms. The dissociation energy is 4.48 ev. This corresponds to an energy just higher than the 14th excited vibrational state.

The ground state of the molecule is not at zero energy. Rather, as already encountered in the case of photons (section 4:13), there is a characteristic positive shift, roughly equal to half the energy difference between the ground state and first excited state. This displacement is characteristic of ground vibrational levels.

7:5

Hydrogen molecules are a major constituent of interstellar space and are the predominant constituent of dark clouds in the galactic plane. Unfortunately, the total mass present is only indirectly inferred.

First, there is the difficulty of detecting the gas: Hydrogen is a symmetric dipole molecule, and as we already saw in section 6:13, symmetric configurations can at best radiate if they have a quadrupole moment. Hydrogen does have such a moment, but the transition probability for this type of transition is many orders of magnitude less than for the more usual dipole radiation of unsymmetric molecules. For this reason molecular hydrogen is hard to observe. This is true both of the vibration and of the rotation spectrum.

Though molecular hydrogen can be dissociated even by ultraviolet photons insufficiently energetic to ionize hydrogen atoms and, therefore, able to pass through neutral atomic gas without much hindrance, H_2 is well shielded inside dark clouds by an abundance of interstellar dust. The dust absorbs the ultraviolet photons and thus shields the hydrogen molecules. Inside such clouds, the vibrational spectrum of H_2 can be excited in shocked regions where temperatures rise to $\sim 2000°K$. At these temperatures, colliding molecules can excite each other vibrationally. Once excited, H_2 can undergo a radiative transition to a lower vibrational state, with probability 10^{-6} sec^{-1}, meaning that the molecule typically remains excited for about a week before radiating its excitation energy away. The photon energy then corresponds to the energy difference between two of the levels in Fig. 7.7 and is well defined. The spectrum of the radiation observed, therefore, identifies the molecular hydrogen uniquely, and also tells us the state of excitation of the gas (He67)*.

Normally, the temperature inside the molecular clouds is only 20–50°K. At these temperatures, the hydrogen can at best be excited to low-lying rotational states (Fig. 7.8), and there the lifetime before radiating to a lower state is measured in years if not centuries. Pure rotational emission of H_2 is, therefore, rarely observable. Instead, the presence of the molecular hydrogen is inferred from the observation of rotational transitions of CO which, though far less abundant than H_2, radiates readily. CO emission, however, is not an entirely reliable tracer of H_2, because the CO emission can readily be reabsorbed by the layers of CO through which the radiation must pass to escape a cloud.

PROBLEM 7–3.   If a system of particles has a *moment of inertia*

$$I = \sum_j m_j r_j^2 \tag{7–14}$$

about its axis of rotation, show that the energy and angular momentum are related by

$$\mathscr{E} = \frac{\omega L}{2} = \frac{L^2}{2I} \tag{7–15}$$

Fig. 7.8 Rotational energy level diagram of H_2 for the ground vibrational state. Alternate states have nuclear spins aligned: parallel, ortho (0), or antiparallel, para (P).

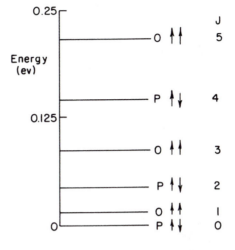

provided we are talking about a classical, rigid, nonrelativistic rotator. ω is the angular frequency of rotation. From this, show that the energy carried away by each quantum of radiation would have to be

$$\delta\mathscr{E} = \hbar\omega \tag{7–16}$$

due to quantization $\hbar$ of angular momentum. In quantum theory where the total angular momentum is given (see section 7:7), by $\hbar[J(J + 1)]^{1/2}$, show that the energy of each state is

$$\mathscr{E} = \hbar^2 \frac{J(J + 1)}{2I} \tag{7–17}$$

and the energy of the quantum released in the transition $J \rightarrow J - 1$ is

$$\delta\mathscr{E} = \hbar^2 \frac{J}{I} \tag{7–18}$$

For massive objects rotating rapidly, show that this is equivalent to the classical formula.

PROBLEM 7–4. If an interstellar molecule has a rotational energy kT in thermal equilibrium with surrounding gas (section 4:18), say $T \sim 100°K$, what is the range of frequencies it will radiate if typical values of atomic weights are 10^{-23} g and typical molecular radii are 2 Å. Make use of some of the expressions obtained in the previous problem to show that radiation may be expected in the far infrared or submillimeter region.

PROBLEM 7-5.   Set up an expression for the probability that a molecule will be found in an excited rotational state J characterized by moment of inertia I about a given spin axis, when it is in thermal equilibrium with a gas at temperature T. From this, convince yourself that a molecule in a cool interstellar cloud ($T \sim 100°$K) cannot be excited into very high rotational states.

PROBLEM 7-6.   If interstellar grains are in thermal equilibrium with the surrounding gas ($T \sim 100°$K), and have typical radii, 10^{-5} cm with typical masses 10^{-15} g, at what frequency could they be expected to radiate away angular momentum? This process is, in fact, not impossible because small inhomogeneous grains can be expected to have appreciable electric dipole moments. Observations would, however, have to be made from above the atmosphere, and terrestrial broadcast bands as well as interstellar plasma absorption might interfere.

PROBLEM 7-7.   It has been suggested that massive cosmic objects might exist having such high angular momenta, that contraction to high density becomes impossible. In such a situation the object might slowly cool down without ever becoming a star because its central temperature does not get high enough to sustain nuclear reactions.

Such an object might, however, be capable of losing angular momentum through systematic emission of circularly polarized radiation.

An object like this might also emit gravitational radiation. Since the graviton carries away twice the angular momentum of photons, see how the formulas of Problem 7-3 would change if applied to gravitons. Try the same thing for neutrinos that carry away only half the angular momentum of photons. Under normal conditions, in any of these cases, the probability for the emission of a quantum is quite small as the discussion of transition probabilities in the next sections will show. If electromagnetic radiation were in fact emitted, it would not be transmitted by the interstellar medium, because the expected rotational frequency ω for massive objects would be small. Why not? What would happen to the energy? (Re 71)

7:6 THE INFORMATION CONTAINED IN SPECTRAL LINES

An excited atomic system, left to itself, will spontaneously jump into a state of lower energy. The mean time required before such a transition occurs varies from one specific situation to another, and depends on such factors as the symmetries of the states, the dimensions of the system, and so on (see section 7:7). If the system has a total probability P of leaving the excited state in any unit time interval, then its total life in the excited state is $\delta t = 1/P$. The energy of the state therefore cannot be determined with arbitrarily great accuracy. The limited time in the excited state implies that the energy can only be determined to an accuracy $\delta\mathscr{E}$ given by the uncertainty principle

$$\delta\mathscr{E} = \frac{\hbar}{\delta t} \qquad (7\text{--}19)$$

Improbable transitions therefore have a narrow natural line width. The accuracy to which the transition energy between two states i and k can be determined depends on the lifetime both of the upper and the lower states; the total frequency width of the line δv, then, is the sum of the widths of the two levels. This total width is usually denoted by

$$\delta\omega = \frac{(\delta\mathscr{E}_i + \delta\mathscr{E}_k)}{\hbar} = \gamma = 2\pi\delta v \qquad (7\text{--}20)$$

γ is called the *natural line width* for the transition.

We will show in section 7:8 that the spontaneously decaying atom emits radiation with a spectral distribution or *line shape*

$$I(\omega) = I_0 \frac{\gamma}{2\pi} \frac{1}{(\omega - \omega_0)^2 + \gamma^2/4} \qquad (7\text{--}21)$$

The intensity drops to half the maximum value at frequencies $\omega_0 \pm \gamma/2$.

In astronomical sources the natural line width is seldom directly observed; but deviations from this width can give us a great deal of information, and it is useful to list the various line broadening effects:

(a) Doppler Broadening

This effect is of random motions of the atoms or molecules whose emission is observed. For small velocities the frequency shift of the radiation is roughly proportional to the line of sight velocity component v_r (see equation 5–44):

$$\Delta\omega = \omega_0 \frac{v_r}{c}, \qquad v_r \ll c \qquad (7\text{--}22)$$

There are two kinds of motion that can contribute to the Doppler broadening— the thermal velocity of the emitting atoms within a cloud, and the turbulent velocities peculiar to the clouds that are superposed along a line of sight. Sometimes these two effects can be resolved.

We can, for example, observe the absorption of stellar radiation by interstellar sodium atoms. Sodium absorbs very strongly in the yellow part of the visible spectrum, in a pair of lines known as the sodium D lines at 5890 and 5896 Å. If these lines are examined at very high resolution, so that shifts of the order of one part in 10^6 are detected, then measurements of velocities as low as $\sim 3 \times 10^4$ cm sec^{-1} can be distinguished. What we then observe is a series of discrete broadened absorption lines due to individual clouds absorbing light along the line of sight, and a definite characterizing width for individual lines representing a given cloud. Whether this individual line width is entirely due to thermal motions, or also partially due to turbulent motions on a smaller scale within each cloud, cannot immediately be recognized (Ho 69).

PROBLEM 7–8. If atoms of mass m move with velocities determined by Maxwell-Boltzmann statistics, the probability of observing a given line of sight velocity v_r is proportional to $\exp\left(-v_r^2 m/2kT\right)$ (see equation 4–56). Show that the line shape for thermal Doppler broadening therefore should be of the form

$$I(\omega)\, d\omega = I_0 \exp - \left[\frac{mc^2(\Delta\omega)^2}{2\omega_0^2 kT}\right] d\omega \tag{7–23}$$

See also Problem 4–27 and show how the line width can immediately be related to the temperature, through equation 7–22.

In general the width at half maximum, for Doppler broadening

$$\delta = \omega_0 \left[\frac{2kT(\ln 2)}{mc^2}\right]^{1/2} \tag{7–24}$$

is much greater than the natural line width γ. However, it drops off exponentially and, therefore, much faster than the natural line width. The observed wings of very strong lines, for example *Lyman-α* in interstellar space, therefore generally are due to natural line width and to the other causes listed below (Je 69).

(b) Collisional Broadening

In the relatively dense atmospheres of stars, atoms or ions often suffer a collision while they are in an excited state. Because any given collision may induce a transition to a lower state, the collective effect of such collisions is to increase the total transition probability. Thus, if the spontaneous transition rate were γ and the number of collision induced transitions in unit time is Γ

$$\frac{\delta\mathscr{E}}{\hbar} = \gamma + \Gamma \tag{7–25}$$

The emitted line has the spectral intensity distribution of the natural line shape, except that γ is replaced by $\gamma + \Gamma$.

(c) Other Types of Broadening

There are other effects due to interactions with neighboring atoms that can cause shifting and splitting of states through the influence of electric fields (Stark effect), through resonance coupling between atomic systems, and so forth. These processes all lead to line broadening, but at low densities their effects are small.

PROBLEM 7–9. To obtain a better feel for the relative importance of the Doppler and collision line widths for visual spectra obtained from stellar atmospheres, show that: If the atmosphere has a density n, thermal velocities $(3kT/M)^{1/2}$, and collision cross section σ,

$$\frac{\Gamma}{\delta} \sim \frac{n\sigma c}{\omega_0} \tag{7-26}$$

where factors of order unity are neglected. For visible light $\omega_0 \sim 10^{15}$ and collision cross sections have typical dimensions of atoms, 10^{-16} cm^2. $n \ll 10^{21}$ cm^{-3} in normal stellar photospheres and, hence, $\Gamma \ll \delta$.

7:7 SELECTION RULES

In section 7:2 we had indicated that the interaction of an atomic system with radiation obeys certain conservation principles. Compliance with these principles requires that certain transitions be forbidden, while others are allowed. The rules that tell us about the permitted transitions are called *selection rules*. We had already seen roughly how these rules come about. Here we will examine the question in somewhat geater depth.

When any two atomic systems combine to form a bigger system, the addition of angular momentum takes place so that, along any arbitrarily chosen direction z, the final angular momentum J_{zf} is the sum of the two initial angular momenta along that direction.

$$J_{zf} = J_{z1} + J_{z2} \tag{7-27}$$

where subscripts 1 and 2 refer to the two individual initial systems.

Because the precise measurement of the z-component precludes a simultaneous precise measurement of the transverse components, equations of the form (7–27) do not exist for the x and y directions. The equation is true whether we interpret the symbols J_{zi} as the z-component of the angular momentum or only as the quantum number for this angular momentum component which, when multiplied by $\hbar$, represents the actual angular momentum component. In what follows below, we will interpret the symbol J_{zi} as a quantum number. The values that these quantum numbers can take on are zero, half integer, or integer.

A second statement can be made about the addition of the squares of the angular momentum values. The (angular momentum)^2 is also a precisely measurable quantity that can, however, be measured simultaneously with J_{zi}. This second statement is somewhat more complicated. Nevertheless, reduced to its essentials it says that the allowed values of (angular momentum)^2 always take on numerical values

$$\text{(angular momentum)}_i^2 = J_i(J_i + 1)\hbar^2, \qquad i = 1, 2, \dots, f, \qquad (7\text{–}28)$$

where the relationship between J_i and J_{zi} requires that J_{zi} takes on values

$$J_{zi} = J_i, J_i - 1, J_i - 2, \dots, 1 - J_i, -J_i \qquad (7\text{–}29)$$

We had said that the z direction can be arbitrarily chosen. Let us choose it to represent the direction along which a photon approaches the atomic system. In the scheme used here, we can assign subscript 1 to the photon, 2 to the initial atomic state, and f to the atomic state after photon absorption. The choice of the z direction then shows that

$$J_{zf} = J_{z2} \pm 1 \qquad (7\text{–}30)$$

since $J_{z1} = 1$. This tells us that an atomic system with half integer values of J_{zi} must have J_i half integer, and that no transitions to integer values of J_{zi} or J_i are possible through photon absorption or emission. Similarly a system with integer angular momentum quantum numbers always maintains those properties under photon absorption.

Let us still try to understand why equation 7–28 takes on the particular form it does. We know that the maximum value that J_{zi} can take on is J_i. For this particular state, equation 7–28 states that there is an additional amount of angular momentum $J_i\hbar^2$ to be associated with the transverse angular momentum components. These components can therefore never be zero, unless J_{zi} itself is zero also and, in that case, they must be zero. The transverse angular momentum contribution comes about because the uncertainty principle does not permit a simultaneous precise measurement of two or more angular momentum components.

We note that the quantum numbers J_{zf} and J_{z2}, alternatively, may be taken to correspond to the magnetic quantum numbers we previously labeled m. Equation 7–30 then gives a selection rule which states $\Delta m = \pm 1$ when the direction of photon emission is along the magnetic field lines. This is why only two Zeeman shifted lines are observed along that direction. Along a direction of emission perpendicular to the field, equation 7–30 still is true, but the association of J_{z2} with m is then no longer valid. A division of photons into groups having $J_{z1} = \pm 1$, that is, in terms of left- or right-handed polarized light, would then mix the contributions from different magnetic energy levels m. This would happen because the photons from the various levels m, viewed along that direction, are plane polarized; as already stated in section 6:12, plane polarized light can be considered as a superposition of left- and right-handed polarized components. Hence, viewed along the direction perpendicular to the field, Δm may have values 0, as well as ± 1, even though (7–30) still is obeyed.

Equation 7–30 leads to one other selection rule, which is very important: It is impossible for an atomic system to undergo transitions between angular momentum states whose values are zero through absorption or emission of one photon. It is easy to see that this must be true. If $J_2 = 0$, then $J_{z2} = 0$ also, and similarly if $J_f = 0$, $J_{zf} = 0$. But by (7–30) both these z-components cannot be zero simultaneously. Hence the selection rule as stated must be true. This rule is absolutely inviolable, no matter whether electronic or vibrational transitions are involved. It is always true!

$$J = 0 \not\rightarrow J = 0 \qquad (7-31)$$

In quantum mechanics, selection rules like these are linked to the symmetry of the atomic system. When the system has sufficiently complex symmetries, the selection rules also become correspondingly complex. We have only shown one or two of the simplest selection relations here, but it is worth remembering that even the more complex appearing rules actually are basic symmetry statements, which become relatively simple when viewed in terms of the appropriate symmetry. The angular momentum selection rules discussed here are based on the rotational symmetry of atomic systems.

Equation 7–31 holds true only for transitions involving a single photon. In the very improbable two photon transitions, it is possible for such a transition to occur, and the angular momentum carried off by the individual photons is then oppositely directed. Such transitions are possible in tenuous nebulae in interstellar space where atoms in an excited state of zero angular momentum can exist undisturbed for long periods of time (Va 70, Sp 51b). In laboratory systems, where pressures are higher, such excited states normally become de-excited through atomic collisions.

An interesting feature of angular momentum quantization is that the existence of quantized states for all matter implies a lack of interactions with radiation having nonquantized angular momentum. Whatever fields, electric, magnetic, weak or strong nuclear, gravitational, or others that may exist in the universe, should therefore have associated radiation that is quantized in terms of half integer spin angular momentum $\hbar$, or multiples thereof. For example, when gravitational waves are discovered and examined, we are confident we will find them to have quantized spin angular momentum. The current prediction is that their spin should be $2\hbar$—twice that of photons (Gu 54).

7:8 ABSORPTION AND EMISSION LINE PROFILE

In estimating the amount of natural interstellar hydrogen along the light path between an ultraviolet emitting star and the earth, we need to know both the shape and the total strength of the Lyman-α absorption line. These two pieces of information will permit us to determine the amount of hydrogen in terms of the observed absorption line width. Intrinsically the calculation of line strength and shape is a quantum mechanical problem. Classical theory, however, permits us to calculate the line shape on the basis of a harmonic oscillator model. The model also yields the right order of magnitude for the line strength. However, we must take care not to take this classical model too seriously, because by itself it does not lead to quantized energy states for atomic systems.

We will first derive an expression for the emission line profile using semi-classical methods.

We start with equation 6–88 for the total energy, $I = 2e^2\dddot{r}^2/3c^3$, radiated by a charged oscillator per second. We can see that this intensity corresponds to a force

$$\mathbf{F} = \frac{2}{3}\frac{e}{c^3}\dddot{\mathbf{d}} \tag{7–32}$$

because the average work done by that force, in unit time, is then

$$\langle \mathbf{F} \cdot \dot{\mathbf{r}} \rangle = \left\langle \frac{2e}{3c^3}\dddot{\mathbf{d}} \cdot \dot{\mathbf{r}} \right\rangle = \left\langle \frac{2}{3c^3}\frac{d}{dt}(\dot{\mathbf{d}} \cdot \ddot{\mathbf{d}}) - \frac{2}{3c^3}\ddot{\mathbf{d}}^2 \right\rangle = I \tag{7–33}$$

Here the term containing $\dot{\mathbf{d}} \cdot \ddot{\mathbf{d}}$ vanishes because d and $\ddot{d}$ are exactly out of phase in simple harmonic motion. The *damping force F* is small compared to the harmonic force; the oscillation in other words lasts over many cycles. We can therefore write the equation of motion as

$$m\ddot{\mathbf{r}} = -m\omega_0^2\mathbf{r} + \frac{2}{3}\frac{e^2}{c^3}\dddot{\mathbf{r}} \tag{7–34}$$

This equation is very much in the spirit of (6–107), except that we have a damping (instead of a harmonic driving) force here. Since the damping is weak, the motion is almost harmonic and we can approximate

$$\dddot{\mathbf{r}} = -\omega_0^2\dot{\mathbf{r}} \tag{7–35}$$

We then rewrite (7–34) as

$$\ddot{\mathbf{r}} = -\omega_0^2 \mathbf{r} - \gamma \dot{\mathbf{r}} \quad \text{with} \quad \gamma = \frac{2}{3}\frac{e^2 \omega_0^2}{mc^3} \ll \omega_0 \tag{7–36}$$

which has the approximate solution:

$$\mathbf{r} = \mathbf{r}_0 e^{-\gamma t/2} e^{-i\omega_0 t} \tag{7–37}$$

since $\gamma \ll \omega_0$.

The oscillating dipole thus sets up an oscillatory field of the form

$$\mathbf{E}(t) = \mathbf{E}_0 e^{-\gamma t/2} e^{-i\omega_0 t} \quad \text{(RP)} \tag{7–38}$$

where only the real part enters into physical consideration. This is not mono-chromatic anymore because it changes with time; and only time invariant oscilla-ting fields can be strictly monochromatic. A time dependent change in intensity affects the frequency spectrum. The total field is now written in terms of an integral over the entire range of frequency components:

$$\mathbf{E}(t) = \int_{-\infty}^{\infty} E(\omega) e^{-i\omega t} d\omega \tag{7–39}$$

By a theorem from *Fourier theory*, an integral of this form can be inverted to give

$$E(\omega) = \frac{1}{2\pi} \int_{-\infty}^{\infty} E(t) e^{i\omega t} dt \tag{7–40}$$

If we now introduce the field (7–38) into this equation and note that $E(t)$ is de-fined only for time $t \geq 0$, we can readily integrate to obtain

$$E(\omega) = \frac{1}{2\pi} \frac{E_0}{i(\omega - \omega_0) - \gamma/2} \quad \text{(RP)} \tag{7–41}$$

We then can obtain the spectral line intensity (see Fig. 7.9):

$$I(\omega) = |E(\omega)|^2 = I_0 \frac{\gamma}{2\pi} \frac{1}{(\omega - \omega_0)^2 + \gamma^2/4}$$

$$I(\nu) = I_0 \left(\frac{\Gamma}{2\pi}\right) \left[(\nu - \nu_0)^2 + \frac{\Gamma^2}{4}\right]^{-1}, \qquad \gamma \equiv 2\pi\Gamma \tag{7–42}$$

where I_0 is the total intensity integrated over all frequency space and $\omega = 2\pi\nu$, $\omega_0 = 2\pi\nu_0$:

$$I_0 = \int_{-\infty}^{\infty} I(\omega) d\omega = \int_{-\infty}^{\infty} I(\nu) d\nu \tag{7–43}$$

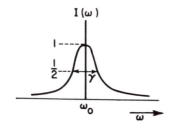

This type of line shape is sometimes called the *Lorentz profile*, and Γ and γ are called the *natural line width*. This is the full frequency width at half maximum. We have not yet shown that the absorption and emission profiles are the same. That question will, however, be taken up in section 7:9.

7:9   QUANTUM MECHANICAL TRANSITION PROBABILITIES

Much astrophysical information is to be obtained from the intensity of absorption or emission lines. The strengths of the lines define the number densities of given atoms, ions, or molecules in a source—or along the line of sight to a source—and the ratio of the strengths of lines can be used to determine the excitation temperature of gases through application of the Saha equation (section 4:16).

However, in order to obtain this information in useful form, we must first relate the intensity of the spectral absorption or emission lines to the number density of atoms or ions in different energy levels; and we can only do that if we know the transition probability between states of the system.

Very roughly the transition probability depends on three factors: (a) on the symmetry properties of the atomic system, (b) on its size in relation to the wavelength to be absorbed or emitted, and (c) on the statistics of the radiation field. The first of these factors includes the selection rules discussed in section 7:7, the statements about charge-to-mass ratio made in section 6:13 and other similar restrictions. The second factor represents the relative probability for dipole, quadrupole, and higher multipole radiation; that is size dependent as seen, for example, in (6–93). The third factor, to be discussed now, depends only on the radiation field and is quite general for any transition regardless of the atomic or nuclear system involved.

If we want to compute the probability that an atomic system will undergo a transition from some state i to another state j, this probability will be proportional to the number of ways in which a change in the photon field can occur. For example, the probability for emission of a photon with radial frequency ω

is (see equation 4–65a) proportional to

$$\frac{\omega^2 \, d\omega \, d\Omega}{(2\pi c)^3} = \frac{v^2 \, dv}{c^3} \, d\Omega \tag{7–44}$$

for photons polarized in one particular sense. Here $d\Omega$ is the increment of solid angle. This factor, considered in isolation, makes transitions in the optical domain, where $\omega \sim 3 \times 10^{15}$, much more probable than transitions, say, in the radio region at $\omega \sim 3 \times 10^9$.

Equation 7–44 holds only for *spontaneous photon emission*. In general, the emission probability is proportional to $n_\omega + 1$, where n_ω is the ambient density of photons per phase cell that already have the momentum and polarization characterizing the photon to be emitted. This preferential emission of photons along directions of photons already present in the vicinity of the atom is called *stimulated* or *induced emission*. It is the exact opposite of ordinary *photon absorption*. The number of absorptions is again proportional to $n_\omega + 1$, if n_ω is taken to be the number density per phase cell of photons left after the atom has reached the upper state. We therefore see that the *transition probability* $P(\omega, \theta, \phi)$ per unit solid angle and frequency range quite generally obeys the relation

$$P(\omega, \theta, \phi) \, d\Omega \, d\omega = [n(\omega, \theta, \phi) + 1] \frac{\omega^2}{(2\pi c)^3} \, d\Omega \, d\omega \tag{7–45}$$

Here $n(\omega, \theta, \phi)$ is the probability per unit frequency range that a photon state is occupied when the atomic system is in its upper energy state and ω is the mean transition frequency. Let us now return to the factors (a) and (b) mentioned earlier. These factors have to be evaluated quantum mechanically. In general, the result of such calculations is a *matrix* with elements U_{ij} giving the *transition amplitude* between any two states i and j of the atomic system. The actual transition probability between these two states is proportional to $|U_{ij}|^2$.

The prescription for obtaining the transition probability per unit solid angle is to multiply the product $|U_{ij}|^2 \, P(\omega, \theta, \phi)$ by the numerical factor $2\pi/\hbar^2$. Thus

$$\text{Transition probability per unit time} = \frac{2\pi}{\hbar^2} |U_{ab}|^2 \, [P(\omega, \theta, \phi)] \, d\Omega$$

$$= \frac{2\pi}{\hbar^2} |U_{ab}|^2 \, [n(\omega, \theta, \phi) + 1] \frac{\omega_{ab}^2}{(2\pi c)^3} \, d\Omega \tag{7–46}$$

Since the energy of the states is quite narrowly defined $n(\omega, \theta, \phi)$ will normally not change appreciably over the bandwidth of the line. The matrix elements already include the integration over the frequency bandwidth that appeared explicitly in equation 7–45. The transition probability (7–46) thus includes an

integration over the entire frequency range ω and, specifically, includes consideration of strongly absorbed or emitted photons at the line center as well as the less
readily absorbed and emitted photons in the line wings. More precisely stated,
it includes consideration of the line shape (7–42).

We still need to relate the quantum mechanical transition probability to
equation 6–86 which expressed the intensity I absorbed by an oscillating dipole in
terms of the second time derivative of the dipole moment d.

$$I = \frac{2}{3c^3} \ddot{d}^2 = \frac{2e^2 \ddot{r}^2}{3c^3} \tag{6–86}$$

Since **r** has the time dependence (7–37), equation 6–86 is readily rewritten as

$$I = \frac{2e^2 \omega^4}{3c^3} \langle r^2 \rangle = \frac{32\pi^4}{3c^3} e^2 v^4 \langle r^2 \rangle \tag{7–47}$$

where the brackets $\langle \ \rangle$ indicate a time average. The intensity I is related to the
spontaneous transition probability of quantum mechanics; and we must therefore
set $n = 0$ in equation 7–46, if a comparison with (7–47) is to be made. In the
dipole approximation the matrix element U_{ab} will make a contribution

$$|U_{ab}|^2 = 2\pi\hbar\omega_{ab}e^2|r_{ab}|^2 \sin^2 \theta \tag{7–48}$$

where an integration has been carried out over the possible directions of polariza-
tion. The physical meaning of $e^2 |r_{ab}|^2$ will be discussed below. The total intensity
now is given by the product of the transition probability (7–46) and the photon
energy $\hbar\omega_{ab}$:

$$I \, d\Omega = \hbar\omega_{ab} \cdot \frac{2\pi}{\hbar^2} \frac{\omega_{ab}^2}{(2\pi c)^3} \cdot 2\pi\hbar\omega_{ab}e^2|r_{ab}|^2 \sin^2 \theta \, d\Omega \tag{7–49}$$

where θ represents the angle between the vector **r** and the direction of propagation of the emitted radiation. Integrating over all angles of emission, we obtain
the total intensity of spontaneously emitted radiation

$$I = \frac{4}{3} \frac{e^2}{c^3} \omega_{ab}^4 |r_{ab}|^2 = \frac{64}{3} \pi^4 \frac{e^2}{c^3} v_{ab}^4 |r_{ab}|^2 \tag{7–50}$$

We see that the formula obtained quantum mechanically is almost the same as
the classical expression. We only have to replace the time average $\langle r^2 \rangle$ by
$2|r_{ab}|^2$ if we wish to obtain identical forms. This connection is consistent with
the correspondence principle. We note, however, that $e^2 |r_{ab}|^2$ is not an exact
quantum mechanical analogue to the mean square dipole moment. For each
individual state of the atomic system, a or b, the dipole moment would be given
by expressions involving the diagonal matrix elements er_{aa} or er_{bb}, respectively.

7:9

The quantities er_{ab}, instead, denote a property that is influenced both by the initial and the final state of the system. They have no exact classical analogue and we therefore need not be surprised that a factor 2 appears in equation 7–50. In fact, we had no reason to expect complete identity of classical and quantum mechanical forms. After all, the quantum theory of radiation is supposed to go a step beyond the classical results in order to provide an understanding of abrupt transitions. Its results must therefore differ from those of classical theory in some essential form.

Thus far we have only a formal solution that does not yet allow us to estimate the strength of an emission or absorption line. We can however still make use of equation 7–36 for that purpose. We note that γ^{-1} is a time constant, so that γ taken by itself is equivalent to a transition probability. By setting the value for γ equal to the transition probability (7–46), we therefore are able to estimate U_{ab} and also an absorption cross section for radiation. We write

$$\gamma = \frac{2}{3} \frac{e^2 \omega_{ab}^2}{mc^3} = \frac{2\pi}{\hbar^2} \frac{\omega_{ab}^2}{(2\pi c)^3} \int |U_{ab}|^2 \left[n(\omega, \theta, \phi) + 1 \right] d\Omega \qquad (7\text{--}51)$$

For spontaneous emission, $n(\omega, \theta, \phi)$ can be set equal to zero. The value for γ that is used here has of course been derived on the basis of a dipole radiator model, and the integral on the right-hand side of equation 7–51 will therefore contain the same factor 2/3 that already came up in the evaluation of the classical expression 6–85.

$$\int |U_{ab}|^2 \, d\Omega = \frac{2}{3} |U_{ab}|^2 \, 4\pi \qquad (7\text{--}52)$$

Hence

$$|U_{ab}|^2 = \frac{\pi e^2 \hbar^2}{m} \qquad (7\text{--}53)$$

This information suffices for us to find the atomic system's absorption cross section for radiation.

Let this cross section be

$$\sigma = \int \sigma(\omega) \, d\omega \qquad (7\text{--}54)$$

and let the atomic system be surrounded by an isotropic photon gas of density $n'(\omega, \theta, \phi)$. The total number of photons absorbed can therefore be expressed as

$$n'(\omega, \theta, \phi) \, \sigma c \, d\Omega = \frac{2\pi}{\hbar^2} |U_{ab}|^2 \left[n(\omega, \theta, \phi) + 1 \right] \frac{\omega_{ab}^2}{(2\pi c)^3} \, d\Omega \qquad (7\text{--}55)$$

Here the left side represents the number of photons per unit frequency range of a continuous spectrum intercepted by the cross-sectional area in unit time, and the right-hand side gives the probability for absorption of a photon, as expressed in (7–46). We can cancel the photon densities in equation 7–55 if we follow the procedure of letting $n(\omega, \theta, \phi)$ stand for the fractional number of photon states occupied when the atomic system is in its upper state. Because we have taken n' to represent the number density of photons present before absorption, that is, when the atomic system still is in its lower state, it is clear that

$$n'(\omega, \theta, \phi) = [n(\omega, \theta, \phi) + 1]\frac{\omega_{ab}^2}{(2\pi c)^3} \tag{7–56}$$

The factor $\omega_{ab}^2/(2\pi c)^3$ appears because n is a number density per phase cell while n' is a density per unit volume of normal, three-dimensional configuration space. From (7–53) and (7–55) it then follows that

$$\sigma = \frac{2\pi^2 e^2}{mc} = 2\pi^2 r_e c \tag{7–57}$$

where

$$r_e \equiv \frac{e^2}{mc^2} \tag{6–168}$$

The cross section (7–57) has precisely the value we would obtain by classical means if we modified equation 6–107 to include a radiative reaction force (see equation 7–32)

$$F_{rad} = \frac{2}{3}\frac{e^2}{c^3}\dddot{r} \tag{7–58}$$

representing the force on the moving charge due to its emission of radiation.

A series of remarks now is in order:

(1) The cross section obtained here holds only for atomic systems for which an oscillating charged dipole represents a satisfactory description. This must be strongly emphasized! Each type of atom or molecule has its own structure and therefore will interact with photons in its own way. However, an essential feature shared by many atomic systems is that electrons are bound to a nucleus or core. In a stable quantum state the electron then resists the efforts of an applied electromagnetic field to move it from its equilibrium position or, more accurately, from its equilibrium orbital distribution within the atomic system.

In this respect, the electron behaves as though it were harmonically bound to the more massive core. This justifies the use of the classical dipole approximation as a guide to the quantum treatment. However, it does so only for atoms or molecules having a dipole moment and for wavelengths long compared to the atomic dimensions. The limitations that held for classical radiators therefore hold equally well in the quantum limit. That was already pointed out in section 6:13, but perhaps it is worth stating again.

(2) No atom behaves precisely like a classical harmonic oscillator. Its cross section, therefore, is not precisely that given in (7–57). We can define an *oscillator strength f* that represents the actual absorption strength of a given line in units of $2\pi^2 e^2 (mc)^{-1}$. A value $f = 1$ represents an absorption equal to that of the classical dipole.

(3) As already noted in equation 7–54, the cross section of the atomic system varies with frequency. The frequency distribution is of the form (7–42), as indicated earlier.

PROBLEM 7–10. Show that if the absorption cross section is

$$\sigma_{ab}(\omega) = \frac{2\pi e^2}{mc} f_{ab} \frac{\gamma/2}{(\omega - \omega_{ab})^2 + (\gamma/2)^2} \tag{7–59}$$

$$\sigma_{ab}(\nu) = \frac{2\pi e^2}{mc} f_{ab} \frac{\Gamma}{2} \left[(\nu - \nu_{ab})^2 + \left(\frac{\Gamma}{2}\right)^2 \right]^{-1}, \qquad \Gamma = \frac{\gamma}{2\pi}$$

the total cross section obtained in (7–57), multiplied by an oscillator strength f_{ab}, is obtained on integrating over all frequencies.

---

(4) The identity of absorption and emission cross sections is already implied in the form which equation 7–46 takes. This is true both of the magnitude of the absorption and also of its spectral distribution. When no radiation field at all appears to be present, that is, $n(\omega, \theta, \phi) = 0$, we still have the vacuum field or zero level photon population present; and this field induces the spontaneous emission of radiation, discussed in more detail in section 7:10.

(5) The numerical values associated with various kinds of transitions also are of interest. The absorption cross section (7–57), corresponding to unit oscillator strength, has a value $\sigma \sim 0.17$ cm^2 sec^{-1}. This cross section, when multiplied by the radiative flux *in unit frequency interval*, gives the total amount of radiation absorbed by an atom. Sometimes it is useful to know the maximum absorption at the line center. The peak absorption cross section is then a useful quantity to know.

PROBLEM 7–11. Show that the maximum absorption cross section has the value $\sigma(\omega_{ab}) = (3\lambda^2/2\pi) f_{ab}$, so that the apparent size of the atom at resonance is roughly a factor of two lower than the wavelength squared, for $f_{ab} \sim 1$. λ is the wavelength of the radiation.

PROBLEM 7–12.   What fraction of the radiation in an emission or absorption line lies within the bandwidth γ defined by the natural line width?

PROBLEM 7–13.   Show that the spontaneous transition probability is roughly $\gamma \sim (5\lambda^2)^{-1}$ in cgs units. It therefore has a value of 10^8 sec^{-1} for visible light.

In a different spirit we can use (7–50) to write the transition probability w as

$$w \sim \frac{e^2}{c^3} \frac{\omega_{ab}^3}{\hbar} |r_{ab}|^2 \sim \frac{e^2}{c^3\hbar} \left(\frac{me^4}{\hbar^3}\right)^2 \left(\frac{\hbar^2}{me^2}\right)^2 \omega_{ab} \sim \left(\frac{e^2}{c\hbar}\right)^3 \omega_{ab} \sim \frac{1}{(137)^3} \omega_{ab} \quad (7\text{–}60)$$

where we have made use of equation 7–3 for a hydrogenlike atom to roughly estimate the radiated frequency ω_{ab} and have set the Bohr radius (7–4) equal to $|r_{ab}|$. The *fine structure constant*

$$\alpha = \frac{e^2}{\hbar c} \sim \frac{1}{137} \quad (7\text{–}61)$$

taken to the third power then appears in the last element of equation 7–60. The transition probability for visible radiation is of order 10^8 sec^{-1} and, correspondingly, we can see from (7–60) that it should be of order 10^{11} sec^{-1} for X-rays, $\sim 10^{14}$ sec^{-1} for γ-radiation, and $\lesssim 10^4$ sec^{-1} for radio waves. Interestingly (7–60) is independent of the mass of the emitting particle. It does not have to be an electron, but can be an ion. Of course the lifetime of the state is just the reciprocal of the transition probability.

The magnitude for oscillator strengths can vary greatly. For the hydrogen Lyman series we have values Lα(0.42), Lβ(0.08), Lγ(0.03), Lδ(0.01), and so on. Occasionally f values are slightly larger than 1.0, or at the other extreme values of 10^{-10} or even less can also occur. The oscillator strengths must therefore be evaluated individually for any given atom or molecule and will depend strongly on the structural properties of the atomic system.

The f values for different transitions in an atom or molecule are not independent. In particular, a given atom cannot have an arbitrarily large number of strong absorption or emission lines: If we sum f values for all possible transitions, between all possible states in an atom or ion, we should obtain a number equal to the total number of electrons in the atom. If the atom has strongly bound inner electrons, then the sum should equal the number of the more weakly bound valence electrons. This is the *Thomas-Kuhn sum rule* (A163)*. For hydrogen the sum of all f values should equal unity.

7:9

PROBLEM 7–14. For many years astronomers believed that atoms could be repelled by sunlight, to a sufficient extent to account for the long comet tails we observe.

For a small molecule, having an oscillator strength $f = 1$ and a mass $m = 5 \times 10^{-23}$ g, calculate the ratio of solar radiative repulsion to gravitational attraction. For comet tails the observed repulsive acceleration corresponds to an effective ratio of the order 10^2 to 10^3. Assume that all of the sunlight is roughly evenly distributed between 4×10^{-5} and 7×10^{-5} cm wavelengths. Does it appear likely that radiation produces this repulsion?

Magnetohydrodynamic forces are currently thought primarily responsible for the acceleration of tail constituents. In contrast, as seen from problem 4–17, radiation pressure does exert major forces on interplanetary dust grains, and is responsible for propelling grains away from the comet nucleus, into a dust tail.

---

(6) Atomic systems that do not have a dipole moment can at best undergo transitions through quadrupole or magnetic dipole radiation. The transition probabilities for such processes are of order $(r/\lambda)^2$ smaller, where r is a typical dimension of the atomic system. This is consistent with what we found in section 6:13, since $r/\lambda \sim r\omega/c \sim v/c$. From (7–60) we also see that $(r\omega/c)^2 \sim (1/137)^2$ for atoms. In rough agreement with these estimates, we find that actual transition probabilities for magnetic dipole and for quadrupole transitions are of order 10^3 and $1 \sec^{-1}$ (He50).*

7:10 STIMULATED EMISSION, COHERENT PROCESSES, AND BLACKBODY RADIATION

Stimulated emission is not a mechanism in the same sense as, say, electric dipole, quadrupole, or synchrotron emission, which we discussed in Chapter 6. It is a process that will work through the aid of any one of these mechanisms. Stimulated emission can occur whenever an electromagnetic wave of frequency $v = \mathscr{E}/h$ impinges on a particle in an excited state at energy $\mathscr{E}$ above some other state.

The electromagnetic wave or photon at frequency v can then stimulate or induce the emission of an additional photon that has exactly the same polarization, direction of propagation, and frequency as the stimulating photon. The characteristics of this newly formed photon are indistinguishable from those of the stimulating photon in the sense already discussed in sections 4:11 to 4:13, and we say that the radiation is *coherent*.

Let us first show the role that *stimulated* (sometimes called *induced*) *emission* plays in the process of blackbody emission. Again, blackbody radiation is not a mechanism. It is a process that depends on the existence of radiative mechanisms, and we can obtain a blackbody spectrum through a wide variety of mechanisms. That is why the appearance of a characteristic blackbody spectrum in an astrophysical source can only tell us something about the surface temperature of the source—nothing about the physical processes that actually give rise to the observed radiation.

Blackbody radiation always is the result of a succession of emission and absorption processes. There are two basic requirements: First, the temperature of absorbing particles must be constant in the vicinity of the emitting surface so that the photons emanating from the surface are in thermal equilibrium with particles at one well-defined temperature; second, for this equilibrium to become established, we require that the assembly of absorbing particles at constant temperature be large enough so that a succession of absorption and re-emission steps occur before energy escapes from the surface of the assembly.

The number density $n(\mathcal{E})$ of particles in an excited energy state $\mathcal{E}$ is given by the Boltzmann distribution (4–47) in terms of n_0, the number density in a lower state:

$$n(\mathcal{E}) = n_0 e^{-\mathcal{E}/kT} \qquad (7\text{–}62)$$

To determine the equilibrium conditions between photons and particles we can proceed in a way that is different from the approach taken in Chapter 4. We can look for the conditions for which the number of photons absorbed by the assembly of particles is just equal to the number emitted per unit volume, because those are the conditions in equilibrium. To analyze this situation we will consider only the transitions occurring between two given energy states of the particle. The presence of other states will not alter the conclusions.

A particle in the lower of the two states can absorb a photon and transit to the upper energy state. In the upper state, the particle can either spontaneously emit a photon or else it can be induced to emit a photon through stimulation by radiation of appropriate spectral frequency. In equilibrium, the sum of the induced and spontaneous downward transitions must equal the number of upward transitions. Let the probability of emitting a photon of frequency v be $A(v)$ in unit time interval. Let the probability of absorbing a photon be $n(v, T) cB(v)$ where $n(v, T)$ is the photon density at temperature T and frequency v, and $B(v)$ is a cross section per unit frequency; then it is easy to see that the probability for stimulated emission for a given excited particle equals the probability for absorption by some other particle in the lower state. As illustrated in Fig. 7.10, this is a consequence of time reversal symmetry that holds for all electromagnetic processes.

We are now ready to write the equation for equilibrium between absorption and emission of photons in the frequency range dv around frequency v.

$$n(v, T) B(v) n_0 c \, dv = [A(v) + n(v, T) cB(v)] n(\mathcal{E}) \, dv \qquad (7\text{–}63)$$

7:10

Fig. 7.10 In part (*a*) a photon of frequency *v* stimulates the emission of a similar photon, while the particle energy drops by an amount $\mathcal{E} = hv$. In (*b*), the time-reversed process takes place. The particle transits to the higher energy state $\mathcal{E}$ by absorbing a photon. As discussed in the text, spontaneous emission of radiation (*c*) can be considered caused by radiation oscillators in the ground state and corresponds to the time-reversed process of absorption of energy from a singly excited radiation oscillator (*d*).

Combining this with equation 7–62 gives

$$n(v, T)\, dv = \frac{A(v)/B(v)\, c}{e^{hv/kT} - 1}\, dv \tag{7–64}$$

If we considered spontaneous transitions as events produced by the ground state of the radiation field, which was discussed in section 4:13, then we can set $A(v)$ equal to the number density of radiation oscillators multiplied by the transition probability per unit time $B(v)$. Basically this is equivalent to stating that all emission processes are induced, that the probability for emission is proportional to the sum of all populated radiation oscillator states, that the ground state of each oscillator ($n = 1$) is always populated and that some radiation oscillators containing what we have called photons are in higher states n. From the phase space enumeration of the number density of radiation oscillators at frequency v (equation 4–65a) we then see that

$$A(v) = \frac{8\pi v^2}{c^2}\, B(v) \tag{7–65}$$

To relate these coefficients to earlier work, we note that (7–50) gives

$$\frac{I}{hv} = \frac{64}{3h}\pi^4 \frac{e^2 v^3}{c^3}|r_{ab}|^2 = \int A(v)\, dv = A_{ab} \tag{7–66}$$

PROBLEM 7-15.   According to the correspondence principle, the transition probability should be related to the classical radiation intensity in the limit of large atomic systems. In the ionized regions of interstellar space, transitions often occur between highly excited states of atomic hydrogen (Ka59, Hö65). Show that the correspondence argument leads to

$$\frac{d\mathscr{E}}{dt} = h\nu A_{n,n-1} = \frac{\omega^4 e^2 a_n^2}{3c^3} \tag{7-67}$$

where a_n is the Bohr radius in the nth state. Show that this gives

$$A_{n,n-1} = \frac{64\pi^6 m_e e^{10}}{3c^3 h^6 n^5} = \frac{5.22 \times 10^9}{n^5} \tag{7-68}$$

PROBLEM 7-16.   Show that $B(\nu)$ differs from $\sigma_{ab}(\nu)$, (7–59) only by a factor of 2. Derive a relation between A_{ab} and f_{ab}.

We see now, that

$$n(\nu, T)\, d\nu = \frac{8\pi\nu^2}{c^3} \frac{d\nu}{e^{h\nu/kT} - 1} \tag{7-69}$$

This corresponds to equation 4–71 for blackbody radiation, and shows that the blackbody process depends heavily on the concept of stimulated emission.

The process we described is stable and self-regulating. If $n(\nu, T)$ is lower than the value given in equation 7–69, the spontaneous emission exceeds the sum of absorption and stimulated emission processes; and in this way the population of photons becomes increased until it reaches the value given by (7–69). Conversely, if $n(\nu)$ is high, absorption will lower it back to the equilibrium value.

It is worth noting that the *Einstein Coefficients* $A(\nu)$ and $B(\nu)$ are sometimes defined slightly differently—for example, in terms of emitted or absorbed energy. Throughout this section, we also have taken the statistical weights g_n, g_{n-1}, discussed in section 4:16 to be unity. If that is not true, equations 7–62 and 7–63 must be modified, but (7–69) remains unaltered.

7:11 STIMULATED EMISSION AND COSMIC MASERS

Let us ask what would happen if the relationship between n_0 and $n(\mathscr{E})$ was not given by the Boltzman relation (7–62). For small deviations from the thermal equilibrium the photon bath would tend to cause n_0 and $n(\mathscr{E})$ to come back to equilibrium. But if the number of particles in the upper state starts to exceed the number in the lower state, then an entirely different process comes into play.

Clearly, this situation can never come into existence under conditions of thermal equilibrium, because $\exp(-\mathscr{E}/kT)$ is always less than unity for positive values of $\mathscr{E}$ and T. A *population inversion*, $n(\mathscr{E}) > n_0$, can therefore only be brought about by an artificial process. Sometimes one describes a population inversion as a state of negative temperature, for then the exponential term in equation 7–62 can exceed unity. However, this is primarily a descriptive term and does not define any physical process.

To see what happens if population inversion is brought about, we note that the probability for stimulated emission always exceeds the probability of absorption in this case. In any given transition a radiation oscillator is therefore more likely to rise to a higher energy state than to a lower one. As the radiation propagates through the assembly of particles, it therefore becomes amplified. Moreover, since the emitted photons have the same characteristics as the stimulating photons, the amplified radiation is coherent. In the laboratory the process described here corresponds to a *maser*. When the process occurs on a cosmic scale we therefore talk about maser processes. A maser which operates at optical frequencies is called a *laser*.

Cosmic maser action will be maintained provided that the pumping of energy into the assembly of particles, that is, the rate of excitation of the upper levels can keep up with the downward spontaneous and induced transitions, and maintain the density of particles in the upper energy state $n(\mathscr{E})$ greater than that in the lower state n_0.

The pumping process can take several forms. We might have a very energetic photon excite particles into an energy state $\mathscr{E}'$ from which the transition probability to the ground state is low and the probability for transition to a metastable state with energy $\mathscr{E}$ is high. This type of maser is called a *three-level maser*. Another means for producing a population inversion can come about chemically. Suppose that a molecule is formed in the interaction of two atoms and that it is formed in a high energy state. If the formation rate is sufficiently high, a population inversion can be maintained and maser action can set in.

In select circumstellar clouds and interstellar regions, OH radicals and/or water vapor molecules H_2O have certain energy states pumped up to population inversion. The pumping action is not understood, and it is curious that different regions of space show a variety of differing OH levels inverted. We therefore seem to have a number of different pumping mechanisms that evidently come into play under differing conditions.

Cosmic masers emit extremely intense coherent radiation. Since all induced photons travel along the same direction, they appear to come from an improbably compact region (Fig. 7.11). The radiation reaching our telescopes arrives contained in a well-defined, extremely small, solid angle.

The smallness of the observed solid angle is misleading. It may not represent the actual size of the cloud, but might represent only the dimension of the region

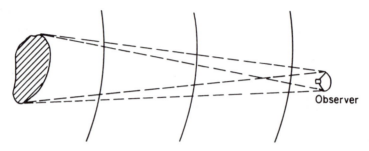

Fig. 7.11   The observer may not see the entire cloud of particles (dashed line) in coherent radiation. He may see only that portion in which the coherent wave originated. That can be a much smaller region, possibly a volume as small as λ^3, where λ is the wavelength of the maser radiation.

in which the coherent radiation originated. This volume can be seen, from phase space arguments, to equal h^3/p^3, or $c^3/v^3 = \lambda^3$, where λ is the wavelength of light. Some masers have by now been resolved by long baseline radio interferometers, and this indicates that masing may be initiated over larger regions in space.

Several types of masers are known. There are OH masers whose positions coincide with those of Mira variable stars. Such a maser's luminosity is roughly $10^{-4} L_\odot$ and it shows variability, over periods of months, synchronized with the star's pulsations. We also recognize OH and H_2O masers associated with dust clouds in or near HII regions. The H_2O masers can have luminosities up to $\sim L_\odot$, and show variability over periods of weeks. These variations are far too slow to be due to scintillation—refraction of radiation by moving clouds in interplanetary or interstellar space. A number of other molecules like SiO, also form interstellar masers.

Some types of masers may be pumped by strong infrared radiation fluxes exciting molecules into population inversion in the masing regions.

One characteristic of the interstellar masers is that the radiation is highly polarized. As already explained, stimulated emission always involves formation of photons with the same sense of polarization as the stimulating photon. This means that all the photons derived from a given progenitor will have identical polarization and the radiation is 100% polarized.

To realize how quickly the intensity of a beam increases as it traverses a cloud in which the population is inverted, we note that for a gain g per interaction, and for an optical depth n in the cloud, the outgoing beam will have an intensity of

$$N = g^n$$

Suppose that the gain is 1.1—that is, the probability for stimulated emission is 10% higher than for absorption. After 100 mean free paths, the total number of photons in the beam will be 10^4. After 200 such successive absorption-emission processes, the number would have reached 10^8, and so on. An emitting region, therefore, need not have an opacity in excess of several hundred in order to emit extremely bright maser radiation. This radiation is all concentrated into a very narrow spectral bandwidth and in this narrow spectral range it therefore has the equivalent brightness of a fantastically hot body, with T up to 6×10^{13}°K (Ra71).*

7:12 STELLAR OPACITY

In the interior of a star, where matter is highly ionized, the interaction of photons with matter often determines the rate at which energy is transported through the star.

To be sure, the photon interaction tells us nothing about neutrino energy transport, which may be very important in some stellar models. The neutrinos, generally, simply escape, leaving the central regions of the star at the speed of light. In contrast to photons, they normally do not affect stellar structure except in the limited sense that the nuclear reactions in the center of a star heat the stellar material by a reduced amount whenever neutrino production occurs. We then speak about *neutrino energy losses*.

In the absence of such losses, three mechanisms act to transport energy from the center of the star, where nuclear reactions occur, out to the periphery where radiation is emitted into space. The transport can occur through convection of stellar matter, through conduction of heat by a degenerate electron gas, or through transfer of radiation. These processes will be discussed in Chapter 8. Each one plays an important role in different types of stars. Here, we will only be concerned about the factors that determine the rate at which photons transport energy.

A γ-photon liberated in a nuclear reaction almost immediately yields its energy to the stellar material through ionization of neutral or partially ionized atoms, or through collisions with electrons. So strong is the interaction of radiation with matter, that energy initially liberated at the center of a star normally requires tens of thousands of years before finally escaping at the stellar surface. The star is highly opaque to radiation, and it is interesting to see how the opacity originates, since many features of stellar structure and evolution depend on this physical property.

We will be interested in four distinct types of interaction of radiation with matter: (a) Thomson or Compton scattering of radiation by free electrons (6:14, 6:20); (b) *free-free absorption* or emission (section 6:16); (c) *bound-free interactions* in which an electron undergoes a transition between a bound and a free state; (d) *bound-bound* transitions, that is, the excitation or de-excitation of atoms or ions by photons.

7:12

In order to compute the mean opacity of stellar matter, we will have to proceed in three steps. First, we will need to know the interaction cross section of radiation with matter for each of the four processes. This gives us the opacity due to the individual interactions through a simple proportionality relationship. However, the total opacity of stellar matter is not just the sum of the individual opacities, and a suitably chosen mean opacity must be computed, properly weighting the individual contributions made by processes (a) to (d) and also taking induced emission into account. Stimulated emission decreases the opacity because the energy transport rate is increased.

The contributions of the various processes to the opacity depend very strongly on the temperature. In the cool surface layers of a star, where atoms are only partially ionized, the opacity may be dominated by bound-bound and bound-free transitions. At high temperatures where ionization may be nearly complete, the opacity due to free-free interactions becomes dominant. At the highest temperatures where induced emission reduces the opacity due to factors (b) through (d), electron scattering plays a dominant role.

We will let *extinction* denote the amount of radiation eliminated from a beam of light through absorption or scattering. We can then define the extinction $\not{E}$ of a slab of matter of unit thickness, through which radiation passes at normal incidence, as

$$\not{E} = \kappa\rho \tag{7-70}$$

where the *opacity* of the substance is denoted by the symbol κ and ρ represents the density. The opacity for radiation at a particular spectral frequency v is denoted by $\kappa(v)$. Summing over the opacity contributions of processes (a) to (d), at any given frequency, we can therefore write a total opacity $\kappa^*(v)$ at frequency v:

$$\kappa^*(v) = \kappa_e + \left[\kappa_{ff}(v) + \kappa_{bf}(v) + \kappa_{bb}(v)\right]\left[1 - e^{-hv/kT}\right] \tag{7-71}$$

where the subscripts respectively mean electron scattering, free-free, bound-free and bound-bound. κ_e is frequency independent and therefore becomes predominant as the temperature increases. κ^* represents the true opacity with induced emission taken into account.

The proper averaging of $\kappa^*(v)$ over the entire range of frequencies depends on our purpose. In the case of stellar energy transport, which will be discussed in Chapter 8, we effectively need to know the mean free path of radiation as it travels through the star. Since this is inversely proportional to the opacity, we effectively need to average $1/\kappa^*(v)$ over the entire spectral range. This average, however, must still take into consideration that the radiation spectrum is not flat, and that the energy transport rate will therefore also depend on the radiation spectrum defined by the local temperature at any given point of the star. We will consider this later, in Chapter 8. For the present we will only show how

$\kappa^*(\nu)$ depends on processes which occur on an atomic scale, and how the individual opacities depend on atomic interaction cross sections for radiation.

(a) Scattering by Free Electrons

At temperatures sufficiently low for photon energies to lie well beneath the electron rest mass energy, that is, $T \ll mc^2 k^{-1} \sim 10°K$, relativistic effects can be neglected and the scattering cross section is simply the Thomson cross section.

$$\sigma_e = \frac{8\pi}{3}\left(\frac{e^2}{mc^2}\right)^2 = 6.65 \times 10^{-25} \text{ cm}^2 \tag{6-102}$$

This is frequency independent. At the centers of highly dense stellar cores, the temperature may however become large enough so that the Klein-Nishina cross section for Compton scattering (6–167) gives a more accurate representation, and a frequency dependence then does exist. At the density of the sun, $\rho \sim 1 \text{ g cm}^{-3}$, the number of electrons per cubic centimeter is of order 10^{24}, so that the mean free path of radiation between electron scattering events is only of the order of 1 cm. If n_e is the number of electrons in unit volume, the opacity for scattering is given by

$$\kappa_e \rho = \sigma_e n_e \tag{7-72}$$

(b) Free-free Interactions

This process was discussed for tenuous plasmas in section 6:16, but the same theory describes the denser plasmas inside stars. We note that the classical expression (6–137) must unwittingly contain the induced emission factor $[1 - \exp(-h\nu/kT)]$ which at long wavelengths approaches $h\nu/kT$. If we, therefore, divide (6–137) by this factor and also by the number densities, we obtain an absorption coefficient per ion, for unit density of ions and electrons. It will have the form

$$\alpha_{ff} = \frac{8}{(6\pi)^{1/2}} \frac{Z^2 e^6}{cm^2 h \nu^3 v} \ln[\ldots] \tag{7-73}$$

where we have set $v = (3kT/m)^{1/2}$ and have assumed that the argument of the logarithmic function has the same character as in (6–137). The actual, quantum mechanically correct, result is (To47)

$$\alpha_{ff} = \frac{4\pi e^6}{3\sqrt{3} \, chm^2} \frac{Z^2}{v} \frac{g_{ff}}{\nu^3} \tag{7-74}$$

7:12

where Z is the effective charge of the ion considered and g_{ff} is called the Gaunt factor. It contains the logarithm in (6–137) and is of order unity for most cases of interest.

(c) Bound-free Absorption

Quantum mechanically, one can also compute an absorption coefficient for bound-free transitions, when only one electron per atom is active in absorbing radiation. This has the form (Cl68)*

$$\alpha_{bf} = \frac{64\pi^4 m e^{10} Z^4}{3\sqrt{3}\, ch^6 n^5} \frac{g_{bf}}{v^3} \tag{7-75}$$

where n is the principal quantum number and g_{bf}, the Gaunt factor, again is of order unity and only mildly depends on n and v. This equation of course only holds when the photon energy exceeds the ionization energy χ_n in the nth state,

$$hv > \chi_n \sim \frac{2\pi^2 m e^4 Z^2}{n^2 h^2} \tag{7-76}$$

(d) Bound-bound Transitions

These have cross sections already discussed in the last section. They depend strongly on the actual structure of the individual atom, and do not give rise to a continuum absorption cross section as do the factors (a) to (c). As discussed in the next section, these cross sections play an important role in determining the radiative transfer rate through a stellar atmosphere; but they do not play a significant role in the stellar interior, where processes (a), (b), and (c) dominate radiative transfer rates (Chapter 8).

The opacity of low density ionized matter also is a measure of the radiant power emitted from unit volume at given temperature T. If the chemical composition of the plasma corresponds to the cosmic abundance (Fig. 1.7, Table 1.2), Figure 7.12 shows the radiated power; it assumes that self-absorption by the plasma can be neglected. At high densities, where that assumption no longer is valid, the plasma would simply radiate with a brightness characteristic of any blackbody at temperature T.

We note that for these low densities forbidden line emission dominates. *Forbidden transitions* are those for which dipole (and sometimes higher multipole) radiative transitions are not allowed by the *selection rules*. When a plasma is

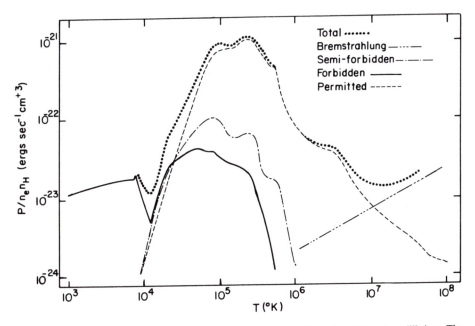

Fig. 7.12 Radiated power for unit volume of a low density ionized gas in collisional equilibrium. The power indicated is for electron and hydrogen concentrations of $n_e = n_H = 1 \text{ cm}^{-3}$. n_H represents the total number density for hydrogen atoms and protons. To obtain the total power radiated from unit volume, the ordinate would have to be multiplied by $n_e n_H$. The plasma considered has a chemical composition typical of cosmic sources (see Table 1.2, Fig. 1.7). (After D.P. Cox and E. Daltabuit, Co71a.)

at sufficiently low densities so that collisions are rare, a forbidden radiative transition with a correspondingly low transition probability may imply a metastable lifetime of seconds or years for the excited metastable atom. Since collisions are rare in the low density medium, these transitions may take place nevertheless. As the gas density increases collisions start dominating the de-excitation of the atom and the forbidden lines disappear.

If we know the lifetimes of a number of different metastable atoms in a hot interstellar nebula, we can often conclude a great deal about temperature and density conditions, by studying the forbidden lines.

In stars and dense stellar atmospheres, collisions between atoms and ions are frequent and no forbidden lines are expected; but even the earth's upper atmosphere is tenuous enough so that forbidden oxygen lines appear in auroral spectra.

7:12

7:13 CHEMICAL COMPOSITION OF STELLAR ATMOSPHERES—THE RADIATIVE TRANSFER PROBLEM

In order to determine the abundance of various chemical elements in the atmospheres of stars, we must be able to correctly interpret their absorption or emission spectra. This interpretation is a complicated process. First, it depends on the correct choice of a model of the stellar atmosphere. By this we mean that we have to choose an effective temperature T_e, a value for the star's surface gravity and a parameter ξ_t representative of the turbulent velocity in the atmosphere.

In interpreting individual Fraunhofer (absorption) lines in terms of the number density n_i of a given atom or ion, i, in a given state we find that the theory always yields expressions proportional to $n_i f g$, where f is the oscillator strength of the transition and g represents the statistical weight of the lower energy level—the level from which the transition takes place. For hydrogen, helium, and other one- and two-electron ions, the f values can be quantum mechanically computed. For such complex spectra as those of iron, however, f values must be obtained through laboratory experiments.

Another important parameter required for an abundance determination when the absorption line is very strong is the *damping constant* γ, which represents broadening of the line due to the intrinsically finite lifetime of the states and due to the shortening of this life through collisions with electrons and atoms.

We can define an *equivalent width* W_λ of a Fraunhofer line. It represents the total energy absorbed in this line, divided by the energy per unit wavelength emitted by the star in its continuum spectrum around wavelength λ. Figure 7.13 shows this relationship.

For very weak lines the amount of radiation absorbed and, hence, the equivalent width, depends linearly on the abundance n_i and on the product $g f n_i$. As W_λ approaches the Doppler width due to thermal and turbulent motion, the absorption line becomes saturated, and the *curve of growth* (Fig. 7.14), which represents the growth of W_λ with increasing material traversed, flattens out. For still stronger lines absorption in the wings of the lines becomes possible. Here the parameter γ (equation 7–51) determines the amount of radiation that is absorbed.

In determining the abundance of various chemical elements in stars, we have to keep in mind that the population of different atomic or ionic energy states depends quite critically on the atmospheric temperature and also to some extent on the surface gravity that determines the pressure. The Boltzmann and Saha equations are applied in these computations, on the assumption that the atmosphere is in thermodynamic equilibrium. Often, the f values for a given transition are not well enough understood; but in some such situations we can at least

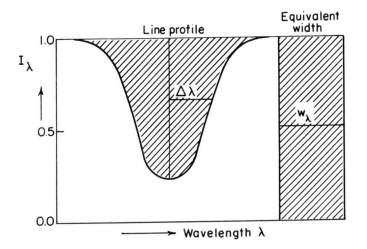

Fig. 7.13 Profile and equivalent width W_λ of a Fraunhofer line. The intensity of the continuum has been made equal to 1. The area under the line profile is equal to that of a completely "black" strip in the spectrum of width W_λ, usually measured in milliangstroms (after A. Unsöld, Un69).

obtain an idea of the relative abundances of an element in a given star compared to its abundance in the sun.

To relate quantitatively the total flux from a star or nebula to its chemical and physical properties, we proceed in the following way: The *intensity* $I(v)$ of radiation at spectral frequency v changes as it crosses a layer of matter of thickness dx. There is a loss of intensity through absorption and a corresponding gain through emission. For normal incidence on the layer, the total intensity change is

$$\frac{dI(v)}{dx} = -\kappa(v)\,\rho I(v) + j(v)\,\rho \qquad (7\text{–}77)$$

where the first term represents extinction (see equation 7–70) and $j(v)$ is the emission per unit mass. $j(v)$ may strongly depend on the radiation intensity itself as in the case of strong scattering or induced emission. When $j(v)$ is entirely due to induced emission, the opacity $\kappa^*(v)$ in (7–71) is the difference between $\kappa(v)$ and $j(v)/I(v)$.

Equation 7–77 can be rewritten as

$$\frac{1}{\rho\kappa(v)}\frac{dI(v)}{dx} = -I(v) + J(v) \qquad (7\text{–}78)$$

where $J(v) \equiv j(v)/\kappa(v)$ is called the *source function* and (7–78) is called the *equation of transfer* (Ch50)*.

In section 8:7 we will discuss the transfer of radiation from the center of a star to its periphery. It will then be necessary to consider not only normal incidence on a layer, but also incidence at other azimuthal angles θ. For arbitrary angles of incidence (θ, ϕ) we can express the energy density of radiation as

$$\int \rho(v)\, dv = \frac{1}{c} \iint I(v, \theta, \phi)\, d\Omega\, dv, \qquad (7\text{--}79)$$

where $I(v, \theta, \phi)$ is the intensity in the direction (θ, ϕ).

The radiative flux depends on the intensity in a related way:

$$F = \int F(v)\, dv = \iint I(v, \theta, \phi) \cos\theta\, d\Omega\, dv \qquad (7\text{--}80)$$

If we consider $I(v, \theta, \phi)$ to be a distribution function that specifies the angular distribution of radiation at frequency v, then $\rho(v)$ and $F(v)$ involve the zeroeth and first moments of this function. The second moment leads to the radiation pressure

$$P = \int P(v)\, dv = \frac{1}{c} \iint I(v, \theta, \phi) \cos^2\theta\, d\Omega\, dv \qquad (7\text{--}81)$$

This relation follows from the discussion of sections 4:5 and 4:7. The radiation pressure will be important in the theory of stellar structure, where hydrostatic equilibrium requires a balance between gravitational forces and pressure gradients. In some stages of stellar evolution, notably in stages leading to planetary nebulae, these gradients depend more strongly on radiant than on kinetic gas pressures. Radiant pressures also play a role in determining the atmospheric structure, particularly of giant and supergiant stars.

Let us still describe the factors that determine the shape of a spectral absorption or emission line seen in a star's atmosphere. We have already discussed factors that lead to broadening of a line. However, we still should mention that for gas at a given temperature T, the emission line intensity $I(v)$ will normally not exceed the blackbody intensity at that temperature and at frequency v. Stimulated and spontaneous emission will therefore tend to increase the brightness of an emission line in its wings, as radiation is transferred through the star's atmosphere. The center of the line may already have become saturated—reached its peak intensity—close to the surface of the star. This effect will lead to emission line broadening. Similarly absorption lines become broadened on passage through the cool outer portions of a star's atmosphere because absorption in the wings becomes increasingly probable as more matter is traversed. We talk about a

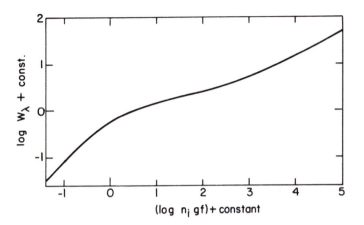

Fig. 7.14 Curve of growth showing increasing absorption with increasing amount of material traversed in a stellar atmosphere. The equivalent width W_λ is plotted against $n_i g f$ where n_i is the abundance of an element in a particular energy state, f is the oscillator strength of the Fraunhofer line, and g is the statistical weight of the absorbing state (after A. Unsöld, Un69).

curve of growth for a spectral line. By this we mean a plot of the equivalent width W_λ (Fig. 7.13) against the product $n_i f$ of the number of atoms n_i in a column of unit area through the atmosphere, and the oscillator strength of the transition f. Sometimes, as in Fig. 7.14, the curve of growth is a plot of W_λ not against $n_i f$ but against some other specified function—in this case $\log n_i g f$, where g is the statistical weight of the absorbing state.

ANSWERS TO SELECTED PROBLEMS

7-1. $E_n = -\dfrac{p_n^2}{2\mu} = -\dfrac{n^2 h^2}{2 r_n^2 \mu} = -\dfrac{Ze^2}{2 r_n}$, $\quad E_n = -\dfrac{Z^2 e^4 \mu}{2 n^2 h^2}$.

$$\text{Number of states} = \frac{\text{phase space volume}}{\text{volume of unit cell}} = \frac{16\pi^2 n^2 h^2 \cdot h}{h^3} \propto n^2.$$

7-2. $r \sim \dfrac{h}{\sqrt{p^2}}$ where $\dfrac{p^2}{2M} = \mathbb{V}_0 - \mathscr{E}_b$.

7-3. $\mathscr{E} = \dfrac{1}{2} \sum_i m_i v_i^2$

$$I = \sum m_i r_i^2$$

$$\text{and } \mathscr{E} = \frac{\omega L}{2} = \frac{L^2}{2I}$$

$$\delta\mathscr{E} = \frac{L}{I}\delta L = \omega h.$$

Since $L = h\{J(J + 1)\}^{1/2}$,

$$L = \sum m_i v_i r_i$$

$$L = \omega I, \quad \omega L = \sum m_i v_i r_i \omega$$

$$\mathscr{E}_J = \frac{L^2}{2I} = \frac{h^2(J + 1)J}{2I} \quad \text{and} \quad \delta\mathscr{E} = \mathscr{E}_J - \mathscr{E}_{J-1} = h^2 \frac{J}{I}.$$

For rapidly rotating massive objects $hJ \sim L$ and $\delta\mathscr{E} = L\,\delta L\,/I$.

7–4. $\frac{1}{2}I\omega^2 = \frac{3}{2}kT, \quad T = 100°\text{K}$

$$I \sim \frac{2}{5}mr^2 \sim \frac{2}{5}(10^{-23}\,\text{gm})(2 \times 10^{-8})^2$$

$$\omega \sim 5 \times 10^{12}\,\text{sec}^{-1}, \quad v = \frac{\omega}{2\pi} \sim 8 \times 10^{11}\,\text{Hz}, \quad \lambda \sim \frac{3}{8}\,\text{mm}.$$

7–5. In (4:18) we stated that the rotational excitation probability would be proportional to a Boltzmann factor. Its form is

$$\exp\left(-\frac{h^2 J(J + 1)}{2IkT}\right)$$

which is small for large J and low T.

7–6. $E = \frac{1}{2}I\omega^2, \quad I = \frac{2}{5}(10^{-15})(10^{-10})\,\text{g cm}^{-2}$

$$= kT \sim 1.4 \times 10^{-14}\,\text{cgs}$$

$$\therefore \omega \sim 8 \times 10^5\,\text{sec}^{-1}$$

$v \sim 10^5$ Hz, which is slightly below the AM radio band.

7–7. $\delta\mathscr{E}_{\text{grav}} = 2h\omega, \quad \delta\mathscr{E}_{\text{neutrino}} = \frac{1}{2}h\omega$ and results of Problem 7–3 are applied directly.

7–8. Probability $\propto \exp\left(-\frac{v_r^2 m}{2kT}\right)$

$$f(v_r) = \left(\frac{m}{2\pi kT}\right)^{3/2} \exp\left(-\frac{mv_r^2}{2kT}\right)$$

from 7–22, $v_r = \dfrac{\Delta\omega}{\omega_0} c$

$$I(\omega) = I_0 \exp\left(-\frac{mc^2 \, \Delta\omega^2}{2\omega_0^2 kT}\right)$$

From Problem 4–27 $\langle v_r^2 \rangle = \dfrac{kT}{m} = \dfrac{\langle \Delta\omega^2 \rangle}{\omega_0^2} c^2.$

7–10. $\sigma_{ab}(\omega) = \dfrac{2\pi e^2}{mc} f_{ab} \dfrac{\gamma/2}{(\omega - \omega_{ab})^2 + (\gamma/2)^2}$

$$\int_{-\infty}^{\infty} \frac{(\gamma/2)\, d\omega}{(\omega - \omega_{ab})^2 + (\gamma/2)^2} = \pi$$

$$\therefore \ \sigma = \int_{-\infty}^{\infty} \sigma_{ab}(\omega)\, d\omega = \frac{2\pi^2 e^2}{mc} f_{ab}$$

7–11. $\sigma_{\max} = \dfrac{2\pi e^2}{mc} f_{ab} \dfrac{1}{\gamma/2}$

where $\dfrac{\gamma}{2} = \dfrac{e^2 \omega_{ab}^2}{3c^3 m}, \quad \omega_{ab} = \dfrac{2\pi c}{\lambda}$

$$\therefore \ \sigma_{\max} = \frac{3}{2} \frac{\lambda^2}{\pi} f_{ab}.$$

7–12. $I_1 = \dfrac{I_{ab}}{\pi} \displaystyle\int_{\omega_{ab}-\gamma/2}^{\omega_{ab}+\gamma/2} \dfrac{(\gamma/2)\, d\omega}{(\omega - \omega_{ab})^2 + (\gamma/2)^2} = \dfrac{I_{ab}}{2}$

$I_2 = \dfrac{I_{ab}}{\pi} \displaystyle\int_{-\infty}^{\infty} \dfrac{(\gamma/2)\, d\omega}{(\omega - \omega_{ab})^2 + (\gamma/2)^2} = I_{ab}$

$I_1/I_2 = 1/2.$

7–13. $\gamma = \dfrac{2}{3} \dfrac{e^2}{mc^3} \omega_{ab}^2 \sim \dfrac{2}{3} \dfrac{e^2}{mc^3} \dfrac{4\pi^2 c^2}{\lambda^2} \sim \dfrac{1}{5\lambda^2}.$

7–14. $F_{\rm rad} = \displaystyle\int \sigma(\omega) \dfrac{I(\omega)\, d\omega}{c} = \dfrac{\sigma}{\Delta c} I_{TOT} = \dfrac{\sigma}{c} \dfrac{L_\odot}{4\pi r^2 \Delta}$

where Δ is the frequency bandwidth.

$$F_g = \frac{GM_\odot m}{r^2}, \qquad \sigma = \frac{2\pi^2 e^2}{mc} f \sim \frac{2\pi^2 e^2}{mc}$$

$$\therefore \frac{F_{\text{rad}}}{F_{\text{grav}}} = \frac{(\sigma/\Delta c)(L_\odot/4\pi r^2)}{GM_\odot m/r^2} = \frac{\sigma L_\odot}{\Delta 4\pi c G M_\odot m}.$$

7-15.   From (7-47) $\dfrac{d\mathscr{E}}{dt} = \dfrac{2e^2\omega^4 \langle r^2 \rangle}{3c^3} = \hbar\omega A_{n,n-1}$ (by 7-47 and 7-66)

$$\mathscr{E}_n = \hbar\omega = \frac{m}{2} \frac{e^4}{\hbar^2} \left\{ \frac{1}{(n-1)^2} - \frac{1}{n^2} \right\}$$

for n large $\omega \cong \dfrac{me^4}{\hbar^3} \dfrac{1}{n^3}$.

Take $a_n^2 = (n^2\hbar^2/me^2)^2 = 2\langle r^2 \rangle$ from (7-2) and (7-5), to obtain $A_{n,n-1} = 64\pi^6 me^{10}/3c^3 n^5 h^6$.

7-16.   In equations 7-63 and 7-65 we called the number of photons of a given polarization absorbed per particle per second $n(v, T) cB(v)$. In defining σ, we talked about photons of either polarization, so that $\sigma(v) = 2B(v)$.

$$\frac{I}{hv} = \int A(v)\, dv = A_{ab} \qquad (7\text{-}66)$$

$$I(\omega) = \frac{I_{ab}}{\pi} \frac{\gamma/2}{(\omega - \omega_{ab})^2 + (\gamma/2)^2} \qquad (7\text{-}42)$$

$$\sigma(\omega) = \frac{2\pi e^2}{mc} f_{ab} \frac{\gamma/2}{(\omega - \omega_{ab})^2 + (\gamma/2)^2} \qquad\qquad (7\text{-}59)$$

$$\therefore A(\omega) = \frac{1}{\pi} \frac{A_{ab}(\gamma/2)}{(\omega - \omega_{ab})^2 + (\gamma/2)^2} \, , \; A(v) = \frac{8\pi v^2}{c^2} B(v)$$

$$\therefore A(\omega) = \frac{2\omega^2}{\pi c^2} B(\omega)$$

$$= \frac{\omega^2}{\pi c^2} \sigma(\omega)$$

$$\therefore A_{ab} = \frac{8\pi^3 e^2 v^2}{mc^3} f_{ab}.$$

8

Stars

8:1 OBSERVATIONS

We do not know how stars are formed, nor just how they die. But we think we understand the structure of the most commonly found stars and the mechanisms by means of which they generate the energy we see as starlight.

How sure can we be of this understanding? How correct is the theory of stellar structure? Such questions are difficult to answer. Many of the most important stellar processes go on deep in a star's interior, while the observations available to us mainly concern themselves with surface characteristics. Conditions in the star's central regions must therefore be inferred, and our evidence is indirect.

Generally the merit of a theory is judged by the number of unrelated observations it can explain, and when the observations are indirect, a larger than usual body of data is desirable. For the theory of stellar structure and evolution, we have several different classes of observations:

(a) We have measurements on the masses of a variety of stellar types. However, the number of precision measurements is small. Each such measurement involves the detailed analysis of a stellar binary system (see section 3:5).

(b) We can determine the luminosity of a star quite accurately provided the star is near the sun where its distance is easily measured and interstellar extinction is negligible. In the past few years, bolometric magnitudes have become well established as satellite observations have increasingly provided far-infrared, far-ultraviolet and X-ray fluxes. These observations suggest frequent sizeable deviations from blackbody behavior.

(c) The surface temperature of stars can be obtained in three different ways.

(i) We can observe a color temperature (section 4:13);

(ii) we can determine the effective temperature if the star's angular diameter is known (section 4:13); or

(iii) we can observe the strengths of spectral lines representing transitions between various excited atomic states. Since the relative population of excited states is governed by the (temperature dependent) Boltzmann factor, the temperature can be computed directly, provided the relevant transition probabilities are known. These probabilities can be calculated or, preferably, observed in the laboratory.

These three independent techniques yield a satisfactory estimate of stellar surface temperature, but the temperature in the interior of stars remains unobserved.

(d) We can determine the angular diameter of a star by interferometric means (section 4:12). It can also be obtained indirectly if both the apparent magnitude and the temperature of the star are known. But we can determine the linear diameter of the star only if we know both the angular diameter and the star's distance. In some cases eclipse data from binaries can also yield stellar diameters.

The luminosity, diameter and surface temperature of a star are interrelated and involve no more than two independent parameters, provided the star's spectrum is simple enough to allow the assignment of a single representative temperature.

(e) The chemical makeup of stars is spectroscopically determined. The abundance of the different elements obtained in this way refers only to the surface layers of the stars. Using current observational techniques, we cannot directly verify conjectures about the composition in the stellar interior.

We find that the abundance of elements on the surfaces of stars varies from one type of star to the next. Normally, hydrogen is by far the most abundant constituent. By mass its concentration lies close to 70%. Helium, the next most abundant element, has a concentration of 25 to 30% by weight. Neon, oxygen, nitrogen, and argon follow in order of decreasing abundance; together they account for about 2% of the total mass. Carbon, magnesium, silicon, sulfur, and iron are next; each of these has an approximate abundance of the order of one part per thousand, by weight.

One task of a theory of stellar evolution must be the correct prediction of the abundance of chemical elements in different types of stars. The interior of stars is the most plausible place for heavier elements to be formed from the relatively pure hydrogen—or more probably from a hydrogen-helium mixture—that seems to have been the main constituent of the Galaxy at the time the earliest stars formed in globular clusters.

The relative abundances of the various isotopes of different elements are repeatedly found in similar ratios in stars, in the interstellar medium, in meteorite fragments, and in the earth's crust. The similarity of these ratios cannot be accidental, and the detailed explanation of the hundreds of known abundance ratios provides a severe task for the theory of stellar evolution.

8:1

(f) For a limited number of stars we have measurements on the surface magnetic fields. Peculiar stars of spectral type A are the best studied members of this group. Their high fields, which may be of order 10 kilogauss in strength, are relatively easy to measure. White dwarfs too are expected to have strong fields. $B \sim 10^5$ to 10^8 gauss may not be unusual, and a detailed study of such objects is likely to be valuable (Ke70). Eventually theories of stellar evolution will have to take magnetic fields into account. Current theories have not yet incorporated magnetic effects to any great extent.

(g) The surface rotational velocities of stars can be statistically determined from a study of different spectral classes. While projection effects preclude an analysis of the rotational velocity of individual stars, the statistical study of large numbers of stars has indicated that the young O and B stars have extremely high rotational velocities, that these velocities slowly drop until late stars of spectral type A are reached, and that redder stars of types G to M have very low rotational velocity (see Fig. 1.9 and Table A.4). We are only just starting to consider effects of rotation on stellar evolution.

(h) Similarly, a statistical study of stellar velocities in our galaxy (see Table A.6) tells us that the origins of stars of differing spectral types must be dissimilar. Presumably these stars were formed at different epochs in the Galaxy's life. A final theory of stellar evolution will have to consider such age differences for stars, and will have to take into account that the chemical composition of the interstellar gas from which stars are formed must change as the Galaxy ages.

(i) For a number of different stars, notably K-giants, O-stars, and nuclei of planetary nebulae, we now have data on mass loss. For the first two of these spectral types, the information comes from spectral measurements of gas outflow. In the case of planetary nebulae, we actually see the accumulation of ejected gas. Such evidence is important both for studying unstable states of stars, and for understanding the chemical changes that the interstellar gas may suffer when material, which has undergone nuclear reactions in stars, is returned to interstellar space. Nova and supernova ejecta also yield important data in this respect. Unfortunately we do not as yet have enough information to judge whether violent but infrequent explosive events dominate or merely contribute to the cycling of matter between stars and interstellar space.

(j) For the sun, we also have sparse data on internal rotation rates (Di67b) from solar oblateness studies; and we know of a lack of neutrino emission (Da68). For other stars such data are completely lacking. Similarly, we have information on solar cosmic ray and X-ray emission; X-ray data also exist by now for an appreciable number of other stars. In each of these cases, however, it is unclear how the circumstellar regions from which such radiation reaches us are affected by conditions prevailing inside the stars.

(k) Finally, a very important body of statistical information is contained in the Herzsprung-Russell and color magnitude diagrams (see Figs. A.3, 1.4, 1.5, and 1.6).

The distribution of stars within quite narrow confines on a H-R diagram sets a relatively detailed standard that must be met by any theory of stellar structure and evolution. The theory must prohibit the appearance of stars with characteristics corresponding to the empty regions of the diagram. Moreover, the relative density of stars in the populated portions of the H-R diagram must also be explained by such a theory. Finally, the theoretical evolution of a model star must always take it from one well-populated part of the diagram to another, without straying into an empty region or spending much time in a sparsely populated domain. For, if a correct model star strayed in this way, we would surely observe similar stray stars in nature.

The detailed structure of the H-R diagram then provides us with a truly severe observational criterion that must be met by theories of stellar evolution. An acceptable theory must explain the significance of the main sequence, the existence of the red giant and horizontal branches, the meaning of the variable turn-off point that has the main sequence joining the red giant branch at differing locations in different groups of stars, and a variety of other features.

We find then that we know too little about stellar masses and diameters to have these parameters play a hypercritical role in the theory of stellar structure, but the Hertzsprung-Russell diagrams that are available for a large number of different stellar groups and populations, and the tables of chemical abundances compiled for many astronomical objects, provide a wealth of observational detail against which to gauge the merit of our theories.

The purpose of the present chapter is to outline the main ideas involved in our current theories of stellar structure, and to show the extent to which these theories fit observations (Bu 57)*.

8:2 SOURCES OF STELLAR ENERGY

We have shown how one determines the overall characteristics of stars—their radii, masses, and luminosities. Knowing these, we can now ask ourselves what might cause stars to shine so brightly? Clearly they shine because they are hot. But why are they hot? What is the source of energy that can heat a star and replenish the energy that it loses so readily in the form of starlight?

Before these problems can be discussed, we may want to answer a somewhat different question: How much energy does a typical star radiate away in the course of its life? Here we may want to know the average luminosity of the star and its age at death—whatever form that death might take. It is not easy to decide how old a given star is, because stellar ages are far greater than the few hundred years during which reliable astronomical observations have been carried out; but two pieces of information are useful.

First, stars like the sun, occupy positions on the lower main sequence of the Hertzsprung-Russell diagram, and are found not to have noticeably changed either in brightness or in color since photographic techniques became well established around the turn of the century.

Second, the sun must be older than the earth which, as judged from the abundance of the radioactive uranium isotope U^{238} and its decay products, is more than four aeons old. The sun is thought to be of the order of 4.5 æ old. From paleontological evidence, we surmise that the temperature of the sun cannot have varied a great deal in the past 3×10^9 y during which life has existed on earth. Fossil remains that we find today indicate that liquid water must have been present on earth during this entire interval. Had the sun been somewhat cooler or hotter during this epoch, the oceans might have frozen or evaporated away, and the observed early forms of life would have died out.

We can therefore assume that, to rough approximation, the sun has radiated at its present rate for about five aeons. Since its luminosity is $L_\odot = 4 \times 10^{33}$ erg sec^{-1}, the total radiated energy emitted by the sun thus far is $\sim 6 \times 10^{50}$ erg. Because the solar mass is $M_\odot = 2 \times 10^{33}$ g, this amounts to an an energy-to-mass ratio of 3×10^{17} erg g^{-1}.

We ask ourselves whether this much energy could be made available by chemical reactions, or else through slow gravitational contraction which, as seen from equation 4–136, yields radiant energy in quantities that are of the order of the increase in absolute value of the potential energy gained by the contracting body.

Neither of these sources turns out to be adequate. The energy yield of chemical reactions normally does not exceed 100 kilocalories, or about 4×10^{12} erg per gram. If the sun had to depend on chemical sources, its total age would be no greater than $\sim 5 \times 10^4$ years. This is factor of 10^5 too short.

If we assume, for purposes of a rough estimate, that the sun has the same density throughout, its total potential energy would be

$$\mathbb{V} = -\int_0^{R_\odot} \left(\frac{4\pi}{3}\right) \rho r^3 \frac{G}{r} (4\pi \rho r^2)\, dr = -\frac{3}{5} \frac{M_\odot^2 G}{R_\odot} \tag{8–1}$$

which amounts to $\sim 2 \times 10^{48}$ erg. This corresponds to 10^{15} erg per gram, still two or three orders of magnitude short of the required energy. Even a hundredfold density increase at the center of the star could not change this result significantly.

We cannot, on these grounds alone, rule out a very dense central core with $\rho \sim 10^{15}$ g cm^{-3} and radius $R \sim 10^5$ cm: Approximately the right amount of gravitational energy would then be available. But while this source of energy seems to be important, for very compact stars, it appears to play no significant role for normal stars.

The only remaining source of energy involves nuclear reactions. The high abundances of hydrogen and helium found in the universe strongly suggest that hydrogen is transmuted into helium at the centers of stars. The energy released per gram of hydrogen is very large in this reaction.

We note that the mass difference between four hydrogen atoms and one atom of helium is

$$4m_H - m_{H_e} = 0.029m_H \qquad (8-2)$$

The transmutation of hydrogen into helium therefore includes a mass loss of the order of 7×10^{-3} g for each gram of converted hydrogen. Since the energy given off in the annihilation of mass m is mc^2, this amounts to an energy liberation of 6×10^{18} erg/g—ample compared to the amount required, even if only a fraction of a star's hydrogen content is converted into helium (Be39).

If we now ask about the lifespan of stars on the main sequence and about the rate at which stars are born, we can proceed in the following way: Let us first assume that we know the lifespan τ_i for a given type, i, of main sequence star. Let the number density in the Galaxy be ϕ_i for this kind of star. We can then define a birthrate function—usually called the *Salpeter birthrate function* ψ_i

$$\psi_i = \frac{\phi_i}{\tau_i} \qquad (8-3)$$

giving the rate of star formation in unit volume of the Galaxy. For disk population stars the formation rate will of course be high only in and near the Milky Way disk, while the birthrate will be negligible in the halo.

We can also obtain a very rough estimate of the age of a star when it moves off the main sequence. Suppose that a certain fraction of the stellar mass $f(M)$ needs to be exhausted of hydrogen before the star moves onto the red giant branch. If the initial composition of the stellar material contains a fraction (by mass, not by number of atoms) X, in the form of hydrogen available for nuclear conversion, the energy $\mathscr{E}$ liberated by the star while it still resides on the main sequence is

$$\mathscr{E} = 6.4 \times 10^{18} f(M) XM \text{ erg} \qquad (8-4)$$

The numerical factor gives the energy in ergs liberated by one gram of hydrogen converted into helium. The time taken to expend this energy is just the energy $\mathscr{E}$ divided by the star's luminosity L. Now Fig. 8.1 shows that the mass-luminosity relation for main sequence stars is roughly $L = L_\odot (M/M_\odot)^a$, where $3 \lesssim a \lesssim 4$.

Fig. 8.1 Mass-luminosity diagram for main sequence stars. The dashed and solid lines are different theoretical fits to the data represented by solid points. (Reprinted with permission from the Royal Astronomical Society.)

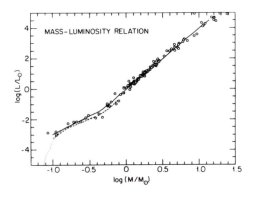

Taking $a \sim 3.5$, the star's life τ on the main sequence becomes

$$\tau = 6.4 \times 10^{18} X f(M) \left(\frac{M_\odot}{M} \right)^{5/2} \frac{M_\odot}{L_\odot} \tag{8-5}$$

$f(M)$ is of order 15 % for stars with solar composition for which $X \sim 0.7$. Inserting these two numerical values, making use of the mass-luminosity relation once more, and converting the time scale to years we find

$$\tau \sim \left(\frac{L_\odot}{L} \right)^{5/7} \times 10^{10} \text{ y} \tag{8-6}$$

The sun should therefore have a total lifespan of 10^{10} y, while O stars, which are some ten thousand times brighter, should survive only a few million years.

8:3 REQUIREMENTS IMPOSED ON STELLAR MODELS

Granted that sufficient energy is available from *hydrogen burning* (8–2) and perhaps from other nuclear reactions, we still need to investigate whether the hypothesis of nuclear energy conversion also fits all the other characteristics observed in stars. These are:

(a) Conditions inside stars must be compatible with adequate nuclear reaction rates. These rates provide for an energy generation budget of the order of the observed luminosity of stars. The energy released at the star's surface must further be predominantly in the form of visible, ultraviolet or infrared radiation since most of a normal star's radiation is actually emitted at these wavelengths. If a predominant fraction of the generated energy were channeled elsewhere, say into neutrino emission, then we would still be faced with the problem of accounting for the visual starlight.

(b) We will find that nuclear reaction rates depend on the temperature, density, and chemical composition of the matter in stars, and it will be important to discover whether the values of these parameters, needed to maintain a given luminosity, are compatible with stable stellar structure.

Pressure equilibrium, for example, must be maintained throughout the star. But this pressure is determined by two factors. First, the pressure in any region is determined by the local temperature, density, and chemical composition. The relationship between these quantities is summarized in an equation of state such as the ideal gas law or some similar expression. Second, the local pressure must be just able to support the weight of material lying overhead—matter at larger radial distance from the center of the star. This is called the condition of *hydrostatic equilibrium.*

If the temperature and density are too high, the local pressure becomes too large and the star expands. If the pressure is too low, the star will contract. We will see that any appreciable deviation from pressure equilibrium leads to a readjustment that takes no more than about an hour. A star that lives for many *aeons* must therefore be very close to pressure equilibrium throughout, unless the star pulsates.

(c) The energy generated at the center of a star must be able to reach the surface within a time small compared to the age of the universe or the evolutionary age of the star; otherwise, the whole life of the star would have to be described by transient conditions, and the stable characteristics of main sequence stars could not be explained.

(d) The temperature at any given distance from the center of the star must not only lead to the correct pressure (condition b), it must also be compatible with adequate energy transfer rates to assure that the luminosity just equals the rate of energy generation (condition c).

(e) At the center of the star the *luminosity* must be zero. This means that there is no finite outflow of energy, no mysterious source pouring out energy from an infinitesimal volume about the center of the star.

At the same time, there can be no more than an infinitesimal mass enclosed in an infinitesimal volume about the center of the star. These two requirements impose boundary conditions on the differential equations implied by requirements (a) to (d).

(f) At the surface of the star the pressure and temperature can usually be taken to be very small compared to values found in the central regions. This follows from the equation of state and from condition (b) which required pressure balance throughout the star. It is a statement of the fact that stars have high internal pressures and that the boundary between star and surrounding empty space is relatively sharp. Nevertheless, some caution has to be observed in applying this last condition; and differences will arise between early spectral-type stars—where energy is transported in the surface layers primarily by radiation—and late spectral types whose surfaces are convective.

8:4 MATHEMATICAL FORMULATION OF THE THEORY

The requirements described above can be summarized in a number of differential equations. In giving this formulation we will find it convenient to follow a procedure slightly different from that of section 8:3.

(a) The change of pressure dP on moving a distance dr outward from the center of a star is

$$dP = -\frac{\rho G M(r)}{r^2} dr \qquad (8-7)$$

where ρ is the local density and $M(r)$ is the mass enclosed by a surface of radius r. This increment of pressure is produced by the gravitational attraction between $M(r)$ and the mass $\rho\, dr$ enclosed in the incremental volume of height dr and unit base area. G is the gravitational constant.

(b) The change of mass $dM(r)$ on moving a distance dr outward from the center is

$$dM(r) = 4\pi r^2 \rho\, dr \qquad (8-8)$$

(c) The change of luminosity $L(r)$ within an increment dr at distance r from the center of the star is

$$dL(r) = 4\pi r^2 \rho \varepsilon\, dr \qquad (8-9)$$

where ε is the energy generation rate per unit mass.

(d) In general, this generation rate is a function of the local density ρ, temperature T, and the mass concentrations X_i of elements i. Hydrogen and helium mass concentrations are usually labeled X and Y, respectively:

$$\varepsilon = \varepsilon(\rho, T, X, Y, X_i) \qquad \text{where} \qquad i = 1, \ldots, n \qquad (8-10)$$

when n elements other than hydrogen and helium are present in significant amounts.

(e) The local pressure is related to the temperature, density, and chemical composition. We will find it convenient to write this in the form

$$P = P(\rho, T, X, Y, X_i) \qquad i = 1, \ldots, n \qquad (8-11)$$

8:4

because it will facilitate comparison of pressures derived from expressions 8–7 and 8–11. The right side of this equation is a general form for an equation of state which often is well approximated by the ideal gas law (4–37).

(f) Next, the temperature gradient must be related to the parameters that assure a stable luminosity profile throughout the star. Two possibilities arise here:

(i) If the star has a low opacity κ,

$$\kappa = \kappa(\rho, T, X, Y, X_i), \qquad i = 1, \ldots, n \tag{8–12}$$

so that light can travel long distances within the star before being absorbed or scattered, then no large temperature gradients can arise. In this case the transfer of energy is achieved by radiation alone. The photons are emitted, scattered, absorbed, re-emitted many times; and their energy and number density changes as they diffuse through the star in a complex random walk that eventually takes them from the center to the star's surface. There they start on their long journey through space.

(ii) If the opacity is high, this random walk may be excessively slow. The center of the star then becomes too hot and the stellar material starts to convect. A convective pattern of heat transfer sets in and, as we will see below, the temperature gradient is given by the so-called *adiabatic lapse rate* which depends on the ratio of heat capacities of the material, $\gamma = c_p/c_v$ (see section 4:18).

Corresponding to these two alternatives, we can derive temperature gradients of the form

$$\frac{dT}{dr} = F_1[\kappa, L(r), T, r] \qquad \text{for radiative transfer} \tag{8–13}$$

or

$$\frac{dT}{dr} = F_2(T, P, r, \gamma) \qquad \text{for convective transfer} \tag{8–14}$$

(g) The two boundary conditions implied by (e) and (f) in section 8:3 are

(i) at $r = 0$, $M(r) = 0$, and $L(r) = 0$ \hfill (8–15)

(ii) at $r = R$, $T \ll T_{\text{central}}$, and $P \ll P_{\text{central}}$ \hfill (8–16)

where R is the star's radius. For purposes of computing hydrostatic pressures, the relations (8–16) are tantamount to writing

$$T(R) \approx 0, \qquad P(R) \approx 0 \tag{8–17}$$

Equations 8–7 to 8–17 constitute the foundations of the theory of stellar structure. We will examine them in greater detail throughout much of this chapter.

8:4

One point, however, is particularly interesting and should be mentioned now. The equations state nothing about the physical source of the generated energy. The overall structure and appearance of the star can therefore give no clue about whether nuclear reactions indeed are responsible for stellar luminosities, or which particular reactions predominate at any given evolutionary stage. We have to derive this information by indirect means—mainly by looking at the debris ejected by stars that become unstable, or by spectrally analyzing stars whose surfaces expose matter previously evolved at the center.

8:5 RELAXATION TIMES

Suppose we could artificially change the temperature or pressure within a star. After this perturbation stopped, the star would again relax to its initial temperature and pressure equilibrium.

The time required to relax is called the *relaxation time*. We will find that the relaxation time in response to a pressure change is very much faster than the time required to re-establish temperature equilibrium.

(a) We first wish to estimate the time required to reach pressure equilibrium. Let the perturbed pressure $P_p(r)$ differ from the equilibrium pressure $P(r)$ by a fractional amount f

$$P_p(r) - P(r) = fP(r) \tag{8–18}$$

This pressure acts on a mass $M - M(r)$ lying at radial distance greater than r, with a force $F = 4\pi r^2 fP(r)$. As a result, that material moves with an acceleration

$$\ddot{r} = \frac{4\pi r^2 f P(r)}{M - M(r)} \tag{8–19}$$

We suppose that a displacement Δr amounting to a fraction g of the total radius R is required to relieve the pressure difference

$$\Delta r = gR \tag{8–20}$$

Then the time required to obtain this displacement with the acceleration given in equation 8–19 is of the order of

$$\tau_P \sim \left(\frac{2\Delta r}{\ddot{r}} \right)^{1/2} = \left[\frac{gR[M - M(r)]}{2\pi r^2 f P(r)} \right]^{1/2} \tag{8–21}$$

Let us compute the approximate value of τ_P. We can estimate $P(r)$ and $M(r)$ by assuming a uniform density throughout the star, and considering a star with

8:5

one solar mass $M = M_\odot = 2 \times 10^{33}$ g contained in one solar radius $R = R_\odot = 7 \times 10^{10}$ cm. The density then is $\rho \sim 1$, and from (8–7)

$$P(r) = -\int_R^r \frac{4\pi}{3} \rho^2 r G \, dr = \frac{2\pi}{3} \rho^2 G (R^2 - r^2) \tag{8–22}$$

Let us choose $r \sim R/2$, then

$$P\left(\frac{R}{2}\right) \sim 10^{15} \text{ dyn cm}^{-2}$$

$$M(R) - M\left(\frac{R}{2}\right) \sim 2 \times 10^{33} \text{ g}$$

and

$$\tau_P \sim 5 \times 10^3 \sqrt{\frac{g}{f}} \text{ sec}$$

For small perturbations, g/f can be expected to be of order unity and the relaxation time is of the order of an hour.

We note that the speed of propagation of pressure information is roughly $(P/\rho)^{1/2}$ (see equation 4–31). This speed is of order 3×10^7 cm/sec. Pressure information can therefore be conveyed over distances $R_\odot$ in $\sim 2 \times 10^3$ sec, a time comparable to the pressure adjustment time.

PROBLEM 8-1. (a) Show that the temperature $T(R/2)$, under the conditions assumed above, is $\sim 10^7 \, ^\circ$K.

(b) One difference between a planet and a star (Sa70a) is that for planets Coulomb forces on electrons and ions are more important than gravitational forces. The opposite is true of a star. Let $\mathscr{E}_c$ be a typical Coulomb interaction energy. Show that a planet's mass M_p is of order

$$M_p \lesssim \frac{1}{\rho^{1/2}} \left[\frac{\mathscr{E}_c}{G A m_H} \right]^{3/2} \sim \frac{R \mathscr{E}_c}{G A m_H} \tag{8–23}$$

where A is an average atomic mass, measured in atomic mass units, and m_H is the mass of the hydrogen atom.

(c) If $\mathscr{E}_c$ is of the order of binding energies of solid material, about 10^{-11} erg, show that Jupiter, which consists largely of hydrogen, lies near the upper limit of the mass range for planets.

(b) Next we wish to estimate the time taken to transport heat from one point within the star to another. If the transport process is radiative, the time can be computed as a random walk process; we only need to know the mean free path of radiation, and that is given (see 7:12) by the opacity of the material κ. When a beam of n photons passes through a layer of thickness dl, a fraction dn will be absorbed or scattered by the material. The loss of photons from the beam can then be expressed as

$$dn = -n\kappa\rho\, dl \qquad (8\text{--}24)$$

where ρ is the density of the material. We note that we have not gone into detail about the scattering process. Some processes strongly favor light scattering into a forward direction. Such scatterers make a medium much less opaque than isotropic scattering centers. We will assume here that the scattering is isotropic. Alternatively, we could count a photon as being lost from the beam only after a large number of collisions has increased its angle with the original direction of propagation significantly, say to 90°, so that all memory of the original direction is lost. We made a similar assumption about electron scattering in (6:16). We wish to calculate the mean free path of the photons under such conditions. Integrating equation 8–24 we obtain

$$n = n_0 e^{-\kappa\rho l} \qquad (8\text{--}25)$$

The mean distance $\langle l \rangle$ traveled by a photon before it is absorbed or strongly scattered is then

$$\langle l \rangle = -\frac{\displaystyle\int_0^{n_0} l\, dn}{n_0} = \int_0^{\infty} l\kappa\rho e^{-\kappa\rho l}\, dl = \frac{1}{\kappa\rho} \qquad (8\text{--}26)$$

For a star like the sun, $\kappa\rho$ is of order unity, and the mean free path is of the order of a centimeter. To traverse a distance of the order of the solar radius $R \sim 10^{11}$ cm, we would require 10^{22} steps, which would cover a total distance $\sim 10^{22}$ cm. The total time taken is $\sim R^2\kappa\rho/c$ sec when we do not count the time required between absorption and reemission. The time constant therefore is at least of the order of 10^{11} sec, several thousand years.

(c) Energy can also be transported by convection, provided high enough thermal gradients can be set up. A buoyancy force then accelerates a hot blob of matter upward and returns cooler matter down toward the center of the star. For nondegenerate matter we can take $\Delta\rho$, the density difference between hot and cold material, to be

$$\Delta\rho \sim \frac{\rho}{T}\Delta T \qquad (8\text{--}27)$$

the upward force on unit volume of the hotter material is

$$F(r, \rho, \Delta T) = \frac{M(r)G}{r^2}\frac{\rho}{T}\Delta T \qquad (8\text{--}28)$$

8:5

which leads to a convective motion accelerated at a rate

$$\ddot{r} = \frac{M(r)G}{r^2} \frac{\Delta T}{T} \qquad (8\text{--}29)$$

If the blob travels a distance of the order of one-tenth of a solar radius, the time required is

$$t = \left[\frac{2R_\odot}{10\ddot{r}} \right]^{1/2} \sim 3 \times 10^6 \text{ sec} \sim 1 \text{ month} \qquad (8\text{--}30)$$

for $M_r \sim 10^{33}$ g, $T \sim 10^7$ °K, and $\Delta T \sim 1$°K.

This is a rather fast transport rate. It sets in whenever radiative heat transfer is too slow to maintain thermal equilibrium within the stars. We will return to this stability problem in section 8:9, where we will also justify the choice of $\Delta T \sim 1$°K.

(d) If the center of a star is degenerate, the electrons can readily transport heat at a rate much faster than is possible by either radiative or convective means. This comes about since the electrons cannot give up their energy to other particles: All the lower electron energy states already are filled and there is no space for another electron that is about to lose energy. The mean free path for electrons therefore becomes extremely long, and heat transport proceeds swiftly. In the limiting case an electron could traverse the entire degenerate region and not lose energy until it reaches the nondegenerate surroundings. If the span in question amounts to a distance of the order of $R_\odot/10^2$, the traversal times at $T \sim 10^7$ °K would be of the order of one second. This represents the thermal relaxation time for the degenerate core of a star.

8:6 EQUATION OF STATE

The equation of state needed to define the pressure in terms of temperature, density, and composition depends on whether (i) conditions at the center of a star are nondegenerate or degenerate, and (ii) the temperature is high enough to involve relativistic behavior.

(a) Nondegenerate Plasma

At the high temperatures found within stars all but the heaviest elements are completely ionized. Under these conditions the electrons and ions are far apart compared to their own radii since electrons and bare nuclei have radii of order

8:6

10^{-13} cm. The ideal gas law can therefore be expected to hold:

$$P = nkT \qquad (4\text{--}38)$$

where n is the number of particles in unit volume.

In Table 8.1 we enumerate the contribution to the number density by the various particles.

Table 8.1 Number Densities

	Number of Ions	Number of Electrons
Hydrogen	$\dfrac{X\rho}{m_H}$	$\dfrac{X\rho}{m_H}$
Helium	$\dfrac{Y\rho}{4m_H}$	$\dfrac{Y\rho}{2m_H}$
Others	$\dfrac{Z\rho}{\langle A \rangle m_H}$	$\dfrac{Z\rho}{2m_H}$

The symbols X, Y, Z, represent the concentration, by mass, of hydrogen, helium, and heavier elements, respectively. $\langle A \rangle$ is the mean atomic mass of the heavier elements. In the last column of the table the number of electrons contributed by the heavier elements is obtained on the hypothesis that the number of electrons per atom is $\langle A \rangle / 2$. This is a fairly good approximation for the less massive elements. The number of ions contributed by the heavier elements amounts to a negligibly small fraction of the total population—only about one part per thousand. The total number density of particles to be inserted in the ideal gas law relation, therefore, is roughly

$$n = \frac{\rho}{m_H}\left[2X + \frac{3}{4}Y + \frac{1}{2}Z \right] \qquad (8\text{--}31)$$

and the equation of state reads

$$P = \frac{\rho k T}{m_H}\left[2X + \frac{3}{4}Y + \frac{1}{2}Z \right] \qquad (8\text{--}32)$$

At first we might think that P represents the total pressure; but it does not. It is only the pressure contribution due to particles. A further pressure, due to the

8:6

presence of electromagnetic radiation, must be added to the particle pressure to give the total pressure. This is true both for the nondegenerate and the degenerate situation.

We had already found in section 4:7 that the radiation pressure has a value numerically equal to one-third of the energy density. Inside the star that density is aT^4; the refractive index is practically unity, and the relationship of Problem 4–21 reduces to equation 4–72. Hence

$$P_{Rad} = \frac{aT^4}{3} \tag{8–33}$$

The equation of state for nondegenerate matter now reads

$$P_{Total} = \frac{\rho k T}{m_H} \left(2X + \frac{3}{4}Y + \frac{1}{2}Z \right) + \frac{aT^4}{3} \tag{8–34}$$

(b) Degenerate Plasma

The maximum number of electrons that can occupy unit volume is (see section 4 :11)

$$n_e = \frac{8\pi}{3} \frac{p_0^3}{h^3} \tag{8–35}$$

where p_0 is the momentum corresponding to the Fermi energy. The number density of electrons also can be written as

$$n_e = \left(X + \frac{1}{2}Y + \frac{1}{2} Z \right) \frac{\rho}{m_H} \equiv \frac{1}{2}(1 + X) \frac{\rho}{m_H} \tag{8–36}$$

since

$$X + Y + Z = 1 \tag{8–37}$$

The pressure contribution due to isotropically moving electrons is then given through equations 4–27, 4–28, and 4–30, as

$$P_e = \int_0^{p_0} \int_0^{2\pi} \int_0^{\pi/2} 2n_e(p)\, p \cos\theta\, v \cos\theta \sin\theta\, d\theta\, d\phi\, dp \tag{8–38}$$

$$= \frac{1}{3} \int_0^{p_0} \frac{8\pi p^2}{h^3} pv\, dp \tag{8–39}$$

where $n_e(p)$ is the number density of electrons having momenta in the range p to $p + dp$.

(i) In the nonrelativistic case $v = p/m_e$ and the electron pressure is

$$P_e = \frac{8\pi}{15} \frac{p_0^5}{m_e h^3} \tag{8-40}$$

Substituting for p_0 from (8–35) and (8–36), equation 8–40 becomes

$$P_e = \frac{h^2}{20 m_e m_H} \left(\frac{3}{\pi m_H} \right)^{2/3} \left(\frac{(1 + X)}{2} \rho \right)^{5/3} \tag{8-41}$$

(ii) In the relativistic case $v \sim c$, and equation 8–38 integrates to

$$P_e = \frac{2\pi c}{3 h^3} p_0^4 \tag{8-42}$$

Using the same device to eliminate p_0, we obtain

$$P_e = \frac{hc}{8 m_H} \left(\frac{3}{\pi m_H} \right)^{1/3} \left(\frac{1 + X}{2} \rho \right)^{4/3} \tag{8-43}$$

To obtain the total pressure we need to add the pressures P_i contributed by individual ions. These normally are nondegenerate, as was pointed out in section (4:15).

$$P_i = \left(X + \frac{1}{4} Y \right) \frac{\rho k T}{m_H} \tag{8-44}$$

Finally we have to add the radiation pressure from equation 8–33 to obtain

$$P_{Total} = P_e + P_i + P_{Rad} \tag{8-45}$$

8:7 LUMINOSITY

We have estimated the time required for a star to recover from a thermal perturbation when a range of different conditions prevails within the star. However, we still have to ask ourselves "When does each of these different conditions predominate? Under what circumstances is radiative heat transfer dominant? When is convection a major contributor, and what are the conditions that favor degeneracy?" These are the problems we must look at next. When we obtain an answer we will also be able to quantitatively express the rates of energy transfer that add to give the total luminosity of a star.

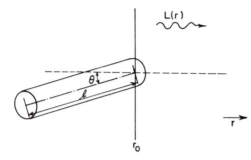

L(r)

Fig. 8.2   Illustration to show relation between luminosity and temperature gradient.

r₀

The total flux at radial distance $r = r_0$ from the star's center is the difference between the outward and inward directed energy flow. Let the temperature at r_0 be T_0. Radiation passes through the surface r_0—which can be assumed to be plane since the radiation mean free path is very small compared to r_0—at all values of azimuthal angle θ (Fig 8–2). The unattenuated flux originating at a distance l along direction θ, which would pass through the surface in unit time, would be

$$aT^4(l, \theta) \cdot c \cos \theta \cdot \frac{2\pi \sin \theta \, d\theta}{4\pi} \tag{8–46}$$

where

$$T(l, \theta) = T_0 - \frac{dT}{dr} l \cos \theta \tag{8–47}$$

$c \cos \theta$ represents the cylindrical volume from which radiation crosses unit area of the surface in unit time, and $2\pi \sin \theta \, d\theta/4\pi$ gives the fraction of the total solid angle at (l, θ) containing the radiation that will pass through the appropriate unit area, at r_0.

In actuality, radiation from (l, θ) does not reach r_0 unattenuated. A photon originating at l only has a probability $\pi(l)$ of reaching r_0:

$$\pi(l) = \kappa\rho e^{-\kappa\rho l} \tag{8–48}$$

This follows from (8–24) and also gives the proper normalization

$$\int_0^\infty \pi(l) \, dl = 1 \tag{8–49}$$

The radiant flow through unit area is now formally given as

$$F(r_0) = \int_0^\infty \int_0^\pi a \left(T_0 - \frac{dT}{dr} l \cos \theta \right)^4 \cdot c \cos \theta \cdot \frac{2\pi \sin \theta \, d\theta}{4\pi} \pi(l) \, dl \tag{8–50}$$

8:7

However, to obtain the actual radiative flux we must decide what value of κ to use in equation 8–48. Equation 8–24 did not take into account stimulated emission which, as explained in section 7:12, is important. On the other hand, if we use $\kappa_*(v)$ from (7–71) we have to properly average over all frequencies to arrive at a suitable mean opacity. This will be done in section 8:8 below.

PROBLEM 8-2. The luminosity $L(r)$ at any radial distance within the star is

$$L(r) = 4\pi r^2 F(r) \qquad (8-51)$$

Show by integration that to first order

$$L(r) = -\frac{16\pi ac}{3\kappa\rho} r^2 T^3 \frac{dT}{dr} \qquad (8-52)$$

8:8 OPACITY INSIDE A STAR

In section 7:12 we had already discussed the four sources of opacity: electron scattering, free-free transitions, free-bound transitions, and bound-bound transitions. However, we have not yet indicated how to compute the mean opacity obtained from these four contributing factors. It is that opacity that has to be used in expression 8–52.

Two factors enter into consideration of the mean opacity to be used in this expression. First, we have to average over all radiation frequencies; but clearly those frequencies at which the radiation density gradient is greatest should receive greater weight in the averaging process if the opacity is to give an accurate assessment of the radiative transfer rate. Secondly, those frequency ranges in which the opacity is smallest potentially make the greatest contribution to energy transport. We therefore will be more interested in averaging $1/\kappa(v)$ rather than $\kappa(v)$.

Let us write (8–52) in its more fundamental form, involving energy density $\rho(v)$ of radiation at frequency v and temperature T (see equation 4–71).

$$L(r, v) = \frac{-4\pi r^2}{3\rho\kappa^*(v)} c \frac{d\rho(v)}{dr} \qquad (8-53)$$

Here we have defined a contribution $L(r, v)$, at frequency v, to the total luminosity $L(r)$ at r; and we have set the total energy density U equal to the blackbody energy density. $\kappa^*(v)$ is the opacity at frequency v that takes account of stimulated

8:8

emission:

$$L(r) = \int_0^\infty L(r, v)\, dv \quad \text{and} \quad U = \int_0^\infty \rho(v)\, dv = aT^4 \qquad (8\text{--}54)$$

We can neglect bound-bound transitions since they play a negligible role in the stellar interior. Equation 7–71 therefore simplifies to

$$\kappa^*(v) = \left[\kappa_{bf}(v) + \kappa_{ff}(v) \right] (1 - e^{-hv/k}) + \kappa_e \qquad (8\text{--}55)$$

and we can define a mean opacity

$$\frac{1}{\kappa} = \frac{\displaystyle\int_0^\infty \frac{1}{\kappa^*(v)} \frac{d\rho(v)}{dT} \frac{dT}{dr}\, dv}{\displaystyle\int_0^\infty \frac{d\rho(v)}{dT} \frac{dT}{dr}\, dv} \qquad (8\text{--}56)$$

called the *Rosseland mean opacity*, in which (4–71) can be substituted for $d\rho(v)/dT$. The Rosseland mean opacity, as can be seen from (8–53), does indeed favor the frequencies important to the transfer process, by using the energy density gradient $d\rho(v)/dr$ as a weighting function for $1/\kappa^*(v)$, which is a measure of the mean free path at frequency v. The opacity at any frequency is the sum of contributions from bound-free (bf) and free-free (ff) transitions, and from electron scattering (e).

$\kappa_{bf}(v)$ and $\kappa_{ff}(v)$ themselves are sums over the opacity contributions of the individual states of excitation n, of the various atoms and ions A present at radial distance r, in the star

$$\kappa_{bf}(v)\rho = \sum_{A,n} \alpha_{bf} \left(\frac{X_A \rho}{A m_H} \right) N_{A,n} \qquad (8\text{--}57)$$

$$\kappa_{ff}(v)\rho = \sum_A \int \alpha_{ff} \frac{X_A \rho}{A m_H} n_e(v)\, dv \qquad (8\text{--}58)$$

Here $X_A \rho / A m_H$ is the number density of atoms of kind A, X_A is the abundance by mass of atoms or ions with mass number A, m_H is the mass of a hydrogen atom, and $N_{A,n}$ is the fraction of these atoms or ions in the nth excited state. $n_e(v)$ is the number density of electrons in a velocity range dv around v. α_{bf} and α_{ff} are the atomic absorption coefficients defined in (7–74) and (7–75). As shown in (7–72),

$$\kappa_e \rho = \sigma_e n_e \qquad (8\text{--}59)$$

where the right side is the product of the electron number density and the Thomson (or—at high energies—the Compton) scattering cross section.

8:8

To evaluate $N_{A,n}$ we make use of the *Saha equation* (4–105) which, for high ionization, leads to

$$N_{A,n} = n^2 \left[n_e \frac{h^3}{2(2\pi m_e kT)^{3/2}} e^{\chi_n/kT} \right] \qquad (8\text{–}60)$$

where we have considered that most of the ions are in the $r + 1$st ionization state. We can understand this equation in the following way:

χ_n is the energy needed to ionize the atomic species A from the nth excited state; m_e is the electron mass. Using a Bohr atom model this energy is (7–5)

$$\chi_n \sim \frac{2\pi^2 e^4 m_e}{h^2} \frac{Z'^2}{n^2} \qquad (8\text{–}61)$$

where Z' is the effective charge of the ion considered. Equation 8–61 assumes that all the excited atoms of a given species A will be in the same state of ionization at radial distance r from the star's center. In our present notation this means that in equation 4–105, $n_r/n_{r+1} = N_{A,n}$. We note that $N_{A,n}$ is proportaional to n^2. This is because the statistical weight g_r—the number of sublevels—of the nth excited state is $2n^2$ (see Problem 7–1). From section 4:16 we also have $g_e = 2$. Similarly, the ion can also exhibit two states of polarization $g_{r+1} = 2$. But for any given final state there are only two possible combinations of polarization, $g_{r+1}g_e = 2$.

Making use of equation 7–75 for α_{bf}, with χ_n from (8–61) substituted into this expression, we can now obtain

$$\kappa_{bf}(v) = \frac{2}{3} \sqrt{\frac{2\pi}{3}} \frac{Z'^2 e^6 h^2 \rho (1 + X) Z}{c A m_H^2 m_e^{1.5} (kT)^{3.5}} \left[\frac{1}{n} \frac{\chi_n}{kT} e^{\chi_n/kT} \left(\frac{kT}{hv} \right)^3 g_{bf} \right] (8\text{–}62)$$

Here Z is the metal abundance, by fraction of the total mass. We have summed (8–57) only for these constituents, since hydrogen and helium do not contribute significantly to the bound-free transitions. The summation over states has been neglected, since the lowest state n usually contributes most. We also have used an electron density from Table 8.1:

$$n_e = \tfrac{1}{2}(X + 1) \frac{\rho}{m_H} \qquad (8\text{–}63)$$

Equation 8–62 can be considerably simplified, if approximate values of the opacity suffice. For example, we can restrict our attention to those levels for which $\chi_n/kT \sim 1$, $hv/kT \sim 1$, since this makes use of the frequencies and ionization potentials that will contribute most to the opacity.

Constituents which would be ionized at lower temperatures, $\chi_n \ll kT$, already are almost fully ionized and have too few bound electrons to be effective, while those with higher χ_n values absorb too few of the photons present. Similarly the photons of frequency $v \sim kT/h$ are weighted most favorably by the Rosseland mean.

For most elements we can also choose a typical value $Z'^2/A \sim 6$.

With these approximations we obtain *Kramer's Law of Opacity* for bound-free absorption:

$$\kappa_{bf} = 4.34 \times 10^{25} Z(1 + X) \frac{\rho}{T^{3.5}} \frac{\langle g_{bf} \rangle}{f} \tag{8-64}$$

where $\langle g_{bf} \rangle$ is the mean Gaunt factor, which is always of order unity, and f contains correction factors—all of order unity also—which arise because of the approximations we have made. For free-free transitions, we can similarly obtain expressions (Sc58b)* :

$$\kappa_{ff} = \frac{2}{3} \sqrt{\frac{2\pi}{3}} \frac{e^2 h^6 (X + Y)(1 + X)\rho}{cm_H^2 m_e^{1.5}(kT)^{3.5}} \frac{1}{196.5} g_{ff}$$

$$\tag{8-65}$$

$$= 3.68 \times 10^{22} \langle g_{ff} \rangle (X + Y)(1 + X) \frac{\rho}{T^{3.5}}$$

where $\langle g_{ff} \rangle$ is the mean Gaunt factor (7–74). We note that if we had taken $\kappa(v)$ in equation 6–137 and substituted into (8–56) for $\kappa^*(v)$, we would have obtained an opacity expression proportional to $e^6 n^2 c^{-1}(m_e kT)^{-1.5}$ and a weighted mean function proportional to v^{-2} that would be proportional to $h^2(kT)^{-2}$. This is just the dependence found in (8–65). For electron scattering, (8–59) combined with the number of free electrons (8–63), yields

$$\kappa_e = \frac{4\pi}{3} \frac{e^4}{c^4 m_H m_e^2}(1 + X) \sim 0.19(1 + X) \tag{8-66}$$

Electron scattering is the main contributor to the opacity at low densities and high temperatures, where the interaction between electrons and ions is weakened.

Figure 8.3 shows the relative importance of scattering and absorption processes for different densities and temperatures. At high densities, where electrons become degenerate, heat is transferred most rapidly through conduction by these electrons.

Figure 8.4 shows the opacity as a function of temperature in stars whose composition is similar to that of the sun.

Thus far we have discussed radiative transfer only in the interior of a star. However, the equations of radiative transfer also play a dominant role in the transport of energy through stellar atmospheres (section 7:13).

8:8

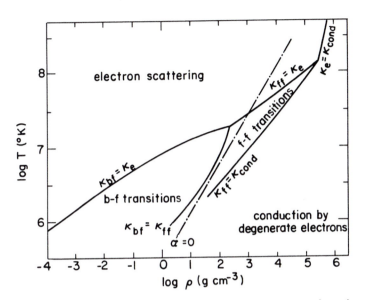

Fig. 8.3 Opacity as a function of density and temperature in a star of population I composition. The diagram is divided into four regions characterized by different mechanism of energy transport. The sources of opacity that dominate these mechanisms are electron scattering, bound-free transitions, free-free transitions, and the effective opacity that would describe the energy transport by degenerate electrons. The dashed line shows where the degeneracy *parameter* α (see equation 4-93) equals zero (after Hayashi, Hoshi, and Sugimoto, Ha62c).

PROBLEM 8-3. Using equations 8–7, 8–52, and the ideal gas law, show that the luminosity of stars should be roughly proportional to M^3.

We find, in reality, that main sequence stars more nearly obey the *mass-luminosity relation* (Fig 8.1):

$$L \propto M^a \qquad 3 \lesssim a \lesssim 4 \qquad (8\text{--}67)$$

Presumably this relation holds in main sequence stars because radiative transfer dominates there, while convective transfer (section 8:9 below) is more important in the giants, and degenerate electron transfer dominates in compact stars and in compact stellar cores.

However, radiative transfer is always present, even when these other processes dominate. The total energy transfer rate is then the sum of all the different transfer rates.

8:8

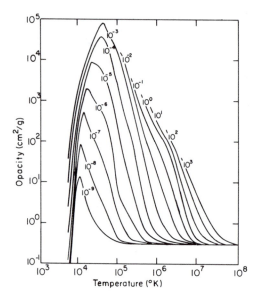

Fig. 8.4 Opacity for stars whose composition is similar to that of the sun. Each curve represents a different density value ρ, measured in g cm^{-3}. (After Ezer and Cameron, Ez65. With permission of the editors of *Icarus. International Journal of Solar System Studies*, Academic Press, New York.)

8:9 CONVECTIVE TRANSFER

Let us establish the conditions under which the temperature gradient becomes so large that the medium starts to convect, and the spherically symmetrical temperature distribution about the stellar center becomes unstable.

Consider an element of matter at some density ρ'_1 and pressure P'_1 surrounded by a region with exactly the same characteristics (ρ_1, P_1) (see Fig 8.5):

$$\rho'_1 = \rho_1 \qquad P'_1 = P_1 \tag{8–68}$$

The element is then moved to a new position, subscript 2, where its final pressure P'_2 equals the ambient pressure P_2:

$$P'_2 = P_2 \tag{8–69}$$

Using (8–30) we found that a convective motion of this kind is fast compared to the time required for radiative heat transfer, and we can therefore consider the process to be adiabatic. Equation 4–123 then implies that

$$\rho'_2 = \rho'_1 \left(\frac{P_2}{P_1} \right)^{1/\gamma} \tag{8–70}$$

Fig. 8.5 Convective outward motion of a low density "bubble." When thermal gradients become too high, such convective motion sets in and becomes the dominant vehicle for heat transport.

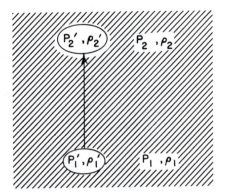

For the highly ionized plasma in a star, the ratio of heat capacities is $\gamma = c_p/c_v = 5/3$.

If the initial displacement of the element was upward, and we find that $\rho_2' > \rho_2$, then the element will be forced down toward its initial position, 1; the medium is then stable. However if $\rho_2' < \rho_2$, then the initial displacement leads to further motion along the same direction—upward—and the medium is unstable. Convection can then set in.

The condition for stability therefore is

$$\rho_2 < \rho_2' = \rho_1' \left(\frac{P_2'}{P_1'} \right)^{c_v/c_p} = \rho_1 \left(\frac{P_2}{P_1} \right)^{c_v/c_p} \tag{8-71}$$

where we have made use of expressions 8-68, 8-69, and 8-70. This can be rewritten as

$$\frac{d\rho}{\rho} < \left(\frac{P + dP}{P} \right)^{c_v/c_p} - 1 = \frac{c_v}{c_p} \frac{dP}{P} \tag{8-72}$$

In terms of radial gradients this becomes

$$\frac{1}{\rho} \frac{d\rho}{dr} < \frac{c_v}{c_p P} \frac{dP}{dr} \tag{8-73}$$

which, with the ideal gas equation 4-37, leads to the stability condition

$$\frac{dT}{dr} > \frac{T}{P} \left(1 - \frac{c_v}{c_p} \right) \frac{dP}{dr} \tag{8-74}$$

Both dP/dr and dT/dr have negative values. The right side of (8-74) is called the adiabatic temperature gradient, and we conclude that stability will prevail when

8:9

the absolute value of the temperature gradient is less than that of the adiabatic gradient. When the absolute value of dT/dr becomes larger than the absolute value of the adiabatic gradient, instability sets in and heat is transferred by convection.

To compute the heat transfer rate we have to know four quantities: the velocity v of the buoyant element, its heat capacity C, its density, and the temperature differential ΔT between the element and the surroundings to which it finally imparts this temperature. The heat transport rate per unit area is then

$$H = C\rho v \, \Delta T \tag{8–75}$$

Here C is the heat capacity under the assumed adiabatic conditions. Using the acceleration given in equation 8–29 and assuming transport over a distance one-tenth of a solar radius, the mean velocity v is of order $\left[\ddot{r} \dfrac{R_\odot}{10} \right]^{1/2}$

$$
\begin{aligned}
H &\sim C\rho \left[\frac{GM(r)}{Tr^2} \frac{R_\odot}{10} \right]^{1/2} (\Delta T)^{3/2} \\
&\sim C\rho \left[\frac{GM(r)}{Tr^2} \right]^{1/2} \left(\frac{d\,\Delta T}{dr} \right)^{3/2} \left(\frac{R_\odot}{10} \right)^2
\end{aligned}
\tag{8–76}
$$

The distance $R_\odot/10$ is chosen somewhat arbitrarily since we do not know how to estimate the cell size. Convective theories are currently seeking to solve that problem. We take $d\,\Delta T/dr$ to be the difference between the actual and the adiabatic gradient. The equations we have obtained hold equally well for the upward convection of hot matter and downward convection of cool material. For a given gradient $d\,\Delta T/dr$ we can now obtain the order of magnitude of H if the heat capacity is known. We have not yet discussed the equation of state, although we have assumed an ideal gas law above. For a completely ionized plasma the heat capacity is roughly known, even though the process described here proceeds neither at constant pressure nor at constant volume. We will, however, not be far wrong in taking $2RT$ per gram, where R is the gas constant (see equation 4–34).

We now wish to see at what gradients the convective flux exceeds radiative transfer. This can be done by checking the value of $d\,\Delta T/dr$ at which the total convective flux equals the luminosity. With $r \sim R_\odot/2$, $M(r) \sim M/2$, $C \sim 2 \times 10^8$ ergs/g, $\rho \sim 1$ g cm^{-3}, $T \sim 10^7$ °K, and $L \sim 10^{34}$ erg sec^{-1}, we have

$$L = 4\pi R_\odot^2 H \sim 4\pi R_\odot^2 C\rho \left(\frac{GM(r)}{Tr^2} \right)^{1/2} \left(\frac{R_\odot}{10} \right)^2 \left(\frac{d\,\Delta T}{dr} \right)^{3/2}, \quad \frac{d\,\Delta T}{dr} \sim 10^{-10} \tag{8–77}$$

The average temperature gradient for a star is of order $T_c/R \sim 10^7/10^{11} \sim 10^{-4}$, where T_c is the central temperature. The required excess gradient is only of the order of one millionth of the total gradient. Over a distance $\Delta r \sim R_0/10$, the excess temperature drop corresponds to $\sim 1°K$, the figure we had previously used in establishing the time constant for convective transport in equation 8–29 and 8–30.

We have now dealt with all the differential equations discussed in section 8:4; but we still have to derive the energy generation rate through nuclear reactions taking place at the center of a star. This is done in the next section.

8:10 NUCLEAR REACTION RATES

The nuclear reactions that take place in stars are largely reactions in which two particles approach to within a short distance, become bound to each other, and at the same time release energy. These *exergonic* processes are the ultimate source of energy for the star.

Let us look into the various factors that determine the reaction rate. We will assume that two kinds of particles are involved and will label them with subscripts 1 and 2, respectively. The reaction rate then is proportional to:

(i) the number density n_1 of nuclei of the first kind,
(ii) the number density n_2 of nuclei of the second kind,
(iii) the frequency of collisions, which depends on the relative velocity v with which particles approach each other, and
(iv) on the velocity dependent interaction cross section $\sigma(v)$ which normally is proportional to $1/v^2$. However, in order for a reaction to occur, the Coulomb barrier, which bars positively charged particles from approaching a nucleus, must be penetrated. This makes the reaction rate proportional to
(v) the probability $P_p(v)$ for penetrating the Coulomb barrier; that has an exponential form

$$P_p(v) \propto \exp - \left(\frac{4\pi^2 Z_1 Z_2 e^2}{hv} \right) \tag{8–78}$$

Here $Z_1 e$ and $Z_2 e$ are the nuclear charges.

Once the nuclear barrier has been penetrated, there is a probability P_N for nuclear interaction. This is insensitive to particle energy or velocity, but does depend on the specific nuclei involved. We therefore introduce a factor proportional to
(vi) P_N the probability for nuclear interaction. For the interaction of two protons, this interaction is known from theory. For all other reactions laboratory

data has to be used to evaluate the probability. The rate of the process further is proportional to

(vii) the distribution of velocities among particles. This can be assumed to be Maxwellian since the nuclei normally are not degenerate. Equation 4–59 gives

$$D(T, v) \propto \frac{v^2}{T^{3/2}} \exp - \left(\frac{1}{2} \frac{m_H A' v^2}{kT} \right) \tag{8–79}$$

where $A' = A_1 A_2 / (A_1 + A_2)$ is the reduced atomic mass, measured in atomic mass units.

We can now write down the overall reaction rate in unit volume

$$r = \int_0^\infty n_1 n_2 v \sigma(v) \, P_p(v) \, P_N D(T, v) \, dv \tag{8–80}$$

This integral is readily evaluated because of the narrow range of velocities in which the product of P_p and D is high. Outside this velocity range the integrand is too small to make a significant contribution to the integral. We proceed in the following way. The integral in equation 8–80 has the form

$$\int_0^\infty v \exp - \left(\frac{a}{v} + bv^2 \right) dv \tag{8–81}$$

The integrand has a sharp maximum at the minimum value of the exponent. We take the derivative of the exponent with respect to v and equate to zero. This gives the value v_m

$$v_m = \left(\frac{a}{2b} \right)^{1/3} = \left(\frac{4\pi^2 Z_1 Z_2 e^2 kT}{h m_H A'} \right)^{1/3} \tag{8–82}$$

To evaluate the integral, however, we still need to estimate the effective velocity range over which the integrand is significant. For order of magnitude purposes it will suffice to take a range between points where the value of the integrand has dropped a factor of e. This happens at v values for which

$$\left(\frac{a}{v} + bv^2 \right) - \left(\frac{a}{v_m} + bv_m^2 \right) = 1 \tag{8–83}$$

Since the deviation from v will be small, we set

$$v = v_m + \Delta \tag{8–84}$$

and substitute in equation 8–83. Terms linear in Δ drop out, but the quadratic

8:10

terms yield

$$\left(\frac{a}{v_m^3} + b\right) \Delta^2 = 3b\,\Delta^2 = 1 \qquad (8\text{–}85)$$

$$\Delta = \pm\sqrt{\frac{1}{3b}} = \pm\sqrt{\frac{2kT}{3A'm_H}} \qquad (8\text{–}86)$$

The integral (8–80) can now be readily evaluated. First, however, we would like to lump all the proportionality constants into a single constant B and relate velocity to temperature everywhere.

We note that

$$\Delta \propto T^{1/2} \qquad (8\text{–}87)$$

and that the integrand is proportional to

$$v_m \cdot \frac{1}{v_m^2} \cdot \frac{v_m^2}{T^{3/2}} = \frac{v_m}{T^{3/2}} = T^{-7/6} \qquad (8\text{–}88)$$

where we have made use of the relation (8–82). This means that $r \propto T^{-7/6}\Delta \propto T^{-2/3}$

We can set

$$n_1 = \frac{\rho_1}{m_1} = \frac{\rho}{m_1}X_1 \qquad \text{and} \qquad n_2 = \frac{\rho}{m_2}X_2 \qquad (8\text{–}89)$$

where X_1 and X_2 are the concentrations and m_1 and m_2, the masses of nuclei of species 1 and 2. Absorption of factors m_1 and m_2 into the proportionality constant B then yields the reaction rate

$$r = B\rho^2 X_1 X_2 T^{-2/3} \exp - 3\left(\frac{2\pi^4 e^4 m_H Z_1^2 Z_2^2 A'}{h^2 kT}\right)^{1/3} \qquad (8\text{–}90)$$

Thus far we have developed an estimate of reaction rates without much thought about the individual reactions involved, the required temperatures and densities, and the resulting energy liberation rate. We now return to these points.

We first ask ourselves how much energy would be needed for two particles to interact. It is clear that a nuclear reaction can only take place if the particles approach to within a distance of the order of a nuclear diameter $D \sim 10^{-13}$ cm. However, since both nuclei are positively charged, they tend to repel each other and the work required to overcome the repulsion is

$$E = \frac{Z_1 Z_2 e^2}{D} \sim 2 \times 10^{-6} Z_1 Z_2 \text{ erg} \sim Z_1 Z_2 \text{ Mev} \qquad (8\text{–}91)$$

This might lead one to think that temperatures of the order of 10^{10} °K would be required for nuclear reactions to proceed. This is far higher than the 10^7 °K temperature we had estimated in Problem 8–1.

Two factors allow the actual reaction temperature to be so low. First, a small fraction of the nuclei with thermal distribution $D(T, v)$ has energies far above the mean. Second, two particles have a small but significant probability of approaching each other by tunneling through the Coulomb potential barrier rather than going over it. This probability is quantum mechanically determined and is included in the function $P_p(v)$.

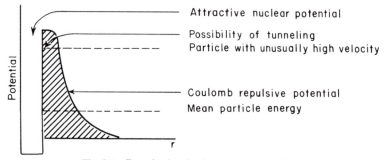

Fig. 8.6   Energies involved in nuclear reactions.

These two factors suffice to allow nuclear reactions to proceed at mean energies some 10^3 times lower than those employed to produce nuclear interactions in the laboratory. The main difference is that in the laboratory speed is essential, while the star is in no hurry. In the laboratory, we want high reaction rates so that results may be obtained within a few minutes or, at most, hours. In contrast, a reaction having a probability of transmuting a given particle in the course of 10 billion years is sufficiently fast to produce the luminosities characterizing many stars like the sun. This prolongation of the available time by a factor of $\sim 10^{14}$ is the prime difference that permits low temperature generation of energy and transmutation of the elements at the centers of stars at cosmically significant rates.

8:11   PARTICLES AND BASIC PARTICLE INTERACTIONS

A number of basic particles are involved in most nuclear reactions taking place in stars. We list their properties in Table 8.2.

The spin of a particle tells us the type of statistics it obeys. Integral spin implies obedience to the *Bose-Einstein* statistics, while a half integral spin labels a particle as a *Fermion*.

8:11

Table 8.2 Some Particles That Take Part in Many Stellar Nuclear Reactions.

Particle	Symbol	Rest Mass		Charge esu	Spin	Mean Life Mean	Class
		g	Mev				
Photon	γ	0	0	0	1		Photon
Neutrino	v	0	0	0	$\frac{1}{2}$		Lepton
Anti-neutrino	$\bar{v}$	0	0	0	$\frac{1}{2}$		Antilepton
Electron	e	9×10^{-28}	0.511	-5×10^{-10}	$\frac{1}{2}$		Lepton
Positron	e^+	9×10^{-28}	0.511	$+5 \times 10^{-10}$	$\frac{1}{2}$		Antilepton
Proton	$\mathscr{P}$	1.6×10^{-24}	938.256	$+5 \times 10^{-10}$	$\frac{1}{2}$		Baryon
Neutron	$\mathscr{N}$	1.6×10^{-24}	939.550	0	$\frac{1}{2}$	6.4×10^2	Baryon

A number of basic conservation laws govern all nuclear reactions:

(a) Mass-energy must be conserved (section 5:6).

(b) The total electric charge of the interacting particles is conserved.

(c) The number of particles and antiparticles must be conserved. A particle cannot be formed from an antiparticle or vice versa. But a particle-antiparticle pair may be formed or destroyed without violating this rule. In particular:

(d) The difference between the number of *leptons* and *antileptons* must be conserved (*conservation of leptons*); and

(e) The difference between the number of *baryons* and *antibaryons* must be conserved (*conservation of baryons*).

With these rules in mind we enumerate some of the most common nuclear reactions found in stars:

(i) Beta Decay

A *neutron* as a free particle, or as a nucleon inside an atomic nucleus can decay giving rise to a proton, an electron, and an *antineutrino*:

$$\mathscr{N} \rightarrow \mathscr{P} + e^- + \bar{v} \qquad (8\text{--}92)$$

This reaction often is *exergonic* and can proceed spontaneously. When (8–92) proceeds in reverse, we speak of inverse beta decay.

$$\mathscr{P} + e^- \rightarrow \mathscr{N} + v \qquad (8\text{--}92a)$$

8:11

(ii) Positron Decay

Here a proton gives rise to a neutron, positron, and neutrino. This process is *endergonic*—requires a threshold input energy—since the mass of the neutron and positron is considerably greater than the proton mass.

$$\mathscr{P} \rightarrow \mathscr{N} + e^{+} + \nu \tag{8-93}$$

In principle all these reactions could go either from left to right or right to left; but normally the number of available neutrinos or antineutrinos is so low that only the reaction from left to right need be considered.

(iii) $(\mathscr{P}, \gamma)$ Process

In this reaction a proton reacts with a nucleus with charge Z and mass A, to give rise to a more massive particle with charge $(Z + 1)$. Energy is liberated in the form of a photon, γ:

$$Z^{A} + \mathscr{P} \rightarrow (Z + 1)^{A+1} + \gamma \tag{8-94}$$

A typical reaction of this kind involves the carbon *isotope* C^{12} and nitrogen isotope N^{13}

$$C^{12} + H^{1} \rightarrow N^{13} + \gamma \tag{8-95}$$

(iv) (α, γ) and (γ, α) Processes

In processes of this kind an *α-particle*—helium nucleus—is added to the nucleus or ejected from it. The excess energy liberated in adding an alpha particle is carried off by a photon. The energy required to tear an alpha particle out of the nucleus also can be supplied by a photon. These two processes are particularly important in nuclei containing an even number both of protons and neutrons— the *even-even nuclei*. These nuclei are especially stable and play a leading role in the processes that lead to the formation of heavy elements.

(v) $(\mathscr{N}, \gamma)$ and $(\gamma, \mathscr{N})$ Processes

Such processes involve the addition or subtraction of a neutron from the nucleus. A photon is emitted or absorbed in the reaction and assures energy balance.

8:11

8:12 ENERGY GENERATING PROCESSES IN STARS

A variety of different energy generating processes can take place in stars. We will enumerate them in the succession in which they are believed to occur during a star's life.

(a) When the star first forms from the interstellar medium it contracts, radiating away gravitational energy. The amount of energy available from this process was already computed in (8–1). During this stage no nuclear reactions take place.

(b) When the temperature at the center of the star becomes about a million degrees, the first nuclear reactions set in. From the discussion of section 8:10 it is clear that these reactions will not set in sharply as some given temperature value is exceeded. The temperature is not a threshold in this sense, even though threshold energies are involved. Instead, we can think of a critical temperature at which reactions will proceed at a certain rate. We will choose to define the critical temperature T_c as that temperature at which the mean reaction time becomes as short as five billion years. Because of the rapid increase in reaction rates with temperature, the reactions will become completely exhausted in a very short time (on the scale of a billion years) if T_c is exceeded. The first reactions to occur are those that destroy many of the light elements initially in the interstellar medium and convert them into helium isotopes. We list the reactions and the energy released in each reaction (Sa55). Note that this energy is carried away by photons or neutrinos, but we have not specifically shown this here:

$$
\begin{array}{llll}
D^2 + H^1 \rightarrow He^3 & 5.5 \text{ Mev} & & \\
Li^6 + H^1 \rightarrow He^3 + He^4 & 4.0 & & \\
Li^7 + H^1 \rightarrow 2He^4 & 17.3 & & \\
Be^9 + 2H^1 \rightarrow He^3 + 2He^4 & 6.2 & \left.\right\} \text{ two-step} & (8\text{–}96) \\
B^{10} + 2H^1 \rightarrow 3He^4 + e^+ & 19.3 & \left.\right\} \text{ reactions} & \\
B^{11} + H^1 \rightarrow 3He^4 & 8.7 & &
\end{array}
$$

These reactions have lifetimes of the order of 5×10^4 y at respective temperatures $\sim 10^6$, 3×10^6, 4×10^6, 5×10^6, 8×10^6, and 8×10^6 °K. At temperatures ranging from about half a million degrees to five million degrees, these reactions would take place rapidly as the star contracts along the *Hayashi track* (Fig. 8.11)— a fully convective stage in the star's pre-main sequence contraction. Because these temperatures are so low, the elements burn up everywhere including the surface layers of the stars, where they can be destroyed because convection takes place.

With few exceptions these light elements are only found in small concentrations in the surface layers of stars. However, large concentrations of, say, lithium can be found in some stars; these are called lithium stars and constitute a puzzle. The theory of stellar evolution seeks to explain such anomalies by providing coherent ideas on the origin of the chemical elements.

None of the reactions listed in (8–96) contribute a large fraction of the total energy emitted by stars during their lifetime. However, they are of interest in connection with the theory of the formation of elements; and the low abundance of these chemical elements in nature provides one test of the accuracy of our notions.

(c) When the temperature at the center of the star reaches about 10 million degrees, hydrogen starts burning (Be39). The reactions and mean reaction times for any given particle are given below for $T = 3 \times 10^7 \,°$K. The amount of energy liberated in each step is also given.

$$
\begin{array}{llll}
H^1 + H^1 \to D^2 + e^+ + \nu, & 1.44 \text{ Mev}, & 14 \times 10^9 \text{ y} \\
D^2 + H^1 \to He^3 + \gamma, & 5.49 \text{ Mev}, & 6 \text{ sec} & (8\text{–}97) \\
He^3 + He^3 \to He^4 + 2H^1, & 12.85 \text{ Mev}, & 10^6 \text{ y}
\end{array}
$$

The first and second reactions have to take place twice to prepare for the third reaction. Not all of the energy liberated contributes to the star's luminosity. Of the energy liberated in the first step 0.26 Mev is carried away by the neutrino and is lost. The total contribution to the luminosity is therefore 26.2 Mev for each helium atom formed.

This set of reactions is the main branch of the *proton-proton reaction*. Other branches are shown in Fig. 8.12. Hydrogen burning can also take place in a somewhat different way, making use of the catalytic action of the carbon isotope C^{12}. This set of reactions comprises the *carbon cycle* or a more elaborate scheme sometimes called the *CNO bi-cycle*, since carbon, nitrogen, and oxygen are all involved; the CN portion is energetically the more significant (C168)*. The reaction times are given for 15×10^6 and $20 \times 10^6 \,°$K.

	Mev	$15 \times 10^6 \,°$K	$20 \times 10^6 \,°$K	
$\to C^{12} + H^1 \to N^{13} + \gamma$	1.94	$\sim 10^6$y	$\sim 5 \times 10^3$y	
$N^{13} \to C^{13} + e^+ + \nu$	2.22	15 min		
$C^{13} + H^1 \to N^{14} + \gamma$	7.55	2×10^5y	2×10^3y	
$\to N^{14} + H^1 \to O^{15} + \gamma$	7.29	2×10^8y	10^6y	(8–98)
$O^{15} \to N^{15} + e^+ + \nu$	2.76	3 min		
$N^{15} + H^1 \to C^{12} + He^4$	4.97	10^4y	30y	
$N^{15} + H^1 \to O^{16} + \gamma$	12.1	4×10^{-4} of $N^{15}(\mathscr{P}, \alpha) C^{12}$ rate		
$O^{16} + H^1 \to F^{17} + \gamma$	0.60	2×10^{10}y	5×10^7y	
$F^{17} \to O^{17} + e^+ + \nu$	2.76	1.5 min		
$O^{17} + H^1 \to N^{14} + He^4$	1.19	2×10^{10}y	10^6y	

8:12

Fig. 8.7 Nuclear energy generation rate as a function of temperature (with $\rho X^2 = 100$. $X_{CN} = 5 \times 10^{-3} X$ for the p-p reaction and carbon cycle, but $\rho^2 Y^3 = 10^8$ for the triple-α process) (Sc58b, from Martin Schwarzschild, *Structure and Evolution of the Stars*, copyright © 1958 by Princeton University Press, p. 82.)

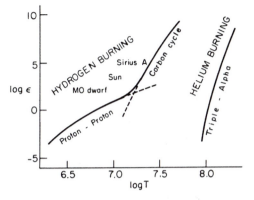

Many of the reaction times still are somewhat uncertain. The second part of the cycle occurs about 4×10^{-4} times as often as the first, since the $N^{15}(\mathscr{P}, \alpha) C^{12}$ reaction is about 2.5×10^3 times more probable than the $N^{15}(\mathscr{P}, \gamma) O^{16}$ reaction.

In the decay of the N^{13} particle 0.71 Mev is carried off by the neutrino; and in the O^{15} decay 1.00 Mev is lost on the average. The total energy made available to the star per helium atom formed is therefore only 25.0 Mev, slightly less than the energy available from the proton-proton reaction. The reaction rates given here are for total concentrations of the carbon and nitrogen isotopes amounting to $X_{CN} \sim 0.005$. The relative predominance of the proton-proton reaction and the carbon cycle as a function of temperature is given in Fig. 8.7. The C^{13} formed in the CN cycle can act as a source of neutrons as can other particles with mass number $4n + 1$. We will see that in reactions (8–102) and (8–103) below. In these, Ne^{21} can be produced as follows:

$$Ne^{20} + H^1 \rightarrow Na^{21} + \gamma \qquad 2.45 \text{ Mev} \qquad 10^9 \text{y at } 3 \times 10^7 \, ^\circ K$$
$$Na^{21} \rightarrow Ne^{21} + e^+ + v \qquad 2.5 \qquad 23 \text{ sec} \qquad (8\text{–}99)$$

The initial formation of Ne^{20} is discussed in paragraph (e) below. The hydrogen-burning reactions we have discussed contribute the energy given off by the star during its long stay on the main sequence. Once the hydrogen at the center of the star is largely depleted, helium burning can set in as described immediately below. In general, hydrogen burning will continue in a shell surrounding this depleted core.

(d) When the hydrogen burning phase of a star is completed, no further nuclear energy generating processes may be available for some time and the star slowly contracts (Fig 8.8). Its central temperature rises continually as a result, until at about $10^8 \, ^\circ K$, helium burning sets in (Sa52). In this process three alpha particles are transmuted into a carbon nucleus. Two steps are involved:

$$He^4 + He^4 \rightarrow Be^8 + \gamma \qquad -95 \text{ kev}$$
$$Be^8 + He^4 \rightarrow C^{12} + \gamma \qquad +7.4 \text{ Mev} \qquad (8\text{–}100)$$

8:12

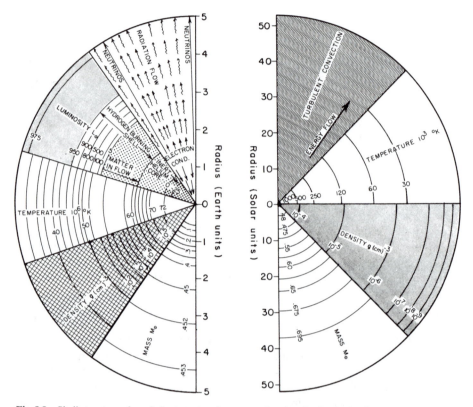

Fig. 8.8 Shell structure of a red giant star in whose central regions hydrogen has become depleted (after Iben, Ib70). The section on the left is a blown up version of the little disk in the center of the drawing at the right. ("Globular Clusters Stars." Copyright © 1970 by Scientific American, Inc. All rights reserved.)

The first reaction is *endergonic*. Energy has to be supplied to make it proceed. The Be^8 nucleus is unstable, and decays back into two alpha particles. An equilibrium is set up between alpha particles and Be^8 particles in which the concentration of Be^8 is quite small, of the order of 10^{-10} times the concentration of alpha particles. This particular abundance is determined by the lifetime of the metastable Be^8, the density and energy (temperature) of the helium and by the magnitude of the (negative) binding energy, -95 kev.

(e) The star's core does not stay in the helium burning phase for a very long time, because the available amount of energy is small ($\sim 10\%$), compared to the energy generated in the hydrogen-burning phase. At higher internal temperatures a succession of (α, γ) processes may set in to form O^{16}, Ne^{20}, and Mg^{24}. This type of process is called the *α-process*. After depletion in the core, helium burning may continue in a shell surrounding the depleted core. This shell is surrounded by a hydrogen-burning shell.

(f) At higher temperatures yet, 10^9°K, reactions may take place among the C^{12}, O^{16}, and Ne^{20} nuclei. At this stage there would be no supply of free helium, but these particles can be made available through a (γ, α) reaction. The densities at this stage are of the order of $\rho \gtrsim 10^6$ g/cc. A typical reaction is

$$2Ne^{20} \rightarrow O^{16} + Mg^{24}, \qquad 4.56 \text{ Mev} \qquad (8-101)$$

Mg^{24} can capture alphas to form Si^{28}, S^{32}, A^{36}, and Ca^{40}.

That this process actually takes place may be partly confirmed by the relatively large natural abundance of these isotopes compared to isotopes of the same substances, or neighboring elements in the periodic table (Fig. 8.10).

This chain eventually terminates in the iron group of nuclei. These are the most stable of all the elements, because their masses per nuclide are at a minimum. During the time that an equilibrium concentration is being reached between these even-even nuclei, the expected temperature and density are

$$T \sim 4 \times 10^9 \text{ °K}, \qquad \text{and} \qquad \rho \sim 10^8 \text{ g/cc}$$

This process is called the *equilibrium* or *e-process*. The α and e processes probably occur rapidly—perhaps explosively.

(g) In a second generation star—one that has formed from interstellar gases containing appreciable amounts of the heavier elements—we may find that Ne^{21} is produced. At high temperatures, in the helium core, we can then have the exergonic reaction

$$Ne^{21} + He^4 \rightarrow Mg^{24} + \mathcal{N}, \qquad 2.58 \text{ Mev} \qquad (8-102)$$

Similarly from the carbon cycle there will be some C^{13} available and we may have the reaction

$$C^{13} + He^4 \rightarrow O^{16} + \mathcal{N}, \qquad 2.20 \text{ Mev} \qquad (8-103)$$

taking place.

These neutrons are preferentially captured by the heavy nuclei, particularly those in the iron group, and these can then be built up into heavier elements yet. There are of the order of a hundred C^{13} and Ne^{21} nuclei available for each iron group element and, hence, an abundance of neutrons is at hand. Elements as heavy as Bi^{209} can be built up in this way. The chain only ends at Po^{210}, which is unstable and undergoes α-decay. In addition, light nuclei such as Ne^{22} also can be built up and, with the exception of the even-proton-even-neutron nuclei, most particles with $24 \leqq A \leqq 50$ are believed to have been built up through neutron capture. This neutron process is *slow*; it is therefore called the *s-process*.

8:12

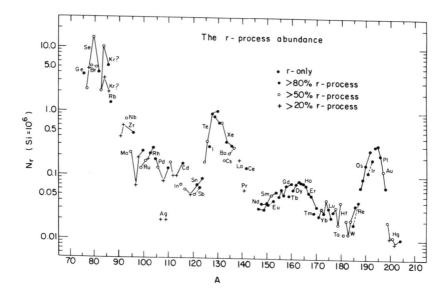

Fig. 8.9 *r*-process abundances estimated for the solar system by subtracting the calculated *s*-process contribution from the total observed abundances of nuclear masses. Isotopes of a given element are joined by lines, dashed lines for even *Z* and solid lines for odd *Z*. All abundances are normalized to silicon, whose abundance would be 10^6 on the same scale. Three peaks, and a broad rare earth bump, are the most prominent features of this plot. The question marks indicate a high uncertainly about the relative roles of the *r*- and *s*-processes in producing krypton. (After Seeger, Fowler, and Clayton, Se65.)

Neutron capture at this stage typically requires several years to several thousand years. This time scale is slow compared to beta decay rates, and only those elements can be built up that involve the addition of neutrons to relatively stable nuclei.

Evidence for a stage with abundant neutrons comes from peaks in the abundance curve at mass numbers $A \sim 90$, 138, and 208. These nuclei have closed shells of neutrons with $N = 50$, 82, and 126.

The stage of stellar evolution in which the *s*-process becomes effective probably is represented by the segment EF in the H-R diagram of Fig. 1.6. In this repeated red giant phase, a series of helium shell flashes may produce a convective behavior in which the helium-rich shell above the helium-burning layer is mixed with the carbon-rich core, and in which eventually even the outermost hydrogen-rich layer can be convected into the core (Sc70). In this situation C^{12} and H^1 (see equation 8–98) can yield C^{13} which produces the required neutrons through interaction with He^4.

(h) In addition to the slow neutron process, it is likely that neutrons can also be added to heavy nuclei in a *rapid process* (*r-process*) that takes place at least in some stars. Some such process is required, in any case, if a theory of the buildup of chemical elements is to explain the existence of elements beyond Po^{210} which α-decays with a half-life of only 138d:

If a star runs out of all energy sources, a rapid implosion can take place on a free fall time scale which as we saw earlier, corresponds to times of the order 1000 sec. Extremely high temperatures then set in and iron group nuclei can be broken up into alpha particles and neutrons; for, in Fe^{56} there are 4 excess neutrons for 13 alphas. All this takes place at temperatures of order $10^{10}\,°K$ and with neutron fluxes of order $10^{32}\,cm^{-2}sec^{-1}$. The *r*-process can build up elements to about $A \sim 260$ where further neutron exposure induces fission that cycles material back down into the lower mass ranges.

In the breakup of iron group elements into helium, the ratio of specific heats γ becomes less than 4/3 so that an implosion occurs (section 4:20). This is accompanied by γ-photon production, pair formation, and electron-positron pair annihilation which at these high pressures gives rise to large neutrino fluxes. The neutrino flux lifts off the outer layers and the rapid neutron process then takes place while the star is again expanding—explosively. It is believed that this is the process involved at least in some types of supernova explosion. A neutron star forms from the central imploding core.

Detailed computations based on neutron capture cross sections and nuclear decay times, both measured in the laboratory, show that many features of the abundance curve in the region between $A = 80$ and $A = 200$ can be explained if the *r*-process occurs (Fig. 8.9). This leads us to believe that the sequence of events described above is at least roughly correct. Two related comments might still be made.

(i) Proton-rich isotopes are relatively rare although they can be produced in ($\mathscr{P}$, γ) processes (sometimes called *p*-processes) or in a (γ, $\mathscr{N}$) reaction. Such nuclei could be produced if hydrogen from the outer layers of a star could come into contact with hot material from the core in convective processes. Generally, however, the *r*-process can account semi-quantitavely for the abundance ratios of many of the heavier elements.

(ii) The uranium isotopes U^{235} and U^{238} might be expected to arise in roughly equal abundances in the *r*-process. However, their present ratio as found in the earth is of the order of 0.0072. This would be expected on the basis of their respective half lives of 0.71 and 4.5 æ if we assume a common time of formation some six aeons ago. This gives us a scheme for dating the origin of terrestrial material. We still face many uncertainties, some of which are illustrated by Problem 10–14 in section 10:13.

A brief stage in which carbon, oxygen, and silicon are successively burned at

8:12

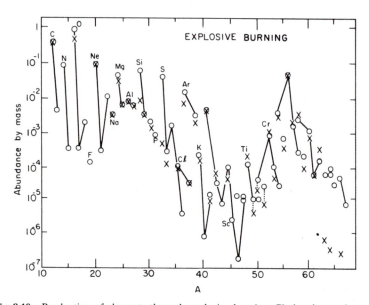

Fig. 8.10 Production of elements through explosive burning. Circles show solar system abundances by mass, crosses are computed values. Explosive burning of carbon at 2×10^9 °K and density $\rho \sim 2 \times 10^5$ g cm^{-3} contributes to atomic masses up to $A \sim 30$. Oxygen burning at 3.6×10^9 °K, at densities of $\sim 5 \times 10^5$ g cm^{-3}, contributes up to atomic mass values $A \sim 50$. Silicon burning at temperatures in the range 4.7 to 5.5×10^9 °K and density 2×10^7 g cm^{-3} accounts for the more massive nuclear abundances shown. This is a composite plot based on the work of Arnett and Clayton (Reprinted with permission from *Nature*, Vol. 227, p. 780, copyright © 1970 Macmillan Magazines, Ltd., Ar70). It is possible that similar abundances could be obtained by Wagoner's explosive nucleosynthesis (see section 8:16) (Wa67). Note that solid lines join different isotopes of the same element.

higher and higher temperatures, during the explosion of a star in the mass range of 20 to 40 $M_\odot$, has been suggested (Ar70) as responsible for the observed abundances of elements in the range $20 \leq A \leq 64$. Initially carbon in the helium depleted core of a star would undergo fusion reactions of the kind:

$$C^{12} + C^{12} \rightarrow \begin{array}{ll} Na^{23} + \mathscr{P} & 2.238 \text{ Mev} \\ Mg^{23} + \mathscr{N} & -2.623 \text{ Mev} \\ Ne^{20} + \alpha & 4.616 \text{ Mev} \end{array} \qquad (8\text{--}104)$$

These reactions would take place at a temperature of 2×10^9 °K. The initial density would be of order 10^5 g cm^{-3}. The reactions are assumed to last for about a tenth of a second, after which the explosion has cooled the stellar matter enough to stop the processes. At higher temperatures, 3×10^9 °K, oxygen also burns and thereafter silicon Si^{28} disintegrates. In this latter process, the silicon splits into seven α's, which are absorbed by other Si^{28} nuclei to form increasingly massive nuclei in the range up to Fe^{56}. These processes will take place if ignition of the nuclear fuel takes place at a temperature of $\sim 5 \times 10^9\,°K$ and a peak density of $2 \times 10^7\,g\,cm^{-3}$ in the helium depleted core. The exact abundance ratios found at the end of the explosive process depend in part on the neutron excess, that is, the fractional number of neutrons in excess of protons in the nuclei initially present. A neutron excess

$$\eta = \frac{(n_{\mathcal{N}} - n_{\mathscr{P}})}{(n_{\mathcal{N}} + n_{\mathscr{P}})} \tag{8--105}$$

of ~ 0.002 gives remarkably good agreement as shown, for example, by Fig. 8.10. It is not clear what stellar configuration should be assumed during this explosion. Are the different substances arranged in concentric shells, or somehow irregularly distributed over the core? These processes are not sufficiently well understood.

8:13 THE HERTZSPRUNG-RUSSELL DIAGRAM AND STELLAR EVOLUTION

We believe that stars form from the interstellar medium. Initially a cool cloud of interstellar gas contracts, giving off thermal radiation in the far infrared. As the contraction proceeds the cloud temperature rises. It is possible that the surface temperature then remains constant for a long period as the star becomes smaller and smaller. This means that the brightness of the star decreases during this stage, but the color remains constant. The *Hayashi track* of the star on a Hertzsprung-Russell diagram is therefore represented by a nearly vertical line. Iben has indicated that this stage is followed by a nearly horizontal leftward motion toward the main sequence (Fig. 8.11). As the light elements are burned, the track may undergo some short-lived changes, but finally the star settles down for a protracted stay on the main sequence.

Actually, as shown in Fig. 1.6, there is a very slight motion through the Hertzsprung-Russell diagram, even during this hydrogen burning phase. Over a period amounting to several aeons, a star moves from A on the initial *zero-age main sequence* to point B. The star becomes brighter and larger. We believe that the sun had a zero-age luminosity of 2.78×10^{33} erg sec^{-1} and a radius of 6.59×10^{10} cm, compared to its current values of 3.90×10^{33} erg sec^{-1} and 6.94×10^{10} cm, 4.5 æ later (St65).

When hydrogen burning is completed at the center of the star, it may still continue in a thin shell surrounding a central core of helium. The core contracts until its temperature becomes sufficiently high to produce helium burning.

8:13

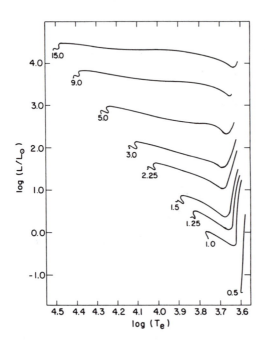

Fig. 8.11 Contraction of stars toward the main sequence. The path of the stars across this Hertz-sprung-Russell diagram proceeds toward the left. The left end of the curves roughly coincides with the main sequence. The star with mass $15M_\odot$ completes the transit shown here in $\sim 6 \times 10^4$ y: the $0.5 \, M_\odot$ star in 1.5×10^8 y. The steep portion on the right is called the *Hayashi track* (Ha66. After Iben, Ib65.)

During this stage the star expands and becomes both brighter and redder. The track *BC* in Fig. 1.6 shows how the star moves upward and to the right in the color-magnitude diagram. This branch of the diagram is the *red giant branch*.

Up to now the theory is well understood. What happens next is not as well known. Somehow, the red giant stars shed some of their mass and eventually end up to the left of the main sequence and below it, in the form of *white dwarfs*.

We believe (Sc70)* that when point *C* in Fig. 1.6 is reached, helium burning through the triple alpha process sets in. This probably occurs in a flash. The star is forced to expand, but it keeps burning helium at a slower rate in the convecting central core, while hydrogen burns in an outer shell. The loop *DE* probably represents this stage, which would last some 10^8y in a star of mass $\sim 1.2M_\odot$. As indicated in section 1:5, a series of less well understood stages then follows in rapid succession. Probably there is a stage of carbon burning and, as already indicated, this may be rapid. We associate these stages in which helium burning goes on in a shell surrounded by an outer hydrogen-burning shell, with the *horizontal branch* stars. These must be short-lived because few of them are observed. The *cepheid variables* may be associated with this stage of nuclear evolution. These stars change period at a rate that changes substantially over a time of the order of a million years.

8:13

The theory of stellar evolution allows us to draw one interesting conclusion about the age of *globular cluster stars*. Using plausible stellar models, we can compute what the hydrogen-burning time scale should be for stars having different masses. A cluster contains stars with a variety of masses; but at any given time only stars having a particular color and magnitude will be about to leave the main sequence to enter the red giant branch. Now since stellar luminosities and masses are related (see Problem 8–3), we can use the *turn-off point* where stars leave the main sequence as an indicator of the mass of these stars that are just completing the hydrogen-burning phase. The age of these stars can then be calculated and this defines the time when the cluster of stars must initially have been formed—assuming of course that all the stars were born roughly at the same time.

Ages of globular clusters derived in this way amount to about 10 aeons. The computed ages are not identical for all different clusters, indicating that they were formed at different epochs and perhaps during different phases of the life of the Galaxy.

The last stage in the evolution of stars with low mass seems to be the *white dwarf stage* in which the stellar interior becomes degenerate and no further contraction takes place. The star then gradually cools down over a long period, but no further nuclear reactions take place.

Massive stars do not go through the white dwarf stage. Their ultimate fate is not known. Perhaps they explode leaving only a small remnant that could become a white dwarf. Perhaps they shed mass nonexplosively. Perhaps they keep contracting until they become degenerate neutron stars. Such stars are described in more detail in section 8:16.

8:14 EVIDENCE FOR STELLAR EVOLUTION, AS SEEN FROM THE SURFACE COMPOSITION OF SOME STARS

As already discussed in section 1:6, the spectra of a wide variety of stars show atmospheric compositions very similar to that of the sun. This is shown in Table 8.3 where the small differences are within the expected errors of the observations and of the data reduction. The table presents the logarithm of the ratios of the number densities of heavier elements relative to hydrogen where, for comparison, $\log n_H$ is set equal to 12.00 in all the stars.

The similarity in abundances for stars of as widely differing ages as a B0 star, which probably formed only a few million years ago, and a red giant or the planetary nebulae, which should be among the oldest objects in the Galaxy and hence seven or more aeons old, indicate that the interstellar medium may not have changed much during the time between formation of these various stars.

Table 8.3 Abundances log *n* for "Normal" Stars Relative to log *n* = 12 for Hydrogen. Compiled from various sources by A. Unsöld (Un69)*.

Atomic number	Element	Abundance: log n					
		Sun		τ Scorpii B0 V	ζ Persei Bl Ib	Planetary Nebulae	Log $(n/n_\odot)$ ε Virginis G8 III
		Goldberg, Müller, Aller (1960, 1967)	Various Recent Sources				
1	H	12.00	12.00	12.00	12.00	12.00	0.00
2	He		11.2	11.12	11.31	11.25	
3	Li	0.90	0.97 $\leq$ 0.38				
4	Be	2.34					
5	B	3.6					
6	C	8.51	8.51 8.55	8.21	8.26	8.7	− 0.12
7	N	8.06	7.93	8.47	8.31	8.5	
8	O	8.83	8.77	8.81	9.03	9.0	
9	F					5.5	
10	Ne			8.98	8.61	8.6	
11	Na	6.30	6.18				+ 0.30
12	Mg	7.36	7.48	7.7	7.77		+ 0.04
13	Al	6.20	6.40	6.4	6.78		+ 0.14
14	Si	7.24	7.55	7.66	7.97		+ 0.13
15	P	5.34	5.43				
16	S	7.30	7.21	7.3	7.48	8.0	+ 0.09
17	Cl					6.5	
18	A			8.8		6.9	
19	K	4.70	5.05				+ 0.10
20	Ca	6.04	6.33				+ 0.10
21	Sc	2.85					− 0.07
22	Ti	4.81					− 0.07
23	V	4.17					− 0.04
24	Cr	5.01					0.00
25	Mn	4.85					+ 0.07
26	Fe	6.80		7.4			+ 0.01
27	Co	4.70					− 0.03
28	Ni	5.77					+ 0.03
29	Cu	4.45					+ 0.06
30	Zn	3.52					+ 0.05
31	Ga	2.72					
32	Ge	2.49					
37	Rb	2.48	2.63				
38	Sr	3.02	2.82				+ 0.02
39	Y	3.20					− 0.17

8:14

Table 8.3 (*continued*)

Atomic number	Element	Sun		τ Scorpii B0 V	ζ Persei Bl Ib	Planetary Nebulae	Log $(n/n_\odot)$ ε Virginis G8 III
		Goldberg, Müller, Aller (1960, 1967)	Various Recent Sources				
40	Zr	2.65					− 0.15
41	Nb	2.30					
42	Mo	2.30					
44	Ru	1.82					
45	Rh	1.37					
46	Rd	1.57					
47	Ag	0.75					
48	Cd	1.54					
49	In	1.45					
50	Sn	1.54					
51	Sb	1.94					
56	Ba	2.10	1.90				− 0.09
57	La	2.03					− 0.08
58	Ce	1.78					− 0.08
59	Pr	1.45					+ 0.37
60	Nd	1.93					+ 0.06
62	Sm	1.62					+ 0.01
63	Eu	0.96					
64	Gd	1.13					
66	Dy	1.00					
70	Yb	1.53					
72	Hf						
82	Pb	1.63	1.93				+ 0.18

This lack of evidence for chemical evolution is somewhat puzzling although the constancy of the chemical makeup of the medium could be attributed to a lack of mixing of the surface layers of the stars with matter from the stellar interior. This would need to be coupled with a return only of surface material to the interstellar medium whenever a supernova or other unstable star ejects matter. Of course (see Table 1.1 or section 1 : 6) some very metal-deficient stars are known to exist, and these exceptional objects do suggest chemical evolution. Clearly the evidence is ambiguous.

In a small fraction of the observed stars, some material from the stellar center does seem able to reach the surface in appreciable quantities. This may happen regularly in some *close binary stars* where surface material systematically flows from one component to its companion, exposing lower layers in which nuclear evolution is considerably advanced. There may also be other processes that permit the convection of matter from the interior to the surface of individual stars (Un69)*.

At any rate overabundances of helium, carbon, or of the metal elements is exhibited by some components of close binaries. For these, the spectroscopically determined abundance ratios are consistent with the processes already mentioned: the burning of hydrogen into helium through the proton-proton reaction or through the CNO cycle, the burning of helium into carbon C^{12}, and the burning of carbon into heavier elements. Evidence for the e-, s-, and r-processes also appears to be accumulating.

The *helium stars* (Table 8.4) show a systematic increase of the helium abundance over stars shown in Table 8.3. In fact, the nominally assumed abundance $n_{He} = 11.61$ represents complete burning of every set of four hydrogen nuclei to produce a helium nucleus, if an initial hydrogen abundance $\log n_H = 12.00$ is assumed. If the initial helium abundance is taken comparable to the values in Table 8.3, $\log n_{He} = 11.2$.

A comparison of the two tables also shows that, although in the star HD 160641 the CNO composition remained almost unchanged, the star HD 30353 shows a reduction of carbon and oxygen relative to nitrogen. This would perhaps indicate that the proton-proton reaction dominates the hydrogen burning in the first-named star, while the second star has undergone the CNO cycle, which has left most of the carbon and oxygen in the form of nitrogen. The equilibrium abundance of nitrogen in this cycle would in any case, be expected to be high because of the slow conversion of N^{14} into O^{15} shown in (8–98). The relatively high oxygen abundance shown by the star's spectrum indicates that other processes might also be at work. This is also true in HD 168476 and HD 124448, where helium burning may have replenished some of the carbon converted in the CNO bi-cycle.

The *carbon stars*, red giants that exhibit anomalously large surface abundance of carbon, either in atomic form or as a component of small radicals, give evidence that the *triple-alpha process* plays an important role. Various types of carbon stars show differing histories. Some have heavy element abundances reduced relative to hydrogen by an order of magnitude. These also exhibit elements, from barium Ba on, which are overabundant relative to iron by an order of magnitude. This overabundance must have been produced by neutron addition.

Interesting information can also be obtained from the ratio of C^{12} to C^{13} as determined from the spectra of diatomic radicals. Slightly shifted from the spectra associated with the C^{12} isotope, one finds a weaker band structure from radicals containing C^{13}. On earth and in the sun the C^{12}/C^{13} ratio is ~ 100. But for the CNO bi-cycle the lifetimes given in section 8:12 indicate an equilibrium ratio of $\sim 5:1$, and in most carbon stars this is the ratio actually observed; other

8:14

Table 8.4 Abundances of Elements in Helium Stars Relative to $\log n_{He} = 11.61$ (after A. Unsöld Un69).

Atomic Number	Element	Abundances: $\log n$				
		HD 160641	HD 168476	HD 124448	BD + 10° 2179	HD 30353
1	H		< 7.1	< 7.8	8.49	7.6
2	He	11.61	11.61	11.61	11.61	11.6
6	C	8.66	9.16	9.01	9.51	6.2
7	N	8.77	8.35	8.38	8.67	9.2
8	O	8.91	< 8.3	< 8.4	< 8.2	7.5
10	Ne	9.42	9.05			8.5
12	Mg	7.61	7.53	7.75	7.2	
13	Al		6.19	6.61	5.8	
14	Si	7.61	7.12	7.21	7.42	7.6
15	P		6.06			
16	S		6.75	7.19		7.8
18	A			6.9		
20	Ca		6.00	6.40	5.91	
21	Sc		4.3			
22	Ti		5.98	6.3		
23	V		4.65			
24	Cr		5.20	4.8		
25	Mn		4.57	4.84		
26	Fe		7.42	7.58		
28	Ni		5.4	5.2		

carbon stars show solar abundances. Where the ratio is low, it indicates that somehow the carbon formed through helium burning is later subject to the CNO bi-cycle. But because the carbon:nitrogen:oxygen abundance ratios found in individual carbon stars frequently are inconsistent with the CNO bi-cycle (Th72) it may be necessary to invoke other explanations.

Another class of stars, the *barium stars*, exhibit an unusually strong doublet due to barium at wavelengths of 4554 and 4934 Å. In general, for such stars elements with atomic number $Z > 35$ are overabundant by about an order of magnitude compared to the solar abundance. In these stars iron with $Z = 26$ is normal, but strontium with $Z = 38$ already is overabundant. These observations are consistent with neutron irradiation.

Similar features hold for the S stars where, again, the great strength of ZrO, LaO, YO bands, and the lines of atomic zirconium, barium, strontium, lanthanum, and yttrium show overabundances in the high Z range. That these elements originate from neutron processes also is consistent with the presence of technetium lines. Tc^{99} has a half-life of 2×10^5y, similar to the life computed for stellar evolution along the red giant branch. This isotope can be produced in the s-process. We can therefore conjecture that the isotope has been formed on the surface of the star within the last few 10^5y, or more recently; or else it was formed in the star's interior during this time and brought out to the surface. Had it, however, been present when the star first formed several aeons earlier, no measurable traces would have remained by this time. This gives direct evidence for current nucleosynthesis in stars, and because we only know of the neutron processes for forming such heavy elements, it supports our belief that the s-process takes place in the stellar interior.

The rare earth element promethium, whose longest-lived isotope has a half-life of only 18 y, has been tentatively identified in the atmosphere of HR465, a star that exhibits spectral lines of many other rare earth elements. This would show that nuclear reactions must be going on to replenish the decaying promethium (A170). Since convection times (8–30) are of the order of a month, this promethium need not have been generated very close to the surface, although that too might be possible. These observations still need confirmation.

8:15 THE POSSIBILITY OF DIRECT OBSERVATIONS OF NUCLEAR PROCESSES IN STARS

Thus far we have described the currently expected sequence of nuclear events that may be going on in the interior of stars, and have identified these events with phases in the life of a star represented by different portions of the Hertzsprung-Russell diagram. Since the sequence of events, the variety of possible reactions, and the number of assumptions required are so large, direct verification of the postulated nuclear reactions would be highly desirable.

The most promising observations that can be made in this respect are measurements on neutrinos emitted in the nuclear reactions. As already indicated *neutrinos* carry off somewhere around 2 to 6% of the hydrogen-to-helium conversion energy, depending on whether the proton-proton reaction or the CN cycle predominates. The neutrinos can escape from a star, virtually without hindrance, because in a 1 cm² column, one stellar radius deep, the neutrino would typically encounter $\rho R/m_h \sim 10^{35}$ nuclei, where $\rho \sim 1\,\text{g cm}^{-3}$, $R \sim 10^{11}$ cm and $m_H \sim 10^{-24}$ g. Since

8:15

the neutrino interaction cross section with nuclei normally is of order 10^{-45} cm^2, only one neutrino in 10^{10} would be intercepted on its way out of the star.

An attempt to directly observe neutrinos from the sun has been carried out by Raymond Davis, Jr. and co-workers at the Brookhaven National Laboratory (Da68). Their experiment was based on the large absorption cross section for neutrinos exhibited by the chlorine isotope Cl^{37}, in the reaction

$$Cl^{37} + v \rightarrow Ar^{37} + e^- \qquad (8\text{--}106)$$

This reaction requires a minimum neutrino energy of 0.81 Mev. The argon isotope Ar^{37} is radioactive and makes itself evident through a 34-day half-life, 2.8 kev *Auger* (*X-ray*) transition which can be recorded. The reaction cross section for this process is only large for high energy neutrinos, however, and so not all of the neutrinos emitted by the sun could be counted in this way. The experimenters figured they would only be able to observe the neutrinos from the decay of the boron isotope B^8, which is formed in very small quantities according to the predictions of nuclear theory. The neutrino given off in that decay can have energies as high as 14 Mev. The scheme, which first gives rise to boron, is shown in Fig. 8.12, together with the probability for the occurrence of each reaction.

In order to detect this process, the group made use of chlorine in the form of 520 metric tons of C$_2$Cl$_4$. The argon produced in this liquid was purged out of the tank with helium gas and trapped at low temperatures to separate it from the helium. The trapped argon was then released into a counting chamber. The cross section for neutrino capture from the boron decay was $\sigma \sim 1.35 \times 10^{-42}$ cm^2,

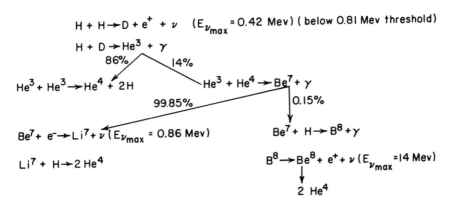

Fig. 8.12 Nuclear reactions leading to the production and decay of B^8 in the sun, showing relative probabilities for the different reactions and the energy of neutrinos produced. The branching ratios are temperature dependent. The ratios shown are expected at temperatures of $\sim 1.5 \times 10^{7\circ}$K at concentrations $X = 0.726$, $Y = 0.26$, $Z = 0.014$ (Ba72).

and the expected number of captures amounted to somewhere between 2 and 7 per day. The actual observed upper limit to the counts attributable to Ar^{37} decay was somewhat lower than that, limiting the flux at the earth to less than 2×10^6 cm^{-2} sec^{-1} neutrinos from this reaction. For the 0.86 Mev neutrino, $\sigma \sim 2.9 \times 10^{-46}$ cm^2 and the contribution from the Be^7 electron capture was small compared to B^8.

This experiment apparently also yielded an upper limit of about 10% for the contribution by the CNO cycle to solar energy production. If it played a bigger role, the neutrinos produced by O^{15} decay in this cycle would have been detected.

After this experiment had been completed, theorists re-examined their predictions on the basis of improved measurements of the sun's composition and nuclear reaction probabilities, but the expected flux was not substantially changed. In the past few years, larger tanks of C_2Cl_4 have come into use and theoretical work also has continued. The newer expected neutrino counting rate, however, still exceeds the observed by a factor of ~ 3 in most calculations. A new type of detector using the gallium isotope Ga^{71} is also planned. It would have the advantage of sensitivity to neutrinos from the proton–proton reaction (Ba85).

8:16 THE POSSIBILITY OF ELEMENT SYNTHESIS IN OBJECTS EXPLODING FROM AN INITIAL TEMPERATURE OF 10^{10} °K

One observation that has not been satisfactorily answered by the theory of nuclear evolution in stars is the detection of some heavy elements in the atmospheres of the earlerst stars that can be found in the Galaxy. The indications are that these stars, which are found as members of *globular clusters*, contain matter that has already undergone some nuclear processing. This would imply that the earliest stars formed within the galaxy were not formed from hydrogen or hydrogen-helium mixtures alone, but must have been formed from matter that already had an admixture of heavier elements. From a cosmological viewpoint this may have important implications. If the earliest stars we know contain heavier elements, then where did those come from? Is it necessary or even possible that these elements were formed during some primordial exploding state of the Universe, or can we explain their presence in other ways?

Before becoming too concerned, we should consider the possibility that the observed heavy elements are no more than a surface contamination. We ask whether there is a possibility that stars might have picked up a sufficient amount of interstellar material, simply on passage through dense interstellar clouds. The present composition of the interstellar material certainly would have a high enough abundance of heavy elements to satisfy observations. However, when the calculations are made (and these take a form similar to Problem 3–8), we find that the *accretion rate*, or capture rate, for interstellar matter, is too small to give rise to the observed metal abundances.

Computations by Wagoner (Wa67) and by others indicate that the observed abundances of the heavy elements in globular cluster stars can be explained by the following hypothesis: At some time before the globular cluster stars were formed, galactic hydrogen collected into one or more massive objects whose central temperatures rose to 10^{10}°K before exploding to spew the material back into the Galaxy. Under the right conditions of expansion following such an explosion, the correct abundance ratios are found. The thought here is that such an explosion might occur on a galaxy-wide scale and perhaps represent a phenomenon of the *quasar* type—very compact and luminous. The material cycled through such a stage would then be available for subsequent star formation in globular clusters. These massive objects, sometimes called *Population III* stars, could comprise an entire galactic mass, be the size of a globular cluster or just a massive star. Either way this idea could explain the presence of heavier elements in the oldest stars within our Galaxy.

This theory might also explain the apparently high amount of helium present in the earliest stars, the halo stars in the Galaxy (Wa71)*, although backers of the standard cosmological model would say that this helium must have been formed during early stages of cosmic evolution when matter was at a high density and temperature (section 12:7).

If the massive object represents the whole universe, in an intially compact state, we can argue that the presently observed 3°K blackbody background radiation, if present primordially, determines the temperature of the earliest evolutionary stages. The rate at which the radius of the universe and hence the temperature and density changed can be calculated. We can then use nuclear reaction rates to determine the composition of the material when the density finally became so low that no further nuclear changes took place (section 12:14).

That composition, of course, depends on the constituents initially present in the universe. In particular, the final neutron-proton ratio will be strongly affected by the number of neutrinos and antineutrinos initially present. If the *electron neutrinos* are very abundant the reaction

$$v + \mathscr{N} \rightleftarrows \mathscr{P} + e^- \qquad (8\text{--}107)$$

will go mainly toward the right and produce a large abundance of protons. When the neutrinos are less dense, two other reactions also play a role

$$e^+ + \mathscr{N} \rightleftarrows \mathscr{P} + \bar{v} \qquad (8\text{--}108)$$

$$\mathscr{N} \rightleftarrows \mathscr{P} + e^- + \bar{v} \qquad (8\text{--}109)$$

8:16

and produce roughly equal neutron and proton densites. For large antineutrino abundances, these reactions run from right to left. The neutrinos considered here of course are electron neutrinos—not muon neutrinos. *Muon neutrinos* influence the development only very indirectly through their contribution to the cosmic expansion rate that is influenced by the total density of matter (section 10:9).

The ratio of protons to neutrons determines the final abundance of elements present. In most cosmological models the evolution stops at mass 7 amu, because nuclei with mass 8 are unstable. Here the triple-α process does not take place because densities are too low. The possible processes are shown in Fig. 8.13.

Detailed calculations for differing initial densities of the various elementary particles predict different abundances of deuterium, He^4, He^3, and Li^7, which might be measurable at the present time by observations of the intergalactic medium. Such observations are very difficult. No such intergalactic gases have been identified to date. As astronomical methods improve, however, identification of these gases should become possible.

8:17   COMPACT STARS

Thus far we have dealt with stars whose densities are roughly comparable to the sun's, except late in life when central portions of these stars become very compact.

We now turn to stars that are orders of magnitude more compact: the *white dwarf* and *neutron stars*. The structure of these stars can be considered from the same general viewpoint that allowed us to understand processes in the interior of ordinary stars. Before proceeding in this direction, we should, however, review one particular argument that we had brought out to demonstrate the importance of nuclear reactions in stellar interiors. In (8–1) we had shown that the potential energy per unit mass of stellar substance is $\sim 3MG/5R$, while the available nuclear energy is of the order of $10^{-2}c^2$, if matter-antimatter annihilation is ruled out. It follows that very compact stars may be able to liberate amounts of gravitational energy in excess of the normal nuclear energies available. This will happen when

$$R \lesssim \frac{MG}{10^{-2}c^2} \sim 10^7 \text{ cm} \qquad (8\text{–}110)$$

for stars of one solar mass. This is still larger than R_S, the *Schwarzschild radius*, at which the conventional expression for potential energy per unit mass would become equal to c^2:

$$R_S = \frac{2MG}{c^2} \qquad (8\text{–}111)$$

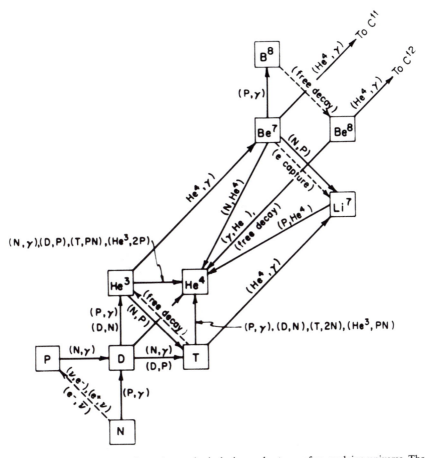

Fig. 8.13 Reactions of importance in nucleosynthesis during early stages of an evolving universe. The exergonic directions are indicated, although rates are often rapid in both directions. The other incoming and outgoing particles, those not shown in squares, are indicated in parentheses. Dashed arrows indicate the beta reactions. Sometimes there are competing reactions leading from one nucleus to another (after R. Wagoner, Wa67).

Since white dwarfs typically have masses of the order of 10^{33}g, the corresponding Schwarzschild radius would be

$$R_S \sim 10^5 \text{ cm} = 1 \text{ km}$$

which is small compared to the white dwarf radii computed below, in section 8:18, but comparable to the neutron star radii ~ 10 km, discussed in section 8:19.

8:17

8:18 WHITE DWARF STARS

We had suggested that the compact stars are so dense that matter becomes degenerate in the interior. At the surface of a white dwarf, the density is lower and no degeneracy will be found in the outer layers of such a star. However, the actual thickness of the nondegenerate layer is small, and we will be able to progress by treating the star as completely degenerate throughout.

To proceed with this computation, we first note that most of the pressure in the stellar interior must be provided by the degenerate electrons. The partial pressure of the nuclei is very small. This comes about because under degenerate conditions, the lowest momentum states of the electron gas are always filled. The more compact the star, the higher the Fermi energy of the electrons, and the higher the electron gas pressure. Since only the electrons are degenerate, as discussed in section 4:15, the nucleon pressure is relatively low. We will neglect it here and take the total pressure to equal the electron pressure, $P = P_e$.

In section 8:6 we had given the electron pressure for a nonrelativistic and for a completely relativistic degenerate electron gas, respectively, as:

$$P = \frac{h^2}{20 m_e m_H} \left(\frac{3}{\pi m_H} \right)^{2/3} \left(\frac{(1 + X)}{2} \rho \right)^{5/3} \quad \text{nonrelativistic} \qquad (8\text{-}41)$$

and

$$P = \frac{hc}{8 m_H} \left(\frac{3}{\pi m_H} \right)^{1/3} \left(\frac{(1 + X)}{2} \rho \right)^{4/3} \quad \text{relativistic} \qquad (8\text{-}43)$$

In general, there will exist an important transition region where the gas is neither highly relativistic nor completely unrelativistic. In that region the pressure can be shown (Problem 8–4) to take the form:

$$P = \frac{8 \pi m_e^4 c^5}{3 h^3} f(x), \qquad \rho = \mu_e \frac{8 \pi m_H m_e^3 c^3}{3 h^3} x^3 \qquad (8\text{-}112)$$

with

$$\mu_e = \frac{2}{1 + X} \qquad (8\text{-}113)$$

The function $f(x)$ is

$$f(x) = \frac{1}{8} \left[x(2x^2 - 3)(x^2 + 1)^{1/2} + 3 \sinh^{-1} x \right] \qquad (8\text{-}114)$$

8:18

where

$$x = \frac{p_0}{m_e c} \tag{8-115}$$

PROBLEM 8-4. For a degenerate relativistic gas, all momentum states (4-65) are filled, and equation 5-30 holds. Using equations similar to (4-27) through (4-30) as a guide, show that

$$P = \frac{1}{3} \int_0^{p_0} pv(p) \, n_e(p) \, dp \tag{8-116}$$

$$= \frac{8\pi}{3 m_e h^3} \int_0^{p_0} \frac{p^4 \, dp}{[1 + (p/m_e c)^2]^{1/2}} \tag{8-117}$$

Setting $\sinh u = p/m_e c$, show that

$$P = \frac{8\pi m_e^4 c^5}{3h^3} \int_0^{u_0} \sinh^4 u \, du \tag{8-118}$$

We see that (8-118) has the same coefficient as (8-112). We can also show by integration that the integral in (8-118) equals the expression 8-114 for $f(x)$.

Small values of $x, x \ll 1$, correspond to the lower density portions where the gas is nonrelativistic, while high x-values correspond to the frequently relativistic central portions of the star.

PROBLEM 8-5. Evaluate $f(x)$ in the limits $x \ll 1$ and $x \gg 1$ and show that equations 8-41 and 8-43 are obtained.

Equation 8-112 is computed on the basis of statistical mechanics and involves no assumptions concerning stars. It is simply an *equation of state* for a partially relativistic degenerate gas, no matter where it may be found. We should note that the pressure is temperature independent in this equation; it only depends on x, which is a measure of the momentum at the Fermi energy of the electron gas; and that is only density dependent. The mathematical problem of computing conditions at the center of the star can therefore be separated into two portions, a hydrostatic and a thermodynamic one.

The hydrostatic equilibrium conditions are the same as those obtained earlier:

$$\frac{dP}{dr} = -\rho \frac{GM(r)}{r^2}, \qquad \frac{dM(r)}{dr} = 4\pi r^2 \rho \tag{8-7, 8-8}$$

To integrate these equations, we assume that we know what the chemical composition of the white dwarf is, because that composition determines the value of μ_e in equation 8–113. We next choose an arbitrary central density ρ_c for the star, and then integrate the hydrostatic equations outward from the star's center until we reach a radius r where the pressure has dropped to zero. In this model computation, that radius represents the surface of the star. The value of $M(r)$ at this radial distance corresponds to the total mass of the star and the value of r represents the actual stellar radius.

Clearly this procedure can be repeated for a range of different central densities, and we therefore obtain a whole family of stellar models with differing central densities. Similarly, we can obtain a new family of models having different chemical composition; but no recomputation is required here, because a change of chemical composition is mathematically equivalent to a simple change of variables. We will show this:

If the initially computed values are denoted by primes, and new variables— corresponding to a new chemical composition—are denoted by unprimed symbols, we find that the required interrelations are

$$P = P'$$

$$\rho = \frac{\mu_e}{\mu_e'} \rho'$$

$$M(r) = \left(\frac{\mu_e'}{\mu_e} \right)^2 M(r') \tag{8–119}$$

$$r = \left(\frac{\mu_e'}{\mu_e} \right) r'$$

We can readily see that substitution of these expressions into equations 8–112, 8–7, and 8–8 leaves the form of these equations unchanged; a change in the chemical composition is therefore equivalent to a change in central density. Consequently we are dealing with a one-parameter family of models, because everything about a given star can be described entirely in terms of an equivalent central density. We present the results of the described computations in the form of Table 8.5.

The argument presented thus far neglects a number of correction factors that act to lower the last few mass values in the table by $\sim 20\%$ (Sc58b). However, we will not be concerned here with factors of that order of accuracy. Instead, we will concentrate on the overall properties of these stars. They are:

(1) The larger the mass of the white dwarf, the smaller is its radius.

(2) For masses comparable to the sun, the white dwarf's radius is a factor of $\sim 10^2$ smaller than $R_\odot$.

Table 8.5 Central Densities, Total Mass, and Radius of Different White Dwarf Models, Taking $\mu_e = 2$ (Negligible Hydrogen Concentration).[a]

$\log \rho_c$	$M/M_\odot$	$\log R/R_\odot$
5.39	0.22	-1.70
6.03	0.40	-1.81
6.29	0.50	-1.86
6.56	0.61	-1.91
6.85	0.74	-1.96
7.20	0.88	-2.03
7.72	1.08	-2.15
8.21	1.22	-2.26
8.83	1.33	-2.41
9.29	1.38	-2.53
∞	1.44	$-\infty$

[a]See text for comments. (After M. Schwarzschild Sc58b.) From *Structure and Evolution of the Stars* (copyright © 1958 by Princeton University Press) p. 232.

(3) There exists an upper limit to the mass—the *Chandrasekhar limit*—above which no stable white dwarf configuration exists—infinite central pressure would be needed to keep the star from further collapse. The actual limit, taking into account the above-cited 20% correction, should be $\sim 1.2M_\odot$.

The reasons for the limit are apparent if we consider that there are quite different relations between central pressure and density in the relativistic and nonrelativistic limits. From (8–41) and (8–43):

$$\text{nonrelativistically:} \quad P \propto \rho^{5/3}, \quad \frac{dP}{dr} \propto \rho^{2/3}\left(\frac{d\rho}{dr}\right) \qquad (8\text{–}120)$$

$$\text{relativistically:} \quad P \propto \rho^{4/3}, \quad \frac{dP}{dr} \propto \rho^{1/3}\left(\frac{d\rho}{dr}\right) \qquad (8\text{–}121)$$

At the same time, the gravitational pressure gradient is

$$\frac{dP}{dr} \propto -\frac{\rho(r)}{r^2}\int_0^r \rho(r')\,4\pi r'^2\,dr' \qquad (8\text{–}122)$$

In the crudest approximation, we can set the density equal to the stellar mass

8:18

divided by the cube of the radius R so that

$$\text{nonrelativistically} \quad \frac{dP}{dr} \propto \frac{M^{5/3}}{R^6}$$

$$\text{relativistically} \quad \frac{dP}{dr} \propto \frac{M^{4/3}}{R^5} \qquad (8\text{--}123)$$

$$\text{gravitational} \quad \frac{dP}{dr} \propto \frac{M^2}{R^5}$$

We note that the dependence of the relativistic pressure gradient on radius has the same power as the gravitational force. Both increase as R^{-5} as the star contracts. This means that once a relativistic white dwarf core is forced to contract by hydrostatic pressure, the counterforce produced through contraction increases at the same rate as the gravitational attraction, and that tends to compress the star even further. There is, therefore, no way in which the star can come into equilibrium. On the other hand, a nonrelativistic gas at the center of a white dwarf can always adjust itself through contraction, until the gravitational forces compressing the star are countered.

We then have the following situation. For small stellar masses, the central pressure is determined more nearly by the nonrelativistic approximation, and the star can find a stable equilibrium position. For more massive objects, the central density becomes so high during contraction that the relativistic regime is reached, and further contraction no longer leads to a situation of equilibrium. The Chandrasekhar limit is therefore symptomatic of the transition from a predominantly nonrelativistic to a predominantly relativistic central core (Ch39)*.

We still must think about the appearance of white dwarf stars in the Hertzsprung-Russell diagram. We recall the definition of the effective temperature of a star

$$L = \sigma T_e^4 (4\pi R^2) \qquad (4\text{--}76)$$

and rewrite this in terms of the solar luminosity and surface temperature:

$$\log \frac{L}{L_\odot} = 4 \log \frac{T_e}{T_{e\odot}} + 2 \log \frac{R}{R_\odot} \qquad (8\text{--}124)$$

If we then use the white dwarf radii and masses from Table 8.5, we can obtain plots of L against T_e as a function of different mass values. Choosing five different representative masses, we obtain the curves shown in Fig. 8.14.

The agreement with observations is satisfactory, and we can feel reasonably confident that the discussion pursued here is at least roughly correct. This is important. For, white dwarfs have a local number density of $\sim 2.5 \times 10^{-2}$ white dwarfs pc^{-3}—10% to 20% of the mass in our neighborhood of the Galaxy (We68).

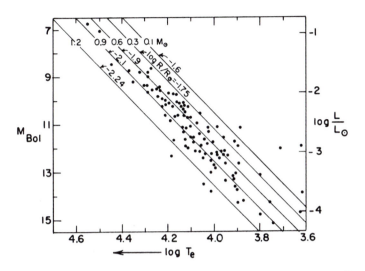

Fig. 8.14 White dwarf Hertzsprung-Russell diagram. Lines of constant radius are shown. Also shown are the masses based on completely degenerate core models containing elements having $\mu_e = 2$ (after Weidemann We68; reprinted with permission from *Annual Review of Astronomy and Astrophysics*, Volume 6, copyright © 1968 by Annual Reviews Inc.).

They are objects we should come to understand well, if we are to learn how stars die.

8:19 NEUTRON STARS AND BLACK HOLES

For some stars the most advanced stage of evolution appears to contain a core that consists of densely packed neutrons. We can imagine the evolution toward this state in the following way (Sa67)*.

The equations in (8–112) are applicable as long as we are dealing with a dense star, but as the star evolves, the value of μ_e changes. As hydrogen becomes depleted, we saw μ_e assumes a value of 2. This holds, for example, for a star in which the major constituent is C^{12}. But as the chemical elements evolve toward the more neutron-rich species, equations 8–36 and therefore 8–113 no longer hold, and for a star rich in Fe^{56} we find $\mu_e = 2.15$. Now, the Chandrasekhar limiting mass is porportional to μ_e^{-2}, as seen from (8–119), so that we can draw a number of curves of mass against central density—as in Fig. 8–15. In these curves, we assume the lowest possible temperature and we show plots for stars of differing chemical composition. Corresponding to each chemical composition is a different Fermi-energy E_F for the electrons at the center of the star, as a direct consequence of the star's changed central density.

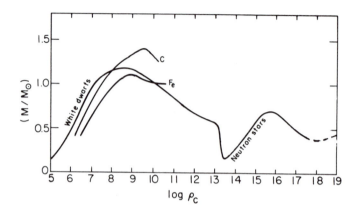

Fig. 8.15   Mass of a cold star, as a function of central density. The full curve is for a representative initial chemical composition and assumes relativistic hydrostatic equilibrium. The curves labeled C and Fe assume pure carbon and iron composition (after Ru71a, Sa67). Negative slopes on the curves indicate regions where there are no hydrostatically stable configurations. The density is given in units of g cm^{-3}. The dashed portion of the curve near central densities of the order 10^{18} g cm^{-3} is very uncertain, because the physical state of matter at these densities is uncertain. (Reprinted with permission of the publishers of The American Mathematical Society, from *Lectures in Applied Mathematics*, copyright © 1967, Volume 10, Part 3, "Stellar Structure Leading up to White Dwarfs and Neutron Stars," p. 33.)

Table 8.6   Density and Electron Fermi Energy at which inverse beta decay becomes energetically favorable (after E.E. Salpeter, Sa67).

	C^{12}	S^{32}	Fe^{56}	Sn^{120}
$\log \rho$ (g cm^{-3})	10.6	8.2	9.1	11.5
E_F (Mev)	13	1.7	3.7	24

Reprinted with permission of the publishers of The American Mathematical Society, from *Lectures in Applied Mathematics*, copyright © 1967, Volume 10, Part 3, "Stellar Structure Leading up to White Dwarfs and Neutron Stars," p. 34.

As the central density increases, for a given composition, the electron Fermi energy always increases up to the point where inverse beta decay takes place and drives the electrons into the nuclei. This is what produces the increasingly neutron-rich elements cited in Table 8.6. The symbolic reaction is

$$\mathscr{P} + e \to \mathscr{N} + \nu \tag{8-125}$$

The reverse reaction cannot take place if the Fermi energy is high enough, because all the electron states into which the radioactive nucleus might decay are already occupied. This gives otherwise unstable nuclei an environmentally induced stability.

The value of μ_e, which is an effective nuclear mass per free electron, also increases during contraction. When the Fermi energy reaches 24 Mev, the density is $\rho \sim 10^{11.5}$ g cm^{-3} and $\mu_e \sim 3.1$. At this stage free neutrons become energetically favorable so that a further increase in density leads to an increased partial density of neutrons, a practically constant density of ions, and a constant electron Fermi energy of 24 Mev.

As the density increases, E_F increases to the point where reaction (8–125) proceeds rapidly and the electrons are driven into the nuclei causing the collapse of the central core, because the electron pressure no longer increases at a sufficient rate during the contraction.

In Fig. 8.15 the curves for stars containing C^{12} and Fe^{56} show a maximum mass at ρ_c values where inverse beta decay first sets in and μ_e increases. At the extreme lower right, the curve for free neutrons is shown. It has a maximum just beyond $\rho_c \sim 10^{15}$ g cm^{-3}.

The reason for this maximum is relatively easy to understand if we compute the mass expected on the basis of a nonrelativistic neutron gas and an extreme-relativistic gas, respectively (Sa67)*.

The virial theorem gives the ratio of pressure P to density ρ in terms of stellar mass as

$$3 \left\langle \frac{P}{\rho} \right\rangle \sim \frac{M}{R} \propto M^{2/3} \langle n^{1/3} \rangle \tag{8–126}$$

where mean values are denoted by brackets, and n is the number density of neutrons. As can be seen from (8–35), (8–40), and (8–42)

$$P \propto n^{5/3} \quad \text{nonrelativistic} \tag{8–40a}$$

$$P \propto n^{4/3} \quad \text{extreme relativistic} \tag{8–42a}$$

Similarly, the ratio of mass density to number density is

$$\langle \rho \rangle / \langle n \rangle \sim m_N \quad \text{nonrelativistic} \tag{8–127}$$

$$\langle \rho \rangle / \langle n \rangle \sim \frac{E_F}{c^2} \quad \text{extreme relativistic} \tag{8–128}$$

because in the extreme case, the rest-mass energy can be neglected. But since $E_F \propto n^{1/3}$, one then has

$$\langle \rho \rangle \propto \langle n \rangle^{4/3} \quad \text{extreme relativistic} \tag{8–129}$$

8:19

and from equation 8–126 we then find

$$M^{2/3} \propto 3 \left\langle \frac{P}{\rho} \right\rangle \langle n^{1/3} \rangle^{-1} \propto \begin{cases} \langle n^{1/3} \rangle & \text{nonrelativistic} \\ \\ \langle n^{1/3} \rangle^{-1} & \text{extreme relativistic} \end{cases} \tag{8–130}$$

This means then that the mass first increases as $\langle n \rangle^{1/3}$ and then decreases as $\langle n \rangle^{-1/3}$ as the density continues to increase.

The cores of massive neutron stars probably contain a variety of *mesons*, *baryons*, and *hyperons* in addition to the neutrons. At these very high densities, general relativistic effects must also be taken into consideration, because, for example, the Newtonian expression for potential energy no longer has meaning. This region is of greatest interest, because potentially it is in these last stages that a star can give off by far its greatest amount of energy by converting a large fraction of its mass into some form of radiation, perhaps gravitational. The star then turns into a *black hole*. However, before we turn to this ultimate form of stellar death, it is worthwhile mentioning a number of important considerations that we have neglected.

At the very high densities involved in neutron stars, the nuclei arrange themselves in a *crystal lattice*. The equation of state and structural properties we have assumed will therefore not be strictly correct. In parts of neutron stars a *superfluid* state may also exist. One model (Fig. 8.16) pictures a solid crust of nuclei floating on a superfluid layer. In summary, a variety of different crystalline or fluid phases may exist at different depths in the star. These will have to be understood more thoroughly before we can claim to understand the structure of neutron stars (Sa70a and Ru71b).

An additional matter of importance is the magnetic field. If a star like the sun were to collapse to a radius of a few kilometers, its magnetic field would be $\sim 10^{12}$ gauss. Is the field really that strong in neutron stars? And if it is, does this strong field permeate the entire interior or does it exist only at the surface? If superconducting effects were present in the core of the neutron star, magnetic fields might be excluded just as they are in laboratory superconductors. But how could that exclusion take place in a rapidly collapsing star?

We also know that an ordinary star which suddenly collapses would have to rotate rapidly if angular momentum is to be conserved. In the region surrounding the neutron star, the rapidly rotating magnetic field is thought responsible for accelerating charged particles to cosmic ray energies and perhaps for radiating away some of the star's rotational energy. The rotation presumably also affects the star's structure, since perfect spherical symmetry no longer holds.

If the Crab Nebula is at all representative, the neutron star, pulsar and supernova phenomena all seem to have a common origin. There may of course exist

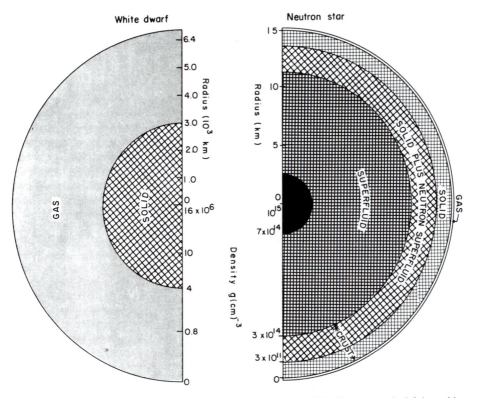

Fig. 8.16 Density and structure of white dwarf and neutron stars. The diagram on the left is a white dwarf model. The shell structure on the right represents a neutron star. Note that the white dwarf radius is 6400 km while that of the neutron star is only 15 km (after Ruderman Ru71b). At the neutron star's center we expect to find mesons and hyperons. (Reprinted with permission from Ruderman, "Solid Stars," copyright © 1971 by Scientific American, Inc. All rights reserved.)

a number of different processes that lead to supernova explosions since a number of different types of supernovae have been observed. Perhaps only some of these lead to the formation of neutron stars and pulsars. Others might be responsible for the formation of heavy elements and their return to the interstellar medium (Ar70) without necessarily also producing a compact remnant. This, in fact, raises the whole question of the interrelationship of supernovae with ordinary and/or recurrent novae, with planetary nebula stars that have shed a great deal of mass and left a hot central star behind, and finally with white dwarfs that might be the remnants of the planetary nebulae. How are all these related? Is the deciding factor between a white dwarf and a neutron star death merely a matter of whether the star's final mass is less or greater than the Chandrasekhar limiting mass (Ch64a)? We do not yet know.

Similarly, we also do not know much about black holes. If a neutron star is massive enough, does it collapse once more into a final state where, as discussed in Chapter 5, it no longer can emit any radiation that will be received outside the star? What happens to the magnetic field when a black hole is formed, and to electric charge? How can we talk about conservation of baryons in the universe when all these particles disappear without a trace, giving off only photons or gravitational radiation and neutrinos as the collapse takes place? What are the correct dynamic equations in such a collapse? Are they the equations of general relativity, or not? Again a thorough discussion is needed; we are just beginning to grope our way (Op39a, Op39b, Ru71a, Pe71, and Ke63).

8:20 VIBRATION AND ROTATION OF STARS

We know from the virial theorem that the absolute value of the potential equals twice the kinetic energy per unit mass. In a star this kinetic energy is represented by the thermal motions of the atomic particles whose speeds are of order of the speed of sound v_s. We can therefore write

$$\frac{GM}{R} \sim v_s^2 \tag{8-131}$$

where G is the gravitational constant, M is the mass, and R the radius of the star. We can now estimate a very rough order of magnitude of the stellar vibration frequency by noting that the period P_{vib} should be comparable to the time it takes to transmit information about pressure changes, across the entire dimension of the star. This time equals $2R/v_s$, and we can write (Sa69)

$$P_{vib}^{-1} = v_{vib} \sim \frac{v_s}{2R} \sim \sqrt{\frac{GM}{4R^3}} \sim \sqrt{G\rho} \tag{8-132}$$

where ρ is the density of the stellar material.

We also note that the maximum rotational frequency is determined by the equilibrium between centrifugal and gravitational forces, since at higher frequencies the star would be torn apart. Nonrelativistically,

$$Rv_{rot,max}^2 = \frac{GM}{(2\pi R)^2}$$

and

$$\tag{8-133}$$

$$\therefore P_{rot,min}^{-1} \sim \frac{1}{\pi}\sqrt{G\rho}$$

8:20

Table 8.7 Approximate Relation Between Stellar Density, Pulsation Period, and Minimum Rotational Period.

Star	Density ρ	P_{vib}	$P_{rot,min}$
Neutron star	10^{15} g cm^{-3}	10^{-4} sec	3×10^{-4} sec
White dwarf	10^7	1	3
RR Lyrae star	10^{-2}	$10^{4.5}$	10^5
Cepheid Variable	10^{-6}	$10^{6.5}$	10^7

We note that the vibrational periods for neutron stars ought to be about a 10th of a millisecond and that the vibration period for white dwarfs should be of the order of a second. Of course these stars can have a wide range of densities, and only representative periods are given in Table 8.7. These can vary by over an order of magnitude. When the Crab Nebula pulsar was discovered, having a period of only 33 milliseconds, it became clear that white dwarfs could not be considered to play a role in the pulsar phenomenon, because the vibration frequencies of white dwarfs would be too low. On the other hand the rotational period of a neutron star could well be several milliseconds. The discovery that the period of the Crab Nebula pulsar is increasing, so that it could have been closer to the minimum expected rotational period of a neutron star, shortly after the supernova explosion in the year 1054 AD, further supports the theory that pulsars are rapidly rotating neutron stars that are losing angular momentum and slowing down as time goes on. Since magnetic pressures would tend to disrupt rapidly rotating neutron stars, the actual minimum rotation period will be somewhat greater than shown in Table 8.7.

Table 8.7 also shows that RR Lyrae variables and Cepheid variables have periods consistent with the very simple vibration picture discussed in the present section. Such stars indeed are pulsating. This can be judged by the periodic Doppler line shifts and color temperature changes. The observed periodicity is the period of that pulsation. The nova remnant DQ Her (Nova Herculis, 1934) has a 71 sec period; and periodic behavior has been seen in some white dwarfs, although these periods are too long to represent fundamental pulsations.

ADDITIONAL PROBLEMS

The following set of problems uses greatly simplified stellar models, mainly to show that we can obtain reasonable order of magnitude estimates of stellar luminosities and lifetimes even without using sophisticated computing methods.

8:20

Reaction rates for the proton-proton reaction and the carbon cycle have the form given in equation 8–90. Schwarzschild (Sc58b) gives the energy generated per unit time and unit mass of matter for the proton-proton reaction:

$$E_{pp} = 2.5 \times 10^6 \rho X^2 \left(\frac{10^6}{T} \right)^{2/3} \exp\left[-33.8 \left(\frac{10^6}{T} \right)^{1/3} \right] \qquad (8\text{--}134)$$

and Clayton (Cl68) gives a similar expression for the CN cycle

$$E_{CN} = 8 \times 10^{27} \rho X X_{CN} \left(\frac{10^6}{T} \right)^{2/3} f \exp\left[-152.3 \left(\frac{10^6}{T} \right)^{1/3} \right]$$
$$(8\text{--}135)$$

where f is an electron screening factor, $f \sim 1$.

For problems 8–6 to 8–8, consider an initial concentration $X_{CN} = 0.005X$ and $Y = 0.12$.

8–6. Consider the following model of a B1 star, that is, a massive young star. Its mass $M = 10M_\odot$ and its radius, $R = 3.6R_\odot$. Its central density is $10 \, \text{g cm}^{-3}$ and the bulk of energy generation takes place within a radial distance $0.15R$. Assume constant density for this region, and a temperature $2.7 \times 10^{7\circ}\text{K}$. Determine the rate of energy generation by the star and its surface temperature, assuming that the star radiates like a blackbody at its given radius. How long can the star exist in its present state without being appreciably changed by the nuclear burning, that is, how soon will it use up the hydrogen in its central core at the present rate?

8–7. Repeat this for a star like the sun in its early main sequence stages, assuming burning in the central region out to $0.2 \, R$. Take a central density of about $55 \, \text{g cm}^{-3}$, and a central temperature $10^{7\circ}\text{K}$; assume hydrogen burning.

8–8. At the present time the concentration of hydrogen, by mass, at the center of the sun is about 70%, throughout a central region out to roughly $0.11 \, R$. The density is about $10^2 \, \text{g cm}^{-3}$ in the center. Take the mean temperature throughout the region to be $1.5 \times 10^{7\circ}\text{K}$ and recalculate the parameters for this star.

8–9. A red giant star whose radius is 100 solar radii, is in an evolutionary state where the inner hydrogen has all been exhausted but helium burning has not yet set in. The main energy source is hydrogen burning that takes place in a thin shell surrounding the inert helium core (Fig. 8.8). Let hydrogen burning take place at a radial distance ranging from 1.8 to 2×10^9cm. The mean density in this layer is about $50 \, \text{g cm}^{-3}$. The temperature is $5 \times 10^{7\circ}\text{K}$. Calculate the above parameters. Take $X_{CN} = 10^{-3}X$, $X \sim 0.5$.

8–10. A white dwarf star is thought to shine by virtue of its stored thermal energy. Assuming its mass to be $0.45 \, M_\odot$, its radius $0.016 \, R_\odot$, and its density roughly uniform throughout (however, see Fig. 8.16). calculate how long the star could radiate at its present rate. Its luminosity is $10^{-3} L_\odot$.

8–11. For a star in which radiative transfer dominates, the energy density at each point is very nearly the blackbody radiation density (at all frequencies v) even though the radiant flux $L(r)$ is predominantly outward directed from the center of the star. Show how closely the radiation density ρ really equals the blackbody radiation density for a given opacity value $\kappa(v)$, by considering the higher order terms in equation 8–50.

8–12. In section 8:2 we argued that the sun's energy must be nuclear, because equation 8–1 did not provide enough potential energy to account for the total solar luminosity over the past aeons. Show that the argument actually was incorrect, by working out the potential energy of a structure in which roughly half the sun's mass is uniformly distributed throughout a sphere of radius $R_\odot$ and half the mass is concentrated in a core of radius 10 km. It is interesting that some years ago, astrophysicists were sure that most of the radiant energy in the universe was directly related to thermonuclear reactions. With the discovery of pulsars, these ideas changed. It is possible that gravitational collapse has contributed at least as much energy as nuclear reactions.

ANSWERS TO PROBLEMS

8–1. (a) $P = nkT$

$$P\left(\frac{R}{2}\right) \sim 10^{15} \text{ dyne cm}^{-2}, \qquad n = \frac{\rho}{m_H} = \frac{1}{1.67 \times 10^{-24}} \text{ cm}^{-3}$$

$$T\left(\frac{R}{2}\right) = \frac{10^{15} \cdot 1.67 \times 10^{-24}}{1.38 \times 10^{-16}} \sim 10^{7\circ}\text{K}.$$

(b) The gravitational potential energy is $\dfrac{GM_p m}{R} \lesssim \mathcal{E}_c$

where m is the mass of an atom, $m = Am_H$.

$$M_p \lesssim \frac{\mathcal{E}_c R}{GAm_H}. \qquad \text{But } M_p \sim \rho R^3$$

$$M_p \lesssim \frac{\mathcal{E}_c M_p^{1/3}}{GAm_H \rho^{1/3}}, \qquad R = \left[\frac{M_p}{\rho}\right]^{1/3}$$

$$M_p^{2/3} \lesssim \frac{\mathscr{E}_c}{GA\rho^{1/3}m_H}, \quad M_p \lesssim \left(\frac{\mathscr{E}_c}{GAm_H}\right)^{3/2} \frac{1}{\rho^{1/2}}.$$

(c) $\mathscr{E}_c \sim 10^{-11}$ erg $R_J = 7.1 \times 10^9$ cm

$\left.\begin{array}{c} \\ \\ \end{array}\right\}$ see Table 1.3

$M_J = 2 \times 10^{30}$ g

while $M_p \lesssim \dfrac{\mathscr{E}_c R}{Gm_H} = 10^{30}$ g

8–2. $F(r) = -\displaystyle\int_0^\infty \kappa\rho e^{-\kappa\rho l}\, dl \int_0^\pi a\left(T - \frac{dT}{dr} l \cos\theta\right)^4 \frac{c}{2}\cos\theta\, d(\cos\theta)$

Expanding, we have to first order in dT/dr,

$$\simeq \frac{4ac\kappa\rho}{3}\frac{dT}{dr} T^3 \int_0^\infty e^{-\kappa\rho l}\, l\, dl.$$

And using (8–26) and (8–51) we have

$$L(r) = \frac{16\pi acr^2}{3\kappa\rho} T^3 \frac{dT}{dr}.$$

8–3. $dP \propto \rho\, \dfrac{M(r)\, dr}{r^2}$ (8–7)

$P = nkT$ so that $dP \propto \rho\, dT$

$\therefore \dfrac{dT}{dr} \propto \dfrac{M(r)}{r^2}$ and $T(r) \propto \dfrac{M(r)}{r}$

$L(R) \propto \dfrac{R^2 T^3}{\rho}\dfrac{dT}{dr}\Big]_{r=R} \propto \dfrac{M^4}{\rho R^3} \propto M^3.$

8–4.   The number of particles incident on a hypothetical surface in unit time and solid angle is

$$n(\theta, \phi, p)\, v \cos\theta\, d\Omega$$

The pressure then is

$$P = \int_0^{p_0} \int_0^{2\pi} \int_0^{\pi/2} 2 \cdot p \cdot \cos\theta\, v \cos\theta\, n(\theta, \phi, p)\, d\Omega\, dp.$$

If the gas is isotropic, $n(\theta, \phi, p) = \dfrac{n(p)}{4\pi}$, $P = \dfrac{1}{3}\displaystyle\int_0^{p_0} pvn(p)\, dp$

where $n(p)$ is given by (4–65) since all states are filled. Now $p = mv\gamma(v)$, which can be solved for v:

$$v = \frac{p}{m\sqrt{1 + p^2/m^2c^2}}$$

$$\therefore P = \frac{8\pi}{3mh^3} \int_0^{p_0} \frac{p^4}{\sqrt{1 + p^2/m^2c^2}} \, dp$$

If $p/mc = \sinh u$ $dp/d\theta = mc \cosh u$ so that

$$P = \frac{8\pi m^4 c^5}{3h^3} \int_0^{u_0} \sinh^4 u \, du.$$

8–5. $f(x) = \dfrac{1}{8}\left[x(2x^2 - 3)(x^2 + 1)^{1/2} + 3 \,\text{arc} \sinh x \right], x = \dfrac{p_0}{m_e c}.$

If $x \ll 1$, $\sinh^{-1} x = x - \dfrac{x^3}{6} + \dfrac{3}{40}x^5 - \cdots$

On expanding

$$8f(x) \approx x(2x^2 - 3)\left(1 + \frac{x^2}{2} - \frac{x^4}{8}\right) + 3x - \frac{x^3}{2} + \frac{9}{40}x^5 - \cdots \approx \frac{8}{5}x^5$$
$$x \ll 1$$

Substitution of $f(x)$ and (8–113) into (8–112) then gives (8–41)

If $x \gg 1$, $\sinh x \sim \dfrac{e^x}{2}$, and $\sinh^{-1} x = \ln(2x)$

$$\therefore f(x)\frac{x^4}{4} \qquad x \gg 1;$$

substitution in (8–112) leads to (8–43).

8–6. We use a temperature $2.7 \times 10^{7\circ}$K and density $\rho = 10 \,\text{g cm}^{-3}$ in equations 8–134 and 8–135:
This gives

$$E_{pp} = 27 \,\text{erg g}^{-1}\text{sec}^{-1} \quad E_{CN} \sim 3 \times 10^3 \,\text{erg g}^{-1} \,\text{sec}^{-1}$$

$\therefore$ The total energy generated is $4\pi/3\rho \, E_{CN}(0.54R_\odot)^3 \sim 7 \times 10^{36}$ erg sec. The total surface area of the star $4\pi(3.6R_\odot)^2 \sim 8 \times 10^{23}$ cm^2. Hence the flux crossing unit area is 0.9×10^{13} erg cm^{-2} sec^{-1} and using the blackbody law

$T^4\sigma$ = Flux across unit area,

$$T \sim 20,000°K$$

The total available energy per hydrogen atom is about 10^{-5} ergs with $n \sim 6 \times 10^{24}$ cm^{-3}. Hence the total energy available per cubic centimeter is about 6×10^{19} erg. The total time during which the star's core can supply energy, therefore, is about 7×10^7y.

8–7. With a temperature $10^7°K$, density 55 g cm^{-3} and hydrogen burning out to $0.2R_\odot$, we find that the proton-proton reaction predominates and yields 190 erg cm^{-3} sec^{-1}. This leads to a surface temperature of about 5000°K.

8–8. Using $X = 0.7$ and $\rho = 10^2$ g cm^{-3} with $T = 1.5 \times 10^7°K$, we again find that the pp reaction predominates, yielding about 2300 erg cm^{-3} sec^{-1}. The total energy generated is $\sim 4 \times 10^{33}$ erg sec^{-1}. This gives a surface temperature of 5900°K.

8–9. At $5 \times 10^7°K$ the carbon cycle predominates. If we assume burning in a thin shell ranging from 1.8×10^9 to 2×10^9 cm, we obtain a carbon-cycle energy generation rate of 4×10^8 erg cm^{-3} sec^{-1} throughout a volume of 10^{28} cm^3. The luminosity therefore is 4×10^{36} erg $\sim 10^3 \, L_\odot$ and the surface temperature at $R = 7 \times 10^{12}$ cm is $T = (L/4\pi R^2\sigma)^{1/4} = 3300°K$.

8–10. A white dwarf can only radiate away the kinetic energy of its ions. Since the electrons are degenerate, they cannot lose energy, and provide the main support against hydrostatic pressures. Just before electron degenerate pressures start predominating, the kinetic energy of the ions is about one-tenth of the white dwarf's potential energy. The virial theorem would predict that half of the energy should be kinetic energy, but there will be two electrons or more for each ion present, and the electrons will have higher energies because of the onset of partial degeneracy. The total available ion energy therefore is of order $0.1 \, M^2G/R$, and the lifetime $\sim 0.1 \, M^2G/RL$. It may still be worth stating how the luminosity can be derived. The nondegenerate outer layers of the white dwarf permit radiative transfer. Using equation 8–52 with $T \sim 0.1 \, MGm_i/kR$ and $dT/dr \sim T/R$, we obtain the luminosity if the opacity is known. The opacity can be computed from the Kramers' expressions, although a look at Fig. 8.4 shows that at high densities the opacity is nearly independent of density and has a very approximate value $\sim 10^8/T$, in the temperature range of interest. The cooling time can therefore be expressed solely as a function of the star's mass, radius, and chemical composition (ion mass). If the luminosity for a star of mass $0.45 \, M_\odot$ is $\sim 10^{-3} \, L_\odot$, and its radius is 1.1×10^9 cm, $\tau \sim 0.1 \, M^2G/RL \sim 1.2 \times 10^{18}$ sec, and the cooling time is of the order of 40 aeons. This problem is discussed more rigorously in (Sc58b).

8–11.   In Problem 8–2, $F(r) = \dfrac{4ac}{3\kappa\rho} T^3 \dfrac{dT}{dr}$ was derived by neglecting higher

order terms, since $\dfrac{dT}{dr} \ll 1$.

The next term is $\dfrac{4ac\kappa\rho}{5} \displaystyle\int_0^\infty e^{-\kappa\rho l} T\left(\dfrac{dT}{dr}\right)^3 l^3\, dl$

$$= \frac{4ac}{(\kappa\rho)^3} \frac{T}{5}\left(\frac{dT}{dr}\right)^3 \int_0^\infty e^{-y} y^3\, dy = \frac{24Tac}{5(\kappa\rho)^3}\left(\frac{dT}{dr}\right)^3$$

$$\rho(v) = \int_0^r 4aT\left(\frac{dT}{dr}\right)^3 \frac{18}{5(\kappa\rho)^2}\, dr$$

is the departure from blackbody energy density in a star.

Hence $\dfrac{d\rho}{dr} = 4a\left[T^3 \dfrac{dT}{dr} + T\left(\dfrac{dT}{dr}\right)^3 \dfrac{18}{5} \dfrac{1}{(\kappa\rho)^2} \right]$

$$= \frac{d\rho(v)}{dr} + \frac{d\rho^*(v)}{dr}.$$

8–12.   For a dense sphere

$$V = \frac{3}{5} \frac{GM^2}{R} \sim 4 \times 10^{52} \text{ erg} \quad \text{for} \quad R = 10^6 \text{ cm}, \quad M = 10^{33} \text{ g.}$$

9

Cosmic Gas
and Dust

9:1 OBSERVATIONS

In this chapter we will try to define a common framework within which most of the processes involving astronomical gas and dust clouds can be understood. Before we can do that, however, we should know something about the temperature, density, state of ionization, and linear dimensions of the clouds and we should summarize the information we have about dust grains.

The methods we use to ascertain the rough quantitative values cited in Table 9.1 differ greatly from one type of medium to another. It may therefore be useful to state qualitatively how this information is derived.

(a) Extragalactic Medium

Very little is known about the extragalactic medium. All the values for particle or field densities are upper limits; for all we know the space between galaxies is completely empty.

An upper limit to the neutral hydrogen density in intergalactic space is set by the lack of an observed Lyman-α absorption continuum for emission from quasars. This argument assumes that the quasars are extragalactic and at distances indicated by their red shifts as computed from Hubble's red-shift–distance relationship. Hubble's relation, however, may only hold for ordinary galaxies. The expected absorption would not correspond to a sharp line because neutral hydrogen lying at different distances between us and a quasar would absorb radiation at differently red-shifted wavelengths. We would therefore expect roughly uniform absorption over the whole wavelength range from the Lyman-α line of the quasar to the local wavelength of Ly-α, 1216 Å.

An upper limit to the number of electrons in intergalactic space can be derived from X-ray background observations on the hypothesis that these electrons are at high temperatures. Their free-free emission then has an appreciable X-ray component. However, if the electron temperatures were low, free-free emission would give lower values of the X-ray background and no X-ray flux would be observed. The precise values of upper limits obtained depend on the model adopted (see Chapter 10) for the expanding universe. We can place another upper limit on the density of the ionized extragalactic component by considering the actual ionization rates available from the known sources of ultraviolet radiation, primarily the quasars. Again, the limit obtained depends on the cosmological model one prefers, but generally the range of number density values is $\gtrsim 10^{-5}$ cm^{-3}.

Upper limits on the extragalactic magnetic field also depend on the assumptions we make. The fields cannot be so great that their pressure would exceed the gas pressure in our Galaxy. Otherwise interstellar hydrogen would actually suffer compression. Neither can the pressure be so great as to compress the extended radio sources that extend out from some galaxies. The characteristics of those sources are very uncertain, however, and therefore yield relatively little information. A coarse upper limit, probably far too high, is $B \lesssim 10^{-6}$ gauss.

(b) Quasistellar Objects, QSO's

Many different models have been constructed to account for observed features exhibited by various quasars. We have to account for the radio, infrared, and X-ray luminosity and for optical line emission. Moreover, it is not clear that all QSO's—quasars, BL Lacertae objects—luminous, compact, extragalactic sources are basically similar in nature. Quasars exhibit strong emission lines while BL Lacertae sources have a bland continuum spectrum.

One model (Ca70b, Gr64), traces the quasar visual line emission of highly excited ions to a nebula having an electron density of order 3×10^6 cm^{-3} and a radius of the order of 1 pc. The emission by relativistic particles in a core having the high magnetic field strength of order 10^5 gauss would then account for much of the remaining radiation. The actual strength of the magnetic field is always uncertain in such computations. First, some of the observed radiation might be produced through inverse Compton scattering so that no field at all would be needed. For the remainder we often assume that synchrotron radiation is important; however, the observed spectrum normally does not cover all wavelengths of interest and the partial information obtained does not suffice to uniquely determine the magnetic field. Basically there are two independent parameters as shown in (6–159): the energy exponent γ of the relativistic particles and the magnetic field strength. If γ is not known, the field strength also remains uncertain.

Faraday rotation within the quasistellar object is a potential means for determining magnetic field strengths, but we obtain no more than an upper limit if the irregularities in the field are unknown (section 6:12). It also is necessary to have an electron density value if the Faraday rotation is to determine a magnetic field strength. That density is only available if recombination line data exist, or free–free emission has been observed (sections 6:16, 6:17, and 7:12).

(c) The Galaxy

In the Galaxy and in some extragalactic objects, neutral hydrogen can be observed through the line absorption or emission of atomic hydrogen at a wavelength of 21 cm. Within the Galaxy we also have Lyman-α absorption of light emitted by O and B stars. These two types of data do not always agree, but the indications are that neutral hydrogen number densities in our part of the Galaxy are of order 0.1 to 0.7 cm^{-3}. Between the spiral arms, the density is lower (Je70) (Ke65).

The electron number density is determined from dispersion data (section 6:11) by using pulsars as sources of radio emission. Different frequency radio waves suffer different time delays (6–58) in traversing the distance between the pulsar and earth.

Except in the case of the Crab Nebula pulsar, the distance to the sources is not well known. Statistically, however, a self-consistent model is obtained if we assume that pulsars cluster around spiral arms (Da69). The electron density obtained then is 0.03 cm^{-3}, averaged over the arm and interarm domains in the local part of the Galactic disk.

Dust grain number densities and radii are computed in the following way. The size of the grains is estimated by the differential extinction of light at different wavelengths. Because red light is less strongly extinguished than blue, we argue that grain sizes should at least be smaller than the wavelength of red light. This argument extends also to ultraviolet wavelengths, with the result that grains are considered to have radii $\lesssim 10^{-5}$ cm. Similar results are obtained from slight coloration effects and polarization effects in reflection nebulae. For any given grain size, an effective extinction cross section can be computed and this, together with interstellar extinction data for stars within a few kpc from the sun determines an approximate dust grain number density for the solar neighborhood.

(d) H II Regions and Planetary Nebulae

The electron temperatures and densities are readily determined from free-free emission observations in the radio domain (section 6:17). Visual and radio recombination line data provide comparable information. This recombination rate can be computed from (7–75) in the following way. Let the recombination cross section for an electron in a gas at temperature T be $Q_n(T)$ where n is the principal

9:1

quantum number of the final state in the hydrogenlike ion. Consider an idealized situation of thermal equilibrium in which the number of ionizations equals the number of recombinations to this level. The recombination rate per electron is, in velocity range dv, proportional to the electron's velocity $v \sim (3kT/m)^{1/2}$; to the cross section $Q_n(v)$ and to the number density $n_e(v)$ of electrons and n_{r+1} of ionized atoms. The ionization rate is proportional to c, to α_{bf} (7–75), to the number density of atoms in the lower ionization state n_r, and to the number density of photons, at a frequency v high enough to produce both ionization and an electron velocity v. If χ_r is the ionizing energy,

$$v = \frac{1}{h}\left(\frac{m}{2}v^2 + \chi_r\right) \tag{9-1}$$

We can then write the equilibrium condition between ionization and recombination, very roughly, as

$$n_e(v)\, n_{r+1} v Q_n(v)\, dv = c \cdot n_r \cdot \frac{8\pi}{c^3} \frac{v^2 \alpha_{bf}(v)}{(e^{hv/kT} - 1)}\, dv \tag{9-2}$$

by making use of the blackbody spectrum (4–71) for the number density of photons $\rho(v)/hv$. Use of the Saha equation 4–105 and the absorption coefficient given by expression 7–75 then leads to the relation

$$\frac{g_{r+1}g_e}{g_r} \frac{[2\pi mkT]^{3/2}}{h^3} e^{-\chi_r/kT} Q_n(v)\, v\, dv = \frac{8\pi}{c^3} \frac{64\pi^4 me^{10}Z^4}{3\sqrt{3}\, h^6 n^5} g_{bf} \frac{dv}{v[e^{hv/kT} - 1]} \tag{9-3}$$

Using the interrelationship (9–1) between variables v and v, and knowing that $g_r \propto n^2$ (see Problem 7–1) we obtain a relation for the recombination rate α_n for unit electron and ion density:

$$\alpha_n = \int_0^\infty v Q_n(v)\, dv$$

$$= \frac{g_r}{g_e g_{r+1}} \int_0^\infty \frac{2^9 e^{10} \pi^5 Z^4 me^{\chi_r/kT}}{h^3 n^5 [6\pi mkT]^{3/2} [e^{(\chi_r + mv^2/2)/kT} - 1]} \frac{d\left(\dfrac{mv^2}{2}\right)}{\left(\dfrac{m}{2}v^2 + \chi_r\right)} \tag{9-4}$$

The fraction outside the integral is $\sim n^2$ (section 8:8).
For visible radiation (9–4) is considerably simplified since kT is small compared to χ_r so that the exponential dependence on χ_r can be neglected. The integral can then be expressed in approximate form (Za54):

$$\alpha_n = \frac{2.08 \times 10^{-11}}{T^{1/2}} \phi(T) \tag{9-5}$$

Table 9.1 Rough Characterization of Gas and Dust Aggregates.

	Representative Object	Density of Neutral Gas n_H (cm^{-3})	Electron Density n_e (cm^{-3})	Radius (cm)
Extragalactic Medium		$< 3 \times 10^{-11}$	$\lesssim 10^{-4}$	$\sim 10^{28}$ cm to cosmic horizon
Quasars	3C273		3×10^6	3×10^{18}
Spiral Galaxy Arm:	Galaxy	0.1 to 0.7 ⎱	0.03	3×10^{20} thick;
Interarm Medium:		$\lesssim 0.05$ ⎰		disk diameter 10^{23}
H II Region	Orion Nebula		10^4	5×10^{18}
Planetary Nebula	NGC6543		6×10^3	10^{17}
Supernova Remnant	Crab Nebula		40	5×10^{18}
H I Cloud	Heiles Cloud I (22^h 29^m5, $74° 58'$)	40 to 125	~ 0.3	10^{19}
Stellar Wind	O star δ Ori	0.14	10^8	at 1 AU from star
Solar Wind			2	at 1 AU from sun
Comet Head	Arend-Roland 1957 III	$n_{molecules}$ $\sim 10^4$		10^{10} cm
Comet Dust Tail	Arend-Roland 1957 III			10^{12} cm long
Comet Ionized Tail	Arend-Roland 1957 III		$n_{ion} \sim 2$ (at 10^{11} cm)	5×10^{12} long

where $\phi(T)$ has a value of 3.16 at 1580°K and 1.26 at 7.9×10^4°K. It therefore does not rapidly change with temperature.

Although the thermal velocities of ions already lead to broadening of spectral lines, we can still determine bulk velocities of turbulent motion when these are high enough to lead to an actual split appearance of spectral lines, or if the thermal broadening contribution can be computed from independent temperature data.

Dust densities in such clouds can be determined by measuring the continuum

Magnetic Field (gauss)	Turbulent or Bulk Velocity $(cm\ sec^{-1})$	Temperature (°K)	Number Densities and Radii of Grains		Remarks
			n_g (cm^{-3})	a_g (cm)	
$\ll 10^{-6}$?		$> 2 \times 10^{4}$°K if the medium produces the diffuse X-ray background	$\lesssim 10^{-15}$	$\sim 3 \times 10^{-5}$ assumed	n_H is dependent on cosmic QSO's and assumed cosmological model (Gu65)
$\sim 10^{5}$? in center	10^8 to 10^9	17,000			velocities from multiple absorption spectra (Gr64) (Ca70b)
$< 10^{-5}$	10^6		10^{-13}	3×10^{-5}	Ly-α and 21 cm n_H data differ (Je70) (Ke65)
	4×10^6	10^4 8400	10^{-9} $\sim 3 \times 10^{-10}$	3×10^{-5} 3×10^{-5}	
3×10^{-4}	10^8	$< 17,000$			(Co70) (Wo57)
$\lesssim 10^{-5}$	10^4 to 10^6	4.5°K to 100°K	10^{-10}	3×10^{-5}	$n_H \sim 4 \times 10^{-6}$ cm^{-3} $n_{H_2} \sim 200$?
	1.4×10^8	10^4°K			(He69) (He68) (Mo67)
3×10^{-5}	4×10^7	10^4 to 10^5°K	$\sim 10^{-13}$ at 1 AU	5×10^{-5}	Wind terminates at 10 to 100 AU; grains orbit the sun; they do not move with wind
	2×10^5				Molecules are: C_2, C_3, CN, NH, CH, OH, NH_2, OH^+, CH^+, and so on.
	10^6		$\sim 10^{-7}$	5×10^{-5}	
$\sim 3 \times 10^{-5}$	10^7				Mainly CO^+ ions, CO_2^+, N_2^+, OH^+, and so on.

radiation from the cloud in the visible part of the spectrum. Much of this radiation is likely to be starlight scattered by dust. Some assumption about particle size must then still be made, and ideally we should also know the chemical composition and physical structure of the grains. Likely candidates are silicates, iron containing minerals, or graphite grains; but no definite knowledge exists. As will be seen in section 9:6, infrared emission from H II regions and planetary nebulae also provide data on dust densities.

9:1

Table 9.2 Absorption and Recombination Coefficients for Hydrogen and Helium.[a]

Atom	Term	a_{v_0}	f	α_n 10,000°K	Q_n
		10^{-18} cm^2		10^{-14} cm^3 s^{-1}	10^{-22} cm^2
H I	1s	6.3	0.436	15.8	32
	2s	15	0.362	2.3	4.7
	2p	14	0.196	5.3	11
	3s	26	0.293	0.8	1.6
	3p	26	0.217	2.0	4.1
	3d	18	0.100	2.0	4.1
	4s	38	0.248	0.4	0.7
	4p	40	0.214	1.0	2.0
	4d	39	0.149	1.0	2.0
	4f	15	0.057	0.6	1.2
	Total			43	88
He I	1s^2 ^{1}S	7.6	1.50	15.9	33
	1s2s ^{3}S	2.80	0.25	1.4	2.9
	1s2s ^{1}S	10.5	0.40	0.55	1.1
	Total			43	88
He II	1s	1.8	0.44	73	150

[a] a_{v_0} is the absorption cross section at the ionization limit; the oscillator strength f is defined in section 7:9; the recombination coefficients α_n and Q_n are defined by (9–4). (After Allen, A164. With the permission of Athlone Press of the University of London, second edition © C.W. Allen 1955 and 1963.)

(e) Supernova Remnants

These remnants can stretch across many tens of parsecs in diameter and often show a circular arc structure. High expansion (Doppler) velocities can be measured by spectral observations; for the Crab Nebula an actual expansion can be observed by a comparison of present-day and 50-year-old photographs. The expansion velocity for the Crab is of the order of 10^8 cm sec^{-1}. For the Crab Nebula strong polarization along a direction perpendicular to the length of the continuum emitting wisps is seen. Assuming that this comes from synchrotron radiation, emitted by highly relativistic electrons spiraling in magnetic fields

9:1

that lie embedded with field lines running the length of the wisps, we can make an estimate of the magnetic field strength. It amounts to about 10^{-4} gauss. Besides relativistic electrons, there also exists lower temperature plasma that can be detected through its Hα emission lines, in the red part of the visual spectrum.

f) Stellar Winds

The stellar wind, in say O and B stars can be detected by observing the Doppler shifted lines of highly excited ions. Initially these were observed in the ultraviolet part of the spectrum through rocket observations (Mo67). Assuming that solar abundance also characterizes the surface material in these stars, it is possible to interpret the observed line strengths in terms of total amounts of matter ejected from these stars. Typically, a massive O star seems to eject matter at a rate of one solar mass in 10^5 to 10^6 y, so that, during the star's lifetime, the ejected mass can amount to an appreciable fraction of the total.

g) Solar Wind

The solar wind density can be measured directly through the use of interplanetary probes that sample the plasma at appreciable distances from the earth where the earth's magnetosphere no longer interferes with observations. Magnetic fields can be measured by magnetometers carried on these spacecraft. Considerable wind variations between quiet and active periods on the sun are observed. The overall wind velocity can reach values of ~ 1000 km sec^{-1} following a solar flare, although during quiet times it does not vary greatly from a mean value around 400 km sec^{-1}. At quiet times the density amounts to a few particles per cubic centimeter. Following a flare the density can increase by an order of magnitude.

h) Comets

These objects have three distinct parts, a roughly spherical head, a long straight tail, and a shorter curved tail. The head, frequently some 10^{10} cm in diameter, contains C_2, C_3, CN, NH, CH, OH, and NH_2 among other molecules, as well as such ions as OH^+, CH^+, and CO^+. Typical molecular densities range around 10^4 cm^{-3}. An extended H_2O and atomic hydrogen envelope around the head has also been discovered through observations from spacecraft. The velocities of the molecules in the head can be measured by Doppler observations and amount to a few km sec^{-1}. The head also includes a solid nucleus that is too small to be directly resolved, except from fly-by spacecraft.

9:1

The long straight tails seen in comets sometimes stretch over a distance larger than an astronomical unit. They are the largest objects in the solar system but their densities are low and the total mass contained is minute. Only ions and no neutral molecules are seen in these tails. The number density of the ions is determined by the intensity of the molecular lines. The f values for the excitation have been computed and we know how to relate the observed brightness of the emission lines to the number of molecular ions in the line of sight and to the rate at which these ions would be placed into excited states by absorption of sunlight.

The shorter curved tails in comets reflect sunlight as judged by the *Fraunhofer* (absorption) *line spectrum* which mimics the sun's. Since the solar lines do not appear broadened in this reflection, the scattering particles must be slowly moving and cannot be electrons, which would have thermal velocities $\sim 10^8$ cm sec^{-1}. We conclude that the tails contain dust. This also is corroborated by the curved shape of the lagging tail. It is curved because the repulsion of the dust by sunlight and the requirement for constant orbital angular momentum about the sun produces an increasing lag for grains at increasing distance from the sun. The sizes of the dust grains can be roughly determined by the rate at which solar radiation pressure pushes them away from the head of the comet. Grains of assorted sizes will follow different paths since the radiative repulsion varies, and it is possible to derive rough estimates of grain sizes at different locations across the spread-out width of the tail. The smallest grains will lie closest to the radius vector pointing away from the sun. The largest grains will be most distant from that axis. From the rough estimate of grain sizes, we can compute the number density of grains as judged from the total scattered sunlight.

Table 9.1 has been constructed to summarize some of the information we have on individual diffuse objects in the solar system, in the Galaxy, and beyond. Within each class listed in the first column, variations in size, density, and so on amounting to orders of magnitude are not uncommon, and we have to be careful not to assume that different members of a class always have identical properties.

9:2 STRÖMGREN SPHERES

In 1939, Strömgren (St39) considered the interaction of a very young star with the interstellar medium. To make matters simple he made two assumptions. First, he suggested that the star lights up rapidly to full strength; secondly, he considered the surrounding medium homogeneous throughout. These two assumptions permitted Strömgren to draw a simple picture of the development of ionized hydrogen regions around massive, ultraviolet emitting stars.

We note that if the star emits a number of photons dN_i capable of ionizing the surrounding gas, then the number of electrons that are stripped off the atoms,

9:2

Fig. 9.1 Schematic diagram of a Strömgren Sphere. H II is the ionized gas. H I is the neutral region and δ the thickness of the separating layer.

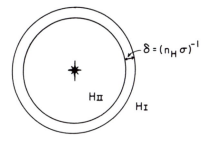

over the same time interval, will also be dN_i if equilibrium is maintained. This assertion is always true in practical cases, since the cross section for ionization by energetic photons is of the order $\sigma \sim 10^{-17}$ cm² and typical gas densities in the vicinity of young stars might be of order $n_H \sim 10^3$ cm⁻³. At these densities a photon can only travel a distance of order $(n_H \sigma)^{-1} \sim 10^{14}$ cm through the neutral medium before it ionizes an atom. But this is a small distance compared to the radii of ionized regions normally observed. These radii range from $\sim 10^{16}$ to $\sim 10^{20}$ cm; hence, practically no ionizing photons can escape through the gas without becoming absorbed.

Although an energetic photon can only travel a short distance in the neutral medium, its path through the ionized gas is very long. Here it occasionally is scattered; but the scattering cross section is relatively small—the Thomson cross section is only $\sim 6.7 \times 10^{-25}$ cm² (6–102)—and we can assume that the ionizing photons proceed undisturbed through the ionized gas immediately surrounding the star, until they hit the boundary region where the neutral gas commences. The thickness of the boundary region also is the mean free ionizing path:

$$\delta = (n_H \sigma)^{-1} \tag{9–6}$$

We will call the neutral clouds, H I regions, and the ionized clouds, H II regions.

We picture an ionizing star, as surrounded by an H II region, which is separated from an embedding H I region by a thin layer of thickness δ. This is shown in Fig. 9.1. If the gas is homogeneous, the surfaces of separation are spherical, and the sphere containing the ionized gas is called the *Strömgren sphere.*

We now ask ourselves, how quickly the sphere becomes established. To do this, we note that the number of atomic particles in a shell of radius R and thickness dR is $4\pi R^2 n_H \, dR$; so that when the star emits dN_i ionizing photons, the radius of the region grows by an amount dR, given by

$$\frac{dN_i}{dt} = 4\pi R^2 n_H \frac{dR}{dt} \tag{9–7}$$

Here we have gone through the additional step of formally dividing both sides by dt, to obtain the time rate of development.

Equation 9–7, however, is only true during the initial stages of growth. It neglects the recombination of ions and electrons that also has to occur in the interior of the Strömgren sphere. For, if an electron and ion recombine to form an atom, a new ionizing photon will be required to separate the two particles. This photon never reaches the boundary R and it will therefore not contribute to the growth of the region. The recombination rate per unit volume is proportional to the product $n_e n_i$, since each electron has a probability of colliding that is proportional to the number of ions it encounters. In addition, the recombination rate per unit volume is also proportional to the recombination factor $\alpha \sim 4 \times 10^{-13}$ cm^3 sec^{-1}, which depends (9–5) on the temperature of the ionized gas, normally $\lesssim 10^{4\circ}$K, and represents the sum over recombination factors leading to all states n (see column 5 of Table 9.2). The full equation, satisfied by the gas therefore, is

$$\frac{dN_i}{dt} = 4\pi R^2 n_H \frac{dR}{dt} + \frac{4\pi}{3} R^3 n_i n_e \alpha \qquad (9–8)$$

During the late developmental stages, this very simple model would predict that dR/dt eventually becomes zero as the sphere grows so large that the star emits photons only just fast enough to keep up with the total number of recombinations. This will happen at an equilibrium radius

$$R_s^3 = \frac{3}{4\pi n_i n_e \alpha} \frac{dN_i}{dt} \qquad (9–9)$$

Equations 9–7 and 9–9 are the extreme cases covered by equation 9–8. They describe the initial growth and final equilibrium value of the radius of the Strömgren sphere as long as the very simplest assumptions are retained.

A number of comments are needed:

(1) Equation 9–8 has to be used in conjunction with some model for ultraviolet emission by stars. If the star's luminosity and temperature are known, then we can readily estimate N_i from the Planck blackbody relation (4–71). In actual fact, the ultraviolet spectrum of a very hot star does not closely approximate a blackbody because absorption by the star's outer atmosphere changes the spectrum of the escaping radiation. This is called the *blanketing effect*. Despite the blanketing effect, however, the blackbody approximation gives roughly the correct magnitude for the number of ionizing photons to be inserted in equation 9–8.

For most stars, ultraviolet observations can only be carried out down to wavelengths of 912 Å, below which photons ionize interstellar hydrogen and are strongly

9:2

absorbed. At such short, extreme ultraviolet wavelengths only a small handful of the nearest stars have been observed.

(2) Many times the recombination of an electron with an ion will yield a photon that is still capable of ionizing another atom. In fact, the only time that this will not be true is if the recombination first takes the atom into a high lying excited state. A recombination directly into the ground state, on the other hand, always yields a photon capable of further ionization. For this reason the second term on the right side of equation 9–8 is an upper limit on the loss of ionizing photons through recombination. Similarly the radius R_s of equation 9–9 is a lower limit for an equilibrium value. The effect turns out to be more important in very dense regions than in tenuous gases surrounding a star. For very dense regions the true R_s value may be more than 10 times greater than that given by (9–9). For values of n_H around 10^4 to 10^5 cm^{-3}, typical of the denser ionized regions normally encountered, the radius R_s is a factor of 2 to 3 higher than predicted by (9–9).

(3) A very quick consideration shows that equations 9–8 and 9–9 cannot be completely correct because they neglect the problem of pressure equilibrium. This can be seen rather easily. For, the ionized region must in any case have at least twice as many particles per unit volume as the nonionized region, that is, at least one ion and one electron for each atom. This means, according to the ideal gas law (4–37), that the pressure on the inner side of the boundary separating ionized from nonionized regions would be at least twice as great as the pressure on the outside; and that only if the temperature was the same on both sides. In practice, the temperature of the H I region is likely to be of order 70°K while the temperature of the H II region normally amounts to ~ 7000°K. The total pressure inside the separating boundary is therefore of order 200 times greater than the pressure outside and expansion must therefore proceed quite rapidly.

(4) If we were to draw the very simplest picture of an expansion, we would proceed by visualizing the process in terms comparable to the inflation of a balloon. If the mass of the surrounding H I region is M, the mass per unit area at the separating surface is $M/4\pi R^2$, and the pressure inside is $2n_i kT_i$. T_i is the temperature of the ionized region, and the factor 2 assumes that the number of ions n_i equals the number of electrons. Neglecting the small gas pressure on the outside of the sphere, we obtain the outward acceleration of the boundary as

$$\ddot{R} = \frac{2n_i kT_i}{(M/4\pi R^2)} \qquad (9\text{--}10)$$

This can be integrated if we first multiply both sides by $\dot{R}$. Then

$$\frac{\dot{R}^2}{2} = \frac{8\pi}{3} \frac{n_i kT_i}{M} R^3 \qquad (9\text{--}11)$$

9:2

which leads to a development time scale of order

$$t \sim \left(\frac{3M}{16\pi n_i k T_i R} \right)^{1/2} \tag{9-12}$$

If we take M roughly equal to one solar mass $\sim 2 \times 10^{33}$ g, $n_i \sim 10^4$ cm^{-3}, $T_i \sim 10^{4\circ}$K, and $R \sim 10^{17}$ cm, we find that

$$\dot{R} \sim 3 \times 10^5 \frac{\text{cm}}{\text{sec}}, \qquad t \sim 3 \times 10^{11} \text{ sec} \tag{9-13}$$

This velocity has to be compared with the random speed of atoms in the cool medium; that only is $\sim (3kT/m_H)^{1/2} \sim 1.5 \times 10^5$ cm/sec at the low temperature of the H I region. The correct dynamics, therefore, cannot be described by equations 9–10 through 9–13 because pressure is normally propagated at roughly the speed of sound in H I regions. If the expansion on the inner edge of the region proceeds faster than the speed of sound, the outer portions of the region will not be aware that a pressure is being exerted at the inner boundary, and will therefore not move. As a result, the quantity of material actually accelerated at any given instant will not have the mass M used in the subsonic approximation (9–10) and the actual velocity $\dot{R}$ will be considerably higher. The equations of supersonic hydrodynamics must therefore be used and these will be explained in section 9:3 below.

(5) Before proceeding to the dynamical treatment of expanding H II regions, it is interesting to point out that equation 9–7 may still hold well for extremely early stages of development, because dR/dt is so high there that the *ionization front*, that is, the region separating ionized and neutral regions, proceeds into the medium at velocities that can be orders of magnitude greater than the speed of sound in the medium. There is then no possibility at all for major instanteous adjustments of density in response to pressure differences between ionized and nonionized regions. This also will be discussed in section 9:3.

(6) It is interesting that the expansion produced by gas pressure brings down the density of ionized material in the H II region and therefore decreases the recombination rate per unit volume. The factor $n_e n_i \alpha$ of the second term in equation 9–8 decreases as R^{-6}, because both n_i and n_e decrease as R^{-3} when only expansion due to excess pressure (in contrast to expansion through further ionization) is involved. This means that the second term on the right of (9–8) is always reduced by pressure induced expansion, thereby giving rise to a higher value for the expansion velocity $\dot{R}$ of the boundary.

(7) Finally, it is important that our whole concept of the development of a Strömgren sphere has been based on a picture in which the central star suddenly brightens and produces ionizing radiation. But that does not at all correspond to the development of massive stars shown in Fig. 8.11. There we had shown that

9:2

a massive O or B star takes some 6×10^4 y to contract to the main sequence and for a good fraction of that time it is bright without emitting much ionizing radiation. Davidson (Da70a) has argued that during the contraction stage, light pressure pushes gas and dust away from the star. This happens because a dust grain with radius a, accelerated by light pressure to a velocity v with respect to the gas, suffers collisions at a rate $n_H v \pi a^2$ with the atoms and, hence, suffers a drag (momentum loss) amounting to a deceleration

$$\dot{v}_d = -\frac{n_H m_H v^2 \pi a^2}{(4\pi/3) a^3 \rho} \tag{9-14}$$

where ρ is the grain's density.

The radiative acceleration, for a star having luminosity L, is

$$\dot{v}_r = \frac{L}{4\pi c R^2} \frac{\pi a^2}{(4\pi/3) a^3 \rho} \tag{9-15}$$

so that equilibrium is established at a velocity

$$v \sim \left[\frac{L}{4\pi c R^2 n_H m_H} \right]^{1/2} \tag{9-16}$$

which has a value of $\sim 1.5 \times 10^6$ for $L \sim 10^{38}$ erg sec^{-1}, $R \sim 10^{17}$ cm, and $n_H \sim 10^4$. This velocity is set up in a time

$$\tau \sim \frac{v}{\dot{v}} = \frac{(4/3) \rho a}{\sqrt{(L/4\pi c R^2) n_H m_H}} \tag{9-17}$$

If $\rho \sim 3$ g cm^{-3} and $a \sim 10^{-5}$ cm, $\tau \sim 2.5 \times 10^9$ sec. From this we see that grains reach equilibrium velocity in a matter of a century, while the contraction of the star to the main sequence takes tens of thousands of years.

The grains drag the gas out to quite large distances through this process. For, a radiative pressure $(L/cR^2 4\pi)$ acting, say, on a column of length R and hence of mass $n_H m_H R$, would produce a mean acceleration of order

$$\dot{v} \sim \frac{L}{4\pi c n_H m_H R^3} \tag{9-18}$$

and for the same conditions chosen above, but with $R \sim 3 \times 10^{17}$ cm, $\dot{v} \sim 10^{-6}$ cm sec^{-2}; in 3×10^4 y a distance of order R would be covered. Davidson therefore argues that when the star begins copious emission of ionizing radiation, most of the gas already has been pushed to large distances. The ionization of course still occurs, but it takes place at the edge of the low density cavity in which the star now finds itself. What may happen then is that the newly ionized gas flows inward to the star, rather than outward away from it. One does not yet know.

9:3 SHOCK FRONTS AND IONIZATION FRONTS

In the preceding section we gave one example of supersonic flow in the vicinity of hot stars. There are many other such examples. The *wind* or stream of gas that continuously sends stellar material out into surrounding space blows at supersonic velocities ranging from a few thousands of kilometers per second for the hottest O stars, down to speeds of the order of 400 km sec^{-1} for stars like the sun.

This fast stream of gas coming from the sun interacts with the earth's *magnetosphere* to give rise to a wide variety of different effects, ranging from the colorful *aurora borealis* to the nuisance of poor radio wave propagation in the broadcast band. The solar wind also gives rise to the long ionized tails of comets.

On a larger scale, supersonic phenomena are encountered in stellar explosions of all kinds, ranging from the small outbursts that regularly occur in flare stars to the explosion of supernovae and the explosive ejection of gas from the nuclei of galaxies.

This brief list shows that supersonic velocities are quite normal in astrophysics. In fact, virtually all phenomena that take place outside stars, or solid bodies like planets and grains of dust, involve velocities greater than the speed of sound.

In this section, we will be concerned with the equations that describe the interaction of an H II region with the surrounding neutral medium; but the treatment is general and can be applied to many other supersonic phenomena in astronomy.

We will assume that a star has suddenly undergone an increase in brightness, that it rapidly ionizes the surrounding medium, and that a *shock front*, or an *ionization front*, or both, move outward into the cool H I region at supersonic velocity. We will see below just how these fronts behave.

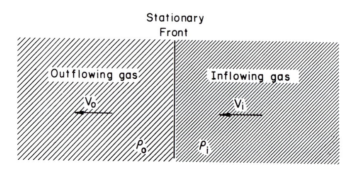

Fig. 9.2 Conditions on the two sides of a shock or ionization front.

There are two ways of considering a front—or dividing region—between the expanding ionized gas and the still unperturbed neutral hydrogen region. We can either consider the front as moving out into neutral gas at some velocity v, or else we can pretend that the neutral gas is moving into a stationary front at velocity, $-v$. After passing through the front, the gas is compressed and possibly ionized; and there will be energy changes—mainly heating.

We will adopt this second point of view, that the front between the two regions is stationary, and we will make a number of demands on the gas flowing through the front, Fig. 9.2.

(a) We require that the mass flow into the front equals the mass flowing out. This is just the continuity condition. If the inflow density and velocity are ρ_i and v_i and the outflow density and velocity are characterized by ρ_0 and v_0, this requirement reads

$$\rho_i v_i = \rho_0 v_0 \equiv I \qquad (9\text{–}19)$$

Here, I is the *mass flow* through unit area in unit time.

(b) We can consider the front to be a surface that absorbs inflowing gas and emits outflowing gas. In that case the pressure of inflowing material would be $\rho_i v_i^2$ even if the inflowing gas had no intrinsic pressure due to random motion of the atoms. $\rho_i v_i^2$ simply is the momentum transferred to the surface per unit area and unit time through absorption of the inflowing particles—and that is what we mean by pressure. Similarly the back pressure due to the seemingly emitted outflowing gas would be $\rho_0 v_0^2$. Since (9–19) has to be satisfied and since v_0 generally differs from v_i, these two pressures will not normally be equal and opposite. We still have two thermal pressures due to the random thermal motions of the gas atoms adding to the overall pressure acting on each side of the front. If the front is not to be accelerated—and we have assumed that there is a constant inflow and outflow velocity here—then momentum conservation requires that the overall pressures on the two sides of the front be equal and oppositely directed.

$$P_0 + \rho_0 v_0^2 = P_i + \rho_i v_i^2 \qquad (9\text{–}20)$$

This is the condition of steady flow.

(c) On passing through the front, the energy content of the gas changes. A number of different sources contribute to the overall energy. For the inflowing gas there is:

(i) kinetic energy due to the bulk flow, $v_i^2/2$ per unit mass flowing into the front;

(ii) internal energy per unit mass U_i (see section 4:18);

(iii) the work done on unit volume as it flows into the surface. Since we are picturing the gas as being stopped—absorbed—at the front, this is equivalent to the gas being compressed down to zero volume as it approaches the front.

The work done in unit time then involves a volume equal to the velocity v_i,

For unit area: work/time $= P_i v_i$ (9–21)

The product on the right of (9–21) is called the *enthalpy* of the gas. Reduced to unit volume, the enthalpy numerically just equals the pressure P_i. Reduced to unit inflow mass, it is P_i/ρ_i.

As it crosses the front, energy Q per unit mass may be fed into the fluid, so that the actual energy gain per unit mass of inflowing material is Q.

In unit time, a mass $\rho_i v_i$ of inflowing material crosses the front. This must contain energy equal to the energy contained in the gas flowing out of the front in unit time except that the outflow energy can be greater by an amount Q. The outflow energy consists of terms similar to those described under (i), (ii), and (iii). We therefore have the energy conservation equation

$$Q + \left(\frac{v_i^2}{2} + U_i + \frac{P_i}{\rho_i} \right) - \left(\frac{v_0^2}{2} + U_0 + \frac{P_0}{\rho_0} \right) = 0 \qquad (9\text{–}22)$$

where we have again used subscripts o to denote outflow. Equations (9–19), (9–20), and (9–22) relating inflow to outflow are sometimes called the *jump conditions*.

PROBLEM 9–1.   Show that the internal energy (section 4:18) can be written as

$$U = c_v T = \frac{R}{\gamma - 1} \frac{P}{R\rho} = \frac{P}{(\gamma - 1)\rho} \qquad (9\text{–}23)$$

This leads to

$$\left[\frac{v_i^2}{2} + \left(\frac{\gamma_i}{\gamma_i - 1} \right) \frac{P_i}{\rho_i} \right] - \left[\frac{v_0^2}{2} + \left(\frac{\gamma_0}{\gamma_0 - 1} \right) \frac{P_0}{\rho_0} \right] = -Q \qquad (9\text{–}24)$$

and since both gases consisting of neutral atoms and gases containing only electrons and atomic ions have γ values of 5/3, we can finally write (9–24) in the form

$$\left[\frac{v_i^2}{2} + \frac{5}{2} \frac{P_i}{\rho_i} \right] - \left[\frac{v_0^2}{2} + \frac{5}{2} \frac{P_0}{\rho_0} \right] = -Q \qquad (9\text{–}25)$$

Equations 9–19, 9–20 and 9–25 describe the motion of a front into a monatomic medium.

Let us first examine the structure of an *ionization front*. We will assume that J ionizing photons are incident on the front per unit area in unit time. The mean energy of these photons is $\chi_r > \chi_0$ where χ_0 is the energy required for ionizing atoms (Ka54)*. Since J is the number of ionizing photons incident on the front

9:3

that divides the ionized from the neutral gas, it follows that J atoms are flowing into the front in unit time and J ions plus J electrons are streaming out.

By referring to (9–19) we see that the mass flow across the front is related to the flux of ionizing photons through the relation

$$I = mJ \tag{9-26}$$

where m is the mass of the neutral atoms. For a pure hydrogen cloud $m = m_H$. We now define the ratio of densities

$$\frac{\rho_0}{\rho_i} \equiv \Psi \tag{9-27}$$

Then, by (9–19) and (9–20)

$$P_0 = P_i - \frac{\rho_i v_i^2 (1 - \Psi)}{\Psi} \tag{9-28}$$

From (9–19), (9–25), and (9–28) we then obtain

$$\left[5\frac{P}{\rho} + v^2 + 2Q \right] \Psi^2 - 5\left[\frac{P}{\rho} + v^2 \right] \Psi + 4v^2 = 0 \tag{9-29}$$

where we have dropped all subscripts, but the pressure, density, and velocity all refer to the inflowing material. We note that the energy supplied at the ionizing front goes partly into ionization and partly into kinetic energy of the particles. The part that goes into kinetic energy on the average is $\chi_r - \chi_0$ per ion pair. Per unit mass of ionized material this relationship corresponds to a mean square velocity

$$u^2 = \frac{2[\chi_r - \chi_0]}{m} = 2Q \tag{9-30}$$

Note that Q in this sense represents only the heating energy, not the energy needed to overcome atomic binding.

This binding energy has not been specifically included in the formalism presented here. Essentially, we have been able to neglect it by concentrating only on those photons having $\chi_r > \chi_0$. For a higher binding energy, fewer photons are available to ionize material.

We also note that the speed of sound in the neutral medium has an adiabatic velocity

$$c = \left[\frac{\gamma P}{\rho} \right]^{1/2} = \left[\frac{5P}{3\rho} \right]^{1/2} \tag{9-31}$$

so that equation 9–29 can now be written entirely in terms of the three velocities

v, u, and c, and in terms of the ratio Ψ

$$[3c^2 + v^2 + u^2]\Psi^2 - [3c^2 + 5v^2]\Psi + 4v^2 = 0 \qquad (9\text{-}32)$$

We can treat this as a quadratic equation in Ψ that can have a pair of coincident roots, two positive roots, or a pair of complex roots depending on whether

$$(3c^2 + 5v^2)^2 \gtreqless 16v^2(3c^2 + v^2 + u^2) \qquad (9\text{-}33)$$

or on whether

$$9(c^2 - v^2)^2 \gtreqless 16v^2u^2 \qquad (9\text{-}34)$$

There are real roots under two conditions. The first condition is that

$$3(c^2 - v^2) \le -4vu \qquad (9\text{-}35)$$

which means that v is greater than some critical speed v_R:

$$v \ge v_R = \frac{1}{3}\left(2u + \sqrt{4u^2 + 9c^2}\right) \qquad (9\text{-}36)$$

This requires that the flux of ionizing photons be larger than a critical value J_R; by (9-19) and (9-26)

$$J \ge J_R = \frac{n}{3}\left(2u + \sqrt{4u^2 + 9c^2}\right) \qquad (9\text{-}37)$$

where n is the initial number density of atoms in the neutral medium. R stands for "rarefied" and D for "dense."

The second condition under which real roots exist is if

$$3(c^2 - v^2) \ge 4vu \qquad (9\text{-}38)$$

which implies velocities of the front less than a critical value v_D and an ionizing flux below J_D:

$$v \le v_D = \frac{1}{3}\left(-2u + \sqrt{4u^2 + 9c^2}\right) \qquad (9\text{-}39)$$

$$J \le J_D = \frac{n}{3}\left(-2u + \sqrt{4u^2 + 9c^2}\right) \qquad (9\text{-}40)$$

The two critical speeds v_R and v_D correspond, for a given ionizing flux, to densities

$$\rho_R = \frac{I}{v_R} \quad \text{and} \quad \rho_D = \frac{I}{v_D} \qquad (9\text{-}41)$$

If the gas ahead of the ionizing front has a value ρ_R or ρ_D, only one possible value of ρ_0 can result, that is, the density in the ionized medium behind the front has a fixed value. Since (9-32) is quadratic in $\Psi = \rho_0/\rho$, we see that for $\rho < \rho_R$ and

9:3

also for $\rho > \rho_D$ there are two possible values which ρ_0 could assume, and for intermediate values of the initial density, $\rho_R < \rho < \rho_D$, there is no permissible value. This means that the ionization front cannot be in direct contact with the undisturbed H I region.

We therefore have the following development of an ionized hydrogen region around a star that suddenly flares up and emits ionizing radiation. Initially, the interface between the ionized and neutral region is very close to the star; the flux J still is very high and well above the critical value J_R. We then have what is called the *R-condition*. The rate at which the front moves into the neutral medium (or vice versa according to the formalism used here) is $v = J/n$. This is just what equation 9–7 stated. However, as the ionization front moves further from the star, the value of J decreases and the front slows down until the critical velocity v_R is reached. This velocity has the approximate value

$$v_R \sim \frac{4}{3} u \qquad (9\text{–}42)$$

because the mean energy of the photons is so high that the excess energy carried off by the ionized particles makes them move at velocities much higher than the speed of sound in the undisturbed neutral medium. The temperature in the ionized medium simply is much higher than in the neutral gas. Typical temperatures in ionized regions of interstellar space are between 5000 and 10,000°K, while H I regions probably have temperatures a factor of 10^2 lower.

When the critical velocity v_R is reached, the ionization front no longer has direct contact with the undisturbed medium. It is now moving slowly enough that a shock front, signaling the impending arrival of the ionized region, precedes the ionization front, and in so doing compresses the medium to a density greater than that of the undisturbed state.

Essentially, this just means that the ionization heats the gas that then expands into the neutral medium fast enough so that a compression wave travels into the neutral gas at a speed exceeding the local speed of sound. A shock front therefore precedes the ionization front into the neutral medium and modifies the density in this medium so that the boundary conditions (9–19), (9–20), and (9–22) once again are satisfied at the ionization front. As the ionization front moves still farther from the star, the velocity with respect to the undisturbed neutral medium drops below the lower critical value v_D. Here a very gradual expansion is going on, no shock is propagating into the neutral medium and the ionization front once again is in direct contact with the undisturbed medium. This is called the *D-condition*.

PROBLEM 9–2.   Show that $v_D \sim 3c^2/4u$. $(9\text{–}43)$

We should still note that the conditions at a normal shock front are identical to those across an ionization front except that the ionization energy is not supplied, that is, $Q = 0$. The equations derived here therefore have a wide range of applications. It is also worth noting that usually a magnetic field is present and that the energy balance and pressure conditions must then also include magnetic field contributions. *Hydromagnetic shocks* are particularly significant because under conditions where interparticle collisions are rare, the magnetic fields are the main conveyors of pressure information throughout the medium. Pressure equilibrium between gas particles is established through mutual interaction via magnetic field compression.

At the interface between H_{II} and H_I regions we sometimes see bright rims that outline the dark, dust-filled regions not yet ionized (Po56). The bright rims generally are located at the edge of the nonionized matter and appear pointed toward the direction of the ionizing star which normally is of spectral type earlier than O9. It is possible (Po58) that these rims occur when ionizing radiation arriving at the H_I region satisfies the D-critical condition and sets up an ionization front that moves into the neutral gas without being preceded by a shock wave.

9:4 FORMATION OF MOLECULES AND GRAINS

One of the puzzles about interstellar grains is their origin. The density in interstellar space is so low that grain formation appears impossible there. To see this, consider the growth rate of a grain. Let its radius at time t be $a(t)$. Interstellar atoms and molecules impinge on the grain with velocity v. If the number density of heavy atoms, having mass m is n, the growth rate of the grain is

$$4\pi a^2 \frac{da}{dt} = \frac{\pi a^2 nmv}{\rho} \alpha_s \qquad (9\text{-}44)$$

where ρ is the density of the interstellar atoms after they have become deposited on the grain's surface, α_s is the sticking coefficient for atoms impinging on the grain, and the left side of (9-44) represents the grain's volume growth. Taking $v \sim \sqrt{3kT/m}$, with $T \sim 100°K$ and $m \sim 20\,amu \sim 3 \times 10^{-23}$ g, $\rho \sim 3$ g cm^{-3} and $n \sim 10^{-3}$ cm^{-3} and the maximum value $\alpha_s = 1$, we obtain

$$\frac{da}{dt} \sim \frac{n\sqrt{3kTm}}{4\rho} \alpha_s \sim 10^{-22} \text{ cm sec}^{-1} \qquad (9\text{-}45)$$

To grow to a size of 10^{-5} cm, the available time would have to be 10^{17} sec $\sim 3 \times 10^9$ y. With more realistic values of α_s, the required length of time increases to an age greater than that of the Galaxy. Here we have neglected the deposition

9:4

of hydrogen on a grain, since pure hydrogen would normally evaporate rapidly.

Of course, there exist regions in space where the number density of atoms like oxygen, nitrogen, carbon, and iron is ~ 1 cm^{-3}. The Orion region is about that dense. If there were no destructive effects there, grains could perhaps form in a time $\sim 3 \times 10^6$ y, if $\alpha_s \sim 1$. Furthermore, if the temperature in a dense cool cloud could become low enough so that hydrogen could solidify on the grains without rapid re-evaporation, the growth rate could be still higher by two or three orders of magnitude.

Thus far we have neglected destructive effects. For example, in H II regions, radiation pressure often accelerates small grains to higher velocities than large grains. Intercollisions of grains can then take place at such high velocities, $\gtrsim 1$ km sec^{-1}, that both of the colliding grains are vaporized. The vapors then have to recondense. In addition, *sputtering* by fast moving protons can knock atoms off a grain's surface after they have become attached. Such destructive effects tend to reverse the growth implied by equation 9–45, or at least will decrease the growth rate. This destruction would be stronger for substances like ice, where molecules are bound weakly, rather than for strongly bound substances like silicates or graphite. These three substances all may be constituents of interstellar grains.

Another—less catastrophic—destructive effect depends on the *vapor pressure* of the material from which the grains are formed. In thermal equilibrium at a temperature T the vapor pressure gives the rate at which molecules or atoms evaporate from a grain's surface. The equilibrium vapor pressure is that pressure of ambient vapor at which the growth rate is just equal to the evaporation rate. This partial pressure, P_{vap}, is related to the vapor density through the equation of state, so that if Dalton's law (4–38) holds, the molecule mass impact rate is

$$nmv \sim \left(\frac{m}{kT} \right)^{1/2} P_{vap} \qquad (9\text{–}46)$$

per unit area. This then must also represent the evaporation rate from the surface. We see that the pressure in the ambient space must exceed the vapor pressure, if the grain is to grow. Hydrogen has a vapor pressure of about 10^{-7} torr at 4°K (Table 9.3). This amounts to about 10^{11} molecules cm^{-3}. Of course, this is a density much higher than any expected for interstellar space. On the other hand, grains are never likely to be cooler than 4°K. This means therefore that hydrogen cannot very well remain on grains, unless it is chemically bound by the presence of other substances or else adsorbed on the basic grain material. For other substances this would not be true. Silicates and graphite, for example, have such low vapor pressures that no appreciable evaporation off grains would occur in periods of the order of the life of the Galaxy. For ice the situation is somewhat more complicated. On approaching close to an individual star, H_2O molecules could evaporate as a grain's temperature rose. Basically this is what happens when a comet approaches the sun from the outer portions of the solar system. The surface warms until water, ammonia, and other ices evaporate. Near most stars sputtering through collisions with atoms is the dominant destroying mechanism; only on very close approach to a star can evaporation be significant.

We believe that grain formation primarily takes place in the atmospheres of very cool grant stars or Mita variables; both eject gas into interstellar space. The atmospheres of these stars are dense so that n in (9–45) is high. Formation must, however, take place in a period as short as a month, because after that, the outflowing gas soon becomes too rarefied. With $n \sim 10^5$ cm^{-3} for the heavier atoms, $T \sim 2 \times 10^{3}$°K, $m \sim 12$ for carbon, it might be possible to form a graphite grain at a rate $da/dt \sim 3 \times 10^{-14}$, a grain with $a \sim 10^{-7}$ cm would form in a month. This would only be a nucleus that could subsequently grow, possibly because radiation pressure would keep it moving slightly faster than the gas flowing out from the star (see equation 9–15).

The fact that Mira variables and cool giants often seem to emit an excessive amount of infrared radiation (Jo67a) would indicate a close association of dust with these stars, and very possibly this is due to the formation and ejection of dust from their atmospheres.

Geisel, Kleinmann, and Low (Ge70) have found that the nova Ser 1970 became brighter in its infrared emission as it dimmed in the visible part of the spectrum. Presumably over a period of weeks, dust formed in the gas ejected by this nova. Since similar behavior was also observed in novae Aql 1970 and Del 1967, novae may generally be responsible for appreciable dust formation.

Finally, the infrared emission from planetary nebulae may indicate the presence of dust; perhaps the ejection process there still permits adequate formation to proceed at rates governed by equation 9–45.

The formation of molecules presents problems similar to dust formation. In the past few years, microwave techniques have been used to discover an increasing number of interstellar molecules, NH_3 (ammonia), CO, H_2O, HCN (hydrogen cyanide), H_2CO (formaldehyde), CN (cyanogen), HC_3N (cyanoacetylene), the hydroxyl radical OH, and many others. To be sure, being smaller aggregates, the molecules form first before the larger grains ever have a chance of forming. However, complicated molecules having several heavy atoms would perhaps not be expected to form readily once smaller stable molecules had already formed. For, these smaller molecules would then not readily accept an additional atom. The theory of molecule formation under the extremely rarefied conditions of interstellar space has, however, not yet been firmly tackled, and little is understood about the formation of molecules as complex as, say, formaldehyde and formic acid. These larger molecules normally are found associated with dense dark clouds in the vicinity of H II regions. It is not clear whether ionized

9:4

or neutral regions, or perhaps only the boundaries of such regions, are most conducive to the formation of molecules.

A number of destructive effects should still be mentioned. Molecules can be destroyed through dissociation—often as a consequence of absorption of, or ionization by, ultraviolet photons. Calculations indicate that the prevalent galactic starlight may destroy molecules like CH_4, H_2O, NH_3, and H_2CO in times of the order of a hundred years, unless the molecules are shielded from the light, inside strongly absorbing dust clouds (St72). Ionization by energetic electrons or cosmic ray particles could produce similar effects. Collisions of interstellar clouds that can produce energetic particles could therefore produce destruction of molecules; however, we could also argue conversely that the high densities at the contact face of two colliding clouds could lead to more rapid formation of molecules. Such competing formation and destruction rates must be compared in a variety of specific settings to see whether the overall situation is conducive or destructive to molecule formation.

9:5 CONDENSATION IN THE PRIMEVAL SOLAR NEBULA

In section 1:7 we presented some current views on the origin of the solar system.

PROBLEM 9-3. Just after the sun first formed, it may have been surrounded by a dense cloud of gas from which the planets eventually condensed. As a first step, small grains probably formed. Suppose the mass was evenly distributed throughout the nebula, that its total mass was equal to twice that of all planets combined (see Table 1.3), that the radius was 10 AU and that the initial abundance was similar to the solar abundance (Table 1.2). Making use of Table 9.3, calculate an approximate distance from the sun at which iron would have condensed. Do the same for carbon. Would water or ice have been able to condense within the nebula? Note that the sun may have been on the Hayashi track (Fig. 8.11) at that time. Assume its luminosity was ten times greater than now.

PROBLEM 9-4. The action of light pressure would have tended to produce homogeneity in the solar nebula: If the outward directed flux amounted to $1.4 \times 10^7/r^2$ erg cm^{-2} sec^{-1} at r astronomical units from the sun, find the orbital velocities of two grains, both orbiting the sun at the distance of Jupiter, both having a grain radius $s \sim 10^{-3}$ cm, one particle having density 2 g cm^{-3}, and the other having density 4 g cm^{-3}. Assume the grains to be spherical and black, absorbing light with a cross section πs^2. Note how the orbital velocity differs as a function of s. The difference is greatest for small s. Grains with large density

9:5

Table 9.3 Relation Between Temperature and Vapor Pressure Compiled from (Ro65), (Du62), and (Le72).[a]

	10^{-11}	10^{-10}	(1 torr $= 1.33 \times 10^3$ dyn cm^{-2}) 10^{-9}	10^{-8}	10^{-7}	torr
C	1695	1765	1845	1930	2030	°K
Fe	1000	1050	1105	1165	1230	

Most solid substances obey a pressure-temperature relationship of the form

$\log_{10} P = A - B/T$,

at low pressures:
P (in torr):

$$\begin{cases} \text{carbon} & A = 12.73 \\ \text{iron} & A = 9.44 \\ \text{NaCl} & A = 7.9 \end{cases}$$

and $B = 4.0 \times 10^{4}$°K
$B = 2.0 \times 10^{4}$
$B = 8.5 \times 10^{3}$

For water the following data are available:

H_2O	7×10^{-9}	3×10^{-10}	7.4×10^{-15}	14×10^{-22}	torr
	143.2	133.2	123.2	90.2	°K

For hydrogen:

H_2	3.1×10^{-7}	8.8×10^{-9}	7.5×10^{-11}	4.5×10^{-13}	torr
	4.0	3.5	3.0	2.6	°K

[a]Parts reprinted by special permission from Rosebury, *Handbook of Electron Tube and Vacuum Techniques*, 1965, Addison-Wesley, Reading Mass. Parts copyright © 1962, John Wiley and Sons. We note that the inner planets consist primarily of low vapor pressure material and that, by and large, the outer planets contain more volatile substances.

and/or size differences would therefore collide more frequently and be destroyed. Grains with nearly identical properties would tend to persist longer.

PROBLEM 9-5. After small particles and chunks were formed through condensation, a second stage of condensation seems to have taken place in which gravitational attraction played a dominant role. Before this time, particles presumably had acquired almost identical low eccentricity, low inclination orbits at any given distance from the sun, and the relative velocities of these grains must have been small. This would have come about because high or low velocity grains would be eliminated preferentially through more frequent destructive collisions with other bodies.

(a) At what size would a body whose density ρ is 3 g cm^{-3} have a gravitational capture cross section that is twice as large as its geometric cross section? Assume

9:5

a relative velocity V_0 for particles to be captured. The result of Problem 3–7 may be useful.

(b) Derive the growth rate of a body with $\rho = 3$ g cm^{-3} moving through a nebula whose density is $\rho_0 = 3 \times 10^{-12}$ g cm^{-3}. Let its relative velocity be $V_0 = 1$ km sec^{-1} and start at a time when its gravitational capture cross section is twice its geometrical cross section.

(c) Show that the mass growth for a spherical gravitating body, whose capture cross section is much greater than its geometric cross section, and whose density $\rho = 3M/4\pi R^3$ has a fixed value, is proportional to $M^{4/3}$ or R^4. More massive bodies therefore have a higher mass capture rate than lower mass bodies whose geometric capture cross section only allows them to capture mass at a rate proportional to R^2.

PROBLEM 9–6.   Suppose that a grain stays spherical as it grows through capture of matter. It moves through the solar nebula at $V_0 = 1$ km sec^{-1}, escapes destructive collisions by chance, and grows from a radius of $\sim 10^{-8}$ cm—one molecule—up to 10 km. If the nebular density is 3×10^{-12} g cm^{-3}, of which a 1% nonvolatile fraction can be captured, and the particle's density is 2.5 g cm^{-3}, show that the growth time is roughly 10^8 y.

9:6 INFRARED EMISSION FROM GALACTIC SOURCES

Infrared emission has now been observed for a variety of different sources. Cool supergiants often seem to be surrounded by dust layers which obscure much of their visible radiation, and these circumstellar clouds emit strongly in the near infrared. At longer wavelengths, principally beyond 10 microns—1 micron = 1 μ = 10^{-4} cm—we find that the galactic plane emits strongly, primarily around H II regions. The Galactic center also is a very strong emitter of radiation. Much of this emission seems to peak around wavelengths of order 100μ although only coarse spectroscopic data exists for this region (see Figs. 6.16 and 9.4).

The near infrared emission from circumstellar dust is relatively easy to explain. Equation 4–78 specifies the temperature that a grain will assume at a given distance R from a star. Some uncertainty always arises because we do not know the emissivity of grains in the visual or in the infrared region. Both the equilibrium temperature, which obeys the relation

$$T = \left(\frac{\varepsilon_a}{\varepsilon_r} \frac{L_\odot}{16\pi\sigma R^2} \right)^{1/4} \qquad (4\text{--}78)$$

and the emission spectrum therefore remain somewhat uncertain. In any case, not all of the grains will be at precisely the same distance from the parent star, and we would therefore expect a rather broad emission spectrum representing emission from grains located at different distances from the star and emitting at different effective temperatures.

The emission process associated with planetary nebulae and H II regions is somewhat different. Here there is relatively little dust. Often the obscuration within the ionized region is not noticeable, and yet strong infrared emission is observed. How can the dust be responsible?

To explain this phenomenon we have to return to the discussion of ionization equilibrium given in section 9:2. We had noted there that, for an equilibrium Strömgren sphere, the recombination rate equals the ionization rate. Each time an electron recombines with a proton to form a hydrogen atom we have two possibilities. Either the recombination leaves the atom in the first excited state $n = 2$, or else it leaves it in a higher level from which a cascade of photon emission processes eventually places the atom into state $n = 2$ or else $n = 1$. Any photon emitted through a transition from $n = 2$ to $n = 1$, or from any higher excited state down to $n = 1$, has a very high probability of being reabsorbed by another hydrogen atom in the $n = 1$ state in the H II region. The photon then wanders through the H II region in a random walk (see Problem 4–4). The mean free path, however, is variable. A photon, successively absorbed and reradiated by atoms moving with thermal velocities, v, becomes Doppler-shifted by $\Delta v \gtrsim v \langle v^2 \rangle^{1/2}/c$, which is large compared to the natural line width, γ, in equation (7–59). The mean free path increases, permitting rapid escape from the region. Nevertheless, the Ly-α pho-

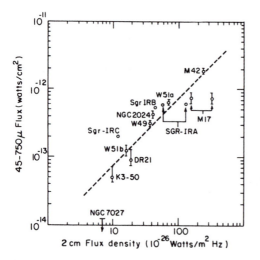

Fig. 9.3 Plot of infrared emission from 45 to 750μ against 2 cm wavelength radio emission from ionized hydrogen regions. (After Harper and Low, Ha71. With the permission of the University of Chicago Press.)

ton's trajectory out of the region is typically several times longer than the cloud radius, and dust has a correspondingly greater chance of absorbing the radiation. We have acted here as though all of the Lyman spectrum photons were Ly-α radiation. This is almost true. A Ly-β photon emitted by one hydrogen atom is likely to be absorbed and to give rise to an Hα photon in an atomic transition $n = 3$ to $n = 2$, succeeded by a Ly-α transition $n = 2$ to $n = 1$.

As a rule of thumb, we can state that for each ionizing photon, the H II region eventually must produce one recombination. That recombination, followed by a succession of emission, reabsorption, and re-emission processes, eventually has to give rise to one photon of the Balmer spectrum, and one Ly-α photon. The oscillator strength for absorption of the Ly-α photon is near unity so that the effective cross section per hydrogen atom is large. Even if the fractional neutral hydrogen density is very low, Ly-α absorption usually is large enough to trap the Ly-α photons in the nebula.

PROBLEM 9–7.   For an H II region in which each ionizing photon has unit optical depth on passing a distance equal to the radius, R, relate n_H, the number density of neutral atoms, to the absorption coefficient a_v listed in Table 9.2.

(a) With reference to Problem 7–11 determine the mean free path at the line center for Ly-α absorption with respect to R.

(b) The absorption bandwidth, γ, usually is small compared to the Doppler frequency shift $\Delta v \sim v \langle v^2 \rangle^{1/2}/c$ where $\langle v^2 \rangle^{1/2}$ is the root mean square velocity in the H II region. For $\langle v^2 \rangle^{1/2} \sim 30$ km sec^{-1}, what is the Ly-α absorption mean free path (see Problem 7–13).

Since each ionizing photon gives rise to a Ly-α quantum of radiation and since this radiation is likely to be absorbed by a grain before it can escape to the boundary of the H II region, we might expect that all the radiation converted into Ly-α would (Kr68) eventually be absorbed by a grain and the energy would be thermally radiated in the far infrared. Most of the ionizing photons given off by a star have an energy less than twice the Ly-α energy. Hence we conclude that more than half the ionizing energy emitted by a star would eventually find its way into infrared radiation. This seems to be at least approximately true. We can make a comparison to the free-free emission from ionized regions observed in the radio domain. This free-free emission can be directly related to the expected recombination line intensities, through equations like (9–5) or (6–141) and (6–139). We can therefore derive a proportionality relationship between the expected free-free radio emission from the H II region and the far infrared flux from grains. Harper and Low (Ha71) have checked this proportionality for a number of regions. The agreement is reasonably good although the infrared

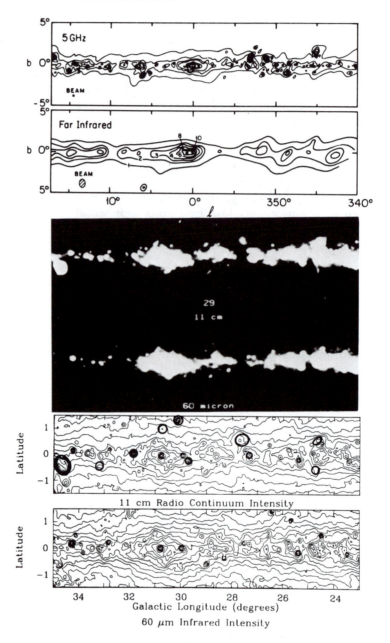

Fig. 9.4 Galactic plane at radio and far infrared wavelengths. The top two panels, respectively, show maps at 5 GHz (6 cm) (Al70a) and 150μ of the Galactic center (Ma79). The central two panels show 11 cm and 60μ photomaps of a small section of the plane. Isophotes of these maps are shown in the lowest two panels. Heavily circled sources on the 11 cm map are supernova remnants. Aside from these, the two central panels are remarkably similar. (Reprinted with permission from *Nature*, Volume 327, p. 211, copyright © 1987 Macmillan Magazines Ltd., Ha87.)

emission observed by them was systematically somewhat high. A number of factors can account for this discrepancy. Figure 9.3 shows Harper and Low's results while Fig. 9.4 gives a comparison of radio and infrared emission maps in the Galactic center region. The correlation is so good that a connection between infrared and radio fluxes seems reasonably well established.

In Seyfert galaxies, where strong infrared emission from nuclei is sometimes observed, the infrared emission mechanism might well be the one we have described. However, infrared radiation can also arise through such processes as synchrotron radiation or inverse Compton scattering, and it is known that the Seyfert galaxies can be strong emitters of X-rays so that highly energetic electrons are likely to be present. We still lack the observational evidence to distinguish between these mechanisms.

9:7 STAR FORMATION

In section 1:4 we presented a number of problems associated with star formation. Three primary processes have to take place if a star is to form: (1) The gas from which a protostar forms has to steadily radiate away energy so that its total energy can continuously decrease—giving rise to an increasingly compact configuration. (2) Angular momentum must somehow be reduced, from the high values associated with distended hydrogen clouds moving in differential rotation about the Galactic center, to the low values observed in stars. (3) The magnetic field observed by Faraday rotation and Zeeman splitting methods must somehow be removed from the contracting cloud of interstellar gas to give the relatively small magnetic field values actually seen at the surface of stars.

To these three requirements we should add one more that apparently is met rather easily: (4) A relatively rapid initial compression of the gas is required to trigger contraction. This compression must be accompanied by cooling, and the total energy loss must be large enough so that the cloud can no longer expand back to roughly its initial diameter. The compression, in other words, has to be strongly inelastic. Some such triggering mechanism is needed because turbulent motions, normally present in the interstellar medium, would otherwise disrupt the contraction process before it had even properly started. These turbulent motions are produced by the expansion of ionized regions and by radiation pressure that acts on the gas via the dust grains (Ha62a).

This particular compression criterion can apparently be met through the strongly compressive shock waves formed around H II regions. Although these expanding regions tend to endow the cool clouds with turbulent motion, they also can produce rapid compression, as we saw in section 9:3. This compression can lead to continued contraction when coupled with strong radiative cooling through such processes as grain emission, radiation from molecular hydrogen that may be abundant in dark clouds (Go63), or perhaps emission from H_2O vapor that has a high dipole moment and can be rotationally excited by low energy collisions.

Criteria (1) and (4) can therefore apparently be met. The shedding of angular momentum and dissociation of magnetic fields, however, still pose unsolved problems. We will only be able to show one or two possible ways out of the dilemma.

Let us consider the angular momentum problem first. We might argue that we overestimated the total angular momentum initially present in the proto-stellar matter. The differential rotation effect might perhaps not be as large as initially estimated.

One way in which the angular momentum due to differential rotation might be low would be to gather only that material which had uniform angular momentum about the Galactic center. If, for example, we assumed all gas particles to be in circular motion about the center, then we would gather material only from a ring at some fixed radius R_c from the center. This ring would have a circumference $2\pi R_c$. The areal density—viewed normal to the Galactic plane—of gaseous matter in the solar neighborhood, is of order 10^{20} atoms cm^{-2}, or $\sigma \sim 10^{-4}$ g cm^{-2}. The length $2\pi R_c$ in our neighborhood is $\sim 2 \times 10^{23}$ cm, so that a ring of width

$$W \sim \frac{M_\odot}{2\pi R_c \sigma} \sim 10^{14} \text{ cm} \tag{9-47}$$

would be needed to form a star of one solar mass. Using the figures for a velocity gradient used in section 1:4, we would find then that the differential velocity would be of order 3 cm sec^{-1} across this width and that the angular momentum per unit mass, once contraction had taken place, would only be of order 3×10^{14} cm^2 sec^{-1}. This is a factor of ~ 3 less than the angular momentum per unit mass of solar material and a factor of ~ 300 less than the corresponding angular momentum figure for the whole solar system. With (dv/dr) decreased 100-fold, mass would have to be gathered only from a shorter segment 600 pc along an arc of radius R_c, and W could be increased to $\sim 10^{16}$ cm, thus increasing the angular momentum per unit mass to roughly 3×10^{16} cm^2 sec^{-1} and a solar system or star with reasonable angular momentum characteristics might thus be formed.

What we conclude is that star formation with reasonable angular momentum characteristics would be possible if stars were formed from segments of thin cylindrical shells of matter initially orbiting about the Galactic center in a narrow range of orbital angular momentum values. Whether such a process is possible on other grounds is an entirely different question and much harder to answer.

First, it is not likely that a gravitational contraction within such platelike segments could proceed (Eb55); gravitational contraction normally is most

effective if the contracting volume is roughly spherical, for then a small compression in volume can amount to a relatively large change in potential energy.

Second, the magnetic field problems are not averted in such a configuration.
If the magnetic field lines were largely normal to the radius R_c—and this might
be favored by shear produced through differential rotation—then contraction
along the magnetic field would be unresisted by the field, but the remaining
contraction perpendicular to the field direction would still necessitate compression of field lines. Here, we are still concerned about compression from a
dimension of order 100 pc down to solar system $\sim 10^{15}$ cm dimensions, and this
alone would produce a field of the order of a gauss for a region 10^{15} cm in diameter.
Compression into an object the size of the sun would then still require compression in two dimensions and give a final field of order 10^9 gauss.

The magnetic field would therefore have to be dissipated in some way or other;
there is no immediately apparent way of avoiding the problem in the way we
avoided some of the angular momentum difficulties by contraction along a
region of constant radial distance from the Galactic center. We would not be
able to wholly avert the magnetic field difficulties even if we took the extreme
case involving the gathering of matter along a complete circular arc around
the galactic center. Somehow the magnetic fields must be dissipated.

We should still mention that many means of avoiding the angular momentum
difficulties have been suggested. It might, for example, be that magnetic coupling
of the contracting matter to external gases produces a viscous drag that slows the
rotation of contracting gases. In this case, the magnetic field would actually help
to overcome one of the difficulties of star formation while, on the other hand,
introducing another stumbling block of its own.

We will discuss magnetic fields in more detail in section 9:9, where we will
speculate on the origins of interstellar magnetic fields. Some of the problems
involved in forming magnetic fields are identical to those encountered in trying
to destroy them.

Literally hundreds of ideas on how stars are formed have been advanced in past
decades. However we still are far from any real solution. Here we have only
presented one or two simple concepts; the actual answer to the problem may turn
out to be much more complex. On the other hand, we might also find that we have
misinterpreted some observational results and that actual conditions in interstellar space are really far more conducive to star formation than we have thus
far believed.

9:8 ORIENTATION OF INTERSTELLAR GRAINS

When the light from stars close to the Galactic plane is analyzed, it is found to
be both reddened and polarized. This has been interpreted (see section 6:15) as

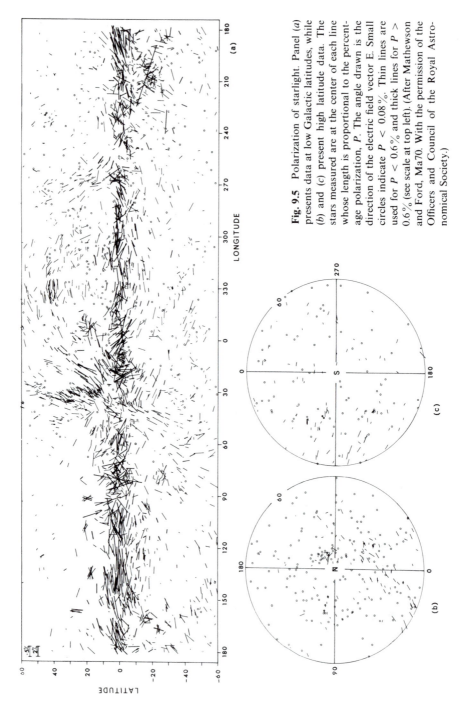

Fig. 9.5 Polarization of starlight. Panel (*a*) presents data at low Galactic latitudes, while (*b*) and (*c*) present high latitude data. The stars measured are at the center of each line whose length is proportional to the percentage polarization, *P*. The angle drawn is the direction of the electric field vector *E*. Small circles indicate $P < 0.08\%$. Thin lines are used for $P < 0.6\%$ and thick lines for $P > 0.6\%$ (see scale at top left). (After Mathewson and Ford, Ma70. With the permission of the Officers and Council of the Royal Astronomical Society.)

due to the alignment of elongated or flattened grains, spinning in such a way that their long axes are close to planes perpendicular to the Galactic disk. The transmitted starlight is polarized in a direction predominantly parallel to the plane (Fig. 9.5).

How can grains become aligned in this way? In Problem 7–6, we showed that grains have angular frequencies of the order of 10^5 Hz. It therefore makes no sense to talk about the orientation of a stationary particle. There is, however, a different kind of orientation involving preferred directions of a grain's angular momentum vector.

This can be illustrated by a simple process. Suppose that a grain is quite elongated, like a stick, and that it is systematically moving through the gas with directed velocity **v**. The gas is so tenuous that individual gas atoms can be considered as colliding with the stick, one at a time. Let us first assume that the gas atoms have no random velocity of their own. The root mean square value of the angular momentum transferred to the stick in any given collision then is

$$\delta L = \left[\frac{1}{a} \int_0^a (mvr)^2 \, dr \right]^{1/2} \sim \frac{mva}{\sqrt{3}} \tag{9–48}$$

where a is half the length of the stick, m is the mass of the atoms, and v their approach velocity to the grain. The root mean square (rms) angular momentum after N collisions can be obtained from a random walk calculation (as in equations 4–12, 4–13) and is $N^{1/2}\delta L$. If the density of a grain is ρ, its width is s, and its length $2a$, we can compute the number of collisions N required to appreciably change any initial angular momentum the grain may have. This is

$$N = \frac{M}{m} = 2\rho s^2 \frac{a}{m} \tag{9–49}$$

where M is the grain mass. This equation merely states that the grain will have an appreciable rotational velocity change after it has sustained collisions with a number of atoms N, whose total mass is equal to M. A random angular momentum accompanies this systematic change. The final root mean square angular momentum, $\langle L^2 \rangle^{1/2}$, which a typical grain will acquire, is

$$\langle L^2 \rangle^{1/2} = N^{1/2}\delta L \sim \left[\frac{2m\rho a^3}{3} \right]^{1/2} sv \tag{9–50}$$

If we take m to be the mass of atomic hydrogen, 1.6×10^{-24} g, $v = 10^5$ cm sec^{-1}, $a = 10^{-5}$ cm $= 3s$, $\rho = 1$ g cm^{-3}, we obtain $L \sim 10^{-20}$ g cm^2 sec^{-1}.

PROBLEM 9–8.   Show that the angular velocity of the grain is $\omega \sim 10^6$ rad sec^{-1} in this case, and that $L \sim 10^{-20}$ g cm^2 sec^{-1} also represents the thermal equilibrium value at a temperature of 100°K.

Since we know angular momentum to be quantized in units of $\hbar$ we see that typically the angular momentum quantum number will have a value

$$J \sim 10^7 \hbar \tag{9-51}$$

We note that the direction of the angular momentum acquired in this process must lie perpendicular to v since all impulses on the stick produce changes in angular momentum δL whose directions lie in this plane. The gas systematically streaming past the grain therefore produces a preferred orientation of the angular momentum axis.

Let us now consider, in contrast to the systematic drift velocity of the grain through the gas, that there also is a random velocity u possessed by the atoms in the gas. Then the random walk process will endow the grain with an angular momentum greater by a factor of $\sim [(u^2/v^2) + 1]^{1/2}$ than the value given by (9–48). In this case, the systematic angular momentum will only be of the order of $[(u^2/v^2) + 1]^{-1/2}$ times the random angular momentum. In many cases $u \gg v$, and only slight preferential orientation of angular momentum can take

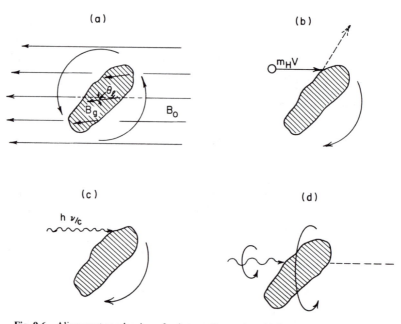

Fig. 9.6 Alignment mechanisms for interstellar grains: (*a*) Process of paramagnetic relaxation; (*b*) alignment by streaming through gas, or through a photon field (*c*). In process (*c*) the photon's linear momentum causes the grain to spin; in process (*d*) the photon's intrinsic spin angular momentum is of importance (see text).

place. However, this may be enough to provide the partial—rather than complete—polarization actually observed. At the very least, it should be a contributing factor.

We still have to examine which orientations of the angular momentum vector will be preferred. To do this, we consider how an individual collision affects an elongated grain. We note that when $v \gg u$, collisions are much more probable if the length of the stick is perpendicular to the direction of **v**. In that configuration more projected area is exposed to the incident gas atoms. The sense of any collisions, then, is to produce a change in angular momentum perpendicular both to the direction of gas flow and to the length of the grain. A rotation about an axis whose direction is along **v** is therefore very unlikely, and it is the absence of angular momentum orientations having this sense that can produce a partially preferred orientation of the sticks with their long axes parallel to the direction of flow. In this process, suggested by Gold (Go52), orientations perpendicular to the direction of flow simply tend to be missing (Fig. 9.6b). This general conclusion holds true even if $v < u$.

Since the sense of polarization is such that light whose electric vector **E** oscillates perpendicular to the plane of the Galaxy is preferentially scattered or absorbed, we assume that the grains also are oriented with their long axes perpendicular to the Galactic plane. They then absorb or scatter light, somewhat like a radio antenna, parallel to the long axis, along which electrons can most readily flow in response to an oscillating E field.

What has been said here about collisions with gas atoms, holds equally well for collisions with photons, Fig. 9.6(c). Here the atomic momentum mv has to be replaced by the photon momentum hv/c if we are to compute δL by means of equation 9–48. The photon effect can be expected to be strong close to bright stars.

For photons Harwit (Ha70) has proposed another effect that will dominate for small grains. We stated in section 7:2 that each photon must have an intrinsic angular momentum $\hbar$. When a photon is absorbed by a grain, *intrinsic angular* momentum $\hbar$ therefore is transferred, Fig. 9.6(d). This means that per photon $\delta L = \hbar$.

We now are in a position to compare these three effects. Per collision, angular momenta:

$$\delta L_g \sim \frac{mva}{\sqrt{3}}, \quad \delta L_e \sim \frac{hva}{\sqrt{3}c} \quad \text{and} \quad \delta L_i \sim \hbar \quad (9\text{–}52)$$

are transferred, respectively, by the gas atoms, extrinsic photon and intrinsic photon angular momentum processes. The last term predominates over the second, if

$$\frac{a}{\sqrt{3}} \lesssim \frac{\lambda}{2\pi} \quad (9\text{–}53)$$

9:8

where $\lambda = c/v$ is the wavelength of the photons. The evidence from light scattering suggests that a $\sim 5 \times 10^2$ Å $= 5 \times 10^{-6}$ cm, so that the intrinsic effect of photons dominates over the extrinsic for visible light. In that case $mva/\sqrt{3} \sim 10^{-29}v$ and the effect of individual atoms is greater for all systematic velocities v in excess of 10^2 cm sec^{-1}, provided that gas collisions are as frequent as photon collisions. This, however, is never true, as we will now show.

If the interaction cross section for gas and photons is taken to be the same, the number of collisions per second is going to be equal to nv for the gas and Nc for photons. N is the number density of starlight—mainly visible and near infrared—photons. Typical values for the gas density n range from 1 to 10^3 cm^{-3} in interstellar space.

The systematic velocities v are harder to assess. In a collision between gas clouds, a grain may find itself streaming through gas with a systematic velocity v as high as 10^6 cm sec^{-1}. But this will last only for a time of order

$$t = \frac{M}{2sanvm} \tag{9-54}$$

the time in which a grain suffers collision with an equal mass of gas. For $n = 10$ cm^{-3}, $v = 10^6$ and with the previously used parameters, we have $t \sim 2 \times 10^{11}$ sec. High velocities may therefore occur $< 1\%$ of the time because the lifetime between acceleration or collisions of clouds is of the order of 10^{14} sec, and it is these accelerations that can initiate systematic motion of grains through ambient gas. The radiation effect on the other hand is always present. Within the Galaxy, $N > 0.02$ cm^{-3} so that $Nc \sim 6 \times 10^8$ sec^{-1} cm^{-2} as contrasted to $nv \sim 10^7$. The photon effects dominate if $\sqrt{3} Nch > nmv^2a$ or if $Nhv > mnv^2$.

Thus far we have considered only orienting effects of gas and photons. Next we must consider the randomizing effects. When v is small compared to u, the randomizing angular momentum per atomic collision, that is, the step size in the random walk process is $mua/\sqrt{3}$. The number of collisions per unit time is proportional to the product nu, the gas density times velocity. It also is proportional to the grain cross section, which we will call σ without defining it further. After some time τ, the collisional random walk process will produce a root mean square angular momentum

$$L_g \sim \frac{mau}{\sqrt{3}} [\sigma nu\tau]^{1/2} \tag{9-55}$$

During the same time interval, the number of thermal photons given off by the grain can be shown to be $\sim 1.5 \times 10^{11} T^3$ cm^{-2} sec^{-1}. Each of these has the ability to endow the grain with angular momentum h in some random direction.

9:8

At typical interstellar grain temperatures of, say, 15°K, we then have

$$L_p \sim [5 \times 10^{14} \sigma \tau]^{1/2} \hbar \qquad (9\text{--}56)$$

Putting $a = 5 \times 10^{-6}$ cm, $u = 10^5$ cm sec^{-1}, $n = 10$ cm^{-3}, we find

$$\frac{L_g}{L_p} \sim 2 \times 10^{-2} \qquad (9\text{--}57)$$

Only for large grains, or for dense regions with $n \gtrsim 10^3$, or for a combination of these, is the gas effect equal to that of the re-emitted radiation.

We note now that the random photon effect, if photons are emitted roughly at wavelengths 100 times longer than the absorption wavelength, will make the L_p vector about 10 times larger than the component L_i due to any systematic radiation anisotropy effects. The photon-orienting effect depends on an intrinsic asymmetry in the radiation field to produce alignment. The illumination by photons coming from directions close to the Galactic plane is about 10 times greater than the illumination perpendicular to the plane—as we know the Milky Way is bright only in a narrow band. If the angle that the angular momentum axis of a typical grain makes with respect to some line lying in the Galactic plane is θ, then the average value of $\cos^2 \theta$ becomes of the order $L_i^2/L_p^2 \sim 0.01$. Observed figures indicate that this ratio probably is more nearly 0.02. However, without knowing much about the ratio a/s for actual interstellar grains, we cannot compute an expected ratio accurately, and these figures will have to suffice until we learn more about the structure of grains.

We next come to an effect first proposed by Davis and Greenstein (Da51). In this process, a grain is bombarded by ambient gas in interstellar space and is thus set spinning about an arbitrary axis. We can now postulate that the grain material is paramagnetic. Such materials, when placed into a magnetic field, set up an internal field whose direction is (Fig. 9.6a) parallel to the external field. The internal field cannot, however, change instantaneously. If the grain rotates with angular velocity ω about a direction perpendicular to the magnetic field, the internal field is forced—again at frequency ω—to change its direction relative to an axis fixed in the grain. However, since this readjustment of direction does not proceed instantaneously, a slight misalignment of the internal and external field arises as shown in Fig. 9.6(a). The interaction of the induced internal field with the externally applied field attempts to compel parallelism by opposing the rotational motion. This drag torque is proportional to the external field **B**, to the internal field that also is proportional to **B**, to the grain volume, V and to ω.

$$\text{Torque} = KVB^2\omega \qquad (9\text{--}58)$$

9:8

This represents the situation when the grain is spinning about an axis perpendicular to the direction of **B**. When the grain spins about an axis parallel to **B**, the induced field does not need to change its direction relative to the external field and no drag force arises.

For an arbitrary spin direction, that component of the spin whose axis is perpendicular to the field will therefore be damped out in a time

$$\tau = \frac{I}{KVB^2} \tag{9–59}$$

where I is the moment of inertia about the spin axis. On the other hand, a component whose spin axis is parallel to **B** remains undamped. This damping process is called *paramagnetic relaxation*.

An aspherical grain whose spinning motion is slowed tends to align itself in such a way that the *axis of greatest moment of inertia* becomes parallel to the angular momentum axis. This inertia axis is perpendicular to the long axis of an elongated grain, and the net effect of paramagnetic relaxation is to align elongated grains with their long axes perpendicular to the magnetic field. For substances with which we are familiar in the laboratory, and for grain temperatures that typically might be of order 10°K, in interstellar space at large distances from the nearest stars, the value of K is probably less than 10^{-12}. As we will see, this leads to rather high values of the relaxation time, unless the magnetic field is at least of order 10^{-5} gauss. But such field strengths are roughly a factor of 3 higher than the general field estimated to be present in the Galaxy. And the relaxation times for the more probable fields $\sim 3 \times 10^{-6}$ gauss (Fig. 9.9) are a factor of 10 too long. Jones and Spitzer (Jo67b) have therefore suggested the possibility that cooperative effects between iron atoms in grains might produce what they have termed *superparamagnetism*, to yield a sufficiently high value of K. Experimental evidence for the existence of such substances does not yet seem to exist.

The relaxation time τ must now be compared to the relaxation time produced by random gas collisions and also to the relaxation time due to re-emission of infrared photons. A comparison to the gas effect (9–54) shows that

$$\frac{\tau}{t} \sim \frac{I}{M} \frac{2sanum}{KVB^2} \sim 10 \tag{9–60}$$

where we have taken $I \sim Ma^2/3$, $n = 10$ cm^{-3}, $u \sim 10^5$ cm sec^{-1}, $3s = a \sim 5 \times 10^{-6}$, $V \sim s^2a$, $B \sim 10^{-5}$ gauss, $K \sim 10^{-12}$. The relaxation therefore is of order $\tau/t \sim 10\%$. For more reasonable B values it is ~ 0.01, comparable to the photon alignment. Even then, we have probably overestimated K, and the effect may be still weaker.

We are therefore faced with a situation in which we could account for the alignment magnetically, if the Galactic magnetic field is strong and lies predominantly parallel to the Milky Way plane. If the motion of gas relative to grains were perpendicular to the plane, the Gold process would give the correct orientation, and we know that the photon flux is in the right direction for the photon process. But in each of these mechanisms we have to strain the properties or the expected parameters of the grains, or the gas, or the magnetic field, or a combination of these, to obtain rough quantitative agreement. It is not unlikely that all of these processes play important roles in the alignment, but since none of the processes is entirely free of difficulties, it also is possible that we have overlooked some dominant factor of greater overall importance.

9:9 ORIGIN OF COSMIC MAGNETIC FIELDS

Magnetic fields are known to exist in stars and in the interstellar medium. Stars like the sun have typical surface magnetic fields of the order of one gauss, but in some A stars the surface fields may reach magnitudes of the order of 40,000 gauss. The fields in the interstellar medium of course are much weaker, typically of the order of 10^{-6} gauss. But there are great variations. In some regions of the Galaxy no magnetic fields at all have been determined in measurements that should have detected fields of strength 10^{-6} gauss, while in other regions, quite strong fields exist. In the Crab Nebula supernova remnant, for example, the field strength is believed to be as high as 10^{-4} gauss, and fields of the order of 5×10^{-5} gauss have been observed by Verschuur in Orion (Ve70).

Where does this field come from? Is it primordial? Is its origin a cosmological question dating back to some early stages of the universe? We do not know.

If magnetic fields are not related to the introduction of matter into the universe—either primordially or on a continuing basis—then they should be generated at some subsequent time. We can then envisage two possible alternatives.

(1) Magnetic fields are formed in the interstellar or intergalactic medium and find their way into stars primarily because stars are formed from the interstellar material, or

(2) magnetic fields are formed in stars and the field lines might then be introduced into the interstellar medium as mass is ejected from the stars. This might be consistent with the high strength of the Crab Nebula field. It would also be consistent with the observation that the solar wind carries along magnetic fields. Whether some portion of this field becomes detached from the sun and strays out into the interstellar medium is not known. But many stars have much more massive winds than the sun, and the outflow of magnetic fields may be a customary accompaniment to the outflow of mass.

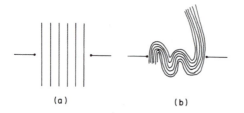

Fig. 9.7 Two magnetic field configurations with the same net flux. Configuration (*a*) has low field strength everywhere. Configuration (*b*) has high field strength in some places. In this figure, the field strength is taken proportional to the number of lines crossing unit length of the abscissa. This would be representative for a situation in which the field lines were embedded in sheets normal to the plane of the paper.

Once a magnetic field exists in very weak form, it can be strengthened by turbulent motion of the gases, and by other types of motion of the medium in which the fields are embedded. The net magnetic flux crossing any given fixed surface cannot be increased in this way, but by folding the field direction many times, local fields of greatly increased strength can be formed without an accompanying high net flux (Fig. 9.7). Turbulent motion therefore obviates the need for strong initial fields. Small, seed magnetic fields can be amplified by turbulent stretching and folding of the field lines.

Let us see how big this effect could be. If a field B_0 initially existed in some location within the Galaxy, the flow of gas at a velocity v could have stretched out field lines maximally at that velocity. Folding the field back on itself also could maximally occur at velocity v, so that the ability of a turbulent motion to amplify the field is limited by the speed of the motions.

Essentially the amplification of the field is given by the ratio of the initial volume $\mathscr{V}_0$ containing the seed fields to the final volume $\mathscr{V}_f$ that would have been obtained through stretching the region through a rectilinear motion, at velocity v

$$\therefore \frac{B_f}{B_0} = \frac{\mathscr{V}_f}{\mathscr{V}_0} \tag{9-61}$$

where B_f is the final magnetic field strength obtained through stretching and folding in a constant volume $\mathscr{V}_0$.

Within the Galaxy explosive velocities of order 10^3 km sec^{-1} are sometimes observed. We can choose this to represent the maximum turbulent velocity. The initial dimension of the Galaxy is ~ 30 kpc along a diameter and ~ 100 pc perpendicular to the disk. If the stretching motion were to go on for 10^{10} y at 10^3 km sec^{-1}, a distance of 10 Mpc would be covered, and a turbulent folding would increase the magnetic field strength respectively by a factor of 300, or 10^5 depending on whether the turbulent motion took place predominantly within the Galactic plane or perpendicular to it.

Since the field in the Galaxy is estimated to have a value of order 3×10^{-6} gauss, at the present epoch, the initial seed fields must at least have had values of the order 3×10^{-11} gauss. There seems no way to escape this conclusion.

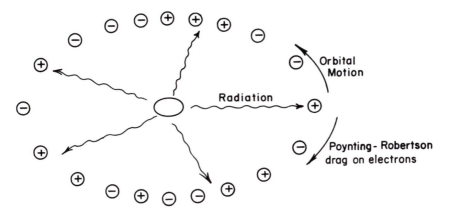

Fig. 9.8   Poynting-Robertson drag on electrons produced by a luminous body acting on an orbiting plasma. The drag on protons is much weaker, and a small net current can therefore be induced.

A primordial field of this magnitude must therefore have been present initially, or else some mechanism must have existed for producing this field. A number of processes have been suggested for setting up such a *seed field* that later could grow in strength through turbulent motion.

One relatively easily understood process could be provided by the Poynting-Robertson effect which could slow down electrons orbiting about a luminous massive object, while leaving protons almost unaffected. The current produced in this way will set up a weak magnetic field (Ca66).

This type of effect applies different forces to electrons and protons and is sometimes called a *battery effect*, because it sets up a current like a battery does. Here we will only show that the Poynting-Robertson effect, Fig. 9.8, would be able to provide sufficient energy for setting up the magnetic field. Some, or perhaps even most of the energy may, however, end up in some form other than magnetic energy. A complete analysis of such effects is very complicated and depends in detail on the interaction of the electrons with protons, on the direct differences in radiation pressure acting on electrons and protons, on the resulting tendency for positive and negative charges to slightly separate along a radial direction from the light source, and so on. A great deal of work remains before we can decide whether battery effects contribute to the generation of interstellar magnetic fields, but the effect discussed here should serve as an illustration of the type of mechanism that could perhaps be effective.

We note that the Poynting-Robertson drag on an electron is large because the Thomson cross section (see equation 6–102) is $(m_p/m_e)^2 \sim 3 \times 10^6$ times larger for electrons than for protons. Here m_e and m_p are the electron and proton mass. The deceleration is even stronger for electrons, being a factor $(m_p/m_e)^3$ times larger than for protons, because for electrons the drag force acts on a smaller mass. From (5–48) we see that the orbital angular momentum L, for an electron in a circular orbit, changes at a rate

$$\frac{1}{L}\frac{dL}{dt} = \frac{L_s}{4\pi R^2}\frac{\sigma_T}{m_e c^2} \tag{9–62}$$

where L_s is the source luminosity, R is the distance from the source, and σ_T is the Thomson cross section. The work dW/dt done on an electron in unit time is equal to the force F acting on it, multiplied by the distance v through which the electron moves in that time.

$$\frac{dW}{dt} = Fv = \left(\frac{1}{R}\frac{dL}{dt}\right)\frac{L}{Rm_e} = \frac{L_s \sigma_T L^2}{4\pi R^4 m_e^2 c^2} \tag{9–63}$$

The largest number of electrons that could be slowed down in this way would be $N \sim 4\pi R^2/\sigma_T$, since if we had more electrons than this, some would shadow the others. Hence the maximum total work that can be done on the clouds is

$$\frac{NdW}{dt} \sim \frac{L_s}{R^2}\frac{L^2}{m_e^2 c^2} = L_s\frac{v^2}{c^2} \tag{9–64}$$

This gives the maximum work that can go into building up a magnetic field over volume $\mathscr{V}$:

$$NW \sim \mathscr{V}\int \frac{d}{dt}\left(\frac{B^2}{8\pi}\right)dt \sim \frac{B_f^2}{8\pi}\mathscr{V} \tag{9–65}$$

where $B^2/8\pi$ (see section 6:10) is the instantaneous magnetic field energy density. For the Galaxy, gas is contained in a disk volume $\mathscr{V} \sim 3 \times 10^{66}$ cm^3 and $B_f \sim 3 \times 10^{-6}$ gauss.

$$\therefore NW = \frac{L_s v^2}{c^2}\tau \sim 10^{54} \text{ erg} \tag{9–66}$$

Let us first see whether the total field could have been produced had the Galaxy at one time been as bright as a quasar for, say, 3×10^6 y. Taking typical quasar internal velocities $v \sim 10^8$ cm sec^{-1}, $\tau \sim 3 \times 10^6$ y $\sim 10^{14}$ sec; we would require $L_s \sim 10^{45}$ erg sec^{-1} at peak efficiency to produce a field of $\sim 3 \times 10^{-6}$ gauss.

This does not seem too unreasonable for producing the entire flux, so that perhaps we would not even need the subsequent turbulent amplification. However, if we want to do the same thing in our Galaxy right now, we find that there are so few electrons that only $\sim 10^{-5}$ of the total flux would be scattered by electrons near the center where velocities are $\sim 3 \times 10^7$ cm sec^{-1}. There $L_s \lesssim 10^{43}$

erg sec^{-1}, so that the overall rate of work done on electrons is decreased by $\sim 10^8$. In 10^{14} sec a seed field of $\sim 3 \times 10^{-10}$ gauss could be formed for the Galaxy—in 10^{10} y, a field of $\sim 2 \times 10^{-8}$ gauss. Such a seed field could be amplified by turbulence as indicated above.

Other processes that could produce similar currents by acting differentially on electrons and ions have been suggested. One such process could work on the basis of viscous drag (Br68).

Such processes could also act to destroy magnetic fields. If the orbiting plasma contains an initial magnetic field, the drag acting on the electrons may be in a direction that produces a current that opposes the magnetic field. Such processes might play a role in decreasing the magnetic fields during protostellar stages.

Dynamo effects, which also can produce magnetic fields, should just be mentioned, for completeness: however, not much is known about the role they may play in setting up cosmic fields. This is a difficult theoretical problem.

9:10 COSMIC RAY PARTICLES IN THE INTERSTELLAR MEDIUM

Cosmic ray particles, mainly high energy electrons and protons, contribute an energy density of about 10^{-12} erg cm^{-3} to the interstellar medium. This compares to a mean starlight density of $\sim 7 \times 10^{-13}$ erg cm^{-3} and a kinetic energy of gas atoms, ions, and electrons ranging from about 10^{-13} erg cm^{-3} in the low density cool clouds, to roughly 10^{-9} erg cm^{-3} in high density H$\,$II regions.

Is there any interaction between the cosmic rays, the gas, and the radiation field?

This interaction, in fact, is quite strong. In this section we will estimate that strength under different conditions. The interaction usually implies an energy loss for a cosmic ray particle.

Such losses can be divided in the following way (Gi69), (Gi64):

(a) Highly relativistic electrons having energies $\mathscr{E} \gg mc^2$, lose energy to the interstellar medium through a number of different processes that sometimes are collectively referred to as *ionization losses*. These comprise (i) the ionization of atoms and ions, (ii) the excitation of energetic atomic or ionic states, and (iii) production of Cherenkov radiation. These effects are not always separable and their interrelationship will be determined in part by the electron energy and in part by the nature of the medium. Neutral and ionized gases give rise to different loss rates. Table 9.4 gives expressions for these and other cosmic ray losses discussed below.

(b) These ultrarelativistic electrons also can suffer *bremsstrahlung* losses. Those occur when electrons are deflected by other electrons or by nuclei of the medium.

Table 9.4 Energy Losses of Cosmic Ray Particles in the Interstellar Medium (After Ginzburg, Gi69).[a]

(a) Ionization Losses for Electrons with $\mathscr{E} \gg mc^2$:

in a plasma:
$$-\frac{d\mathscr{E}}{dt} = \frac{2\pi e^4 n}{mc}\left\{\ln\frac{m^2 c^2 \mathscr{E}}{4\pi e^2 n \hbar^2} - \frac{3}{4}\right\} = 7.62 \times 10^{-9} n\left\{\ln\left(\frac{\mathscr{E}}{mc^2}\right) - \ln n + 73.4\right\} \text{ ev sec}^{-1}$$

in a neutral gas:
$$-\frac{d\mathscr{E}}{dt} = \frac{2\pi e^2 n}{mc}\left\{\ln\frac{\mathscr{E}^3}{mc^2 \chi_0^2} - 0.57\right\} = 7.62 \times 10^{-9} n\left\{3\ln\left(\frac{\mathscr{E}}{mc^2}\right) + 20.2\right\} \text{ ev sec}^{-1}$$

Electron Radiation Losses:

for plasma:
$$-\frac{d\mathscr{E}}{dt} = 7 \times 10^{-11} n\left\{\ln\left(\frac{\mathscr{E}}{mc^2}\right) + 0.36\right\}\frac{\mathscr{E}}{mc^2} \text{ ev sec}^{-1}$$

for neutral gas:
$$-\frac{d\mathscr{E}}{dt} = 5.1 \times 10^{-10} n\frac{\mathscr{E}}{mc^2}\text{ev sec}^{-1}$$

Electron Synchrotron and Compton Losses:
$$-\left[\left(\frac{d\mathscr{E}}{dt}\right)_s + \left(\frac{d\mathscr{E}}{dt}\right)_c\right] = 1.65 \times 10^{-2}\left[\frac{H^2}{8\pi} + \rho_{ph}\right]\left(\frac{\mathscr{E}}{mc^2}\right)^2 \text{ ev sec}^{-1}$$

(b) Losses for Nuclei:

in neutral gas:
$$-\frac{d\mathscr{E}}{dt} = 7.62 \times 10^{-9} Z^2 n\left\{4\left[\ln\left(\frac{\mathscr{E}}{mc^2}\right)\right] + 20.2\right\} \text{ ev sec}^{-1}, \quad \text{if} \quad Mc^2 \ll \mathscr{E} \ll \left(\frac{M}{m}\right)^2 Mc^2$$

$$-\frac{d\mathscr{E}}{dt} = 7.62 \times 10^{-9} Z^2 n\left\{3\left[\ln\left(\frac{\mathscr{E}}{mc^2}\right)\right] + \ln\frac{M}{m} + 19.5\right\} \text{ ev sec}^{-1}, \quad \text{if} \quad \mathscr{E} \gg \left(\frac{M}{m}\right) Mc^2$$

for plasma:
$$-\frac{d\mathscr{E}}{dt} = 7.62 \times 10^{-9} Z^2 n\left\{\left(\ln\frac{W_{max}}{mc^2}\right) - (\ln n) + 74.1\right\} \text{ ev sec}^{-1}$$

where $W_{max} = 2mc^2\left(\frac{\mathscr{E}}{mc^2}\right)^2$ if $Mc^2 < \mathscr{E} \ll \left(\frac{M}{m}\right)Mc^2$

$\qquad\qquad = \mathscr{E}$ if $\mathscr{E} \gg \left(\frac{M}{m}\right)Mc^2$

ρ_{ph} is the energy density of photons, M is the nuclear mass, Z the nuclear charge, $\mathscr{E}$ is the particle energy, n is the density of electrons in the medium, and χ_0 is the ionization energy.

[a]Reprinted with the permission of Gordon and Breach Science Publishers, Inc. New York, London, and Paris.

The deflection amounts to an acceleration that causes the particle to radiate. Again the loss rates differ for a plasma in which hydrogen is predominantly ionized and for a neutral gas.

(c) Synchrotron and Compton losses (see sections 6:18 and 6:20) are inter-related loss rates, respectively, proportional to the energy density of the magnetic and radiation fields. That these two processes can be considered to be similar can be seen from a simplified argument. Imagine two electromagnetic waves— photons—traveling in exactly opposing directions in such a way that their magnetic field vectors are identical in amplitude and frequency and their electric fields are exactly opposite in amplitude, but again at the same frequency. The electric field and the Poynting vector $\mathbf{S}$ both cancel for these two waves at certain times, and we are left with a pure magnetic field whose energy density is equal to the total energy in the radiation field. At this point synchrotron loss should occur, and this synchrotron loss should then be equivalent to the losses from inverse Compton scattering off the two protons of equivalent energy. The combined expression for these losses is given in Table 9.4.

(d) For protons and nuclei in the cosmic ray field, we again have ionization losses given in Table 9.4. Synchrotron and Compton losses should be less than those of electrons by the ratio of masses taken to the fourth power $\sim 10^{13}$. There also are a variety of interactions between cosmic ray nuclei and the nuclei of the interstellar gaseous medium and grains.

Table 9.5 gives these interactions for several different groups of nuclear particles interacting with an interstellar gas composed of 90% hydrogen and 10% helium by number of atoms.

Table 9.5 Cross Sections, Mean Free Paths Λ, and Absorption Paths λ.[a]

Cosmic Ray Particle	Cross Section for Collision	Λ Mean Free Path	λ Absorption Path
$\mathscr{P}$	3×10^{-26} cm^2	72 g cm^{-2}	— g cm^{-2}
α	11	20	34
Li, Be, B	25	8.7	10
C, N, O, F	31	6.9	7.8
$Z \geq 10$	52	4.2	6.1
Fe	78	2.8	2.8

[a]For cosmic ray particles in different groups of elements, interacting with an interstellar medium which consists of 90% hydrogen and 10% helium (in number density of atoms) (see text). (After Ginzburg Gi69. Reprinted with the permission of Gordon and Breach Science Publishers, Inc. New York, London, and Paris.)

9:10

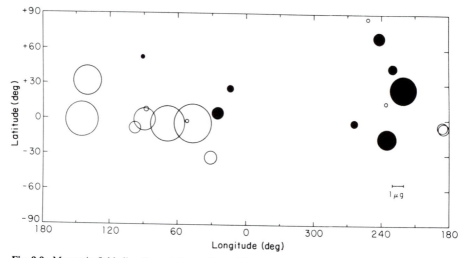

Fig. 9.9 Magnetic field direction at the sun's position, plotted in galactic coordinates. These data actually are mean line-of-sight magnetic field components for pulsars, as judged from their rotation and dispersion measures. For fields greater than 0.3 μgauss the circle diameter is proportional to the field strength. When the field has a direction toward the observer (positive rotation measure), the circles are filled. When they are away from the observer, they are unfilled. The diameter for 1 μgauss is indicated in the figure. The observations are consistent with a relatively uniform field of about 3.5 μgauss directed along the local spiral arm. Note that the directions of greatest field strength are toward longitudes ∼ 60° and ∼ 240°, although there are large variations. These also are roughly the directions of the stellar and gaseous components in the local spiral arm (Bo71). (From Manchester, Ma72b. With the permission of the University of Chicago Press.)

The mean free path Λ gives the distance traveled between nuclear collisions. Essentially a proton travels until it has passed through an effective layer thickness containing 72 g of matter per cm^2. In a cool cloud with density of order 10^{-23} g cm^{-3} this would amount to a distance of order 2 Mpc. Since the cosmic ray particles describe spiral paths in the Galaxy's magnetic field, they traverse such a distance in about 6×10^6 y. Cosmic ray nuclei in the more massive groups suffer collisions more rapidly. Their absorption path length $\lambda = \Lambda/(1 - P_i)$ (where P_i is the probability that the collision will again yield a nucleus belonging to the same initial cosmic ray group) is somewhat longer, as shown in column 4 of Table 9.5.

PROBLEM 9–9.   If the energy loss per collision of a cosmic ray nucleon with a nucleus of the interplanetary medium leads to a loss comparable to the total energy of the nucleon,

$$\left(-\frac{d\mathscr{E}}{dt}\right)_{nucl} = cn\sigma\mathscr{E} \tag{9–67}$$

Fig. 9.10 Proton and alpha particle cosmic ray flux at the earth. At any given energy the proton flux is about 100 times as intense as the electron flux. (Compiled from various sources by P. Meyer, Me69. Reprinted with permission from *Annual Review of Astronomy and Astrophysics*, Volume 7, copyright © 1969 by Annual Reviews, Inc.) (Error bars have been omitted.) At higher energies the flux continues to drop, the flux *J* obeying the power law $dJ/dE \propto E^{-\gamma}$ with $\gamma \sim 2.6$. This appears to be true of all the nuclei of different elements. Their flux energy-curves are quite similar.

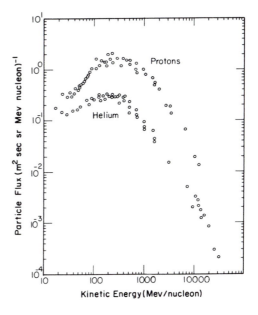

show that this loss dominates the other processes listed in Table 9.4 for cosmic ray nuclei.

PROBLEM 9-10. Using the above loss rate for protons and using the loss rates for cosmic ray electrons having the spectrum shown in Fig. 9.11, calculate roughly how fast the electron and the proton cosmic ray components lose energy and estimate how fast the interstellar medium is being heated by cosmic rays. This cosmic ray heating effect is thought to be important.

The observed flux of cosmic ray protons and alpha particles incident on the earth's atmosphere is shown in Fig. 9.10. Similar data exists for many of the elements. Roughly 90% of the nuclear component of the cosmic ray flux at the top of the atmosphere consists of protons. Alpha particles make up $\sim 9\%$ and the remaining particles are heavier nuclei. Curiously, there is a great excess of Li, Be, B, and He^3 which have a low overall abundance in other cosmic objects. All these constituents are easily destroyed at temperatures existing at the center of stars (section 8:12). We can account for the presence of these elements if they are produced through collisions of carbon, nitrogen, and oxygen cosmic ray particles with hydrogen nuclei in the interstellar medium. The lighter elements then are the *spallation products* of the more massive parent particles. To obtain the amount of these low mass elements observed and also to obtain the correct He^3/He^4 ratio, cosmic ray particles with energies in excess of 1 Gev($= 10^9$ev)

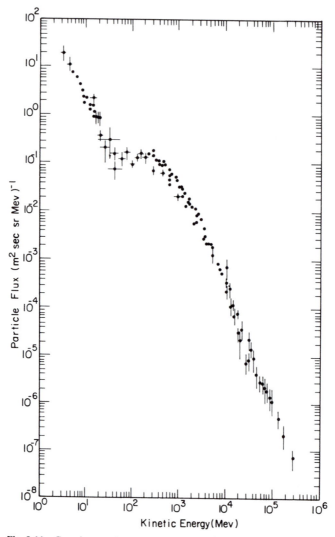

Fig. 9.11 Cosmic ray electron spectrum at the earth. Compiled from various sources by P. Meyer (Me69). Electron and positron abundances are comparable. 1 Gev = 10^9 ev.

would have had to pass through ~ 3 g cm^{-2} of matter (Re68c). This suggests an age of about 2×10^6 y if the particles have been spiraling within the Galaxy all this time. Evidently this represents the mean time taken for cosmic ray particles to diffuse out of the Galactic disk.

9:10

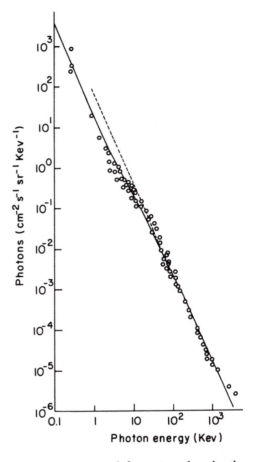

Fig. 9.12 Diffuse X-ray spectrum observed from above the earth's atmosphere. (Compiled from various sources by A.S. Webster and M.S. Longair, We71. With the permission of the Officers and Council of the Royal Astronomical Society.)

We also find that the heavy elements are represented far more abundantly in the cosmic ray flux than in meteorities or in the solar atmosphere. This suggests that these highly energetic particles originate in supernova explosions, pulsars, or white dwarfs, where high concentrations of heavy elements will have been produced during advanced stages of stellar evolution (Co71b).

The cosmic ray flux in the Galaxy appears to be quite steady. Meteorites and lunar surface samples have been analyzed for tracks left by heavy nucleons. The total flux, as well as the relative abundance of heavy nuclei, cannot have drastically changed over the past 5×10^7 y.

Electrons and the somewhat less abundant positrons have fluxes which, at any given energy, amount to about 1% of the proton flux (Figs. 9.10 and 9.11). Interestingly the spectrum of the X-ray flux arriving at the earth (Fig. 9.12) is roughly similar to that of the cosmic ray electrons. This suggests that the X-rays are formed by inverse Compton scattering off electrons by the cosmic millimeter and sub-millimeter radiation flux. However, other mechanisms might also be possible (Po71).

9:11 X-RAY GALAXIES AND QUASARS

A number of galaxies, notably the *Seyfert galaxies* NGC 1275 and NGC 4151, the massive elliptical galaxy and radio source M87, the radio source Centaurus A, and the quasistellar object 3C 273 all are powerful sources of X-ray emission. The X-ray flux from NGC 1275 amounts to 2.4×10^{44} erg sec^{-1} (Gu71), and 3C 273, if at the cosmic distance indicated by its red shift, would emit 1.5×10^{46} erg sec^{-1} (Po71). These objects emit more energy at X-ray frequencies than at all others combined and the flux from NGC 1275 is comparable to the visible flux emitted by normal spirals in the form of starlight.

3C 273 and the two Seyfert galaxies are powerful emitters of infrared energy (K170a) (K170b). It therefore is likely that relativistic electrons in some of these compact sources, inverse-Compton scatter the intense infrared radiation into the X-ray frequency range (We71). Radio emission (Fig. 9.13) would take place by virtue of synchrotron emission given off by the energetic electrons spiraling in their local magnetic field.

The source of infrared emission might be grains thermally radiating energy supplied by the bombardment of relativistic cosmic ray particles.

Whether a given component of the radiation is due to thermal emission, inverse Compton scattering, synchrotron radiation, or some other mechanism, should become apparent when we know more about the brightness fluctuations of these objects in different wavelength ranges. As discussed in section 6:21 some mechanisms of emission have intrinsically faster onset or decay times than others, and the fluctuation rate may therefore permit us to decide which spectral ranges of emission are connected by one and the same emission mechanism and what that mechanism could be.

ADDITIONAL PROBLEMS

9–11. A star composed of hydrogen alone is limited to a maximum brightness. When its luminosity-to-mass ratio exceeds $4 \times 10^4 L_\odot / M_\odot$, its surface layers are blown off. This *Eddington limit* is exceeded in the ejection of *planetary nebula* envelopes. (a) Show that this occurs when the radiative repulsion of electrons exceeds the star's gravitational attraction for protons so that ionized hydrogen no longer is bound. (b) Show that for a star predominantly composed

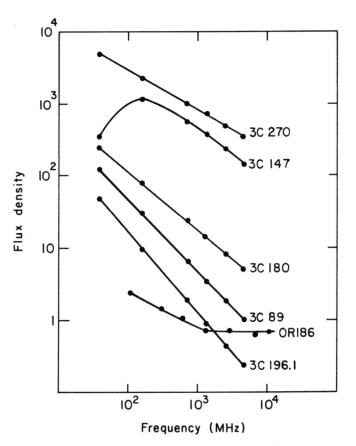

Fig. 9.13 Radio spectra of several extragalactic objects. Note that some have the regular slope discussed in section 6:19. Others show considerable curvature. (Compiled from Kellermann, Pauliny-Toth, and Williams Ke69 and Jauncey, Niell, and Condon, Ja70a). Flux density is measured in units of 10^{-26} wm^{-2} Hz^{-1}. (With the permission of the University of Chicago Press.)

of He^4, C^{12}, O^{16}, or Si^{28}, the luminosity-to-mass ratio can be as high as $8 \times 10^4 L_\odot / M_\odot$ before an outer envelope is blown off.

9–12. The *solar wind* is produced by the high temperature $\sim 2 \times 10^6$°K of the solar corona. The wind velocities, which are ~ 400 km sec^{-1}, are higher than the speed of sound for protons ~ 130 km sec^{-1} for ionized hydrogen at that temperature. Show that this difference can be partly accounted for because

9:11

randomly projected velocities in a confined gas all become projected onto a radial direction if the gas is allowed to freely expand into a much larger volume where collisions between particles no longer are important. A similar process probably plays a role in all stellar winds. Assume equipartition of energy between protons, electrons, and magnetic fields in the corona.

9–13. In a *comet* ionized gas is propelled into a straight tail pointing away from the sun. This seems to happen in the following way. We think that molecules like CO, initially in the comet's head, suffer charge exchange with protons of the solar wind. In this process, which has a very high cross section $\sim 10^{-15}$ cm^2, an electron is transferred from the CO molecule to the proton. The newly formed CO$^+$ ion now is forced to travel along with the magnetic field embedded in the solar wind. The magnetic field is predominantly transverse to the wind direction. Make use of the solar wind data of Table 9.1 to compute the velocity of ions in the comet's tail, if roughly half the protons on any given magnetic line of force undergo charge exchange (Ha62b).

9–14. The *X-ray source* Sco X-1 is believed to be a Galactic object in which plasma with electron density $\sim 10^{16}$ cm^{-3} radiates from a volume of radius $\sim 10^9$ cm at a temperature of $5 \times 10^{7\circ}$K. Several models have been suggested. In one model, gravitational energy of matter falling onto the surface of a white dwarf produces the energy. To account for the fast accretion rate of matter and the high radiation intensity it produces, this matter would have to be falling onto the white dwarf from a binary companion through rapid syphoning of matter onto the white dwarf's surface. Compute the infall rate of matter required and convince yourself that the energies produced in the infall are of the right order of magnitude by comparing gravitational to thermal energies (Pr68). Many other models involving white dwarfs and neutron stars have been proposed. We hope to be able to distinguish between these possibilities by the differing predictions the models make.

ANSWERS TO PROBLEMS

9–1. By (4–117) $c_v = R/(\gamma - 1)$. The ideal gas law states $T = P/R_\rho$. This gives the result (9–23).

9–2. Since $u \gg c$, we can write

$$v_D = \frac{1}{3}\left(-2u + 2u\sqrt{1 + \frac{9c^2}{4u^2}} \right) \sim \frac{2u}{3}\left(-1 + 1 + \frac{9c^2}{8u^2} \right)$$

9–3. (a)

$$\rho_{tot} \sim \frac{4 \times 10^{30}}{\frac{4\pi}{3}(1.5 \times 10^{14})^3} \sim 4 \times 10^{-13} \text{ g cm}^{-3}$$

from Table 1.2

$$\rho_{Fe}/\rho_{tot} \sim 8.9 \times 10^5/4.7 \times 10^{10} \sim 2 \times 10^{-5}; \quad \rho_{Fe} \sim 8 \times 10^{-18} \text{ g cm}^{-3}$$

$$P_{Fe} = \frac{\rho_{Fe}}{m_{Fe}} kT \sim 10^{-11} T, \text{ with } T \sim \left(\frac{L}{16\pi R^2 \sigma}\right)^{1/4} \text{ by (4–78).}$$

For $T \sim 10^3 {}^\circ K$, $P_{Fe} \sim 10^{-8}$ dyn cm$^{-2} \sim P_{Fe,vap}$

Hence iron can condense at distances greater than

$$R \sim \left(\frac{L}{16\pi\sigma T^4}\right)^{1/2} \sim 10^{12} \text{ cm}$$

(b) Carbon will condense ~ 3 times closer to the sun.

(c) The temperature out to 10 AU would have been greater than $T \sim 160^\circ K$. Using the H_2O data in Table 9.3 to derive approximate coefficients A and B for H_2O, we see that the vapor pressure $P_{H_2O} \gtrsim 10^{-7}$ torr $\sim 10^{-4}$ dyn cm^{-2} throughout the nebula. On the other hand, from Table 1.2, assuming all oxygen to be in the form of H_2O,

$$\rho_{H_2O} \sim 1.5 \times 10^{-15} \text{ g cm}^{-3}, \quad P_{H_2O} \gtrsim \frac{\rho_{H_2O}}{m_{H_2O}} kT_{min} \sim 10^{-4} \text{ dyn cm}^{-2}$$

H_2O could therefore have condensed near the edge of the nebula; but since the temperature rises rapidly on approach to the sun, we cannot expect H_2O condensation nearer to the sun than the major planets. We should note, however, that the presence of other substances such as ammonia NH_3 can lower the vapor pressure, since H_2O molecules become more strongly bound.

9–4. For circular motion the balance of forces leads to:

$$r\omega^2 = \frac{3}{4\pi\rho s^3}\left(\frac{4\pi M\rho s^3 G}{3r^2} - \frac{L\pi s^2}{4\pi c r^2}\right)$$

$$\text{velocity} = \omega r = \left[\frac{1}{r}\left(MG - \frac{3 \times 4 \times 10^{34}}{16\pi\rho sc}\right)\right]^{1/2}$$

The velocity difference is

$$\Delta v = \left[\frac{1}{r}\left(MG - \frac{3 \times 4 \times 10^{34}}{32\pi sc}\right)\right]^{1/2} - \left[\frac{1}{r}\left(MG - \frac{3 \times 4 \times 10^{34}}{64\pi sc}\right)\right]^{1/2}$$

for $\rho = 2$ and 4 g cm^{-3}, and $s \sim 10^{-3}$ cm, $\Delta v \sim 10^5$ cm sec^{-1}.

9–5. (a) From Problem 3–7 we see that a body with radius R and density ρ has a total capture cross section equal to twice its geometric cross section if

$$\frac{4\pi}{3}\frac{2G}{V_0^2}\rho R^2 = 1$$

(b) $\quad \dfrac{dM}{dt} = 4\pi\rho R^2 \dfrac{dR}{dt} = \rho_0 V_0 \pi R^2 = \rho_0 V_0 \pi \left[R^2 + \dfrac{2MGR}{V_0^2}\right]$

$$\int \frac{4\rho\, dR}{\rho_0 V_0 \left[1 + \dfrac{8\pi}{3}\dfrac{GR^2}{V_0^2}\right]} = \int dt = \tau$$

Initially $\dfrac{dM}{dt} = 2\rho_0 V_0 \pi R^2 = \dfrac{3}{4}\dfrac{\rho_0}{\rho}\dfrac{V_0^3}{G} \sim 10^{10}$ g sec^{-1}.

(c) For large bodies $\dfrac{dM}{dt} \propto R^4$ or $M^{4/3}$, since $M \propto R^3$.

9–6. Initially $\dfrac{dM}{dt} \sim 4\pi s^2 \dfrac{ds}{dt}\rho = 10^{-2}\rho_0 V_0 s^2$

$$s = 10^{-2}\frac{\rho_0 V_0}{4\pi}t \sim 3 \times 10^{-10}t$$

for a particle with radius s. From Problem 9–5(a) we see that gravitation takes over when $s \sim [3V_0^2/8\pi\rho G]^{1/2} \sim 10^8$ cm, so that gravitation can be neglected for a body with $s \lesssim 10^6$ cm $= 10$ km. The growth time then is

$$t \sim \frac{10^6}{3 \times 10^{-10}} \sim 3 \times 10^{15} \text{ sec} \sim 10^8 \text{ y}.$$

9–7. $n_H R a_v = 1$.

Since most atoms are in the lowest state, we use $a_v \sim 6.3 \times 10^{-18}$ cm^2 so that

$$R \sim \frac{1.6 \times 10^{17}}{n_H} \text{ cm}.$$

(a) For Ly-α, the oscillator strength f is 0.42 so that

$$\sigma \sim \frac{3\lambda^2 f}{2\pi} \sim 3 \times 10^{-11} \text{ cm}^2.$$

This would give an absorption distance

$$\frac{1}{\sigma n_H} \sim \frac{3 \times 10^{10}}{n_H} \text{ cm} \sim 2 \times 10^{-7} \text{ R.}$$

However,

(b) the Doppler shift may make an atom unable to absorb radiation at the central emission frequency of another atom. The mean absorption cross section therefore is $(c\gamma)^2 \sigma / 16\pi^2 \nu^2 \langle v^2 \rangle$ and the mean free path is

$$\sim \frac{16\pi^2 \nu^2 \langle v^2 \rangle}{(c\gamma)^2 \sigma n_H} \sim \frac{3 \times 10^{19}}{n_H} \text{ cm}$$

9–8. For a gas at $T \sim 100°K$, $kT \sim 1.4 \times 10^{-14}$ erg, and the grain that has moment of inertia $I \sim 10^{-26}$ g cm^2 has $\omega \sim 10^6$ rad sec^{-1} to make $kT \sim I\omega^2$. The angular momentum $L \sim I\omega$ therefore has a thermal equilibrium value that also is $L \sim 10^{-20}$ g cm^2 sec^{-1}. This is no coincidence. The "random" angular momentum acquired in $N = M/m$ collisions becomes the "systematic" angular momentum to be altered by the next generation of N collisions. These collisions endow the grain with a random angular momentum of the same magnitude as its initial value, but oriented in some other arbitrary direction.

9–9. Table 9.5 shows a collision mean free path of order 20 g cm^{-2}, which for 10^{-23} g cm^{-3} gives a path of order 2×10^{24} cm and a life $\sim 6 \times 10^{13}$ sec.

$$\therefore \frac{d\mathscr{E}}{dt} \sim 1.6 \times 10^{-14} \mathscr{E} \text{ sec}^{-1}.$$

The loss rates in Table 9.4 are of order $10^{-6} Z^2$ ev sec^{-1}, so that collisional losses dominate for all nuclei with energies in excess of $\sim 10^9$ ev.

9–10. For electrons typical losses from Table 9.4(a) for the most significant part of the energy range covered in Fig. 9.11 are 10^{-6} ev sec^{-1}. At $\sim 10^8$ ev the life is $\sim 10^{14}$ sec. Taken together with the result of Problem 9–9, this indicates a cosmic ray energy loss to the interstellar medium of 10^{-12} erg cm$^{-3}/10^{14}$ sec or $\sim 10^{-26}$ erg cm^{-3} sec^{-1}.

If integrated over a gas-containing volume of 10^{66} cm^3 in the Galaxy, this would indicate an eventual radiation loss of order 10^{40} erg sec^{-1}. The total luminosity of the Galaxy is 10^3 to 10^4 times higher; but only about 10% of this luminosity may contribute to heating the interstellar medium; and in the darkest clouds, where radiation does not readily penetrate, cosmic ray heating may be a dominant factor.

9–11. The radiative repulsion of an electron at distance R from a star is

$$\frac{L\sigma_T}{4\pi R^2 c} \sim 1.8 \times 10^{-36} \frac{L}{R^2}$$

where σ_T is the Thomson cross section (6–102).
The gravitational attraction for a proton is $m_H(MG/R^2)$.
For the sun, this ratio of repulsion to attraction is

$$\frac{\sigma_T L_\odot}{4\pi c M_\odot G m_H} \sim (3 \times 10^4)^{-1}.$$

For a star with $L/M \sim 3 \times 10^4$ greater than the sun, electrons are repelled and pull the protons along. For He^4, C^{12}, and so on each electron pulls a proton plus a neutron, and twice this luminosity-to-mass ratio would be needed.

9–12. The total energy per proton in the corona is $3kT/2$. Because of the magnetic field, protons and electrons will be moving together in the solar wind expansion, and the energy of electron random motion can be transferred into expansion velocity of the protons. The total magnetic energy can also decrease, at the expense of proton velocity, since $B \propto r^{-2}$ and the energy that is proportional to $B^2 r^3$ will be proportional to r^{-1}. The three sources of energy, protons, electrons, and magnetic fields, now supply energy $9kT/2$ for each hydrogenic mass m_H streaming away from the sun. For $T = 2 \times 10^{6\circ}K$, we have $v \sim (9kT/m_H)^{1/2} \sim 4 \times 10^7$ cm sec^{-1}.

9–13. If half the protons undergo charge exchange, then the momentum carried by the others must be shared with an equal number of CO^+ ions. These are 28 times as massive as a proton. Very little momentum is transferred by the proton that exchanges charges so that only the remaining protons supply momentum. A total slowing down of ~ 29 times should occur. For an initial velocity of 400 km sec^{-1} for protons, the final ion velocities would be ~ 14 km sec^{-1}. Actually observed velocities are higher, implying either less complete charge exchange—more protons per ion—or an additional pressure from solar wind protons piling into the more slowly moving plasma, and accelerating it away from the sun through pressure transfer by the magnetic field embedded in the proton–CO^+ plasma.

9–14. The energy of a proton freely falling onto a white dwarf star's surface is $mMG/R \sim 10^{-7}$ erg for $R \sim 10^9$ cm. This is $\sim 10^5$ ev and X-rays up to this energy can therefore be given off by the protons. The actual energy at $5 \times 10^{7\circ}K$ is 5×10^3 ev. The plasma radiates like an optically very thin gas as can be guessed from an extrapolation of Figs. 8.3 and 8.4 and equations 8–64 and 8–65.

10

Structure of the
Universe

10:1 QUESTIONS ABOUT THE UNIVERSE

In preceding chapters we discussed the appearance of stars and stellar systems and we looked in some detail at the immediate surroundings of the sun, the one star to which we have easy access. Now we want to examine the environment in which the sun, the stars, and the stellar systems are embedded. We want to learn about the properties of the universe.

The first questions we would impulsively ask are:

(1) What is the shape of the universe?
(2) How big is it?
(3) How massive is it?
(4) How long has it existed?
(5) What is its chemical composition?

These are some of the simplest and most basic questions we would normally ask about any object and we will find it easy to prescribe conceptually simple observational tests for determining the corresponding properties of the universe. In practice, however, it is very hard to follow these prescriptions, and, for this reason most of the above questions have thus far received only partial answers.

We can approach the problem in two distinct ways. The first is the observational approach. We attempt to observe what the universe is "really" like. The other approach is synthetic. We construct hypothetical models of the universe and see how the observations fit the models. This second procedure might at first glance seem superfluous. It might seem that all we need are observations. That is not so.

Any observation has to be interpreted and can only be interpreted within the framework of theory, even if that theory consists of nothing more than the prejudices that constitute common sense. Common sense itself implies a model. It is three dimensional; time measurements can be completely divorced from distance measurements; bodies obey the Newtonian laws of motion; there are laws of conservation of energy and of momentum. We can attempt to see how far common sense can take us in cosmology. We will find that it is quite useful at times, but it can lead to great misconceptions if applied uncritically.

10:2 ISOTROPY AND HOMOGENEITY OF THE UNIVERSE

If we look out into the universe as far as the best available telescopes allow, we find that no matter which direction we look essentially the same picture presents itself. We find roughly the same kind and number of galaxies at large distances in all directions. There may be statistical variations occasionally, but they appear to be random. The general coloration of galaxies also is the same whichever direction we look. The only systematic color differences we detect are those associated with distance, but the universal red shift of the spectra of distant galaxies does not appear to be dependent on direction.

Strictly speaking, all this is true only when we look at fields of view outside the plane of our own Galaxy. The Milky Way absorbs light so strongly that we always have to make allowances for its presence.

Independence of direction is called *isotropy*. As far as we can tell the universe is isotropic. There are no indications of any preferred directions, except for the flow of time (see section 10:15 below).

Next we take into account all those effects associated with distance. We ask ourselves whether conditions at large distances from us appear to be different from local conditions. Is the universal red shift the only effect we see, or are there other distance dependent factors? If the red shift indeed is the only effect, then we can postulate an expanding model to explain all observations. The red shift is taken to be a Doppler shift caused by the recession of distant galaxies. We imply that if it were not for this velocity produced shift, distant parts of the universe would appear identical with our local environment. In such a model no structural differences exist in different parts of the universe and the universe might be taken to be *homogeneous*.

At this point of the discussion we should stop to reconsider. Our argument is not strictly correct. We have forgotten to take into account the fact that the universe is very large and that we obtain all our information by means of light signals that sometimes take billions of years to reach us. A distant galaxy we view today appears not as it would to a local observer stationed near that galaxy, but rather as it would have looked to such an observer many aeons ago when the galaxy was younger. Therefore, would we not expect that distant galaxies would appear younger and younger, the farther away we looked?

10:2

Not necessarily! And this is the place where theoretical models begin to become important. We have to consider the possible existence of two entirely different models, an *evolving* model and a *steady state* model. These models will have differing histories.

In most evolving models, matter starts out densely confined. At some more or less narrowly defined stage, galaxies are formed. They recede from each other giving rise to a cosmic expansion and to the observed red shift. In this model more distant galaxies should consistently appear younger. Thus far we have not yet stated how the appearance of galaxies changes with age or how to tell an old galaxy from a young one. No one really knows. But since energy is continually being emitted in the form of starlight, we would expect in time some changes to take place in the appearance of galaxies. We conclude that the distant galaxies should at least appear "different" from those nearby. If some such difference, barely suggested by available observation, could be firmly established, we would have strong evidence in support of an evolving universe.

A steady state model takes a different view: The picture presented here is one of a universe that has always existed and will continue to exist. As distant galaxies stream away from us, new matter is created everywhere and the newly created matter gives rise to new galaxies. Through this replenishment the density of matter can always be kept constant. Any depletion due to the cosmic expansion is exactly compensated by the creation of new matter.

In the steady state model two galaxies created at approximately the same time will find themselves receding from each other at ever increasing speeds. As this separation occurs new, younger galaxies are formed in the intervening space. These younger galaxies themselves will recede from each other making room for succeeding generations of galaxies. Any chosen volume of space will, in this way, contain galaxies of varying ages. There will be relatively many young galaxies and decreasingly many old galaxies, because the old galaxies have had time to drift far apart and therefore the number density of old galaxies must be low.

In a steady state universe, the assortment of galaxies in a given volume will be roughly the same at epochs separated by many aeons. There will always be a mixture of young and old galaxies occupying any given volume and the ratio of these galaxies will remain the same. For this reason it does not matter whether we view a distant region today, or several aeons from now. Apart from an increasing red shift, it will always look roughly the same even though the individual galaxies occupying that region will no longer be the same.

The important consequence of this idea is that the view of a distant region observed by us today should, in the steady state model, be roughly the same as the view that a local observer within that region would experience. Consequently the delay introduced by the travel time of light should not affect the age distribution of galaxies observed in a given region of space. That age distribution should be identical with the age distribution an observer sees in the local region around

10:2

him. The steady state theory therefore predicts that distant galaxies generally look just like nearby galaxies; the evolutionary theories, on the contrary, state that there should be differences in the appearance of galaxies at differing distances. Such differences are not firmly established as yet, but this may only be because telescopes cannot yet yield accurate observations of regions so distant that differences would show up.

In principle, we can observationally distinguish between steady state and evolutionary models by looking for age differences in distant galaxies. In practice, however, we do not yet have the means for reliably accomplishing this comparison.

Another test for distinguishing steady state and evolutionary models can also be described. In a steady state universe the mean intergalactic distance never changes from one epoch to another, or from one location to another. In evolutionary models the opposite is true. We usually postulate that most galaxies were formed at a particular epoch in an evolutionary universe. The intergalactic distances should therefore have been closer in the past than they are now. By observing the separation between distant galaxies we would be measuring their separation many aeons ago, at a time when these galaxies were still close together. The number density of galaxies at a large distance should, therefore, be different from the nearby density. Number density counts have been attempted both by visual and radio-astronomical techniques. The latest radio-astronomical results are controversial. They may, or may not, indicate a small change of number density with distance. This is still debated and may not be very significant. A well-established difference in number density would rule out the possibility that the universe can be described by a steady state model.

It is clear that observations on the apparent homogeneity of the universe can be very important; both the numbers of galaxies and the types of galaxies observed in different regions can yield information about the past history and further evolution of the cosmos. But an understanding of the observations can only be achieved by careful study of different cosmological models. The fact that we have never seen matter created and that this process violates common sense is unimportant. Rejection would only be warranted if the required creation rate had actually been ruled out observationally. We will see, however, that a steady state model appears unable to account for the thermal spectrum of the microwave background radiation. The model, therefore, has fallen into disfavor.

10:3 COSMOLOGICAL PRINCIPLE

Some postulates about the universe have to be granted before any theory can be developed. These postulates or axioms must then be shown not only to be internally consistent, but also to be borne out by the postulates' observational consequences.

10:3

One of these postulates is the cosmological principle (Bo52) which can take various forms. The main hypothesis here is that our position in space and time is not unusual. Hence our local physics, and our locally made observations of the universe should not markedly differ from those made by other observers located in different regions of the universe.

The *Perfect Cosmological Principle* states that for any observer located at an arbitrary position, at an arbitrary time in the history of the universe, the cosmos will present exactly the same aspect as that observed by an observer at some other location at the same or even at some completely different epoch. This principle is particularly far reaching and directly leads to the development of the steady state model.

Many cosmologists will not accept this perfect cosmological principle, but favor a more restricted *cosmological principle* stating that at any epoch (suitably defined) observers at different positions in the universe will observe identical cosmic features—except for small local variations.

In a sense all these principles are extensions of the Copernican hypothesis that we should in no way consider ourselves favored observers.

Although the cosmological principle applies only in a statistical sense, since obviously one galaxy looks different from its neighbors, it will nevertheless be very useful when used in conjunction with a number of simple abstract concepts.

The first of these is that of a substratum. The *substratum* in any cosmic model is a matrix of geometrical points all of which move in the idealized way required by the model. Real galaxies will have random velocities measured with respect to the substratum. On the other hand we would expect the mean motion of distant galaxies to be zero as seen by an observer who is at rest with respect to the substratum. We might also expect that the $3°K$ component of the microwave background radiation would appear isotropic to such an observer. A state of rest relative to the substratum can therefore be determined in a number of practical ways. Such a state plays a fundamental role in cosmology and it therefore is useful to call a particle that is at rest with respect to the substratum a *fundamental particle* and an observer who is similarly stationary a *fundamental observer*.

A fundamental observer may have a watch on him. Such a watch, stationary with respect to the substratum, measures *proper time* for the observer. The time measured by locally moving clocks will be different from this time. The proper time of a fundamental observer can be considered to define a *world time* scale that could be used by all fundamental observers for intercomparing measurements. For example, in describing the evolution of a cosmic model we normally think of a *world map* that describes the appearance of the universe at one particular world time. In contrast, we can also think of a *world picture* that is just the aspect the universe presents to a particular fundamental observer at any given time. To see the difference between these concepts, we note that all galaxies are at rest in a world map, but the map may be expanding. On the other hand, in a world picture, distant galaxies would appear to recede from the observer—at least at the present epoch.

10:4 CREATION OF MATTER

The most surprising part of the steady state theory of cosmology is its suggestion that matter is continually being created. It is not being created from something else, the way we might form water from oxygen and hydrogen in the laboratory. This matter is created from nothing!

Or is it? Some theories postulate a new field from which this matter would be created, but thus far this has been done mainly to keep the conservation laws of physics intact. The new field—called the C-field—in that sense is an artifact.

An observer in a steady state universe would expect to see matter created locally and we might wonder whether the rate of creation might be observed directly. We can readily compute what that rate should be. Consider a spherical volume with radius r. This radius is expanding at some rate directly proportional to r. Call this rate Hr.

$$\frac{dr}{dt} = Hr \tag{10-1}$$

The rate at which the volume expands is

$$\frac{d(4\pi r^3/3)}{dt} = 4\pi r^2 \left(\frac{dr}{dt}\right) = 4\pi r^3 H \tag{10-2}$$

If the density of the sphere is to be maintained constant at some value ρ_0 during this expansion, the increased volume must be filled with matter at density ρ_0 so that the rate of matter creation is $4\pi r^3 H \rho_0$ in a sphere of radius r. Dividing by the volume of the sphere, we find the rate of matter creation in unit volume to be $3H\rho_0$. The quantity H is the *Hubble constant*. It is a measure of the expansion of the universe and must be observationally determined. The best present estimate is roughly

$$H = 75 \text{ km sec}^{-1} \text{ Mpc}^{-1} = 2.5 \times 10^{-18} \text{ sec}^{-1} \tag{10-3}$$

This means that an object at a distance of one Mpc (*megaparsec* $= 10^6$ parsec) has a typical recession velocity of 75 km/sec. An object N times this distance has N times that velocity.

An estimate of the value of ρ_0 can be obtained by taking the number density of galaxies (see sections 2:10 and 2:11) and multiplying by a typical galactic mass. The density so obtained is $\sim 10^{-30}$ g cm^{-3}. It is a lower limit to the density since there might exist large amounts of nonluminous unobserved matter in the

10:4

universe. We can now assess the creation rate as

$$3H\rho_0 \sim 10^{-47} \text{ g cm}^{-3} \text{ sec}^{-1} \tag{10-4}$$

If matter is created in the form of hydrogen, this would imply a creation rate of the order of one atom in each one liter volume, every five aeons. At the moment there is no way of measuring such small creation rates.

10:5 HOMOGENEOUS ISOTROPIC MODELS OF THE UNIVERSE

Observations made to date do not indicate that there are any preferred directions or unusually dense regions in the universe. The data are compatible with a homogeneous, isotropic model of the universe, that is, a universe in which there are no select locations or directions.

An observer placed at any location in the universe would see distant galaxies red shifted, in apparent recession, no matter what direction he chose to observe.

In order to construct a model of such a universe, we assume that the red shift indeed indicates a genuine expansion. This assumption has become entrenched in cosmology, primarily through default. When the red shift of distant galaxies was first discovered, a number of explanations were advanced. One by one the competing hypotheses have been eliminated—found incompatible with observations, or unlikely on other grounds. The velocity-induced red shift has been the only hypothesis that could not be discarded. It has survived and may well be the genuine source of the red shift. Still, from time to time the search for alternate explanations continues and probably will continue until the recession hypothesis can be established on a firmer basis.

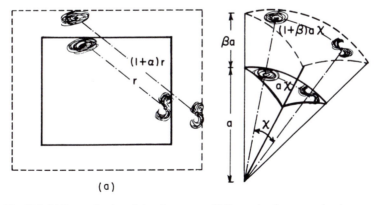

Fig. 10.1 (*a*) Expansion in a flat universe.   (*b*) Expansion in a curved universe.

It is important to have some means of visualizing a model of the universe in which an observer at each point sees all other observers in distant galaxies receding from him. A simple model in two dimensions consists of a rubber sheet (Fig. 10.1). Let spots be painted on this sheet in some random manner. If the sheet is now stretched in length L and width W, by fixed amounts of αL and αW, respectively, then all distances are increased by a fractional amount α. If the spots on the sheet represent galaxies, then a galaxy that initially was at some distance $r = (x^2 + y^2)^{1/2}$ from a given galaxy, will later be at distance $(1 + \alpha) r = \{[(1 + \alpha) x]^2 + [(1 + \alpha) y]^2\}^{1/2}$ where x and y are the components of the separation along the L and W directions, respectively.

A flat rubber sheet is not the only two-dimensional model for an expanding homogeneous isotropic universe. Take a rubber balloon and paint spots on it to represent galaxies. At a given instant let the radius of the balloon be a. Let the angle subtended by two galaxies be χ at the center of the balloon. The distance between galaxies measured along the surface is the arc length $a\chi$. If the balloon is expanded the angle χ remains constant, but the radius increases to some new value say $a' = (1 + \beta) a$, where β is the fractional increase in the radius. The distance between galaxies now is $(1 + \beta) a\chi$, and we note that the fractional increase is independent of χ. This means that if the universe is homogeneous and isotropic at a given instant, an isotropic expansion will keep it that way.

If the time rate of change of β is $\dot\beta$, the recession velocity between the two galaxies is $a\dot\beta\chi$, which increases linearly with angle χ. Dividing the velocity of separation by the distance, we obtain the ratio $a\dot\beta\chi/a\chi = \dot\beta$. We talk about a linear distance-velocity relation because, for increasing separation, the recession velocity increases in proportion to the separating distance.

In section 2:10 and Fig. 2.4, we saw that distant galaxies and clusters of galaxies obey such a relation—at least approximately.

A sphere of radius a is described by the equation

$$x_1^2 + x_2^2 + x_3^2 = a^2 \tag{10-5}$$

where x_1, x_2, x_3 are three mutually orthogonal Cartesian coordinates. An element of length dl on the sphere is given by

$$dl^2 = dx_1^2 + dx_2^2 + dx_3^2 \tag{10-6}$$

Eliminating the coordinate x_3 by means of equation 10-5, we find

$$dl^2 = dx_1^2 + dx_2^2 + \frac{(x_1 \, dx_1 + x_2 \, dx_2)^2}{a^2 - x_1^2 - x_2^2} \tag{10-7}$$

In terms of spherical polar coordinates, we can write dl^2 as

$$dl^2 = a^2 (d\theta^2 + \sin^2\theta \, d\phi^2) \tag{10-8}$$

We can repeat this procedure for a four-dimensional sphere in an exactly analogous way. Here we do not deal with a two-dimensional surface or a three-dimensional space. Rather, we work with a space showing isotropy and homogeneity in three dimensions; and analogously to the three-dimensional approach of equations 10–5 to 10–8, we want to investigate the properties of a three-dimensional *hypersurface* on a four-dimensional *hypersphere*. Problem 10–1 will show that the relation corresponding to equation 10–8 then has the form

$$dl^2 = a^2[d\chi^2 + \sin^2\chi(\sin^2\theta \, d\phi^2 + d\theta^2)] \tag{10–9}$$

PROBLEM 10–1. Show how relation (10–9) is obtained by starting with an equation for a hypersphere

$$x_1^2 + x_2^2 + x_3^2 + x_4^2 = a^2 \tag{10–10}$$

Continue by showing that in terms of three-dimensional polar coordinates, we have

$$dl^2 = dr^2 + r^2 \, d\theta^2 + r^2 \sin^2\theta \, d\phi^2 + \frac{(r \, dr)^2}{a^2 - r^2} \tag{10–11}$$

Then substitute the new variable

$$r = a \sin \chi \tag{10–12}$$

Consider a sphere of radius R in a conventional three-dimensional space. On a two-dimensional spherical surface the distance along the sphere is given by $R\theta$. A circle about $\theta = 0$ on this surface has length $2\pi R \sin \theta$. At increasing distance from the origin, the size of the circle increases to a maximum value $2\pi R$ at a distance $\pi R/2$. After that it decreases and shrinks to a geometric point at the *antipodal position*—at distance πR.

PROBLEM 10–2. Show that on a four-dimensional hypersphere
  (i) the ratio of the circumference of a circle to its radius is less than 2π;
  (ii) the surface area of a sphere is

$$S = 4\pi a^2 \sin^2\chi \tag{10–13}$$

(iii) As the angle χ increases the sphere grows and the surface of the sphere increases reaching a maximum value $4\pi a^2$ at distance $\pi a/2$ before shrinking to a point at distance πa. Show, moreover, that the element (10–9) defines the total volume

$$V = \int_0^{2\pi} \int_0^{\pi} \int_0^{\pi} a^3 \sin^2 \chi \sin \theta \, d\chi \, d\theta \, d\phi \tag{10–14}$$

10:5

so that

$$V = 2\pi^2 a^3 \qquad (10\text{–}15)$$

We can choose a parameter λ

$$\lambda = \frac{1}{a^2} \qquad (10\text{–}16)$$

which defines the *curvature* properties of a space. When the *radius of curvature* is infinite, $\lambda = 0$ and the space has zero curvature. Such a space is called *flat* or *Euclidean*. When $\lambda > 0$, we talk of a space of positive curvature. We can also define spaces of negative curvature for which $\lambda < 0$ if, as in (10–17) below, we replace the right side of (10–10) by $-a^2$. Note that the two two-dimensional universes described above have differing curvature constants. The sheet model is Euclidean. The balloon model has positive curvature.

PROBLEM 10–3.   In a space of negative curvature—a *hyperbolic space*, sometimes called a *pseudospherical space*:

$$x_1^2 + x_2^2 + x_3^2 + x_4^2 = -a^2 \qquad (10\text{–}17)$$

where a is real.
   Show that
   (i)

$$dl^2 = r^2 (\sin^2\theta \, d\phi^2 + d\theta^2) + (1 + r^2/a^2)^{-1} dr^2 \qquad (10\text{–}18)$$

where r can have values from 0 to ∞.
   (ii) Defining $r = a \sinh \chi$ (where χ goes from 0 to ∞).

$$dl^2 = a^2 \{ d\chi^2 + \sinh^2\chi (\sin^2\theta \, d\phi^2 + d\theta^2) \} \qquad (10\text{–}19)$$

Show that the ratio of the circumference of a circle to its radius is greater than 2π.
   (iii) Show that the surface of a sphere is

$$S = 4\pi a^2 \sinh^2\chi \qquad (10\text{–}20)$$

which increases without limit.
   (iv) The volume of the space is

$$V = \int_0^{2\pi} \int_0^\pi \int_0^\infty a^3 \sinh^2\chi \sin\theta \, d\chi \, d\theta \, d\phi \qquad (10\text{–}21)$$

which is infinite.

10:5

To summarize, we note that a space of positive curvature has a finite volume and is closed. Increasing χ beyond a value π returns us to a region already defined by χ values between 0 and π. The space of negative curvature is *open*. The volume of a closed space is finite and given by equation 10–15. The volume of an open space is infinite.

A steady state universe can only exist in a flat space. This has to be so because in a curved expanding space the radius of curvature would be constantly changing and that means that the number of galaxies observed at different distances would continually be changing and would, in fact, be a measure of the evolutionary state (age) of the universe. This does not mean that creation of matter might not be possible in a curved universe; it does mean, however, that such a universe would be observably evolving.

We often see the curvature of cosmological models described by a constant k that can assume values $+1, 0,$ or -1. k is the *Riemann curvature constant* and the values respectively describe universes of positive, zero, and negative curvature. k denotes the algebraic sign of the parameter λ of equation 10–16.

In our balloon model of a universe, a galaxy close to an observer subtends a large angular diameter. At larger distances galaxies of the same size subtend smaller and smaller angular diameters reaching a minimum value when their distance is $\pi a/2$, a is the radius of curvature of the balloon. After that the angular diameter once again increases until it reaches a maximum value of 2π when seen at the antipodal point of the balloon, that is, at a distance πa. An observer can then look in any direction he pleases and see that particular galaxy at one and the same distance from him. Figure 10.2 illustrates these effects.

These effects are found on three-dimensional hypersurfaces in exact analogy. Observations have been made to detect whether there exists a minimum angular diameter for radio-astronomical sources. Thus far these observations have not been successful because quasars intrinsically have minute radii and available techniques are not yet well enough developed to measure these small angles accurately.

10:6 MEASURING THE GEOMETRIC PROPERTIES OF THE UNIVERSE

It is possible, at least in principle, to determine the size and curvature of the universe on the basis of astronomical observations (Ro55)*, (Ro68)*. The simplest quantitative relationship between directly observed quantities and the more abstract geometrical properties of the universe have been derived for models in which complete homogeneity and isotropy pervade. These relationships will be described here; their greatest asset is an independence of the specific dynamics— for example, general relativity—used to describe the evolutionary properties

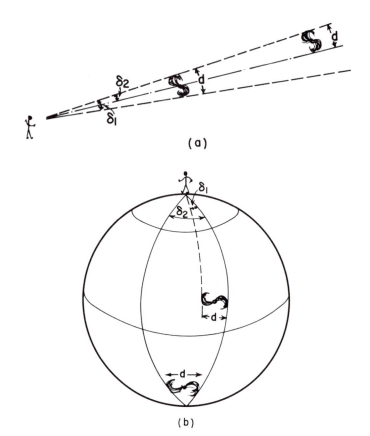

Fig. 10.2 (*a*) Distance–angular-diameter relation in a flat space. (*b*) Distance–angular-diameter relation on the surface of a three-sphere.

of the cosmological models. In effect, we can obtain a geometric description of the universe as it appears at the present world time, without being able to—or, for that matter, needing to—make any statements about how the universe evolved before reaching this state or how it will evolve in the future. At our present stage of understanding, this restricted approach is useful. We do not yet know the size or sense of curvature of the universe; and it is welcome that we could at least obtain that much information, without the additional complication of first needing to know the dynamic laws describing cosmic evolution.

One can show on the basis of group theoretical arguments (Ro33) (Wa34) that the most general metric describing homogeneous isotropic spaces is

$$ds^2 = c^2 \, dt^2 - dl^2 \qquad (10\text{--}22)$$

10:6

the Robertson–Walker metric, for which

$$dl^2 = a^2(t)\{d\chi^2 + \sigma^2(\chi)[d\theta^2 + \sin^2\theta\, d\phi^2]\} \qquad (10\text{-}23)$$

where dl^2 is the metric of a homogeneous, isotropic three-dimensional space. The function $\sigma(\chi)$ has the form $\sin\chi$, χ or $\sinh\chi$, depending on whether the *Riemannian curvature* of the three-space is $k = 1, 0,$ or -1.

In this notation:

(a) the world line of a stationary galaxy is a curve, with χ, θ, and ϕ constant, along which ds measures the world time interval dt;

(b) the world line of any light signal is a *null-geodesic*, meaning that it is characterized by $ds = 0$.

(c) If we choose a specific world time—t = constant, $dt = 0$—we can measure spatial distances within the universe with the aid of the metric $-ds^2$. The curvature of the universe k/a^2 will then be fully determined if we can ascertain the value of k and of $a(t)$. To this end, consider a world diagram representing an observer O located at $(\chi, \theta, \phi) = (0, 0, 0)$ and a galaxy at $(\chi_0, \theta_0, \phi_0)$ = constant. As $a(t)$ changes with time, a constant interval dx, in the auxiliary three-space (10–23) will lead to a changing value of $ds^2 - c^2\, dt^2$. In particular, for a light beam traveling from t_1 to t_0 (Fig. 10.3) we can set $ds = 0$ and then the integration of equation 10.22 along a fixed line of sight (θ, ϕ) yields

$$\int_{t_1}^{t_0} \frac{c\, dt}{a(t)} = \chi \qquad (10\text{-}24)$$

This is the relation between *distance parameter* χ and emission time t_1 as measured by observer O at time t_0. We will keep referring to a *distance parameter* here in-

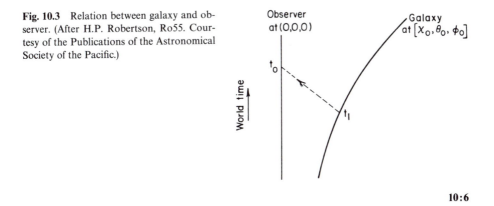

Fig. 10.3 Relation between galaxy and observer. (After H.P. Robertson, Ro55. Courtesy of the Publications of the Astronomical Society of the Pacific.)

10:6

stead of a *distance*. The reason is that it is not quite clear just what we would like to call "distance." Perhaps the *cosmic distance* (see equation 10–12)

$$r = a(t_0)\,\sigma(\chi) \tag{10-25}$$

is a useful measure. This represents the distance in the auxiliary three-space at the world time of observation, t_0.

There are other alternatives. We will see that the apparent luminosity of distant objects makes the quantity $a(t_0)\,\sigma(\chi)\,(1 + z)$ a useful measure of distance. Here z is a measure of the red shift defined by equation 10–26 below and the *luminosity* of a galaxy is derived further on in equation 10–32. This discussion is brought up at the present stage only because it is annoying not to have an exact analogue to all the concepts we normally like to attribute to distance; but in the more general mathematical spaces that are useful in cosmology we cannot expect to have all these properties embodied in a single parameter.

If the light emitted in the interval t_1 to $t_1 + dt_1$ is received between times t_0 and $t_0 + dt_0$, the increment dt_0 can be evaluated either by differentiating equation 10–24 or equivalently by setting $ds = 0$ and replacing $d\chi$ by the constant distance parameter χ in equation 10–22. If the frequency of the emitted signal is v_1 and that of the received signal v_0 then

$$v_0\,dt_0 = v_1\,dt_1$$

because the total number of oscillations in the wave is conserved during propagation. In terms of wavelength $\lambda = c/v$, we have

$$z \equiv \frac{\lambda_0 - \lambda_1}{\lambda_1} = \frac{dt_0}{dt_1} - 1 = \frac{a(t_0)}{a(t_1)} - 1 \tag{10-26}$$

Equation 10–26 defines the measured *red shift parameter* z for radiation emitted at time t_1.

PROBLEM 10–4. Equation 10–26 is not yet in a useful form because we do not know how $a(t)$ varies with time. However, if we make the assumption that $a(t)$ varies regularly, so that a Taylor expansion may be used to determine $a(t_1)$ in terms of derivatives of $a(t_0)$, show that

$$z = \frac{\dot{a}_0}{a_0}(t_0 - t_1) + \left[\left(\frac{\dot{a}_0}{a_0}\right)^2 - \frac{1}{2}\frac{\ddot{a}_0}{a_0} \right](t_0 - t_1)^2 + \dots \tag{10-27}$$

where $\dot{a}_0$ and $\ddot{a}_0$ are the first and second time derivatives of $a(t)$ evaluated at t_0, the time of observation.

We can next ask about the angular diameter δ subtended by a galaxy as observed by O. If the intrinsic, locally measured diameter of the galaxy is D, we can define a parameter distance $d\chi$ across it such that $D = a_1 \, d\chi$. But at world time t_1, when the center of the galaxy has a fixed position (χ, θ, ϕ), equation 10–23 shows that the total parameter length of a circle, drawn transverse to the line of sight and traversing the galaxy's major diameter is $2\pi a_1 \sigma(\chi)$. This length corresponds to the full range of values that θ and ϕ can take. It follows that the linear diameter of the galaxy amounts to a segment $D/2\pi\sigma(\chi) \, a_1$ of a full circle, and that the angular diameter therefore is

$$\delta = \frac{D}{a_1 \sigma(\chi)} \tag{10–28}$$

To convert this into values of $a(t)$ measured at the present epoch, we can still make use of equation 10–26 to obtain

$$\delta = \frac{(z + 1) D}{a_0 \sigma(\chi)} \tag{10–29}$$

The second relation of interest to observational cosmology is the dependence on χ of the number of galaxies $N(\chi)$ whose parametric distance is less than a given value χ, or equal to it. If n is the density of galaxies in the auxiliary three-space defined by the metric dl^2, (10–23), then n is independent of t if evolutionary effects are neglected: in a homogeneous model it is also independent of χ:

$$N(\chi) = 4\pi n \int_0^\chi \sigma^2(\chi) \, d\chi \tag{10–30}$$

where the more general function σ now represents the functions sin and sinh that held for the special case $k = +1$ and $k = -1$ in relations (10–13) and (10–20). Fig. 10.5 illustrates these ideas.

PROBLEM 10-5. Show that (10–30) can be expanded into the series relation

$$N(\chi) = \frac{4\pi n}{3} \chi^3 \left(1 - \frac{k}{5} \chi^2 + \dots \right) \tag{10–31}$$

Both relations (10–29) and (10–31) depend on a knowledge of z, if observations are to be interpreted. However z often is hard to measure at great distances because galaxies there are quite faint. We might therefore prefer to deal with the total observed flux, a readily determined quantity, rather than with the red shift parameters. To do this, we need to know more about the apparent luminosity of distant galaxies as seen by an observer O at our epoch. To determine that we have to make the further assumption that photons are conserved and the energy is related to frequency by the Planck expression $\mathscr{E} = h\nu$ with h a universal constant independent of world time.

If L_1 is the bolometric luminosity of the galaxy at the time of emission, then the apparent bolometric flux $\mathscr{F}$ observed by O is

$$\mathscr{F} = \frac{L_1}{4\pi a_0^2 \sigma^2(\chi)} \cdot \frac{1}{(1+z)^2} \tag{10-32}$$

Here the first term represents the geometrical dilution of radiation, since $4\pi a_0^2 \sigma^2$ is the surface area of an auxiliary three-space drawn about the emitting galaxy at the distance of the observer (equations 10–13 and 10–20). The second term represents the reddening. The term $(1 + z)$ appears squared. One such factor is just due to the decrease in spectral frequency and, hence, the decrease in energy per photon as viewed by the observer. The second factor enters because all conceivable frequencies are diminished, including the rate at which photons emitted by the galaxy arrive at the observer. In unit time interval the observer sees $(1 + z)$ fewer photons than were emitted at the galaxy in unit time. This corresponds to a general slowing down of clocks and again means that less energy arrives at O in unit time.

We now proceed to write the angular diameter relation (10–29) purely in terms of observable quantities. If the flux $\mathscr{F}$ in equation 10–32 is divided by the square of the angular diameter δ (10–29), we obtain

$$\log \mathscr{F} = 2 \log \delta - 4 \log (1 + z) + \log \frac{L_1}{4\pi D^2} \tag{10-33}$$

Observations normally are carried out in a fixed spectral frequency range $\Delta\nu$ in which the radiation detector is sensitive. Let $\Delta\mathscr{F}$ be the rate of reception of photons at O in this spectral range. ΔL_1 is the comparable emission rate for photons in the emission frequency range $\Delta\nu_1$ which, when red shifted, becomes superposed on the received spectral band $\Delta\nu$. Then

$$\log \Delta\mathscr{F} = 2 \log \delta - 3 \log (1 + z) + \log \frac{\Delta L_1}{4\pi D^2} \tag{10-34}$$

We now return to equation 10–32 and convert it into a relation between bolometric magnitudes. We can write

$$m = M_1 + 5 \log \left[\sigma(\chi)(1 + z) \frac{a_0}{10} \right] \tag{10-35}$$

10:6

where a_0 is now measured in parsecs and the division by 10 is necessitated by the fact that absolute magnitudes always refer to the apparent magnitude of an object at a distance of 10 pc. In terms of present values, we replace the magnitude M_1 by its expanded form

$$M_1 = M_0 - \dot{M}_0(t_0 - t_1) + \dots \qquad (10\text{-}36)$$

PROBLEM 10-6. Show that if we further expand in powers of z and χ, relation (10-35) can be written as

$$m = M_0 - 45.06 + 5\log\left(\frac{a_0 z}{\dot{a}_0}\right) + 1.086\left(1 + \frac{a_0 \ddot{a}_0}{\dot{a}_0^2} - 2\mu\right)z + \dots \qquad (10\text{-}37)$$

where

$$\mu = 0.46 \, \frac{\dot{M}_0 a_0}{\dot{a}_0} \qquad (10\text{-}38)$$

is a measure of the change of the magnitude of the galaxy.

The quantity

$$q_0 \equiv -\frac{a_0 \ddot{a}_0}{\dot{a}_0^2} \qquad (10\text{-}39)$$

often appears in cosmology. In equation 10-37 the logarithmic term gives rise to the linear red-shift–distance relation. The Hubble constant

$$H \equiv \frac{\dot{a}_0}{a_0} \qquad (10\text{-}40)$$

which is not constant if $\ddot{a}_0 \neq 0$, allows us to write (equations 10-25, 10-27)

$$cz = a_0 H\chi \qquad (10\text{-}41)$$

in the first approximation, with the consequence that cz can be thought of as the linear velocity of the galaxy and $a_0\chi$ as its distance. If such a linear velocity relation had held throughout the past, the universe would have had to start from a point origin a time $1/H$ ago.

q_0 is sometimes called the *deceleration parameter*. Its observational value is very uncertain. The data do not permit a differentiation between cosmic models

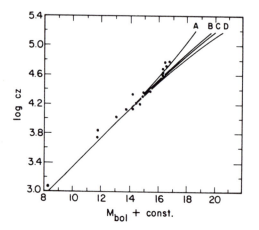

Fig. 10.4 The red-shift–magnitude relation (after Hoyle and Sandage Ho56). Curve A corresponds to $q_0 \sim 2.5$. Curve B is $k = 0$, $q_0 = \frac{1}{2}$. Curve C represents the steady state model (see also Figure. 2.4). (Courtesy of the Publications of the Astronomical Society of the Pacific.)

having distinct q_0 values. We can tabulate the q_0 values in terms of the curvature k of space.

PROBLEM 10-7.   From the definition of q_0, and the fact that in steady state cosmologies $a(t)$ is proportional to e^{tH}, show that for a steady state cosmos, $q_0 = -1$.

Table 10.1.

	k	q_0
For exploding models with zero cosmological constant and pressure (see section 10:9 below)	$\begin{cases} +1 \\ 0 \\ -1 \end{cases}$	$> 1/2$ $= 1/2$ $0 \leq q_0 < 1/2$
Steady state model	0	$q = -1$

The observational data are plotted in Fig. 10.4 (see also Fig. 2.4). No clear distinction between models is possible.

With the information about the apparent magnitude of galaxies, contained in (10–37), we can return to expression 10–31 in an attempt to determine the number of galaxies that would be observed at magnitudes out to a given apparent magnitude m. On replacement of distances by magnitudes in expression 10–31 one can obtain (Ro55)

$$\frac{1}{N}\frac{dN}{dm} = 1.382\left\{1 - (1 - \mu)z + \left[\frac{3}{2} + \frac{kc^2}{5\dot{a}_0^2} + \frac{a_0\ddot{a}_0}{2\dot{a}_0^2}(1 + \mu)\right.\right.$$

$$\left.\left. - \frac{7}{2}\mu + \mu^2 - K\right]z^2 + \ldots\right\} = 0.4\frac{d(\log N)}{d(\log S)}$$

(10–42)

where the expansion is in terms of the red shift z and the meaning of log S is discussed below. The linear term in z depends only on the rate of change of the absolute magnitude of galaxies μ, while the quadratic term depends in addition on the term

$$K \equiv 0.46\frac{\ddot{M}_0}{H^2}$$

(10–43)

In radio astronomy we often talk (section 2:11) about the logarithm of the local flux S obtained from a galaxy, and equation 10–42 then leads to values of the slope $d\log N/d\log S$ characteristic of individual models on a log S-log N plot (Fig. 2.7). Since log S is $(2.5)^{-1}$ times a "radio magnitude" for a radio source, we see that the right side of (10–42) could be written as $0.4\, d(\log N)/d(\log S)$ with K and μ now interpreted as evolutionary parameters in the radio frequency domain.

The relations (10–34), (10–37), and (10–42) all suffer from a common drawback in that separate information such as that inherent in expressions 10–38 or 10–43 must be obtained before meaningful cosmological conclusions can be reached. This problem would not be severe were it not for the fact that in most of the cosmological models considered today, curvature effects of the universe, such as those illustrated in Fig 10.5, apparently occur, if at all, only at such large distances that galaxies, and the stars inside them, presumably evolve very significantly during the time their signals take to reach us. How well we can then define the time derivative $\dot{M}_0$ or $\ddot{M}_0$ is not at all clear. We now know that galaxies often suffer catastrophic structural changes as evidenced, say, by the explosion of material from the nucleus of the galaxy M82 or the extremely powerful radio "jet" of the giant spherical galaxy M87. For many quasars and violently active galactic nuclei, even greater variations in brightness can be expected, and the flux may be emitted in vastly differing spectral ranges at different epochs during their evolution. We

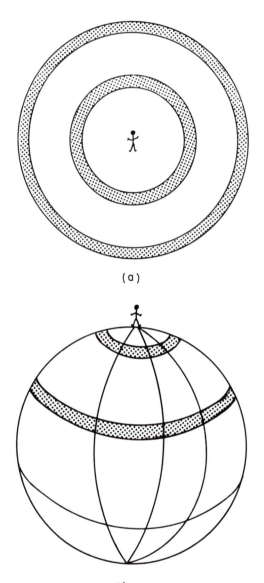

(a)

(b)

Fig. 10.5   (a) Distance–number-count relation in a flat space. (b) Distance–number-count on a spherical surface. Because the circle or surface drawn about the observer at any given distance is always smaller than the corresponding circle or surface in a flat space, the number of galaxies counted at any distance in a spherical universe will also be less than the number counted in a flat or Euclidean universe.

10:6

must bear all this in mind in considering the observational results obtained thus far. In each case strong assumptions about the rate of galactic evolution underlie the correction factors that have to be inserted into the equations derived.

The only exception to this requirement is the steady state model where we can simply assume that locally observed conditions at the observing epoch are and have been characteristic of all portions of the universe for all time. Even here, however, we still have the additional difficulty that observations are carried out in a different spectral range from that at which radiation is emitted by distant objects. For observations in the visual part of the spectrum, we would therefore need to know a great deal about the ultraviolet behavior of sources at least in the observer's vicinity. That information is just becoming available from observations which now can be made through use of orbiting astronomical observatories. It therefore appears that much of the cosmological interpretation of observational data may have to await a more complete understanding of the ways in which local extragalactic objects emit radiation in wavelength ranges not carefully studied to date.

10:7 TOPOLOGY OF THE UNIVERSE

Thus far we have assumed that the universe is *simply connected*—that it has the simplest *topological* structure. In the two-dimensional models, we have talked about spherical surfaces, or planes, or hyperbolical surfaces of negative curvature.

There exist more complicated surfaces some of which can be easily constructed. If we take a rectangular sheet and label the four edges a, b, c, d, as shown in Fig. 10.6(a) we can, first, obtain a cylindrical surface by joining edges a and b.

But there really are several ways of joining a and b. We can give the sheet a twist, as indicated by the arrows, and obtain a Møbius strip as in Fig. 10.6(b).

If edges a and b are joined and edges c and d are joined also, we can obtain a torus or a Klein bottle depending on whether the sheet is given no twist or one twist—Fig. 10.6(c).

The Møbius strip and Klein bottle are *re-entrant*. Starting on the exterior side of a surface, we are able to reappear at the starting point, but on the interior of the surface, without ever crossing an edge or perforating the surface. There is, however, a change in directions. An arrow pointing in a particular direction appears reversed when it returns to its starting point (on the opposite surface of the strip).

These models have by no means been investigated thoroughly. There are many peculiarities. In some re-entrant models, a right-handed glove reappears as a left-handed glove on reaching its starting point after a traversal of the universe.

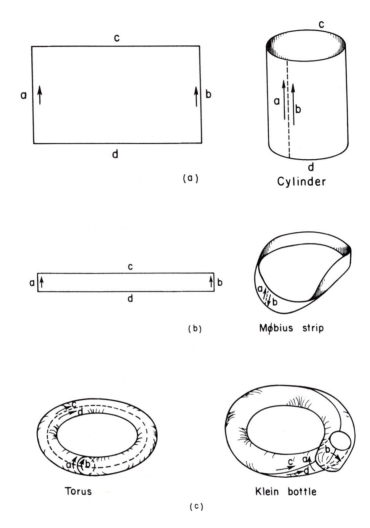

Fig. 10.6

In other models an observer may return to his own past. In still others he might return to his starting point with his arrow of time reversed with respect to his surroundings. Some spaces of negative curvature are not open when complex topological forms are allowed. Clearly much remains to be done in the topological study of the universe (He62).

10:7

10:8 DYNAMICS ON A COSMIC SCALE

In Chapter 3 the mass of galaxies was estimated using the Newtonian laws of physics. The masses of clusters of galaxies computed in this way are roughly consistent with their masses estimated on the basis of brightness. On this scale of events Newtonian dynamics therefore is probably not in error by as much as an order of magnitude. However, there are some features of Newtonian theory that lead to difficulties for phenomena on a large scale:

(a) The propagation time of gravitational signals becomes large so that forces no longer can be assumed to be transmitted instantaneously. The laws of motion should be modified to take this time lag into account.

(b) The laws of special relativity, which are well verified in the laboratory, should hold locally throughout the universe. Again, this feature is not incorporated in Newtonian dynamics and should be.

These and other difficulties are eliminated when general relativistic field equations are used to replace Newton's laws of motion. That does not mean however that general relativity itself might not have drawbacks in treating phenomena on a very large scale. General relativity has been tested out to a scale no larger than the solar system $[o(10^{13} \text{ cm})]$ and it is not clear that the same laws hold on the scale of the universe $o(10^{28})$ cm. Few laws of physics span such large ranges.

General relativity may also run into difficulties in very compact exploding objects—like the universe itself when it was only $\sim 10^{-23}$ sec old—or compact imploding objects like black holes. Bahcall and Frautschi (Ba71b) have pointed out that velocity differences of the order of the speed of light would exist, in such dense exploding or imploding states, across distances of 10^{-13} cm, the size of an elementary particle. This shows that quantum effects should be incorporated into a theory that deals with highly compact states.

10:9 SOME SIMPLE MODELS OF THE UNIVERSE

If we understood the proper dynamics to be used in dealing with phenomena on the scale of the entire universe, we could describe the evolution of different models and give their histories. Despite the fact that the problem of dynamics has not yet been finally solved, we can make the best use of provisional information in order to at least enumerate a number of theoretical models of evolving universes. We can then fit the observed characteristics of the universe to such models and an effort can be made to reject those that do not fit observations.

10:9

(a) Steady State Universe

First described by Bondi and Gold (Bo48) and by Hoyle (Ho48), this universe is flat and provides the same aspect at all times and in all places. The expansion rate is uniform in space and time. Old galaxies and young ones are statistically distributed in some fixed ratio at all distances from an observer.

(b) Static Universe of Einstein

Before the expansion of the universe had been discovered, Einstein (Ei17) proposed a cosmic model based on his general relativistic field equations. This model was static (not expanding). General relativistic dynamics allowed Einstein to calculate the density of such a universe, in terms of its radius, since a static situation will arise for one density value only, if the pressure in the universe is taken to be negligibly small. In fact, most relativistic models assume the pressure to be negligible for dynamic purposes. Observations allow this assumption, and the calculations are greatly simplified if pressure can be neglected. The Einstein universe is spherical ($k = 1$) and has constant radius of curvature a (Fig. 10.7).

In 1930, Lemaitre and Eddington discovered that the Einstein Universe is unstable (Ed30). A small deviation from the perfect conditions postulated by Einstein would result in either a continuing expansion or else an accelerating collapse. They set up a model invoking this instability. It is particularly interesting because galaxies might be able to form at the unstable stage.

(c) De Sitter Model

Shortly after Einstein proposed the static model in 1917, de Sitter (deS17) pointed out that the general relativistic field equations permitted the description of a second model, one that existed in a flat space, $k = 0$. This was an expanding model. At first it had no more than academic interest; but in the late 1920s, Hubble's discovery (Hu29) of the cosmic expansion revived interest in the model. Its main drawback is that the density of such a universe must be zero. However cosmic densities are low anyway and this was not considered to be an overriding difficulty.

(d) Eddington Model

This model of a universe starts out in an Einstein state. It then becomes perturbed by processes involving the formation of the galaxies from an initially uniform distribution of gas, and goes over into a uniform expansion. One difficulty with

10:9

it is the uncertainty whether the formation of galaxies should not lead to an instability that would give rise to contraction rather than expansion. The model is interesting because it focused attention on the fact that cosmology is not just a matter of geometry. A model must also be able to account for the physical state of matter found in the universe. Galaxies are likely to have condensed out of an initially uniform gas: If this gas was in rapid expansion, how was it possible to counteract the expansion in order to force the gas to contract into galaxies? We do not know the answer; but Eddington and Lemaître tried to make a plausible guess.

(e) Lemaître Model

Lemaître (Le50) also proposed another model. The universe starts out in a highly contracted state and initially expands at a rapid rate. The expansion is slowed down and brought to a halt in a state that is nearly identical with the Einstein state. Galaxies form at this stage and give rise to a new expanding phase that continues indefinitely (Fig. 10.7).

An interesting feature of Lemaître's model is the high initial density of the universe. We can compute the temperature and pressure that must have existed at that time, and thus establish the nuclear reactions that should have taken place. We obtain information about the chemical composition of matter at early stages of the universe before galaxies had a chance to form. We may expect this composition to be identical with the chemical composition of the surface material in some of the oldest stars observed in the galaxy.

This points out the importance of the chemical characteristics of the universe. A cosmological model must be able to simulate not only the overall density and pressure of the universe, or the existence of galaxies, it also must be able to predict in some detail the chemical composition of matter from which the earliest stars were formed. The chemical composition of later stars need not be predicted in this way, because nuclear reactions take place in the interior of stars so that early generations of stars can produce heavier chemical elements that later are distributed in interstellar space through ejection or explosion. Subsequent generations of stars can incorporate this newly formed matter and their atmospheres will contain these chemical elements.

(f) Friedmann Models

The relativistic cosmological models described so far have had one feature in common. They all involve a nonzero *cosmological constant* Λ in the relativistic field equations to be discussed below (10–44, 10–45). This constant corresponds to a tension in the cosmic substrate so that work has to be done on the universe in order to expand it; alternately work can be derived during an expansion, depending only on the value of Λ. Friedmann (Fr22) sets this constant equal to zero, essentially denying its existence.

There is hope that the value of Λ can be determined in part from a study of the dynamics of galaxies within a cluster. The presence of a Λ term produces changes in the virial equation, so that the kinetic energy of galaxies would no longer need to be exactly half of $-\mathbb{V}$, the potential energy of the cluster of galaxies (Ja70b). Just exactly how readily such effects could be observed with present techniques is still uncertain.

The Friedmann models can have Riemann curvature $k = -1$ or $+1$. Some of the models start in an extremely dense state and continue to expand. Others start out in a dense state, expand, eventually start contracting, and collapse back into the initial dense state. This cycle might repeat itself and such models then oscillate. In principle an oscillating universe could have existed in the indefinite past and might continue to exist into the indefinite future—much as a steady state universe does. Although the oscillating model need not postulate the creation of matter, it must be able to assure the proper chemical composition following a collapse phase.

Models (b) to (f) all have similar mathematical forms: Lemaître (Le31), Robertson (Ro33) and Walker (Wa34) found that in an isotropic homogeneous space the field equations of Einstein reduce to two simple differential equations in the radius of curvature or *scale factor a*. The first of these gives the mass density, including all radiation energy, in terms of the mass density of the vacuum, the expansion or contraction kinetic energy, and a curvature energy:

$$\kappa\rho = -\Lambda + 3\left(\frac{\dot{a}^2 + kc^2}{c^2 a^2}\right) \tag{10-44}$$

The second gives the change in this density on expansion or contraction, including work done against the pressure of the fluid. This is the energy conservation equation.

$$\frac{d\rho}{dt} + 3\left(\rho + \frac{P}{c^2}\right)\frac{\dot{a}}{a} = 0, \quad \text{or} \quad \frac{\kappa P}{c^2} = \Lambda - \left(\frac{2a\ddot{a} + \dot{a}^2 + kc^2}{c^2 a^2}\right) \tag{10-45}$$

Here κ is $8\pi G/c^2 = 1.86 \times 10^{-27}$ cm g^{-1}, where G is the (Newtonian) gravitational constant; κ is sometimes called the *Einstein gravitational constant*. Λ is the cosmological constant, and ρ and P are the density and pressure of the universe. Dots represent differentiation with respect to world time.

PROBLEM 10-8.   For *Einstein's universe a* = constant and $k = 1$.
(i) Show that

$$\Lambda = \frac{1}{a^2} + \frac{\kappa P}{c^2} \tag{10-46}$$

and that the density of the universe is fixed at

$$\rho = \frac{2}{a^2 \kappa} - \frac{P}{c^2} \tag{10–47}$$

We know that $P/c^2 \ll \rho$, at present. If we lived in an Einstein universe, Λ would have to have a value $\sim a^{-2}$, and $\rho \sim 2(a^2\kappa)^{-1}$.

(ii) Show that if $k = 0$ and $P = 0$ a static universe would require $\Lambda = 0$ and $\rho = 0$, leaving a undefined.

PROBLEM 10–9. De Sitter's universe is flat and empty, $k = \rho = P = 0$. Show that the scale factor a of the expanding universe obeys

$$a = a_0 e^{(\Lambda c^2/3)^{1/2}t} \tag{10–48}$$

and the age of the universe judged from the Hubble constant H is

$$\frac{1}{H} = \sqrt{\frac{3}{\Lambda c^2}} \tag{10–49}$$

PROBLEM 10–10. From the field equations with cosmological constant $\Lambda = 0$ and with zero pressure, obtain the relations

$$\frac{\ddot{a}}{a} = -\frac{4\pi G\rho}{3} \quad \text{and} \quad H^2 q_0 = \frac{4\pi G\rho}{3} \tag{10–50}$$

and, from (10–40), thus obtain also

$$(2q_0 - 1) = \frac{kc^2}{H^2 a^2} \tag{10–51}$$

PROBLEM 10–11. Prove the instability of the Einstein universe for $P = 0$. Note that an infinitesimal expansion makes $\rho < 2(\kappa c^2 a^2)^{-1} - P/c^2$, so that even with $\dot{a} = 0$, we have $\ddot{a} > 0$ and the expansion must continue. The proof for an initial contraction is similar.

PROBLEM 10–12. Prove for a Friedmann universe $(\Lambda = 0)$ that

(a) with $k = 1$ and an initially dense universe for which $P/c^2 = \rho/3$, the solution has the parametric form

$$a = b_0 c \sin x, \quad t = b_0 (1 - \cos x) \tag{10–52}$$

where x is a parameter.

10:9

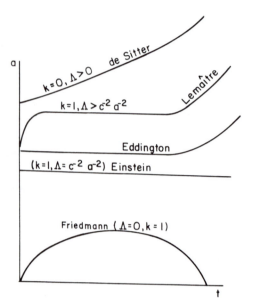

Fig. 10.7 Some cosmological models. The scale of $a(t)$ and of t is not the same for different curves. The only important consideration in this figure is the shape of each curve, rather than its exact dimensions.

(b) For late stages of the universe, with $k = 1$, $P = 0$

$$a = a_0 c(1 - \cos x), \qquad t = a_0(x - \sin x) \qquad (10\text{–}53)$$

Note that x grows monotonically with t, so that $(1 - \cos x)$ eventually must approach zero. The universe first grows, but later on collapses.

(c) For the dense stage, of a *hyperbolic universe* $(k = -1)$

$$a = b_0 c \sinh x, \qquad t = b_0(\cosh x - 1) \qquad (10\text{–}54)$$

(d) For the late stage of a hyperbolic universe

$$a = a_0 c(\cosh x - 1), \qquad t = a_0(\sinh x - x) \qquad (10\text{–}55)$$

10:10 OLBERS'S PARADOX

Suppose that a Euclidean space were uniformly filled with stars. Light emitted by stars in a shell at distance r to $r + dr$ from an observer would be proportional to $4\pi r^2 \, dr$. Of this light, a fraction proportional to $1/r^2$ would be incident on the observer's telescope, since light intensity drops as the inverse square of the distance. From each spherical shell of thickness dr an observer would therefore receive an amount of light proportional to dr alone. On integrating out to infinite

10:10

distance, we find that the light received by the observer should have infinite brightness. This infinity arises only because we have not taken into account the self-shadowing of stars. A foreground star will prevent an observer from seeing a star in a more distant shell, provided both stars lie along the same line of sight. When shadowing is taken into account we find that the sky should only be as bright as the surface of a typical star, not infinitely bright. Of course, that still is much brighter than the daytime sky; and the night sky is fainter still.

To someone who strongly believed in Euclidean space and in the infinite size and age of the universe this would appear paradoxical. Olbers, who advanced this argument in 1826, saw that such a cosmological view could not be held.

If we try to circumvent the argument by introducing curved space, no advance can be made. In such a space the area of a sphere drawn about an observer is of the form of equation 10–13 or 10–20—the surface area $S = 4\pi a^2 \sigma^2(\chi)$ is a function of distance χ alone. The number of stars in a spherical shell is proportional to $S(\chi)\,d\chi$. But the amount of light reaching the observer from that shell is also reduced by a factor $S(\chi)$, and these two factors cancel to give the same distance independence obtained for a flat space.

We could next argue that interstellar dust might absorb the light. But in an infinitely old universe, dust would come into radiative equilibrium with stars and would emit as much light as was absorbed. The dust would then either emit as brightly as the stars, or else it would evaporate into a gas that either transmitted light or else again emitted as brightly as the stars.

As long as galaxies themselves are distributed more or less randomly, this argument for a bright night sky remains valid, and only the overall space density of stars in the universe need be taken into account. That stars aggregate in galaxies, rather than being homogeneously distributed, does not affect the validity of the argument, either.

Unless we wish to suggest that no laws of physics hold for phenomena on such a scale—and then of course we must resign ourselves never to bother with cosmology—we are forced into one of three conclusions (Bo52), (Ha65):

(a) The universe is very young. Stars have not been shining for very long; light from great distances could not yet have reached us.

(b) The constants of physics vary with time. Since these constants affect the rate at which stars emit light, it may be that stars only started shining brightly in recent times.

(c) There are large recessional velocities of stars at great distances, so that their spectral shifts and diminished intensity lead to a lower sky brightness.

The correct explanation appears to be that stars are too sparse in the universe, with a stellar density of only 10^{10} Mpc^{-3}, and that they shine for too short a time, only about $t = 10^{10}$ y. Each star can then be thought of as occupying a volume

10:10

$V_0 = 3 \times 10^{63}$ cm^3. If the star's surface area is $\sigma = 10^{23}$ cm^2 and it radiates for the time t, it could only fill a volume $V_s = c\sigma t = 10^{51}$ cm^3 without diluting the radiation leaving its surface. Since $V_s/V_0 = 3 \times 10^{-13}$, we see that when the radiation of all stars is diluted to fill the entire universe, the radiation density and hence the surface brightness of the sky will be diminished by a factor of 3×10^{12} when compared to the surface brightness of a typical star. Doppler shifts and diminished intensity due to the expansion of the universe are relatively minor factors. The darkness of the sky is largely due to the low density of stars, their youth and their finite energy resources.

10:11 HORIZON OF A UNIVERSE

When a man on a cruising ocean liner wants to determine the distance of the horizon, he need only drop a buoy overboard and determine the buoy's range at the last instant before it disappears over the horizon. If the man is quick enough, he may then be able to shin high up on the ship's mast and briefly see the buoy again before it finally disappears over the horizon a second time. Two points are worth noting.

First, the distance of the horizon depends on the position of the observer. If a preferred horizon distance is to be defined, it should be selected in terms of some fundamental observer placed at some specific height above the ocean surface.

Second, no matter how high above the surface the observer climbs, there is an absolute horizon beyond which he can never see. He cannot see further than halfway to the antipodal point. His absolute horizon divides the surface of the earth into two hemispheres.

An observer placed at a given location in a universe will also be able to define a horizon beyond which he cannot see. In fact, there are a number of ways in which a horizon can be specified. The distance to the horizon may depend on the speed at which the observer is moving, so that a horizon is best defined in terms of a fundamental observer who is at rest with respect to the mean motion of galaxies in his local environment.

W. Rindler (Ri56)* has given a classification of horizons in different cosmological models. He defines three kinds of horizons: an *event horizon*, a *particle horizon*, and finally an *absolute horizon*:

(a) In some cosmological models, galaxies at large distances recede from an observer at ever-increasing speeds. The steady state universe is an example of such a model. In such a universe there will exist a world time t_1 (Fig. 10.8a) at which the distance of a particular galaxy P from the observer A increases at pre-

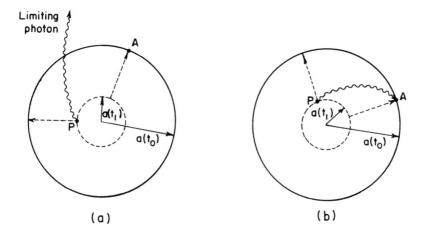

Fig. 10.8 (*a*) *Event horizon.* Light emitted at P will reach A at world time ∞. Events which occur beyond P will lie beyond A's event horizon, as will events at P, after t_1. Events at P before t_1 reach A in finite world time. The limiting photon asymptotically moves parallel to A for increasing values of $t_0 - t_1$. (*b*) *Particle horizon.* Effective locus of light in a four-dimensional expanding spherical universe. A particle formed at P at world time t_1 does not appear on A's horizon until time t_0.

cisely the speed of light.† Before t_1 the galaxy can emit radiation that eventually reaches the observer; but after t_1 radiation emitted by the galaxy cannot reach the observer because the intervening distance is increasing at a rate greater than the speed of light. Events prior to t_1 can therefore be transmitted to the observer whereas events subsequent to t_1 must forever remain hidden from him. Events occurring just before t_1 will appear highly red shifted and dilated in time. The time dilatation dictates that events occuring right at time t_1 will only become accessible to the observer an infinite time after t_1 and of course events occuring after t_1 remain altogether inaccessible to him. It is interesting to note that particles which have at some time been visible to an observer remain so forever although they appear increasingly faint and red shifted. We can now define an *event horizon* as a hypersurface in space and time that divides all events into two classes: those which have been, are, or will be observable to a given observer, and those forever inaccessible to him.

(b) In other cosmological models a different type of horizon is important.

† There is no conflict between such a rapid increase in distance and the special relativistic statement that speeds greater than the velocity of light cannot be attained. Special relativity is valid only as a *local theory* holding in the vicinity of any given point in the universe. By vicinity we mean a domain small enough so that the cosmic expansion can be neglected. Over larger domains where the expansion is significant, a less confined theory is required. One approach is provided by the general theory of relativity.

Take an explosive model in which matter initially is highly compact. At zero time the universe suddenly explodes. Two particles, P and A, which initially were quite far apart may recede from each other at nearly the speed of light. Because of the large separation, light initially emitted by particle P may not reach A for a long time. In particular, light emitted by P at time t_1 may not reach an observer at A until time t_0. Before t_0 the observer is quite unaware of the existence of particle P. After t_0 he can receive messages emitted at P. Essentially particle P enters the observer's horizon at time t_0 (Fig. 10.8b).

We can now define a *particle horizon* for any fundamental observer and for given world time t_0. It is a surface that divides all fundamental particles into two classes: those that have already been observable and those that have not.

It is clear that there can exist models in which both particle and event horizons exist. The Lemaître model is of this kind. Because there is an initial explosion from a compact state, a particle horizon will exist; because there is a subsequent acceleration of galaxies, following the period of galaxy formation in the Einstein state, an event horizon will also come into play.

(c) We may wonder about the distance of the horizon from a moving observer. Clearly, if the observer accelerates himself toward a fast receding galaxy, his event horizon can be extended. A number of results can be proven (Ri56):

(i) In a model without event horizon, a fundamental observer can sooner or later observe any event.

(ii) In a model having an event horizon, but no particle horizon, an observer can be present at any one specified event, provided he is willing to travel and provided he starts out early enough.

Statement (i) depends on the inability of particles to recede at a speed greater than light when no event horizon exists. Statement (ii) hinges on the fact that any given particle must have been within an observer's event horizon at some time in the distant past.

(iii) In a model with both event and particle horizons an observer originally attached to a fundamental particle finds that there exists a class of events absolutely inaccessible to him, no matter how he travels through space. This class of events defines an absolute horizon as shown by the following argument.

Suppose a fundamental observer were placed at some position A in the universe. There can then exist a critical particle P that initially recedes at exactly the speed of light and that enters A's particle horizon at time $t = \infty$. Let the initial distance (10–25) between P and A be D. Next, consider a fundamental observer at a point B situated along the line of sight AP but at a distance D beyond P. Again, P will enter B's particle horizon at $t = \infty$. By moving at the speed of light toward P, observer A would reach P at $t = \infty$. B would be receding at the speed of light relative to P and would, therefore, enter A's particle horizon on A's arrival at P at time $t = \infty$; but all particles beyond B would forever remain

Fig. 10.9 The absolute horizon for an observer at *A* (see text).

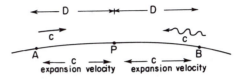

inaccessible to *A*. Position *B* defines an absolute horizon for a fundamental observer at an initial position *A* (Fig. 10.9).

10:12 COSMOLOGICAL MODELS WITH MATTER AND ANTIMATTER

Antimatter is hard to detect at a distance: Antihydrogen gives off a spectrum identical to hydrogen and hence a distant galaxy would look the same whether it was composed of matter or antimatter.

For these reasons we have no way of knowing whether our universe consists only of normal matter—protons and electrons as observed here on earth—or whether there might not be large amounts of antimatter in the universe.

Either way, there are difficulties. If there is no antimatter, or very little, then how are we to account for this cosmic asymmetry? What was it that decided the universe should consist predominantly of protons and electrons, when the probability of forming antiprotons and protons should seemingly be the same. We see no obvious explanation.

We might at first think that protons and antiprotons could have been formed randomly and with equal probability. In that case, just as in a random walk process (Chapter 4) we would expect an excess of either particles or antiparticles. If annihilation on a large scale had subsequently led to the destruction of matter through the reaction

$$\mathscr{P} + \overline{\mathscr{P}} \rightarrow \text{pions} \rightarrow \text{muons} + \text{neutrinos} \longrightarrow \begin{array}{l} \text{electrons} + \text{positrons} \\ \text{gammas} + \text{neutrinos} \end{array} \quad (10\text{--}56)$$

we might now be left with only those particles that had been produced in preeminence, apparently the protons. This idea founders because, in a random walk process the fluctuation—the deviation from equality after N^2 steps—is N. Since the universe now contains $N \sim 10^{78}$ protons, the initial number of protons and antiprotons would have had to be 10^{156}. The destruction of such a vast number of particles would however have produced an overwhelming amount of radiation and there certainly is no evidence for that. The fluctuation hypothesis must therefore be discarded.

10:12

It is still possible to argue that matter and antimatter might somehow become separated into galaxy or star-sized objects that had low probability of interaction. In that case, we would have the problem of explaining how this separation comes about. Matter and antimatter do not seem to repel each other gravitationally (Sc58a, see section 3:7) and some other explanation must be found.

If it really is abundant we might search for cosmic antimatter in two ways. First, cosmic ray particles produced in regions of antimatter might reach the earth from time to time. The fraction of antiparticles in the cosmic ray flux would then be an indicator of cosmic abundance. Unfortunately, the highest energy cosmic ray particles have not been analyzed in this respect. Such experiments are not yet possible. The lower energy cosmic rays do not show antiprotons in abundance, but these particles may have a more local origin and the presence of antigalaxies can therefore not be excluded in this way.

The other way of identifying antiprotons would be through their annihilation at boundaries where matter and antimatter meet. This annihilation would eventually lead to $\lesssim 100$ Mev gamma rays. Thus far we have no good observational data on that radiation. The question therefore still is open and it is not impossible that the universe could consist half of matter, half of antimatter.

A theory combining the ideas of H. Alfvén and O. Klein (K171)* suggests such a state. The primordial substance consisting half of matter, half of antimatter is called *ambiplasma* in this theory. This model of the universe is taken to be composed of isolated "metagalaxies" so that different portions can be separated by cosmic horizons and do not communicate. The theory then postulates that our local portion started out quite large, collapsed gravitationally until the ambiplasma started reacting violently as the density increased and the probability for proton-antiproton annihilation became large. A substantial part of the matter was thus destroyed and this made sufficient energy available for the subsequent expansion of the universe.

Some interesting conclusions can be drawn if we assume only that the contraction did not proceed up to the point where the Schwarzschild radius R_s (8–111) would have been reached. This means that the radius R at the time of greatest contraction (see Figure 10.10) obeys the inequality:

$$\frac{2MG}{R} \lesssim c^2 \tag{10–57}$$

Here M is the mass of the observed portion of the universe.

An argument suggests that R was not much greater than R_s: If we take $M \sim 10^{54}$ g, our best guess from present day observations, then we find $R \sim 10^{26}$ cm at the time of greatest concentration. If much of the matter is to have been destroyed suddenly about the time the radius was R, we must have had an anni-

Fig. 10.10 Radius of the *Klein-Alfvén Universe* as a function of time.

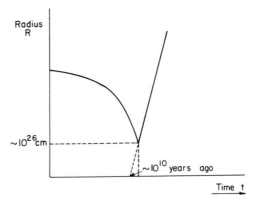

hilation probability of order unity as each particle moved through distance R:

$$n\sigma R \sim 1 \qquad (10\text{–}58)$$

Here σ is the cross section for annihilation and for collisions that randomize the motion and permit eventual annihilation. n is the number density at that epoch:

$$n \sim \frac{M}{R^3 m_H} \qquad (10\text{–}59)$$

m_H is the proton mass. Combining the last two equations gives

$$\frac{M\sigma}{R^2 m_H} \sim 1 \qquad (10\text{–}60)$$

If this is combined with the inquality for the radius, we obtain, after cancelling terms, that

$$\frac{\sigma}{4 M m_H} \gtrsim \frac{G^2}{c^4} \qquad (10\text{–}61)$$

This of course is a verifiable statement, as the authors of the theory point out. Inserting the known values for σ, m_H, G, and c we find that for $\sigma \sim 10^{-25}$ cm^2

$$M \lesssim 10^{55} \text{ g} \qquad (10\text{–}62)$$

which appears to agree with observations. The authors thus feel that a relation between cosmic and atomic quantities has been established and that of course is very interesting.

It is intersting to note that if M were somewhat larger than postulated above.

or

$$M > \frac{c^4 \sigma}{4 G^2 m_H} \qquad (10\text{–}63)$$

10:12

the collapse would go through the Schwarzschild singularity and we would not be here today. We only escape this fate because M is small enough—although barely so.

Even if our portion of the universe had collapsed there might have been others left that remained intact because their masses had been low enough. Hence not everything would be lost in such a universe that has many separate, almost independent portions. A universe in which there were many black holes would conceptually not be too different. It also would have portions between which communication could not be established.

10:13 GALAXY FORMATION

The question of galaxy formation is closely related to the problems of cosmic evolution. The study of galaxies and their past history is therefore not only of intrinsic interest, but also is a means for gaining insight into the type of universe we inhabit.

Two almost unrelated problems actually are involved. First, we have to understand how a sufficiently large amount of matter can be aggregated in a small enough volume to give rise to galaxies. It is of course possible that the galaxies were or are spontaneously formed and that matter, at least for some time, flows out of the volume occupied by the nucleus. This would be acceptable in some forms of the steady state theory. However, usually we have considered galaxies as objects formed from the general cosmic medium, and we have sought inherent instabilities capable of giving rise to galaxies in such a medium. The stability question here is somewhat more involved than in the star formation problem (section 4:20), first, because there are difficulties in defining a suitable potential energy per unit mass of cosmic material and, second, because the rapid expansion of the universe seems to endow the medium with such stability against contraction that sufficiently rapid growth of condensations is hard to understand. We will discuss this problem in some detail below; but first we note that there is a second way in which galaxy formation entails deeper insight into the evolution of the universe. For, on the assumption that enough matter can be brought together by some means or other, we can next ask how this matter behaves once it has aggregated. How does the halo containing old stars form in the galaxy, and how does the galactic disk evolve? If that were understood (see Fig. 10.11a for one possibility) we would better understand the formation of chemical elements and so be able to define a proper age for the Galaxy as shown in section 10:14 below. We would be able to understand at what epoch and perhaps at which cosmic density the initial instabilities set in. Our understanding of dynamic processes on a cosmic scale would therefore be sharpened appreciably.

10:13

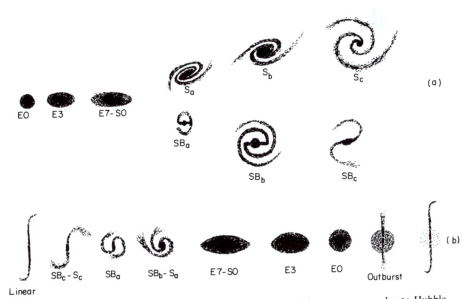

EO   E3   E7-SO   S$_a$ S$_b$ S$_c$ (a)

SB$_a$ SB$_b$ SB$_c$

Linear   SB$_c$-S$_c$ SB$_a$ SB$_b$-S$_a$ E7-SO   E3   EO   Outburst   (b)

Fig. 10.11 Possible evolutionary trends in galaxies: (*a*) An evolutionary sequence due to Hubble. (*b*) A pattern in which a symmetric outburst from a nucleus develops into a barred spiral that winds itself into a regular spiral and then diffuses into an elliptical and eventually spherical aggregate before another outburst takes place. It is also possible that spiral galaxies always remain spiral and that ellipticals always remain elliptical. We do not yet know.

Let us look at some of the alternatives more closely.

(a) Possibilities That Galaxies Are Formed Through Explosions

In the late 1950s Ambartsumian in the Soviet Union suggested that galaxies might be formed, not through the condensation of extragalactic material, but rather through the outflow of material from selected regions. He envisaged galaxies as being formed in pairs through an explosive process. In support of this hypothesis we can cite a number of observational facts.

(i) Galaxies never appear to occur singly. They are only found in pairs or in larger aggregates.

(ii) Some pairs or multiple galaxies are joined by bridges of luminous matter. In a few cases the velocities of the galaxies along the radial direction alone are of the order of several thousand kilometers per second so that it is not likely that these galaxies are gravitationally bound. They seem therefore to have originated recently—perhaps as complete or nearly completely formed galaxies.

(iii) In general, the masses of galaxies that are members of a physically well-

10:13

isolated group or cluster seem to be smaller than the mass that would be required to bind the galaxies gravitationally. The virial expression (3–83) for the relationship between potential and kinetic energy could therefore only be obeyed if there were a great deal of unseen matter between the galaxies binding them together. The total mass needed would be about an order of magnitude greater than the visible mass if Newtonian dynamics still holds on that scale.

(iv) That some large-scale explosive activity takes place in galaxies is shown by studies of such galaxies as M82 (Ly63) and M87. The first of these is found to be ejecting large quantities of hydrogen from its nucleus into interstellar space. The second object is shooting out a jet of gas containing relativistic particles that can be observed through the synchrotron radiation they emit.

(v) Quasars and active nuclei of galaxies can also emit jets, some as long as $\lesssim 1$ Mpc. High velocity absorption lines observed in these objects suggest that matter is being exploded outward at speeds of the order of 10^4 to 10^5 km sec^{-1}. These sources might be representing the formation of galaxies.

If this type of explosive origin is the way in which galaxies are actually formed, see Figure 10.11(b) for example, then much of the difficulty encountered in the condensation theories might be obviated. However, there does not seem to exist enough data at present to permit quantitative estimates of the consequences of such catastrophic formation. We have no good theoretical guides against which to test the hypothesis of explosive origin.

(b) Formation of Galaxies in a Steady State Universe

The formation of galaxies in a steady state universe presents problems if an explosive origin is ruled out. For, in that case, the rate of formation of condensations is completely specified by the value of Hubble's constant H. This comes about because in a steady state universe every feature must reproduce itself—somehow—within a time (see section 10:4)

$$\tau_s = (3H)^{-1} \tag{10–64}$$

To demonstrate the difficulty, we take the most favorable simple case in which a galaxy of mass M is already present and the formation of a new galaxy merely requires that extragalactic matter fall into the galaxy at a rate sufficiently great so that at the end of time τ_s the total gravitationally bound mass be $2M$. Once the mass of a galaxy is doubled, we could envisage a subsequent split into two— the formation of an additional galaxy. Clearly this is more favorable than the spontaneous formation of galaxies in a part of space where no initial attracting "seed" galaxy is present. But the accretion of enough matter before such a fission is a prerequisite.

10:13

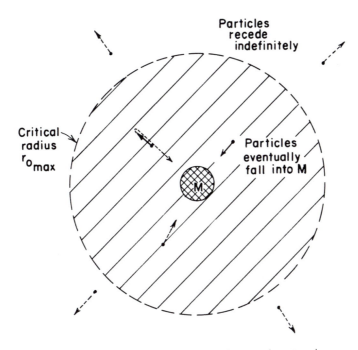

Fig. 10.12 Accretion of matter into a galaxy in a steady state universe.

The most favorable situation will be that in which the external gas has no velocity other than its recession velocity. If there were additional thermal velocities, these also would act to resist gravitational contraction. We therefore assume that the cosmic gas is created at zero temperature. First, we set up a model in which Newtonian dynamics is still obeyed. This can be done because the velocities and masses considered are small. The potential will be central. Close to the mass M a particle will be predominantly under the influence of M. At large distances it will be predominantly under the influence of the cosmic repulsion (Fig. 10.12). Conservation of energy then requires the velocity relationship

$$\left(\frac{dr}{dt}\right)^2 - \left(\frac{dr_0}{dt}\right)^2 = H^2(r^2 - r_0^2) + 2GM(r^{-1} - r_0^{-1}) \qquad (10\text{--}65)$$

where G is the gravitational constant, r is the instantaneous particle distance from M, and r_0 is an initial distance. We imagine the situation in which, before some time t_0, all particles have velocities Hr. At $t = t_0$ the gravitational field

10:13

of M is switched on. For $t > t_0$ we then have

$$\left(\frac{dr}{dt}\right)^2 = H^2 r^2 + 2GM(r^{-1} - r_0^{-1}) \tag{10-66}$$

Second, if r_0 is small, the attractive force of the galaxy is able to decelerate and eventually reverse the initial outflow of matter. Particles, initially at large distances will forever continue to flow away. But there will be some terminal particle initially at some distance $r_{0\,max}$, whose velocity will just be reduced to zero at $t = \infty$. The maximum amount of matter that can possible be gathered into the galaxy in time τ_s then is $4\pi r_{0\,max}^3 \rho/3$, if matter created within the sphere of radius r_0 is neglected. We will return to this point later after computing $r_{0\,max}$.

We do this by noting that particles, initially at distances less than $r_{0\,max}$, will have negative velocities $V \equiv (dr/dt) < 0$ at $t = \infty$, while those initially beyond $r_{0\,max}$ will have $V > 0$. We therefore seek a solution to equation 10–66 that has $V^2 = 0$ and $d(V^2)/dr = 0$ simultaneously.

PROBLEM 10–13.   Show that this happens for

$$r_{0\,max} = \left(\frac{8GM}{27H^2}\right)^{1/3} \tag{10-67}$$

Different versions of the steady state theory propose overall densities of $3H^2/8\pi G$, and $3H^2/4\pi G$, and some versions propose no definite density at all. If we take the larger value $3H^2/4\pi G$, we find that the amount of matter initially contained in the sphere $(4\pi/3) r_{0\,max}^3$ is $8M/27$. This represents the maximum amount of matter that can fall into the condensation of mass M, in time τ_s, no matter what the value of M might be.

If we now add to this all the additional mass that might have been created within $r_{0\,max}$ during time τ_s, we could conceivably double the accreted amount of matter, but we would still have a total mass well below the required amount M. One can show more rigorously that galaxy formation would require a density of at least $\rho_{min} = 15H^2/4\pi G$ for a steady state universe in which galaxies are formed through gravitational forces alone (Ha61). For current estimates of Hubble's constant, this corresponds to a density of about 10^{-28} g cm^{-3} which is two orders of magnitude above the baryon and lepton density estimated to be resident in condensed matter (galaxies). Although not absolutely ruled out by current observations, this density is nevertheless very high. If galaxies form in a steady state universe, through the collapse of intergalactic gas, they probably form nongravitationally.

10:13

(c) Formation of Galaxies in an Evolutionary Universe

The difficulties inherent in steady state universes are also present in evolutionary models. The cosmic expansion makes condensation of matter very difficult if only gravitational self-attraction is considered. A difficulty in the case of evolving universes is that no large-scale condensations initially exist to provide condensation nuclei around which galaxies could form. Such nuclei are possible only in steady state universes since those have always existed. In an evolving universe, however, we start out with a hypothetically uniform medium in rapid expansion and this medium may not be sufficiently unstable to form galaxies.

If we deal with the set of Friedmann universes, then the process of galaxy formation appears to present great difficulties. In a classical paper published in 1946, Lifshitz (Li46) analyzed the stability of this family of cosmic models and showed that instabilities could simply not grow rapidly enough to give us anything like the concentration of matter observed in galaxies. He based his conclusion on an analysis of the most general type of perturbations—seed disturbances—that might grow into larger condensations. His assumption, however, was that these disturbances grow independently of each other. More recently the nonlinear superposition of a number of such perturbations has been considered. There, the possibilities seem more favorable, and perhaps galaxy formation would be possible (Ko69).

Otherwise, evolving universes appear capable of forming galaxies through gravitational condensation only in Eddington-Lemaître models where a sufficiently long sojourn in the Einstein state might permit galaxies to form. In that case, however, formation would only have been possible in the past. The possible formation of galaxies at the present time, as suggested by the observations (i) to (v) in subsection (a) above could not be explained by this means.

Moreover, many cosmologists feel uncertain about whether the cosmic constant Λ could actually be nonzero in the Einstein field equations. It is this constant, of course, that allows the Lemaître universe to go through a quiescent stage before proceeding to a second phase of rapid expansion.

From all this, it is clear that the gathering of enough matter to form a galaxy is difficult in any cosmological model. So stringent are the conditions that perhaps quite severe limitations will be found that all cosmological models must obey if galaxies are to exist at all. One aim of theories of galaxy formation is to ascertain these limitations and thus to further sharpen our understanding of the universe we inhabit.

(d) Formation of Our Own Galaxy

In our own galaxy we do have some additional clues about how the initial birth

10:13

took place. The information is primarily given by the orbital parameters of the very oldest stars in the Galaxy. These stars are deficient in metals, and this gives them an excessively large ultraviolet magnitude relative to more recently formed metal-rich stars. The difference between U and B magnitudes, $\delta(U - B)$, increases with increasing age.

We can now look for the orbital characteristics of stars in the solar neighborhood. This was done by Eggen, Sandage, and Lynden-Bell (Eg62). As indicated in Fig. 10.13, they found that the oldest stars in the Galaxy have highly eccentric, low angular momentum orbits with high velocities perpendicular to the Galactic plane. All this suggest an initial, almost radial collapse toward the center—or

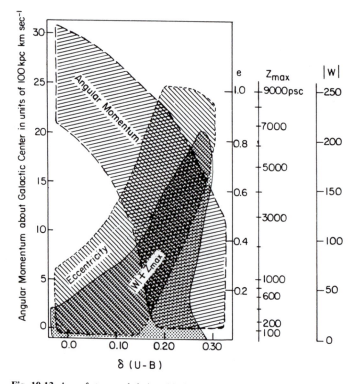

Fig. 10.13 Age of stars and their orbital characteristics: Angular momentum about the Galactic center, eccentricity e of the orbits about the center, the velocity $|W|$ perpendicular to the Galactic plane, and height z_{max} risen above the plane. Large values of $\delta(U - B)$ are associated with old stars in the Galaxy. (After Eggen, Sandage, and Lynden Bell, *Ap. J.*, Eg62. See text.)

possibly an initial, almost radial ejection from the center. As pointed out by Oort (Oo65), the angular momentum per unit mass, measured about the Galactic center, is a factor of eight lower for highly metal-deficient RR Lyrae stars than for disk or spiral arm stars in the neighborhood of the sun. How the disk population could have been formed from matter initially related to the halo population therefore is a real puzzle. Perhaps the origins of the halo and disk stars are quite distinct. Von Hoerner (voH55) found that globular clusters have orbital characteristics similar to the halo stars.

An approximate age for the Galaxy can be obtained on the assumption that *carbonaceous chondrites*, a class of meteorites that often impact on the earth, represent material from which the solar system was formed. We analyze for the content of thorium and uranium isotopes, Th^{232}, U^{235}, and U^{238}. These elements are formed in the *r*-process (section 8:12) in a ratio 1.6:1.6:1. Their α-decay *half-lives* are 1.4×10^{10}, 7.1×10^{8}, and 4.5×10^{9} y, respectively. The present $Th^{232}:U^{238}$ ratio is somewhat uncertain, but the best guess seems to be 3.3:1. The $U^{238}:U^{235}$ ratio is 1:0.007.

PROBLEM 10–14. If 60% of the uranium and thorium was formed in the birth of the Galaxy and 40% was formed continuously between the birth of the Galaxy and the solar system birth $\sim 5 \times 10^9$ y ago, show that the Galaxy would only be $\sim 7 \times 10^9$ y old. If all the metal formation had taken place continuously, at an even rate, the Galaxy would be 20 aeons old. This age still is controversial (Di69).

10:14 DO THE CONSTANTS OF NATURE CHANGE WITH TIME?

How do we know that the speed of light has been the same throughout the history of the universe? Or has it, in fact, always been the same? Does Planck's constant or the gravitational constant, or the charge of the electron change very slowly but yet significantly on a cosmic scale? In fact, can one even answer such questions?

The first person to worry about this problem and to come up with some quantitative indicators was P. A. M. Dirac (Di38). He noted that the constants of nature could be arranged in groups that were dimensionless numbers of order 10^{39} or $(10^{39})^2 = 10^{78}$. In general, such numbers can be constructed by taking the ratio of a cosmic and a microscopic quantity. We do not, of course, expect to get ratios of precisely 10^{39} in each case, but the exponents cluster remarkably closely around the numbers 39 and 78.

Dirac argued that if this was no coincidence, then it indicated that microscopic and macroscopic—atomic or subatomic and cosmic—quantities were interrelated. Since the universe is expanding, then because of the changes in the size of the cosmos, there should be corresponding changes on an atomic scale. In fact, we can see how large this change would be by constructing some dimensionless quantities from the ratio of the radius of the universe to atomic or nuclear lengths:

(a) The radius of the universe is believed to be of order 10^{28} cm.

(b) The radius of the Bohr orbit of an atom is

$$\frac{\hbar^2}{me^2} = 5 \times 10^{-9} \text{ cm} \tag{7-4}$$

(c) The Compton wavelength of the electron is

$$\lambdabar_c = \frac{\lambda_c}{2\pi} = \frac{\hbar}{mc} = 4 \times 10^{-11} \text{ cm} \tag{6-166}$$

(d) The classical electron radius is $\dfrac{e^2}{mc^2} = 3 \times 10^{-13}$ cm $\tag{6-168}$

(e) Nuclear dimensions also are of the order of 10^{-13} cm $\tag{7-7}$

By taking the ratios of (a) to (b) through (e), we get numbers ranging from 10^{36} to 10^{41}.

Dirac argued that since these numbers are dimensionless, they should somehow not change with time. They should stay constant as the universe evolves because a pure dimensionless number has no time or length dependence built into it. Just how the interaction between cosmic and microscopic quantities takes place to keep these dimensionless quantities constant in an evolving universe is not known. Generally, gravitational fields have been considered the only suitable candidates ever since Einstein tried to incorporate Mach's principle into his general relativistic theory of gravitation. In fact, Dirac's hypothesis of constant dimensionless numbers is another version of Mach's principle and its quest to unite the very large-scale behavior of matter with what happens on a local scale. That gravitation may in fact play a role is hinted at by a dimensionless length that can be constructed through use of atomic and gravitational physical constants. We note that none of the ratios (b) through (e) included the gravitational constant G. We can however construct the ratio

(f) $$\frac{\hbar^2}{m^2 m_p G} = 10^{31} \text{ cm} \tag{10-68}$$

Here m_p is the proton mass and this length represents the radius that a hydrogen atom would have if electromagnetic forces were absent and only gravitational forces held the atom together. The ratio of the gravitational to the electromagnetic

Bohr orbit then is just

$$\frac{e^2}{mm_pG} = 10^{39} \qquad (10\text{–}69)$$

This brings up the point that electromagnetic forces and gravitational forces also differ in strength by the same 39 orders of magnitude; for, the ratio of the lengths (b) and (f) happens to be the ratio of electromagnetic force to gravitational force of attraction between proton and electron: $F_E/F_g = e^2/mGm_p$.

If this ratio is to be constant, then we would expect the mass and charge of the electrons to change if cosmic evolution affected the value of the gravitational constant.

What about the ratio, say, of cosmic to atomic or nuclear mass? This is a particularly interesting quantity because the mass of the universe divided by the mass of the proton gives just the number, N, of atoms in the universe. We obtain

$$N = \frac{M}{m_p} = 10^{78} \qquad (10\text{–}70)$$

The puzzling thing about this ratio is that the flow of particles across the cosmic horizon would clearly destroy its constancy on a time scale of the order of an inverse Hubble constant $T = H^{-1} = 4 \times 10^{17}$ sec. Over a period of 10^{10} y, N would change appreciably. Would not that perhaps destroy Dirac's argument that these large dimensionless constants must not change?

There are two ways to answer that. First, we might decide that outflow beyond the cosmic horizon does indeed weaken Dirac's argument. On the other hand, we can also state that Dirac's ideas fit perfectly into a steady state theory of the universe. There the constant replenishment of matter keeps N constant and, in fact, keeps all cosmic parameters constant so that microscopic quantities need not change at all.

This is an attractive feature of the steady state theory because, as we shall see, however crude the observational indications we do possess, they all suggest that the physical constants of nature have not changed with time.

Once we have talked about the dimensionless ratios of forces, masses, and lengths, we have effectively constructed most, if not all, of the independent quantities that are available from the ratios of the parameters: time, length, and mass. After all, we normally express all physical entities in terms of these three basic parameters.

Nevertheless, a couple of other dimensionless numbers will still be presented, partly because of their general interest and partly because all the ratios thus far constructed depend on the electromagnetic and gravitational properties of matter; we have not worried much about strong and weak nuclear interactions.

We therefore might still look at dimensionless constants constructed from the ratios of different times scales:

(a') The cosmic time scale, as already mentioned, is $T = H^{-1} = 4 \times 10^{17}$ sec.

(b')
$$\frac{e^2}{mc^3} \sim 10^{-23} \text{ sec} \qquad (10-71)$$

(c')
$$\frac{h}{mc^2} \sim 10^{-21} \text{ sec} \qquad (10-72)$$

These last two numbers are not entirely independent. They are related through the fine structure constant $\sim 1/137$. Those short times are characteristic of nuclear interactions where time scales of the order of 10^{-22} sec are common. Again, we note that the ratio of the cosmic to microscopic time scales is of order 10^{38} to 10^{41}.

The fact that nuclear dimensions and time scales are not too different from some of the purely electromagnetic ones comes about because of the comparable strength of *strong* and *electromagnetic interactions*. The difference is only about three orders of magnitude, while here we talk about some 39 orders.

The weak interactions have not been considered at all here and, perhaps, they would not even fit into a coherent picture of interactions between cosmos and atom. Perhaps we must not expect too much from a simple idea.

Finally, we can still mention the dimensionless constant that can be formed using only gravitational and cosmic parameters

$$\rho_0 \frac{G}{H^2} \approx 1 \qquad (10-73)$$

where ρ_0 is the density of the universe. This is an observational result. Although it is true that the value of $\rho_0 \sim 10^{-28}$ g cm^{-3}, which would be needed to make the relation exact, is high compared to the estimated density of galactic mass $\sim 10^{-30}$ g cm^{-3}, we must remember that galactic mass may provide only a fraction of the total mass content of the universe. Such a density also represents the value that would be needed to make the universe spherical as, for example, in an Einstein or related model of the universe.

So much for an enumeration of the coincidences that make the dimensionless numbers of nature 10^{39} or small powers (including the zeroeth power) of this number. Many scientists regard them as no more than coincidences. Others feel they are suggestive of underlying basic relationships that should be explained. We will not enter this argument, but rather go on to the observational searches for evolutionary changes in the constants of time that have been inspired by Dirac's ideas.

How fast would we expect the constants to change? We do not know of course, but the inverse Hubble constant H^{-1} might provide a suitable unit of time in terms of which changes should be measured.

If the gravitational action of the universe produces all of the changes on a microscopic scale, it might be interesting to see whether the gravitational constant had appreciably changed during recent aeons. Edward Teller (Te48) first analyzed this question. He looked for climatological changes on earth as a function of time. These would be related to the solar luminosity changes and changes in the earth's orbit resulting from time variations in the constant G. This argument is quite complex, but indicates changes—if any—of $(dG/dt)/G \lesssim 10^{-10}$ per year.

More recent radar observations of the constancy of the orbits of Venus and Mercury (Sh71) indicate similar results and involve no assumptions beyond Newtonian gravitational theory. These results will probably be refined by an order of magnitude in the next few years.

An interesting study on the possible variability of Planck's constant with world time was done by Wilkinson (Wi58). He was interested in the integrated effect of changes in Planck's constant over a period of the order of the age of the earth. The age of the earth and of meteorites can be determined separately from a number of different radioactive decay schemes, some of which involve alpha-particle emission and others beta-decay. These two processes have quite different physical bases, and we would not expect the ages given by beta- and alpha-decay schemes to be the same if the constants of nature varied appreciably.

The evidence cited by Wilkinson comes from a study of *paleochroic haloes*. These haloes are spherical shells observed in rocks that have small inclusions of radioactive material. As the material decays, any alpha particles will give rise to a thin visible shell at the end of the particle's path through the rock where most of the energy is dissipated. Corresponding to individual alpha-particle velocities v, we then obtain individual shells. These shells are easily identified with given α-decay schemes. Two interesting statements can then be made.

(i) The physics of charged particle transit through material, a process that is purely electromagnetic, is invariant over a period of order 2×10^9 y or perhaps slightly more. Otherwise the shells would be diffuse, not thin. This is interesting because the α-decay scheme discussed by Wilkinson involves both electromagnetic and nuclear forces.

(ii) Some alpha-emitting nuclei also have the possibility of emitting through beta-decay. The ratio of these two decay rates is called the *branching ratio*. Wilkinson was able to make the statement that if the branching ratio had increased or decreased by an amount of order 10 over the past 2×10^9 y, we would have found that certain alpha particles that caused haloes should have been absent and others again much stronger. Since no such anomalies were found, any changes taking place over the past few aeons in the many fundamental physical constants involved probably were small.

The search for changes in the fundamental physical constants has only begun. This study may eventually lead to a better understanding of the universe.

10:15 THE FLOW OF TIME

We tend to think that time always increases. But with respect to what? And what do we mean by "increase"?

The simplest answer would be to say that time is that which is measured by a clock. We know how clocks work, and that is that. Of course (section 3:10) there are different types of clocks and we might wish to compare them to see whether they all are running at the same rate or whether there might be, say, a systematic slowing down of one type of clock relative to the others.

Here we have assumed that all possible clocks will always run in one direction only. In that case, however, we would never be able to decide whether time is running "forward" or "backward" because these two directions would be indistinguishable.

For gravitational and electromagnetic processes, we do not know how to define a direction of time's flow. The physics that describes the orbiting of the earth around the sun holds equally whether the earth moves in a direct or retrograde orbit about the sun. Under a time reversal, too, the earth and all the other planets would return along the same orbits that had led them to their current positions in the solar system. But such orbits would be no different from a set of future orbits that could have been predicted from a simple reversal of all velocities involved.

Similarly we could use the orbital motion of an electron in a magnetic field to define time. Here it is interesting that the orbit in which the electron travels is identical to one that a positron would travel if it were going along the same path backward in time.

Both these examples show a basic symmetry that seems to pervade all natural physical processes: that if we reverse the flow of time T, reverse the sign of the electric charge of matter C, and reverse the sign of all positions and motions, (this is also called an inversion) P, then the observed results are indistinguishable from an original process in which none of these reversals or reflections took place. The operation P is called the *parity operation*; C is called *charge conjugation*; and T is the *time-reversal operation*. A fundamental theorem of physics requires that under the combined operations CPT all physical processes remain invariant.

Because of these symmetries, it is apparently impossible for us to know whether we are living in a world in which time is running forward and the universe is expanding, or whether time is running backward and the universe is contracting. These cosmic motions are independent of electric charge so that a charge conjugation would not be noticeable either. We would just assume we were made up of matter, but actually it might be what we currently call antimatter.

How, then, do we actually determine the direction in which time is flowing?

For a long time it was suggested that the second law of thermodynamics defines the direction of time in a unique way. The law states that as time increases, any isolated system tends toward increasing disorder. Light initially concentrated near the surface of a star flows out to fill all space. The reverse never happens. Light-filling space simply does not converge and flow into a single compact object. Such ordered motions, although strictly permitted by a simple time reversal argument, do not happen often. They are possible but highly improbable. The second law of thermodynamics basically states that as time increases, greater randomness comes about because there are many states of a system in which the system is disordered and only few in which it has a high degree of order. If any given state is as likely to occur as any other state, then the chances are that the evolved system will be found in one of the many disordered states rather than in one of the very few ordered configurations.

We could argue, however, that the second law of thermodynamics really is a consequence of the cosmic expansion: An undisturbed system shows no systematic evolution with time. The cosmic expansion provides just the disturbance needed. Because the universe expands, there is always more empty space being created and starlight can flow into this volume to fill it. Distant galaxies are red shifted, the sky looks black, and thermodynamic imbalance is actively maintained. In a static universe, equilibrium could be attained and the direction of time lost.

Let us look at the reverse of all this: If the universe were to collapse, distant galaxies could be approaching us at great speeds and we would observe them highly blue shifted. The night sky would be bright and perhaps light would flow from the night sky into stars rather than the other way around. Under these conditions would the second law still hold? Or would we find that physical processes tended toward greater order as the universe collapsed?

If tendency toward disorder depends on the cosmic expansion, as suggested by T. Gold (Go62), then the flow of time is well correlated with the flow toward order; however, we still have no way of telling whether time is running forward or backward because a reversal of time might be correlated with cosmic collapse, decrease of randomness, blue shifting of galaxies, and so on.

What we think we see in the universe then depends on how we define the flow of time. If we first fall down the steps and then hurt ourselves, the universe is expanding, time flows "forward," and physical systems tend toward randomness.

10:15

If we first hurt ourselves and then tumble upstairs, time is actually running "backward," the universe is contracting and physical systems are becoming more and more ordered. It may just be a matter of definition.

Since we do not like to leave such a fundamental question in such an unsatisfactory state of arbitrariness, we wish that there might be more straightforward ways of determining the direction of the flow of time. A possibility of this sort has come into sight in recent decades.

In 1956 Yang and Lee (Le56) pointed out that parity might be violated in weak interactions. This was swiftly verified in a variety of experiments. It was then realized that invariance under a combination of operations CP seemed to hold universally true. This would mean that reversing the charge on an object would produce a mirror image of its initial physical behavior. In electromagnetic cases this certainly is true. A positron moving in a negative direction traces out the same path as an electron moving in a positive direction at the same velocity through the same electric and magnetic fields. The same rule seemed true also for other types of interactions, including the "weak" interactions of which beta-decay is an example. CP invariance, together with the above-mentioned CPT invariance, indicated that the laws of physics still would remain invariant in time.

In the mid-1960s, however, a group at Princeton University (Ch64b) discovered that violations of CP symmetry occasionally occurred in the decay of neutral K-mesons. This would indicate that time reversal might also be violated in these reactions since otherwise CPT invariance would not hold and the laws of physics then would be in very fundamental difficulties. If time reversal symmetry really were violated, a preferred direction of time could at least be defined, and the discussion of the flow of time might be made easier. A series of experiments searching for time reversal asymmetries has therefore been attempted. Thus far no violations of symmetry have been found.

To show the type of behavior we are searching for, an example may be useful.

If we look at the decay products of the lambda particle (which always has zero charge)

$$\Lambda^0 \rightarrow \mathscr{P} + \pi^- \tag{10–74}$$

into a proton and a negatively charged pion, we are dealing with a weak interaction process. Initially the spin of the lambda particle can be considered to be upward, as shown in Fig. 10.14(a). The proton moves away perpendicular to that spin direction, and its own spin is unknown. It may have components a, b, c along three mutually perpendicular directions as shown. Under time reversal all the spin directions change and so does the direction of the velocity. Magnitudes of the components, however, do not change. Rotating the whole process about the c-axis by 180° gives picture (c), in which the Λ° particle has the same initial spin direction, the proton again comes off in the same direction, but the component

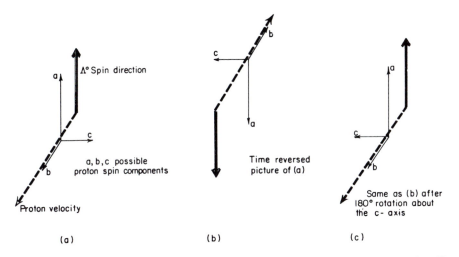

Fig. 10.14 Decay of a Λ^0 particle with spin in an "upward" direction into a proton and pion. The proton velocity vector is shown by a dashed line. Its spin components are indicated by solid lines and the lambda spin direction by iterated arrows. The decay seen in a time reflection transformation is shown in (b) and after a further 180° rotation in (c). If the process is to appear time-reversal invariant, the proton spin component c must have zero value. (Reprinted with permission from Overseth, "Experiments in Time Reversal," copyright © 1969 by Scientific American, Inc. All rights reserved.)

c of the spin is now reversed. Since the 180° degree rotation that has been introduced is only a matter of viewpoint, we would expect that pictures (a) and (c) would be identical if time reversal symmetry held. For this to be true, however, the c component of the proton spin would have to vanish. Time-reversal symmetry in Λ^0-decay would therefore only be preserved if the c component of the proton spin always had a value zero. Experiments in which the proton is scattered off carbon targets allow one to determine the direction of the proton spin. Such experiments thus far have shown the c component to actually be zero, except for a predictable deviation due to the strong interaction of the pion and proton at the moment of decay. This calculable correction can be subtracted from the experimentally observed c values to find that the c component would actually have zero value, as required by time-reversal invariance.

At the same time, one does find violations of parity for the Λ^0 decay, because under such an inversion all linear momenta change directions; but spins, which can be thought of as the vector product of two directed quantities, do not change sign. If parity were to remain unviolated, the proton velocity and the spin component b should always have the same relationship and that could happen only if b were to have the value zero. In actual fact its value is not zero. Sometimes it lies along the velocity vector and sometimes in the opposite direction but one

10:15

of these directions is preferred. Parity is therefore violated in this process, but we do not yet know whether the combination CP is violated also. A complication is that T symmetry is often difficult to check where CP violation is known to occur and vice versa. There appears, however, to be a possibility that T symmetry violations will be uncovered in the next few years and a preferred direction of time might then become established. What is clear is that the time-symmetry problem is one of observational physics and need not necessarily remain in the domain of philosophy.

It also appears that fundamental processes on a nuclear scale seem closely related to important structural features of the universe. We wish we understood the interrelation.

ANSWERS TO PROBLEMS

10–1. $x_4^2 = a^2 - r^2$, $r^2 = x_1^2 + x_2^2 + x_3^2$

$$dx_4^2 = \frac{(rdr)^2}{a^2 - r^2}, \quad dx_1^2 + dx_2^2 + dx_3^2 = dr^2 + r^2\, d\theta^2 + r^2 \sin^2\theta\, d\phi^2.$$

This leads to (10–11) which gives (10–9) on substituting (10–12).

10–2. The radius of the circle is $a\chi = $ constant.
From (10–9) we see that an element at distance $a\chi$ has length $dl^2 = a^2 \sin^2 \chi(\sin^2 \theta\, d\phi^2 + d\theta^2)$:

(i) If we choose $\theta = \pi/2$, as the plane of the circle, the circumference becomes $\oint dl = \int_0^{2\pi} a \sin \chi\, d\phi = 2\pi a \sin \chi$. The ratio of circumference to length therefore is $\chi^{-1} 2\pi \sin \chi \leq 2\pi$

(ii) The area of an element on the sphere is

$$d\sigma = (a \sin \chi \sin \theta\, d\phi)(a \sin \chi\, d\theta).$$

The area of the whole sphere therefore is

$$\iint a \sin \chi \sin \theta\, d\phi a \sin \chi\, d\theta = 4\pi a^2 \sin^2 \chi.$$

(iii) The element of three-dimensional volume suggested by (10–9) is

$$dV = (a\, d\chi)(a \sin \chi\, d\theta)(a \sin \chi \sin \theta\, d\phi) = a^3 \sin^2\chi \sin \theta\, d\theta\, d\phi\, d\chi$$

from which (10–14) and (10–15) follow.

10:15

10-3. (i) $x_4^2 = -a^2 - r^2,$ $\qquad dx_4^2 = -\dfrac{r^2\, dr^2}{(a^2 + r^2)}$

$$dl^2 = dr^2 + r^2(\sin^2\theta\, d\phi^2 + d\theta^2) - \frac{r^2\, dr^2}{a^2 + r^2}.$$

This is equivalent to (10–18). With $r \equiv a \sin h\chi$

(ii) $\quad dl^2 = a^2 \sin h^2\chi\,(\sin^2\theta\, d\phi^2 + d\theta^2) + \dfrac{a^2\,(d\sinh \chi)^2}{1 + \sinh^2\chi}$

but $1 + \sinh^2 \chi \equiv \cosh^2 \chi$ and $d \sinh \chi = \cosh \chi\, d\chi$, so that we obtain (10–19). For $a\chi =$ constant, $\theta = \pi/2$, $dl = a \sinh \chi\, d\phi$, and a circle has circumference $2\pi a \sinh \chi \geq 2\pi a\chi$.

(iii) The element of surface is then seen to be

$$d\sigma = (a \sinh \chi \sin \theta\, d\phi)\,(a \sinh \chi\, d\theta)$$

for a surface $a\chi =$ constant. Integration over all values of θ and ϕ gives (10–20).

(iv) Similarly the volume element $dV = (a \sinh \chi\, d\theta)\,(a \sinh \chi \sin \theta\, d\phi)$ $(a\, d\chi)$ which leads to (10–21).

10-4. $1 + z = a(t_0)/a(t_1)$. A Taylor expansion in $(t_0 - t_1)$ gives

$$1 + z = a_0[a_0 + \dot{a}_0(t_1 - t_0) + \ddot{a}_0(t_1 - t_0)^2/2]^{-1}$$

which, on expanding in $(t_1 - t_0)$ leads to (10–27).

10-5. $\sinh^2\chi = \left(\chi + \dfrac{\chi^3}{6} + \cdots\right)^2 = \chi^2 + \dfrac{\chi^4}{3} + \cdots = \chi^2 - \dfrac{k\chi^4}{3} + \cdots, k = -1.$

$\qquad \sin^2\chi = \left(\chi - \dfrac{\chi^3}{6} + \cdots\right)^2 = \chi^2 - \dfrac{\chi^4}{3} + \cdots = \chi^2 - \dfrac{k\chi^4}{3} + \cdots, k = 1.$

$$N(\chi) = \frac{4\pi}{3}\, n\chi^3\left(1 - \frac{k}{5}\chi^2 + \cdots\right).$$

10-6. On inverting (10–27), we have

$$\Delta \equiv (t_0 - t_1) = z\,\frac{a_0}{\dot{a}_0}\left[1 - z\left(1 - \frac{\ddot{a}_0 a_0}{2\dot{a}_0^2}\right)\right].$$

The integral (10–24), then becomes

$$\chi \sim \int_{t_0 - \Delta}^{t_0} \frac{c\, dt}{a_0 - \dot{a}_0(t_0 - t)} \sim \frac{c}{a_0}\left[\frac{z a_0}{\dot{a}_0} - \frac{z^2}{2}\frac{a_0}{\dot{a}_0}\left(1 - \frac{\ddot{a}_0 a_0}{\dot{a}_0^2}\right)\right].$$

With $\sigma(\chi) \sim \chi$ we now have

$$5 \log \left(\frac{\sigma(\chi) a_0 (1 + z)}{10 \text{pc}} \right) = 5 \log \left\{ \left[\frac{a_0 z}{\dot{a}_0} \right] \left[\frac{c}{10 \text{pc}} \right] \left[1 + \frac{z}{2} \left(1 + \frac{a_0 \ddot{a}_0}{\dot{a}_0^2} \right) \right] \right\}$$

To first order $\dot{M}_0(t_0 - t_1) = - \dot{M}_0 \dfrac{z a_0}{\dot{a}_0}$ and $\log_{10}(1 + A) = \dfrac{A}{2.303}$.

$A \ll 1$ so that $m = M_0 - 45.06 + 5 \log \dfrac{a_0 z}{\dot{a}_0} + \dfrac{2.5}{2.303} \left(1 + \dfrac{a_0 \ddot{a}_0}{\dot{a}_0^2} - 2\mu \right) z.$

10-7. $q_0 = - \dfrac{a_0 \ddot{a}_0}{\dot{a}_0^2}.$

$\dot{a}_0 = a_0 H,$ $\ddot{a}_0 = a_0 H^2$ $\therefore q_0 = -1.$

10-8. (i) $\Lambda = (a^2)^{-1} + \dfrac{\kappa P}{c^2}$ since $\dot{a} = \ddot{a} = 0$ in (10–45).

(10–47) then follows from (10–44) and (10–45).

(ii) The result follows from substitution of the given values in (10–45) and (10–44).

10-9.   (10–44) and (10–45) yield

$$\Lambda = \frac{3 \dot{a}^2}{c^2 a^2} = \frac{2 a \ddot{a} + \dot{a}^2}{c^2 a^2}$$

which has the solution (10–48). The age is $a/\dot{a} = H^{-1} = \sqrt{3/\Lambda c^2}.$

10-10. (10–44) and (10–45) give $-2a\ddot{a} = \dot{a}^2 + kc^2 = \dfrac{\kappa \rho c^2 a^2}{3}$ so that

$$-\frac{\ddot{a}}{a} = \frac{4\pi \rho G}{3} = -\frac{\dot{a}^2}{a^2}\left(\frac{\ddot{a}a}{\dot{a}^2}\right) = H^2 q_0 \text{ and } 2q_0 - 1 = \frac{-2a\ddot{a}}{\dot{a}^2} - 1 = \frac{kc^2}{\dot{a}^2} = \frac{kc^2}{H^2 a^2}.$$

10-11.   Initially, by (10–44), (10–45) $6\ddot{a} = ac^2[2\Lambda - \kappa \rho]$. If ρ decreases because of a disturbance, $\ddot{a} > 0$, the universe expands, which decreases ρ further, and so on (Ed30).

10-12. (a) $\dfrac{\kappa P}{c^2} = - \left[\dfrac{2a\ddot{a} + \dot{a}^2 + c^2}{c^2 a^2} \right] = \dfrac{\kappa \rho}{3} = \dfrac{c^2 + \dot{a}^2}{c^2 a^2}$

so that initially $\ddot{a}a = -(c^2 + \dot{a}^2).$
We try the solution (10–52) that gives

$$\dot{a} = c \cot x, \qquad \ddot{a} = \frac{-c}{b_0 \sin^3 x}$$

which satisfy the differential equation above.

(b) The trial solution (10–53) satisfies the equation

$$2\ddot{a}a = -(\dot{a}^2 + c^2)$$

which follows from (10–45) at this stage of evolution. We see this since

$$\dot{a} = c \sin x [1 - \cos x]^{-1} \qquad \ddot{a} = -c[a_0(1 - \cos x)^2]^{-1}$$

(c) The equation to be satisfied now is $\ddot{a}a = c^2 - \dot{a}^2$ which, following the above procedures, is satisfied by (10–54).

(d) Similarly (10–55) satisfies $2\ddot{a}a = -\dot{a}^2 - c^2$ for the late stages of a hyperbolic universe.

10–13. The two requirements imposed on (10–66) are

$$V^2 = 0 = H^2 r^2 + 2GM\left[\frac{1}{r} - \frac{1}{r_{0\,\text{max}}}\right]$$

and

$$\frac{dV^2}{dr} = 0 = 2H^2 r - \frac{2GM}{r^2}; \qquad r^3 = \frac{GM}{H^2}.$$

Substituted into the first relation this gives $r = \dfrac{3r_{0\,\text{max}}}{2}$ and hence $r_{0\,\text{max}}^3 = \dfrac{8}{27}\dfrac{GM}{H^2}$.

10–14. For continuous formation of Th^{232} and U^{238} at rates dn_T/dt and dn_U/dt, respectively, since a time t in the past, the present abundance ratio R should be

$$R = \frac{\displaystyle\int_0^t \frac{dn_T}{dt} 2^{-t/\tau_T}\, dt}{\displaystyle\int_0^t \frac{dn_u}{dt} 2^{-t/\tau_U}\, dt}$$

where τ_T and τ_U are the half-lives, in aeons.

$$\therefore R = \frac{14}{4.5} 1.6 \frac{[1 - e^{-0.69t/\tau_T}]}{[1 - e^{-0.69t/\tau_U}]} = 3.3$$

$\therefore t \sim 20$ æ, but this does not agree with the $\text{U}^{238}:\text{U}^{235}$ ratio.
If 60% of the material had formed at time t and 40% between t and 5 æ ago, the ratio would be

$$R = 1.6\left[0.6\,\frac{e^{-0.69t/\tau_T}}{e^{-0.69t/\tau_U}} + 0.4\,\frac{\tau_T}{\tau_U}\left[\frac{e^{-3.45/\tau_T} - e^{-0.69t/\tau_T}}{e^{-3.45/\tau_U} - e^{-0.69t/\tau_U}}\right]\right] = 3.3$$

which gives $t \sim 7$ æ.

11

Life in the Universe

11:1 INTRODUCTION

Since prehistoric times, men have always wondered about where they came from and where life originated. As it became apparent that the earth was just one planet orbiting the sun, that the sun was just one star among some $\sim 10^{11}$ in our galaxy, and that the Galaxy itself was only one such object among $\sim 10^{11}$ similar systems populating the universe, it became clear that life on other planets, near some other star, in some other galaxy was possible. The cosmological principle (section 10:3) also makes this idea philosophically attractive.

It would suggest that life is some very general state of matter that must be prevalent throughout the universe. The probability of finding some form of life, however primitive, on other planets either within the solar system or around nearby stars seems very high from this point of view. Nevertheless, it has not been possible to make unequivocal predictions about where life should exist, mainly because we do not yet understand the thermodynamics of living organisms and what different forms life may take.

11:2 THERMODYNAMICS OF BIOLOGICAL SYSTEMS

In thermodynamics one distinguishes between three types of systems. *Isolated systems* exchange neither energy nor matter with their surroundings. *Closed systems* exchange energy but not matter, and *open systems* exchange both matter and energy with their surroundings. Biological systems always are open; but in carrying out some of their functions, they may act like closed systems.

The processes that go on in living systems are also characterized by a certain

11:2

type of time dependence. Some physical processes could take place equally well whether time runs forward or backward. If we viewed a film of a clock's pendulum, we would not be sure whether the film was running forward or backward. Only if the film also showed the ratchet mechanism that advances the hands of the clock, would we be able to tell that the film was running in the right direction. The pendulum motion is reversible but the action of a ratchet is an *irreversible process*. Life processes are invariably irreversible.

In an irreversible process, *entropy* always increases. Entropy is a measure of disorder. If a cool interstellar grain absorbs visible starlight and re-emits the radiation thermally, it does so by giving off a large number of low energy photons. In equilibrium the total energy of emitted photons equals the energy of the absorbed starlight; but the entropy of the emitted radiation is larger. The increased entropy is a measure of the disorder associated with a large number of low energy photons moving in unpredictable, arbitrary directions. The initial state of a single photon carrying a large amount of energy is more orderly and, hence, characterized by a lower entropy.

Biological systems thrive on order. They convert order in their surroundings into disorder. In so doing, however, they also increase their own internal degree of order. The entropy in the surroundings increases, the internal entropy can decrease, but the total entropy of system plus surroundings always increases. The second law of thermodynamics, which states that the overall entropy change— of the entire universe—in any process is always positive, is therefore not violated.

It may seem strange that biological systems can increase their internal order in this way; but actually we encountered a similar process in the alignment of interstellar grains (section 9:8). We saw there that anisotropic starlight arriving at a grain primarily from directions lying within the plane of the Galaxy, tended to orient the grain so it was spinning with its angular momentum axis lying in the plane. An oriented set of grains shows greater order than randomly oriented dust; and the decrease in the grains' entropy is produced through the absorption of low entropy, anisotropic starlight and emission of high entropy isotropic infrared radiation.

These interstellar dust grains are in a state of *stationary nonequilibrium*. Such a state is characterized by transport of energy between a source at high temperature (the stars) and a sink at low temperature (the universe). There is no systematic change of the system in time, although statistical fluctuations in the orientation, angular momentum, and other properties of the grains do take place.

We hope that the study of stationary nonequilibrium processes will lead to a better understanding of the behavior of biological systems (Pr61). For, when a plant absorbs sunlight—photons whose energy typically is ~ 2 ev—and thermally re-emits an equal amount of energy in the form of 0.1 ev photons, it is acting like a stationary nonequilibrium system and, in fact, a wide range of biological processes seem to fit into this pattern.* Pendulum clocks also are stationary systems. Low entropy energy in the form of a wound-up spring is irreversibly turned into high entropy heat. As Schrödinger pointed out (Sc44), living organisms and clocks have a thermodynamic resemblance.

Nonequilibrium characterizes virtually every astrophysical situation since energy is always flowing out of highly compact sources into vast empty spaces. Any biological system stationed near one of these sources could make good use of this energy flow. It would therefore seem that the conditions necessary for the existence of life in one form or another would be commonplace. Maybe life does abound; but perhaps it is of a form we do not yet recognize.

Fred Hoyle (Ho57) has speculated that interstellar dust clouds might be alive. From a thermodynamic viewpoint the situation would be ideal. We know that dust clouds absorb perhaps half of the starlight emitted in a spiral galaxy like ours. The grain temperatures are so low that a maximum increase in entropy can be produced. What is uncertain, however, is whether the grains are not too cold to make good use of the available energy. At the 10 to $20°K$ temperatures that might be typical of interstellar grains, the mobility of atoms within the grains is so low that the normal characteristics we associate with life might be ruled out (Pi66).

A thermodynamically similar scheme has been suggested by Freeman Dyson (Dy60) who has proposed that intelligent civilizations would build thin shells around stars to trap starlight, extract useful energy, and then radiate away heat in the infrared. Perhaps some infrared sources are such objects (Sa66b).

Our experience on earth is that life will proliferate until stopped by a lack of energy sources, a lack of raw materials, or an excess of toxins. It would perhaps be surprising if no form of life had adapted itself sufficiently to make use of the huge outpouring of energy that goes on in the universe and apparently is just going to waste.

A search for unknown forms of life might concentrate on striking examples of nonequilibrium. A Martian astronomer, for example, would find only two pieces of evidence for life on earth. The first is a radio wave flux that would correspond to a nonequilibrium temperature of some millions of degrees. This is produced by radio, television, and radar transmitters. The second is an excess of methane, CH_4, which is very short-lived in the presence of atmospheric oxygen. It is converted into CO_2 and H_2O. Its nonequilibrium concentration, which could be spectroscopically detected from Mars, is rapidly replenished by methane bacteria that live in marshes and in the bowels of cows and other ruminants (Sa70b).

---

*The photochemistry of green plants of course is an enormously varied process that also centers about the buildup of large molecules.

11:2

11:3 ORGANIC MOLECULES IN NATURE AND IN THE LABORATORY

Granted that we do not specifically know how to search for exotic forms of life could we not find indications of extraterrestrial life in the form familiar on earth? All our living matter contains organic molecules of some complexity—proteins and nucleic acids, for example—and we might expect to find either traces of such molecules or at least of their decay products.

The last few years have shown two quite distinct locations in which such complex molecules are found. There may be many more. First, observations of interstellar molecules by means of their microwave spectra have revealed the existence of such organic molecules as hydrogen cyanide, methyl alcohol, formaldehyde, and formic acid (section 9:4). Such molecules as $HC_{11}N$ also exist in interstellar space, and it is quite likely that much larger and more complex molecules will be found in the next few years.

Second, an analysis of a meteorite—a *carbonaceous chondrite*—that fell near Murchison in Australia on September 28, 1969 showed the presence of many hydrocarbons and of 17 amino acids, including six that are found in living matter (Kv70). One such amino acid was alanine. It has the form

$$
\text{Alanine:} \qquad
\begin{array}{c}
\overset{\displaystyle O}{\overset{\|}{}} \\
CH_3-CH-C-OH \\
| \\
NH_2
\end{array}
\qquad (11\text{–}1)
$$

All *organic acids* are marked by the group of atoms

$$
\begin{array}{c}
\overset{\displaystyle O}{\overset{\|}{}} \\
-C-OH
\end{array}
$$

and *amino acids* contain the additional characterizing *amino group* NH_2.

Contamination by terrestrial amino acids seems to be ruled out by three features of these observations.

(a) Alanine can occur in two different forms. One in which the CH_3, NH_2, COOH, and H surrounding the central carbon atom are arranged in a configuration that causes polarized light to be rotated in a left-handed screw sense. The other in which polarized light would be rotated in the opposite sense. These are respectively labeled L- and D-alanine. The symbol L stands for levo—left—and D for dextro—right.

All amino acids can be derived from alanine. If derived from L-alanine such an acid is called an L-amino acid, and if derived from D-alanine, a D-amino acid. All amino acids found in proteins are L-amino acids. Although not all of them rotate light in a left-handed screw sense, they all can be structurally derived from L-alanine.

The Murchison meteorite showed D- and L-forms in essentially equal abundances. These amino acids are therefore very unlikely to have been biogenic contaminants.

On earth, amino acids are overwhelmingly in the left-handed form. Why that should be so is a mystery, because chemically speaking the right- and left-handed forms are equally probable. They are simply mirror images of each other. Perhaps evolutionary considerations have played a role. It might be that primitive life existed in a racemic mixture—having both L and D forms—and that the L-form won out in a competition for the raw materials essential to life.

It may in fact be impossible for racemic life to exist in an effective way. The search for nutrients would be inefficient. A bolt in search of a nut is more readily satisfied if all nuts and bolts have a right-handed thread; trying to match nuts and bolts from a racemic mixture would be extremely vexing.

(b) A second distinctive feature of the Murchison material was that the ratio of carbon isotopes C^{13} to C^{12} was about twice as high as normally found in terrestrial material. This too indicated that contamination could be ruled out.

(c) Finally some of the amino acids found in the material consisted of nonprotein amino acids. This could not have been a contaminant.

Biogenic molecules and molecules needed for the existence of life therefore seem prevalent elsewhere. They occur naturally not just on earth.

We still have to ask how these molecules arise. Is their fabrication simply achieved under normal astrophysical conditions? The answer to this seems to be "Yes."

A series of experiments that had their foundations in the work of Miller (Mi57a, Mi59) has shown that amino acids and other molecules found in living organisms can be produced artificially if mixtures of gases such as ammonia NH_3, methane CH_4, and water vapor H_2O are irradiated with ultraviolet radiation, subjected to electrical discharges or shocks, to X-ray, γ-ray, electron- or alpha-particle bombardment. Thus far these molecules always have been produced in racemic mixtures. Since all the gases that are used in the experiment are atmospheric constituents on a number of planets and are expected universally on cosmic abundance grounds, it seems that solar ultraviolet, X-ray, and cosmic ray bombardment, occasional irradiation by nearer supernova explosions, and other natural sources of irradiation should have been able to produce molecules of biological interest within planetary atmospheres.

This should be true not only in the solar system, but in other similar systems.

11:3

Perhaps the planets around other stars have had sufficiently similar histories so that life would be expected there too.

Although such molecules are readily formed by energetic bombardment, they are also readily destroyed by it. A biogenic molecule formed in the atmosphere might therefore be destroyed unless it were rapidly removed to a safer place. On earth, rain could have washed such molecules out of the atmosphere and into the oceans where they would be shielded from destructive irradiation by a protective layer of water.

We note that the conditions for forming life—or highly ordered biogenic molecules—are those that seem thermodynamically favorable (section 11:2). There is a source of low-entropy energy in solar ultraviolet or cosmic ray irradiation and a possiblity of converting this energy into a higher entropy form through collisions with atmospheric molecules or through radiation at long wavelengths.

11:4 ORIGINS OF LIFE ON EARTH

Before we can make a rough guess about the origins of life on earth, we should know something about the earth's atmosphere during the aeons immediately following the birth of the solar system.

Initially, the atmosphere seems to have contained no molecular oxygen. The most prevalent atmospheric molecules probably were those strongly reduced by the presence of abundant amounts of hydrogen: methane, ammonia, water, and ethane.

Over one or two aeons, the atmosphere became less rich in hydrogen. Free oxygen appeared, perhaps as hydrogen was freed from water vapor through ultraviolet dissociation in the upper atmosphere. Such hydrogen atoms may have escaped the atmosphere altogether although how that happened does not seem well established (Va71).

The forms of life that would be formed under the earliest reducing conditions would be anaerobic. Presumably the very first organism to be formed found itself in a rich environment of large organic molecules (Op61a, b) that had been built up by ultraviolet irradiation and other bombarding mechanisms discussed in section 11:3. Such an organism could feast and procreate at will, until the supply of organic molecules dwindled. Those organisms which obtain energy by breaking down pre-existing molecules are called *heterotrophs*. Clearly they would be at a disadvantage compared to *autotrophs*, organisms which in addition could also make use of energy in other forms; autotrophs that make use of sunlight are called *photoautotrophs*. The autotrophs probably soon took over. Initially they must have been anaerobes, but as hydrogen kept escaping at the top of the atmosphere and oxygen became more prevalent, the anaerobes came to be at a disadvantage

11:4

compared to aerobes which form the basis on which all the higher organisms of today have evolved. When the oxygen concentration in the atmosphere became roughly one percent of its present abundance, respiration should have become a more efficient process than fermentation and the aerobes may have originated at that time.

Living organisms naturally suffer mutations in their genetic makeup—that is, in the code that defines the makeup of the progeny. The mutation rate can be artificially increased through X-ray and other destructive bombardment. The aerobes probably arose from anaerobes through such mutative processes. Being able to make use of atmospheric oxygen, they soon became the dominant form of life. The anaerobes nowadays can proliferate only under conditions where atmospheric oxygen is somehow excluded.

The balance between stability and mutability seems to be particularly important in forms of life that succeed. Without mutability an organism cannot adapt to changes in its environment; but without some stability, higher forms could not evolve either. In Darwin's theory of survival of the fittest, these fittest are likely to be produced through occasional mutations of rather stable forms. The death of individuals seems essential, in order that life forms may evolve. Yet, for life to evolve optimally, each fit individual should attempt to survive—resist death. Presumably there is an optimal eugenic life span. This will vary from species to species. Some male spiders are devoured immediately after mating. For men, a longer life span must be desirable since they are needed to help rear the young.

We think that an eventual grouping of small organisms into larger ones led to multicellular forms and eventually to the higher forms of life encountered today. Interestingly, irreversible thermodynamics should play a role not only in the metabolism of life, as in section 11:2. It is likely that the growth of more highly organized forms of life through mutations and consequent changes in metabolic forms and metabolic rates can also be described using the methods of irreversible thermodynamics.

The next few years should show major advances in the thermodynamic approach to theoretical problems of life, in laboratory studies of the formation of ever more complex biological forms and in our searches for extraterrestrial evidence for life-supporting molecules.

11:5   COMMUNICATION AND SPACE TRAVEL

If life exists elsewhere in the universe, perhaps it also shows intelligence. If it is intelligent, perhaps it has organized into a civilization. How should we exchange information with it and how would others be likely to get in touch with us? (Sh66) (Dr62)

11:5

This is a problem in communications. How does one most effectively send messages over large distances? How does the enormous time lag between sending and reception of electromagnetic signals affect the problem of communication? These questions are actively being studied, but no single optimum way has yet been discovered. Much depends on what we would like to do best.

If you like to travel, perhaps a rocket journey at relativistic speeds would suit you. But then you must decide how to stay alive during the long trip. Suggestions have been made for deep-freezing spacemen who would undertake the journey. Thus far, nothing much bigger than a frog has been successfully frozen and revived, and it is not clear whether the technique could be developed for large mammals. Unmanned spaceflight or flights in which several generations would pass before landing also are possibilities.

Alternately, we might be willing to restrict ourselves to communicating through transmission of radio or visible signals, or perhaps infrared or X-ray messages. Is there any one electromagnetic frequency that is optimal? If such a frequency is found, we still must ask ourselves whether its characteristics are optimal only because of our particular technological resources, or whether there is some more fundamental reason for choosing this particular means of communicating. It is clear that if we choose the wrong frequency for transmitting our signals, no one is likely to receive them. We also are likely to miss messages sent by other civilizations, if we do not know to which frequency we should tune our receiver. We cannot tune in on all frequencies because that might require an insurmountable financial effort. We have to second-guess the correct frequency in order to make the initial communication contact more probable.

What about tachyons? In section 5:12 we mentioned these particles that would travel faster than light. Clearly, if tachyons existed they would have many desirable properties. They could travel at millions of times the speed of light and would therefore make meaningful two-way conversations a real possibility. Moreover, tachyons require only low transmission energy (5–56) and might therefore be very economical. Finally, tachyons would apparently free us from the limitations imposed by cosmic horizons (section 10:11). The one disadvantage might be that tachyons seem not to readily interact with normal matter—otherwise they should probably have been detected by now. Construction of suitable transmitters and receivers might therefore be difficult.

There are an apparently endless set of questions to be answered before an effective means for communicating with other civilizations can exist. We need only note that if tachyons could be readily produced and received, other civilizations would probably use them to the exclusion of all other means of communicating. Yet we do not even know whether tachyons can exist.

To show some of the questions that need be considered it may be worth thinking about the following two problems.

11:5

PROBLEM 11-1. A spaceship slowly accelerates on its voyage from earth to a distant galaxy. As it accelerates to ever higher speeds, it suffers collisions with interstellar gas and dust, with photons criss-crossing space, with magnetic fields, and with cosmic ray particles. Estimate the effects of these and other possible particles and fields of interstellar and intergalactic space on the momentum of the spaceship, electric charges deposited on the ship and the effect of these charges, the abrasion and ablation effects on the hull of the ship, heating effects, and so on. What are the most serious limitations? Almost everything discussed in Chapters 6 and 9 bears on this problem.

PROBLEM 11-2. The rate at which messages can be transmitted and received is normally proportional to the area of the transmitter A and to the solid angle Ω subtended by the receiver at the point of transmission. Let us now assume that the transmitted particles or waves have a momentum range Δp, and that the number of message bits that can be transmitted per unit time—the *bit rate*—equals the number of phase cells contained in the beam sent out during that period (4–65).

(a) For an electromagnetic wave show that the bit rate is

$$\text{Photon bit rate} = A\Omega \frac{v^2}{c^2} \Delta v \qquad (11-2)$$

where v is the frequency, Δv is the bandwidth of the transmitted beam, and the antenna only transmits photons of one polarization.

(b) For a tachyon system, show that if (4–65) is applicable,

$$\text{Tachyon bit rate is} \qquad \left| \frac{A\Omega}{h^3} m^3 c^4 \frac{\Delta N}{N^3} \right| \qquad N \gg 1 \qquad (11-3)$$

where we have assumed that transmission occurs for tachyon velocities ranging from $V = Nc$ to $V = (N + \Delta N)c$, where N is a large number. If the tachyon mass is of the order of the electron mass and the radiation frequency is that of visible light, show that the tachyon bit rate for $N \lesssim 10^7, \Delta N \sim 0.5N$ is several orders of magnitude greater than the electromagnetic bit rate. Show that for $N \sim 10^8$, however, the bit rate and energy expenditure would be comparable to visible light. Equation 5–56 is useful in tackling this problem.

PROBLEM 11-2 ANSWER

This problem is highly speculative, particularly in view of some of the difficulties cited in Chapter 5:

We assume that phase space arguments determine the distinguishability of tachyons and that the number of distinguishable tachyons transmitted per unit time determines the bit rate. For a receiver with area A and solid receiving angle Ω, the volume from which tachyons are received per unit time is ANc, where N is the tachyon speed measured in units c, the speed of light. The momentum space volume occupied by these tachyons is $\Omega p^2 dp$, per mode of polarization. The number of distinguishable tachyons incident on the detector in unit time (here referred to as bit rate) therefore would be

$$\left| \frac{ANc\,\Omega p^2 dp}{h^3} \right|$$

We make use of the relativistic expression

$$\mathscr{E}^2 = p^2 c^2 + m^2 c^4 = m^2 c^4 (1 - N^2)^{-1}$$

relating energy $\mathscr{E}$ and rest mass m to momentum and velocity. This leads to the (imaginary) momentum value

$$p = \frac{N}{\sqrt{1 - N^2}} mc$$

and the bit rate obtained for a velocity range dNc reads

$$\left| \frac{A\Omega}{h^3} m^3 c^4 \frac{dN}{N^3} \right| \qquad \text{for} \qquad N \gg 1$$

The corresponding expression for electromagnetic radiation is

$$\frac{A\Omega v^2 dv}{c^2}$$

where dv is the frequency of the radiation. If we take the frequency to be that of visible light, and take m to be an electron mass, the tachyon bit rate is seen to be many magnitudes greater than the electromagnetic bit rate, as long as N remains less than about 10^7 and $dN/N \sim dv/v$. At that speed, the energy per tachyon would be about $10^{-7} mc^2$ corresponding to about 0.1 ev while the visual radiation would require a transmission energy about an order of magnitude higher.

If $N \sim 10^8$, the bit rate and energy expenditure per message is comparable to that for visible light, but communication across the universe can be achieved in times of the order of 100 years.

A question about the stability of tachyons has recently (Be71) been raised. If they exist but are unstable they would not be suitable information carriers. Clearly our ideas on tachyons still are speculative.

12

Cosmic Origins

In this final chapter, we pool all the resources previously developed to probe back to the earliest times in the existence of the universe, the Galaxy and the solar system. In this quest, we hope to clarify the origin of the chemical elements, identify processes that led to the formation of galaxies and stars, and come to an understanding of the factors that played a role in the formation of the earth and planets. To this end, we start with an investigation of what might be learned from an analysis of the background radiation reaching us from the most distant parts of the universe.

12:1 BACKGROUND RADIATION

Background measurements occupy a peculiar position in astronomy. Frequently, these observations tell us little about an individual phenomenon or group of phenomena. Instead, they provide a set of useful bounds on the possible behavior of cosmic sources, and allow us to set reasonable limitations on the evolution of the universe during past epochs. By examining the background radiation reaching us isotropically from the sky in all wavelengths, we can place useful limits on the energy budget available for cosmic processes.

These potential uses of background information are important, but we must first unscramble radiation reaching us from a variety of highly diffuse sources, some of which are truly isotropic, while others show spatial structure on a large scale. The diffuse radiation that appears to reach us from the largest distances across the universe is the microwave background flux exhibiting a 2.7°K blackbody spectrum. It appears to be truly isotropic, though the earth's motion through the universe lends the impression of a small dipole asymmetry (section 5:9).

We also expect eventually to observe a hitherto undetected flux originating somewhat closer to us, which was emitted during the earliest stages of galaxy

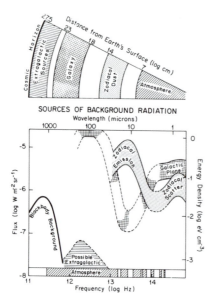

Fig. 12.1 Sources of background radiation showing their expected distance and characteristic spectrum at wavelengths ranging from visible light to microwave frequencies.

formation. Since we do not yet know the epoch at which galaxies first formed, we do not know how redshifted that radiation might be, and therefore do not know just where in the spectrum to search for this component. If the galaxies and their stars formed relatively recently, in an era characterized by a redshift z in the range $z = 5 \rightarrow 10$, this background component would lie in the near infrared. If the earliest galaxies formed around $z = 100$, the emission would be found in the wavelength range from 10 to 100μ.

Within our Galaxy the aggregate emission of billions of red stars provides a further, faint near-infrared background. Also, heated dust clouds emit strongly in the mid and far infrared. Both of these sources are concentrated toward the Galactic plane from which diffuse X-rays and γ-radiation also emanate. Finally, within the solar system, interplanetary dust grains that scatter near-infrared sunlight and re-emit absorbed solar energy at longer wavelengths provide a foreground flux against which all other astronomical sources must be viewed. This zodiacal component is concentrated toward the ecliptic plane. Fig. 12.1 provides a schematic presentation of the expected background contribution from these components. Fig. 12.2 gives an overview of the observed isotropic background radiation at all frequencies at which observations exist. Fig. 12.3 shows the anisotropy of the microwave background radiation, and relates it to the Galaxy's motion with respect to the radiation, in the spirit of section 5:9, as well as with respect to other galaxies.

12:1

Fig. 12.2 The isotropic sky flux versus frequency ν. The left-hand scale gives the observed flux denoted by the solid curves and arrows showing upper limits. The stippled line gives the corresponding radiation density in space and corresponds to the right-hand scale. Upper limits (horizontal bars with downward point-pointing arrows) refer to the left-hand scale.

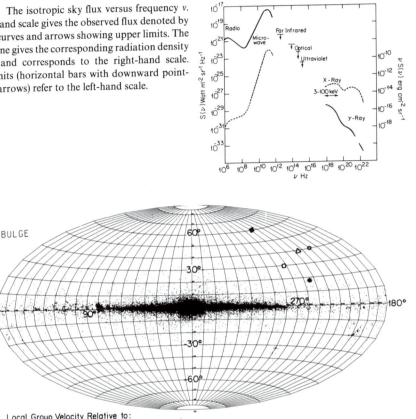

Fig. 12.3 Motion of the local group of galaxies through the microwave background radiation and with respect to various groups of distant discrete sources, superposed on a chart of the Galaxy's far-infrared-emitting giant stars. Velocities are given only where the spectral shift is known in addition to the direction of motion. Other direction estimates come from galaxy observations that assume their distributed radiation to also trace the anisotropy in the mass distribution in Problem 12–14. (Chart due to Beichman, Be87). (Reproduced with permission from the Annual Review of Astronomy and Astrophysics, © 1987 by Annual Reviews Inc.)

12:2 BACKGROUND CONTRIBUTIONS OF DISCRETE SOURCES

We can ask ourselves to what extent a large number of discrete sources filling a field of view could mimic a uniform background. We saw in Chapter 10 that an individual discrete source at distance $a_0\sigma(\chi)$ (defined by (10–25)), whose luminosity at emission is L_1, produces an integrated flux at the observer, given by equation (10–32).

$$\mathscr{F} = \frac{L_1}{4\pi a_0^2\sigma^2(\chi)(1+z)^2} \tag{10–32}$$

where L_1 is the source luminosity locally observed at time t_1, and the quotient on the right represents the uniform spread of redshifted radiation over a surrounding sphere of area $4\pi a_0^2\sigma^2(\chi)$. Our receiving telescope lies on this spherical surface and is pointed toward the source. The factor $(1+z)^{-2}$ in the equation represents the product of redshift and time dilation terms.

If we deal with a nearly monochromatic source whose luminosity L_e is $L_e = L(\nu_e)\,d\nu_e$, then the flux received is

$$\mathscr{F}(\nu_r)\,d\nu_r = \frac{L(\nu_e)\,d\nu_e}{4\pi a_r^2\sigma^2(\chi)(1+z)^2} \tag{12–1}$$

which leads, through

$$\frac{d\nu_e}{d\nu_r} = (1+z), \tag{12–2}$$

where subscripts e and r respectively refer to emission and absorption times, to the expression

$$\mathscr{F}(\nu_r) = \frac{L(\nu_e)}{4\pi a_r^2\sigma^2(\chi)(1+z)} \tag{12–3}$$

This is the spectral line flux to be expected from individual distant quasars and galaxies. To calculate the background radiation received due to a distribution of distant galaxies, we have to integrate over all galaxies in the distance interval χ to $\chi + d\chi$, from which radiation reaches us within the frequency interval $d\nu_r$.

Suppose now that the number density of such galaxies is n_G per unit volume as measured at the time of emission. Using (10–13) or (10–20), we note that the total number of galaxies viewed in the distance interval χ to $\chi + d\chi$ and in solid angle increment $d\Omega$ becomes

$$n_G\,dV = n_G a_e^3\sigma^2(\chi)\,d\chi\,d\Omega \tag{12–4}$$

On combining this with (12–3) we see that the flux received per unit solid angle should be

$$\left.\begin{aligned}\mathscr{F}(\nu_r) &= \int_0^\chi \frac{n_G a_e^3 L(\nu_e)}{4\pi a_r^2(1+z)}\,d\chi = \int_0^\chi \frac{n_G a_e L(\nu_e)\,d\chi}{4\pi(1+z)^3} \\[2mm] &= \frac{c}{4\pi}\int_0^t \frac{n_G L(\nu_e)}{(1+z)^3}\,dt = \frac{c}{4\pi}\int_{z_{min}}^{z_{max}} \frac{n_G L(\nu_e)}{(1+z)^3}\left|\frac{dt}{dz}\right|\,dz\end{aligned}\right\} \tag{12–5}$$

where we have made use of equation (10–24). Integrating over the frequency spectrum using (12–2) yields the integrated flux per unit solid angle

$$
\left.
\begin{aligned}
\mathscr{F}_r &= \int_0^\infty \mathscr{F}(v_r)\, dv_r = \frac{c}{4\pi} \int_{z_{\min}}^{z_{\max}} \frac{n_G L_e}{(1+z)^4} \left|\frac{dt}{dz}\right| dz \\
&= \frac{c}{4\pi} \int_0^\infty \int_{z_{\min}}^{z_{\max}} \frac{n_G(L,z)L(z)}{(1+z)^4} \left|\frac{dt}{dz}\right| dz\, dL(z)
\end{aligned}
\right\}
\tag{12–6}
$$

where the last expression holds for a number of galaxies $n_G(L,z)$ having a luminosity distribution, $L(z)$, that may depend on redshift.

This same question may be considered from a different viewpoint by taking the universe to be isotropic and homogeneous. Suppose the universe is divided into individual cubicles separated by totally reflecting walls which expand along with the universe. The radiation accumulated due to past emission by sources within each cubicle, as measured today, would then be precisely the same as though no walls had existed at all. If at some past epoch characterized by redshift z, the sources within the cubicle had emitted a total energy increment $\Delta\varepsilon$, reflection off the expanding walls would by now have redshifted this energy to a value $\Delta\varepsilon(1+z)^{-1}$. If that cubicle currently corresponds to unit volume—which means that it measured $(1+z)^{-3}$ at the time of emission—then the current energy density is just $\Delta\varepsilon(1+z)^{-1}$ and the integrated flux per solid angle crossing unit surface within the volume or reflected at unit surface area at the walls of the enclosure is just

$$
\mathscr{F}_r = \frac{c}{4\pi} \frac{\Delta\varepsilon}{(1+z)}
\tag{12–7}
$$

This corresponds to equation (12–6), as we can see by considering that we are only interested here in the number density per unit volume as measured today, which is $n_G/(1+z)^3$, and that $\Delta\varepsilon/(1+z)$ corresponds to $\int n_G L_e (1+z)^{-4} |dt/dz|\, dz$ when integrated over a sufficiently short interval Δz.

Equation (12–7) can be employed to place constraints on the amount of matter that could have been converted into radiation at different epochs. Upper limits on the observed background radiation can be used to obtain upper limits for the total radiation energy density $\Delta\varepsilon$ that could have been generated in the universe at any given epoch, z. The ratio $\Delta\varepsilon/\rho_0 c^2$ then provides an upper limit to the fraction of the cosmic mass that could have been converted into radiation at that epoch. Here, ρ_0 is today's baryonic, or atomic mass density of the universe, which has not yet been well determined. We now examine this energy constraint somewhat further.

12:3 COOL EXPANDING UNIVERSE

Let us begin by calculating the energy density of radiation liberated in the formation of heavy elements in stars. If we suppose that all of the elements heavier than hydrogen were formed in stars, we find that the bulk of the energy is liberated just in the formation of helium. The elements heavier than helium are sufficiently rare, and the added binding energy of the nucleons in complex nuclei is sufficiently small, so that heavier elements contribute little to the overall energy budget.

We observe that a fraction $\varepsilon \sim 0.25$ of the mass of atomic matter is in the form of elements heavier than hydrogen. The liberated energy is 0.029 hydrogen masses m_H per helium nucleus formed. While the energy liberated in the formation of helium from hydrogen goes partly into the production of neutrinos and partly into electromagnetic radiation, the bulk of the energy is produced as radiation in both of the major production schemes: the proton–proton reaction and the carbon–nitrogen–oxygen [CNO] cycle. For a contemporary matter density, ρ_0, we therefore expect the radiation density in the universe to be

$$\rho_{rad} \sim \frac{0.029\varepsilon\rho_0 c^2}{4(1+z)} \sim \frac{1.6 \times 10^{18}\rho_0}{(1+z)} \leq \frac{3 \times 10^{-12}}{(1+z)} \text{erg cm}^{-3} \qquad (12\text{–}8)$$

where we have taken $\rho_0 \leq 2 \times 10^{-30}$ g cm^{-3} and an energy injection epoch, z. The dependence on $(1+z)^{-1}$ is readily understood if we consider the ratio of photon-to-nucleon number densities to remain constant during cosmic expansion. Then the energy per photon decreases with increasing redshift as $(1+z)^{-1}$, while the mass–energy for nucleons remains constant. The microwave background corresponding to a temperature of 3°K provides a radiation density of 6×10^{-13} erg cm^{-3}. We therefore see that if the hydrogen-to-helium conversion energy had been recently injected into the universe and had then been rapidly thermalized, it would be possible to account for the observed radiation density, ρ_{rad}. On the other hand, a background due to stellar sources could not have been generated through a hydrogen-to-helium conversion at earlier epochs than $z \sim 4$, because then the red shifted stellar conversion energy would not have sufficed to produce the currently observed background radiation. Instead, we would have to consider other alternatives—perhaps the collapse of massive objects—with annihilation of matter in primordial black holes as the source of radiation.

Energy considerations apart, we also need to explain the background's thermal spectrum. One possible way of subsequently thermalizing the spectrum of the background could have been through the absorption of starlight by dust. The energy could then have been re-emitted in the mid- or far-infrared, and further redshifted to the millimeter-wavelength range where the microwave background radiation is now observed (Rees (Re78)). The argument pursued is slightly different from that sketched earlier in this section. Instead of considering the mass density ρ_0 of the universe directly observed, we consider the possibility of an early, pre-galactic generation of supermassive, $\simeq 10^6 \; M_\odot$ objects, called Population III stars. These stars would have rapidly liberated their energy, evolving into a final dark state where now they can be discerned only through their massive presence—contributing to the binding energy of galaxies and galaxy clusters, but otherwise invisible.

One difficulty with dust as a thermalizer of radiation is the inefficiency with which dust emits at long wavelengths. One would think that dust re-emission would have left its imprint on the background radiation in marked deviations from a blackbody spectrum; but those are not observed. An alternative would be thermalization of radiation through scattering off electrons in a universe that was highly ionized at these early epochs.

The column density of electrons required to produce optical depth unity is the reciprocal of the Thomson scattering cross section, $\sigma_e^{-1} = 1.5 \times 10^{24}$ cm^{-2}. If all the electrons in the universe were evenly distributed today, the column depth would only be $\sim 10^{22}$ cm^{-2} out to the horizon. At epochs $z \geq 10$, however, Thomson scattering could have been significant if all the matter in the universe had been ionized. One can imagine Population III objects lighting up in such a plasma; their radiation would be effectively scattered. But Thomson scattering changes the photon energies only by a fractional amount $(\gamma(v) - 1)$, as seen from equations (6–160) and (6–161). This is of order kT_e/m_ec^2, where $T_e \sim 3(1 + z)°$K is the approximate electron temperature at epoch z. The number of scatterings required for thermalization would then need to be of order $(\gamma(v) - 1)^{-1} \sim m_ec^2/3(1 + z)k \sim 2 \times 10^9/(z + 1)$. In section 12:4 we will see that the time available for scattering at such an epoch would only be $\sim [(H(1 + z)^{3/2}]^{-1}$, where H is today's Hubble constant. The number of scatterings per unit time would then be

$$\frac{(1 + z)^{1/2}m_ec^2H}{3k} \sim n_e\sigma_ec \qquad (12\text{–}9)$$

so that $n_e \sim 2.5 \times 10^5(1 + z)^{1/2}$ cm^{-3}. Such a high density could only have existed at $z \geq 3 \times 10^4$, where the temperature would have been $T \geq 10^5°$K. In turn, that would drive us toward a hot primordial universe, discussed in greater detail in the next sections. At any rate, the idea that the microwave background could be produced by discrete sources has not gained wide acceptance in view of the difficulties just cited.

12:4 EXTRAPOLATING BACK TO EARLY EPOCHS

The isotropy of the background radiation and the thermal spectrum observed provide our strongest clues to the structure of the early universe. To see this we have to analyze the observations quantitatively and recall Einstein's general relativistic field equations (10–44) and (10–45).

$$\frac{8\pi G\rho}{c^2} = -\Lambda + 3[kc^2 + \dot{a}^2][c^2a^2]^{-1} \qquad (12\text{–}10)$$

$$\frac{8\pi GP}{c^4} = \Lambda - [2\ddot{a}a + \dot{a}^2 + kc^2][c^2a^2]^{-1} = \frac{2}{3}\Lambda - \left[\frac{8\pi Gp}{3c^2} + \frac{2\ddot{a}}{c^2a}\right] \quad (12\text{--}11)$$

where ρ is the instantaneous mass–energy density, P is the pressure, Λ is the cosmological constant, k is the Riemann curvature constant, and dots above the *scale factor*, a, represent differentiation with respect to world time. There is little doubt that both P and Λ must be quite small at the present epoch, though earlier in the development of the universe the pressure could well have approached $\rho c^2/3$, and—if there were certain phase transitions at very early opochs—$|\Lambda|$ might also have been large before those transitions took place. At any rate, if we take $\Lambda = 0$ for the current epoch, we see that the *Hubble constant*—which despite its name can vary with time—is

$$H \equiv \frac{\dot{a}}{a} = \left[\frac{8\pi G\rho}{3}\right]^{1/2} \quad (12\text{--}12)$$

provided we set $k = 0$ in (12–10).

This is true only in a flat space. However, we see the velocity of distant galaxies increasing—as far as we can tell—linearly with distance; this implies that (12–12) must be at least a good approximation. In turn, that tells us that $3kc^2/a^2 \ll 8\pi G\rho$, which means that the scale factor a, interpreted as the radius of curvature, must be sufficiently large to at least give the local semblance that the space we inhabit is flat.

With this in mind, we also obtain from (12–11)

$$\frac{\ddot{a}}{\dot{a}} = -\frac{\dot{a}}{2a}, \quad (12\text{--}13)$$

and, on integrating with respect to world time,

$$\ln[\dot{a}a^{1/2}] = \text{constant} \quad (12\text{--}14)$$

or

$$Ha^{3/2} = \text{constant} \quad (12\text{--}15)$$

This is equivalent to (12–12), since we assumed there that the density of the universe was matter-dominated, $\rho c^2/3 \gg P$, in which case the density ρ is inversely proportional to a^3. For other possibilities see section 10:9.

We may still talk about an apparent age, τ, for the universe given by

$$\tau \equiv \frac{1}{H} = \left[\frac{3}{8\pi G\rho}\right]^{1/2} \quad (12\text{--}16)$$

That age is proportional to $a^{3/2}$,

$$\tau \propto a^{3/2}, \quad \rho \propto a^{-3} \text{ [matter-dominated era]} \quad (12\text{--}17)$$

12:4

We can now examine the evolution of the universe at early epochs when the mass–energy density might have been radiation-dominated. For $\Lambda = k = 0$ and

$$P = \frac{\rho c^2}{3} \text{ [radiation-dominated era]} \qquad (12\text{--}18)$$

equation (12–12) still holds, and we have $\ddot{a} = 0$, though, in contrast to (12–17),

$$\rho \propto a^{-4} \text{ [radiation-dominated era]} \qquad (12\text{--}19)$$

This is readily understood if we consider the number of quanta of radiation to be constant during this era. In that case the number density of quanta is proportional to a^{-3}, but the energy per quantum decreases as a^{-1}, because of the red shift. The product of number density and energy per quantum of radiation is therefore proportional to a^{-4}. This holds equally well for photons and other quanta lacking rest mass. To a good approximation it also holds for massive particles at very high temperatures when the particles travel at speeds close to the speed of light, where their rest mass is small compared to the total mass–energy. From equation (12–12) we then have

$$H = \frac{1}{\tau} \propto a^{-2} \text{ [radiation-dominated era]} \qquad (12\text{--}20)$$

In summary:

$$\left. \begin{array}{l} \tau = H^{-1} \propto a[t]^{n/2} \\ \propto [z+1]^{-n/2} \end{array} \right\} \quad \begin{cases} n = 3, \text{ matter-dominated era} \\ n = 4, \text{ radiation-dominated era} \end{cases} \qquad (12\text{--}21)$$

PROBLEM 12–1. (a) Show that the age, t_0, of the universe, when $k = \Lambda = 0$, is just $(2H)^{-1}$ for a relativistic equation of state, $P = \rho/3c^2$.

(b) At a later epoch, when the pressure has dropped to $P = 0$ and the universe is matter-dominated, show that the world time is related to the Hubble constant by $t_a = 2/3H$, provided $t_a \gg t_0'$, where t_0' marks the epoch when the pressure drops well below the relativistic value, toward zero.

From this we see that the world time t_a is always smaller than τ, though τ is a reasonable measure of the age of the universe. Since the epoch during which the universe has been matter-dominated has been far longer than the earlier, relativistic epoch, the actual age of the universe is close to

$$t_a \sim \frac{2\tau}{3} = \frac{2}{3H} \qquad (12\text{--}22)$$

We now are in a position to discuss the evolution of the microwave background radiation during past epochs. The energy density of radiation decoupled from matter must have been proportional to a^{-4} throughout. Since the microwave radiation is thermal, the density is related to temperature by the blackbody expression

$$\rho \propto T^4 \qquad \text{so that } T \propto a^{-1} \qquad (12\text{–}23)$$

Hence, the temperature must have fallen linearly with increasing radius. Throughout the more recent, matter-dominated era, the blackbody radiation temperature, by (12–21), also will have been proportional to $\tau^{-2/3}$. During an earlier, radiation-dominated epoch it would have fallen as $\tau^{-1/2}$. Hence

$$T \propto \tau^{-2/n}, \qquad (12\text{–}24)$$

where n has the same meaning as in (12–21). Now, it is generally assumed that the universe became optically thin to the blackbody radiation some time around the era during which electrons and protons combined to form atoms. At wavelengths longer than the Lyman-α line, hydrogen atoms scatter radiation far less effectively than electrons, whose Thomson scattering cross section is 6.65×10^{-25} cm^2; hence, matter decoupled from radiation at a time when the wavelength of most photons was substantially longer than 1216 Å, a wavelength roughly 10^4 times shorter than the present peak wavelength of the microwave background photons. That would have happened when the wavelength at the radiation peak was 6000–7000 Å, or ~ 1500 times lower than the wavelength at the current microwave peak, roughly a wavelength of 1 mm. Since the temperature of the radiation is also inversely proportional to the wavelength at the blackbody peak, we conclude that the radiation at that epoch must have had a temperature around 4000°K.

The scale factor, a, also would have been ~ 1500 times smaller than now, and the cosmic particle mass-density $\sim (1500)^3$ times larger than the current value, which may be $\sim 10^{-30}$ g cm^{-3}.

At that stage the universe would have been neither entirely radiation- nor entirely matter-dominated, though it would soon become fully matter-dominated and remain that way. Hence, the age of the universe at decoupling was of the order of

$$t \sim \frac{2}{3}\tau_{\text{decoupling}} = \frac{2}{3}\left[\frac{T}{2.7}\right]^{-3/2} H^{-1} \qquad (12\text{–}25)$$

where we have used the currently observed temperature of the background radiation, 2.7°K. If the present age of the universe is 1.3×10^{10} y, the age at decoupling must have been about 60,000 times less, and the universe would have been 2×10^5 years old. Until just after that era the age of the universe would have been proportional to a^2, because radiation would have dominated the mass density.

12:4

12:5 THE ISOTROPY PROBLEM

Equations (10–24) and (12–12) tell us that light can cross the universe at a rate of

$$\frac{d\chi}{dt} = \frac{c}{a(t)} \qquad (12\text{–}26)$$

The maximum available time for travel at any past epoch is $\sim H^{-1}$, which currently is proportional to $a^{3/2}$. Since the distance to be traversed is $a\chi$, it should be easier to travel a parameter distance χ today, than it ever was in the past.

PROBLEM 12–2. Using equations (12–21) and (12–26) as well as the results of Problem 12–1, show that the time taken for light to cross a parameter distance χ and to return to the starting point is always more than twice the time it took to traverse χ in the first place, whether the universe is in a relativistic phase or is filled with nonrelativistic matter.

As it is, we think we are receiving the microwave radiation from distances so great, that the radiation from the recombination era is only just reaching us now. Since the radiation arrives from all parts of the sky, including portions located in diametrically opposite directions, we must conclude that the mircowave radiation originated during an era when different portions of the universe were too far apart to communicate. Yet the radiation we receive from all directions appears identical.

The puzzle to be solved is how portions of the universe, which appear never to have been in causal contact at any previous epoch, could have generated a radiation bath that is as homogeneous and isotropic as the microwave background we now observe.

This problem has received a great deal of attention (Gu81, Al82) and is further discussed in section 12:9 below.

12:6 THE DENSITY PARAMETER Ω

If we set $\Lambda = 0$ in equation (12–10), we can define a critical density ρ_{crit}

$$\rho_{crit} = \frac{3H^2}{8\pi G} \qquad (12\text{–}27)$$

which is just the density given by (12–12). This density is critical in the sense that it implies, through (12–10) and (12–11), that the acceleration $\ddot{a}$ will approach a value of zero, asymptotically as $\dot{a}$ approaches zero, provided $P = k = 0$:

$$\ddot{a} = -\frac{\dot{a}^2}{2\ddot{a}} \qquad (12\text{–}28)$$

The universe then continues to expand indefinitely, but the velocity of expansion decelerates monotomically. This case is shown in Fig. 10.7.

Right now, the actual density of the universe is hardly known. A minimum density is obtained if we estimate the mass condensed in galaxies and divide it by some estimated volume per galaxy, as judged by the volume occupied by the nearer galaxies. To this, we need to add an estimate of the mass contained in intergalactic matter, partly judged from X-ray emission by hot intergalactic gas. If neutrinos have rest mass, the density may, however, be dominated by the additional mass density of these neutrinos, and then the actual density could be far larger than the density comprised in atomic matter. For these reasons the actual density of the universe may substantially differ from our current best estimate of ρ_{crit} which, for a Hubble constant of 75 km sec^{-1} Mpc^{-1} or an expansion age of 1.33×10^{10} y, takes on a value of 1.1×10^{-29} g cm^{-3}. The directly observed atomic matter density of the universe lies an order of magnitude below this value.

In order to describe the evolution of the universe for different possible density values, one frequently makes use of a *density parameter* Ω, defined by

$$\Omega \equiv \frac{\rho}{\rho_{crit}} \tag{12-29}$$

Equations (12–10), (12–11) and (12–12) with $P = \Lambda = 0$, then give

$$\frac{\ddot{a}}{a} = -\frac{\Omega H^2}{2} \tag{12-30}$$

The *deceleration parameter* defined by

$$q_0 \equiv -\frac{a\ddot{a}}{\dot{a}^2} \tag{12-31}$$

thus becomes

$$q_0 = \frac{\Omega}{2} \tag{12-32}$$

We can now examine the ratio of radiation density and matter density at different epochs characterized by the redshift z with which the radiation reaches us. Equations (12–18) and (12–23) tell us that an observed ratio today corresponds to a larger ratio earlier:

$$\left.\frac{\rho_{rad}}{c^2 \rho_{matter}}\right|_{today} < \left.\frac{\rho_{rad}}{c^2 \rho_{crit} \Omega}\right|_{earlier} \tag{12-33}$$

where we have assumed that the current mass-density of the universe is dominated by matter. For a radiation temperature of 2.7°K, the current radiation density

Fig. 12.4 Density/temperature history of the universe since the universe was one second old (after Longair, Lo78). (Reproduced with permission, © 1978 by D. Reidel Publishing Company, Dordrecht, Holland.)

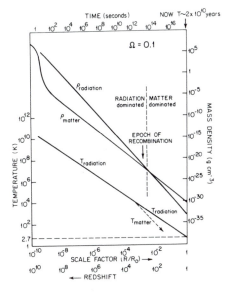

is given by 4×10^{-13} erg cm^{-3}, as in (12–19), equivalent to a mass-density of 4.5×10^{-34} g cm^{-3}. This is a factor of $\sim 2.5 \times 10^4$ smaller than our estimate of the critical density, based on the observed Hubble constant. Figure 12.4 shows a temperature/density history of the universe given by Longair (Lo78) for $\Omega = 0.1$.

12:7 HOT EXPLODING UNIVERSE

The extrapolation backward in time forces us toward the conclusion that the universe must initially have been enormously hot, provided, as discussed in section 12:3, the opacity of the cosmos has always been low over past aeons. If the universe had been opaque recently, the energy for the microwave background could have been generated in the recent past, at rather little expense of energy through conversion of matter into radiation; it could then have acquired its thermal spectrum in passage through the high-opacity medium. In section 12:4 we saw that— barring recent, high opacity—the radiation temperature must have been 4000°K at $z \sim 1500$, so that we can compute the energy density in radiation and show that it must have approximately equalled the atomic matter-energy density at that epoch, obtained from (12–17).

If we now ask how so much radiation energy could have originated in the universe—unless, of course, it was primordially present—we find only matter annihilation to be a suitably energetic source. No other possibilities appear to exist.

To understand such sources of annihilation we must go far back in time. Electron–positron pairs, which would have to exist at very high temperatures, would annihilate at temperatures below

$$T_{ea} \sim \frac{m_e c^2}{k} = 6 \times 10^{9\circ}\text{K} \qquad (12\text{--}34)$$

so that in thermal equilibrium between matter and radiation, electron–positron pair densities rapidly decline below temperature T_{ea}. This energy is transferred to the radiation bath. The temperature T_{ea} corresponds to a redshift $z \sim 2 \times 10^9$. Here k is the Boltzmann constant, not to be confused with the curvature constant conventionally labeled with the same symbol.

Before annihilation, the number of electrons was roughly comparable to, though lower than, the number of photons. After annihilation, the number of electrons equaled the number of baryons, roughly 10^9 times lower than the number of photons in the universe. That ratio then remained virtually unaltered throughout the subsequent evolution of the universe. Today the number density of photons is ~ 500 cm^{-3} and the baryon density is $\leq 10^{-6}$ cm^{-3}.

If matter and antimatter both were present at the earliest epochs, then the baryons currently observed may merely represent the excess of baryons over antibaryons present earlier. The annihilation temperature for protons and antiprotons is

$$T_{pa} \sim \frac{m_p c^2}{k} = 10^{13\circ}\text{K} \qquad (12\text{--}35)$$

That temperature is equivalent to an energy of 931 MeV, well within the range of current accelerator energies; and the physics of matter at these energies is thoroughly understood. We could, therefore, pursue this question further in a search for the original sources of energy in the universe, but we will leave that quest for now and only return to it in sections 12:10 and 12:12 below.

Before electron–positron annihilation, radiation and matter interacted rapidly, since densities were high. Through annihilation, however, the number densities of electrons and positrons rapidly dropped, leading to the drop in matter density seen in Fig. 12.4 at a cosmic age of only a few seconds. Later, just before the era of electron–proton *recombination* to form hydrogen, the matter density was $(1 + z)^3 \sim (T/2.7^\circ\text{K})^3 \sim 3.4 \times 10^9$ times higher than at present, the electron number density was of order 10^3 cm^{-3}, and Thomson scattering optical depth unity was reached in traversing a distance $\sim 10^{21}$ cm. This corresponds to a travel time of $\sim 10^3$ y, and gives an estimate for the time required for thermalization of the photon spectrum through interaction with the electrons. The age of the universe at that stage was $\sim 2 \times 10^5$ y, so that the universe remained opaque until all but $\sim 1\%$ of the electrons had combined with protons to form hydrogen. All record of the

12:7

spectral distribution of radiation at earlier times, therefore, disappears in thermalization, and any deviation from a blackbody shape can have been initiated only after recombination was virtually complete (We77)*.

Between the annihilation era and the era of *recombination* there was one additional epoch of note—the stage at which helium and deuterium formed from protons and neutrons. Helium is formed through a chain of steps initiated by the formation of deuterium. At a temperature of $\sim 10^9 \,^\circ$K, the weakly bound neutron in a deuterium nucleus is still readily dissociated from its companion proton, but the nucleus frequently can survive long enough for an additional proton or neutron to interact with it to form, respectively, He^3 or H^3. Both of these isotopes are more stable than deuterium at these temperatures and can survive to add, respectively, a neutron or proton and form He^4, which is very stable. At temperatures slightly below $\sim 10^9 \,^\circ$K deuterium can also survive without being torn apart through collisions.

Elements heavier than helium could not have been formed in appreciable numbers at this stage, because there are no stable nuclei containing five or eight nucleons; and most of the neutrons present were ultimately incorporated into helium nuclei. The final ratio of helium to hydrogen was determined by the temperature $\sim 10^9 \,^\circ$K, at which hydrogen-to-helium conversion took place. That temperature was sufficiently low to drive interactions between protons, neutrons, neutrinos, antineutrinos, electrons and positrons toward favoring the less massive proton–electron combination over the more massive neutron, because the neutron–hydrogen mass difference is 1.4×10^{-27} g, whose equivalent temperature is

$$T = \frac{[m_N - m_H]c^2}{k} = 9 \times 10^9 \,^\circ \text{K}. \tag{12-36}$$

Hence, at the time the temperature dropped below $10^{10}\,^\circ$K, the balance dictated by the Boltzmann factor $\exp[-(m_N - m_H)c^2/kT]$ shifted significantly to favor protons over neutrons. Neutrons began to decay with a characteristic half-life of ~ 10.6 min. Helium was produced when the ratio of densities n_H/n_N was roughly 6:1. This permitted a final hydrogen-to-helium number ratio of $\sim 12:1$ and allowed a small fractional amount of deuterium and an even smaller amount of lithium and beryllium to also survive. Those processes will be discussed in more detail in section 12:14 below.

12:8 THE FLATNESS PROBLEM

Once we reconcile ourselves to the idea that the universe must at one time have been very compact and hot, a new puzzle emerges. We noted earlier that the universe may very well be flat, that is, have Riemann curvature constant $k = 0$. This is astonishing when one thinks about the alternatives. If we rewrite equation (12–10) in the form:

$$\frac{8\pi G \rho a^2}{3} = kc^2 + \dot{a}^2 \qquad (12\text{–}37)$$

we note that the term on the left has the form of a potential, while the second term on the right has the form of a kinetic energy per unit mass. If we consider conditions in the remote past, when the universe was very young, we see from equation (12–19) that the potential term was proportional to a^{-2}, and therefore very large. $\dot{a}^2$ must therefore also have been very large, and the difference between these two quantities, namely kc^2, would have remained constant. In order for the term involving k in equation (12–37) to be small today, it must have been a fantastically small fraction of the potential or the kinetic energy per unit mass, when the universe was very young. If we go back to a time when the universe was no older than 10^{-35} sec, $a(t)$, according to (12–21), would have been roughly 10^{24} times smaller than at decoupling or about 10^{27} times smaller than today. The two major terms in equation (12–37) would then have had to be 10^{54} times larger than today, and hence their difference, kc^2, would have to have been 10^{54} times smaller than either term, since kc^2 appears to be no larger than either of the other terms today. The delicacy of this balance at early epochs appears to be in need of explanation, unless k always had been precisely zero throughout all ages.

12:9 THE PLANCK ERA

A set of cosmological models which seem to explain both the isotropy and the flatness problems has been proposed in the past decade. These models all have one feature in common, namely that they originate in a highly compact form. Originally their expansion is rather slow, so that communication across major portions of the universe is possible in the earliest moments. Thereafter a phase transition sets in, during which the universe expands at an enormous rate—to such an extent, in fact, that the k term in equations (12–10) and (12–11) becomes negligibly small for all subsequent times. The expansion then decelerates and the universe continues to expand much in the fashion we had deduced from consideration of the microwave background radiation in section 12:7.

Before outlining these ideas in quantitative form, we first must ask about the time span over which the laws of physics as we currently understand them might be valid. Here we are really questioning the range of validity of general relativity. At extreme densities and over extremely short time intervals, that theory needs to be replaced by a quantized theory of gravitation which remains to be invented. However, it is possible to estimate conditions under which these quantum effects will become pronounced.

Suppose we had an extremely compact mass, so compact that it was contained in its own Schwarzschild radius. On quantum mechanical grounds that mass should also be of dimensions comparable to the Compton wavelength. Setting these two dimensions equal to each other with the aid of equations (5–63) and (6–166), we obtain

$$\frac{2mG}{c^2} = \frac{h}{mc} \tag{12–38}$$

which permits us to solve for the mass m in terms of the other natural constants. We define the *Planck mass*

$$m_p \equiv \left(\frac{hc}{2\pi G}\right)^{1/2} = 2.18 \times 10^{-5} \text{ g} \tag{12–39}$$

where a factor of π has still been included, as a matter of convention. If this is inserted in the expression for the Compton wavelength, we obtain the *Planck length*

$$l_p \equiv \left(\frac{hG}{2\pi c^3}\right)^{1/2} = 1.61 \times 10^{-33} \text{ cm} \tag{12–40}$$

where again, by convention, Planck's constant is divided by 2π. The shortest time during which it makes sense to talk about such a mass is the length of time light would take to traverse the Planck length. This is the *Planck time*,

$$t_p = \left(\frac{hG}{2\pi c^5}\right)^{1/2} = 5.38 \times 10^{-44} \text{ sec} \tag{12–41}$$

Over shorter intervals than that, one end of the Planck mass distribution would cease to be aware of the presence of the other end, so that the laws of causality would no longer apply. The Planck time must therefore be considered the earliest time in the existence of the universe for which the Einstein field equations (12–10) and (12–11) could apply. There is no guarantee that they apply that early, but there is no reason to expect that they could apply any earlier.

Finally, dividing the Planck mass by the cube of the Planck length gives us a measure of the density the early universe could have attained. Again, the laws of relativity could not be expected to apply at higher densities than this *Planck density*

$$\rho_p \equiv \frac{2\pi c^5}{hG^2} = 5.18 \times 10^{93} \text{ g cm}^{-3} \tag{12–42}$$

It makes sense to also ask about the temperature that might have existed at the Planck time. We can proceed by considering two possibilities: only bosons being present or only fermions being present. For bosons, having two possible spin values, the density–temperature relation has the form of equation (4–72). For fermions restricted to a single spin value, the energy density in the relativistic extreme is

$$\rho = \frac{4\pi}{h^3} \int_0^\infty \frac{p^3 dp}{e^{pc/kT} + 1} = \frac{7\pi^5}{30h^3c^3} (kT)^4 = \frac{7}{16} aT^4. \tag{12–43}$$

where a is the radiation constant 7.56×10^{-15} erg cm^{-3}°K^{-4}. It should not be confused with the scale factor for which we have used the same symbol.

For a mixture of bosons with statistical weight g_{bi} and fermions with statistical weight g_{fj}, we then find

$$T_p = a^{-1/4} \rho_p^{1/4} \left(\frac{1}{2} \sum_i g_{bi} + \frac{7}{16} \sum_j g_{fj} \right)^{-1/4} c^{1/2} \tag{12–44}$$

The factor $c^{1/2}$ enters because equations (4–72) and (12–43) refer to energy-density, rather than mass-density. When the number of available species is of order unity, the temperature becomes

$$T_p \sim \left(\frac{c^2 \rho_p}{a} \right)^{1/4} \sim 10^{32}°\text{K} \tag{12–45}$$

The mass contained within the present-day horizon of our universe is about 10^{55} g. At the Planck time this mass would have been contained in a volume of order 10^{-39} cm^3, having diameter $\sim 10^{-13}$ cm. That span, however, can only be traversed at the speed of light, in 3×10^{-24} sec, a period which is long compared to the Planck time. We again encounter the difficulty here of causally connecting one end of the universe to the other during these initial states. We should not be surprised at this, because we had set up equation (12–41) with the idea that the Planck time simply connects different portions of a Planck mass causally—not a mass 60 orders of magnitude larger and hence 20 orders of magnitude more extended.

12:10 INFLATIONARY COSMOLOGICAL MODELS

To overcome this causality problem, one may consider the possibility of a phase transition that occurs in the early universe. Roughly 10^{-35} sec after the Planck era, an interval of rapid expansion begins, during which the density and hence the temperature both drop by many orders of magnitude. During this time the scale factor a, the temperature T, the world time t, and the density obey the proportionalities (12–21) to (12–23)

$$a \propto t^{1/2} \propto T^{-1} \propto \rho^{-1/4} \tag{12–46}$$

so that the temperature drops to a value of 10^{28}°K, where particles have energies 10^{15} Gev. One Gev $= 10^9$ ev.

At an energy 10^{15} Gev, *grand unified theories of particle interactions* set strong and weak nuclear, as well as electromagnetic, forces roughly equal to each other. Below that energy the strong and the electro-weak forces no longer remain comparable, symmetry is broken, and a phase transition can set in. In the inflationary models, however, that transition is not allowed to take place. The material is supercooled through an expansion that lasts another 10^{-33} sec. The expansion rate during this phase is dictated by a cosmological constant Λ, which represents the energy density of the *Higgs field*, a scalar field which mediates the interactions between particles. During the Planck era Λ is too small to significantly affect the cosmic expansion. But at a temperature below 10^{15} Gev Λ becomes a dominant factor controlling the expansion. Λ is sometimes considered to be a *false vacuum* and is referred to by that name. After the passage of 10^{-34} sec Λ vanishes and is replaced by the present-day vacuum, in which the constant Λ is immeasurably small, $|\Lambda| \leq 10^{-55}$ cm^{-2}.

During the Higgs era, however, Λ is very large. We can write (12–10) and (12–11) for $k = 0$ as

$$\frac{8\pi G\rho}{3} = -\frac{\Lambda c^2}{3} + \frac{\dot{a}^2}{a^2} \tag{12–47}$$

$$\frac{8\pi GP}{c^2} = \frac{2\Lambda c^2}{3} - 2\frac{\ddot{a}}{a} - \frac{8\pi G\rho}{3} \tag{12–48}$$

and with an equation of state

$$\frac{P}{c^2} = -\rho = -\frac{\Lambda c^2}{8\pi G} \tag{12–49}$$

we obtain

$$H^2 = \frac{\dot{a}^2}{a^2} = \frac{2\Lambda c^2}{3} \tag{12–50}$$

The universe expands exponentially at a rate

$$\dot{a} = ae^{Ht} = a\exp\left(\frac{2\Lambda^2}{3}\right)^{1/2}t, \tag{12–51}$$

but since the cosmological constant corresponds to the energy density term at $t = 10^{-35}$ sec or $\rho \sim 10^{77}$ g cm^{-3}, its value during this period must be of order

$$\left(\frac{2\Lambda c^2}{3}\right)^{1/2} \sim 4 \times 10^{35} \text{ sec}^{-1} \tag{12–52}$$

In the course of 1.5×10^{-34} sec, the universe can therefore expand by a factor of e^{60} or more than 10^{26}. All that time the cosmological constant remains constant; the energy required to keep up the expansion is provided by the release of tension through the equation of state in which the pressure term is just the negative of the energy density. At the end of this expansion, the dimensions of the region encompassing the universe in our current horizon are of the order of 10 cm. This means that the universe we now contemplate has come from a region only about 10^{-25} cm in diameter at the start of the inflationary phase. During the preceding 10^{-35} sec that region could have been traversed by radiation, come to pressure equilibrium and attained homogeneity. There is no guarantee that this actually happened; but a state of bland homogeneity, over spans which ultimately developed into regions larger than the presently observed universe, is considerably less puzzling in this model. Note also that the region $\sim 10^{-25}$ cm in diameter is more than ten orders of magnitude smaller than the size estimated in section 12:9. The difference arises from the hypothesis that most of the mass–energy density observed in the universe today originated when Λ became zero.

A feature of the inflationary models is that they also explain the flatness of the universe, the lack of observed curvature, and hence the value $k = 0$ of the Riemann curvature constant. We only note that any curved surface (Fig. 10.1(b)) when sufficiently expanded will appear flat in a small region around any chosen point. The inflation increases the scale of the universe so enormously and balloons the universe out to such a large extent, that locally, within any horizon defined by the speed of light, the curvature becomes negligible.

Of importance in considering these models is the recognition that the portion of the universe we now see is only an unimaginably small fraction of a larger universe which will forever remain unknown to us—out of touch, beyond physical reach, beyond study by physical means. Since physics normally confines itself to statements about systems that can be examined observationally or through experiment, the proposition that such remote realms of the universe exist, though they could never be observed, breaks with traditional ideas about the range of permissible scientific inference.

PROBLEM 12–3.   Fundamental particle physics suggests that the universe should have been filled with magnetic monopoles in the earliest moments of existence of the universe. *Magnetic monopoles* are carriers of magnetic charge, similar in fashion to electrons which carry a unit of electric charge. These particles, first postulated by Dirac in 1931, were reconsidered by Polyakhov and independently by t'Hooft in 1974 (Po74, Ho74) and are likely to have energies of order 10^{16} Gev. Show that after inflation the universe could have contained fewer magnetic monopoles than one for each galaxy that would eventually form.

12:10

12:11 THE MATTER–ANTIMATTER ASYMMETRY

The inflationary universe suggests ways in which the matter–antimatter asymmetry —the predominance of protons and electrons over antiprotons and positrons— could be explained. One suggestion has been that the phase transition at $\sim 10^{-34}$ sec generates *Higgs bosons*, designated by the symbol X. Higgs bosons can take the form either of charged particles (X_+) or their antiparticles (X_-). If particles and antiparticles have slightly different decay schemes because they violate charge and parity conservation, a lasting asymmetry of matter over antimatter may be introduced. This explanation is quite speculative and, in any case, gives no fundamental justification for postulating the asymmetry.

At any rate, it is interesting, as Problem 12–3 points out, that the number of particles remaining in the accessible universe after inflation could have been quite low. If these few particles had influenced phase transitions in their individual surroundings, they could have produced large-scale domains of matter stretching over regions of the order of clusters of galaxies and separated from each other by similar regions of antimatter. We still have no observational evidence that large portions of the universe are not dominated by antimatter in the way that our region appears dominated by matter.

12:12 THE POST-INFLATIONARY STAGE

Toward the end of the inflationary stage, the predicted phase transition finally sets in. The energy of the false vacuum becomes available much as energy is liberated when liquid water freezes to form ice. The value of the cosmological constant now essentially drops to zero, as already mentioned, and the energy previously residing in the vacuum field is now converted into radiation and matter. The temperature, which had dropped drastically as a result of the inflationary expanion, is again raised to its original value before expansion started. This is easily seen, because during inflation the value of Λ and the energy density it represents remain precisely constant. That energy density is related to the temperature by an expression of the form (12–44), but with the density ρ_p in that expression replaced by $\Lambda c^2/8\pi G$ as in (12–49).

The equation of state is now replaced by the equation of state of a relativistic gas, and the expansion becomes that of a normal relativistic Friedmann model as described in sections 12:4 and 12:6.

12:13 THE ENTROPY OF THE UNIVERSE

The internal energy of radiation at temperature T, in a volume V, is $UV = aT^4V$, according to equation (4–125). Its pressure is jut one-third of the radiation density

$aT^4/3$, (4–43), and we can therefore write the first law of thermodynamics as in (4–125):

$$dQ = T\,dS = dU + P\,dV = 4aT^3V\,dT + \frac{4aT^4}{3}\,dV \qquad (12\text{–}53)$$

where the first equality is a definition of the *entropy* S. While dQ is not an exact differential, dS is.

The second law of thermodynamics asserts that the entropy of a closed system can at best remain constant, but will normally increase during any physical process.

PROBLEM 12–4.   Show that equation (12–53) integrates to

$$S = \frac{4aT^3V}{3} \qquad (12\text{–}54)$$

for radiation.

For an adiabatic expansion, $dS = 0$ and the product

$$VT^3 = \text{constant} \qquad \text{(adiabatic process)} \qquad (12\text{–}55)$$

It is now easy to define an entropy per baryon in the universe. The entropy is strongly dominated by the radiation. The baryons—because of their scarcity—contribute only negligibly. The entropy per baryon, then, is just the entropy per unit volume divided by the number of baryons in the same volume

$$\frac{S}{Vn_B} = \frac{4aT^3}{3n_B} = 10^{-14}\frac{T^3}{n_B}\,\text{erg}^\circ\text{K}^{-1} \qquad (12\text{–}56)$$

or, since the photon number density (equation 4–72) is $20T^3$ cm^{-3},

$$\frac{S}{Vn_B} = 5 \times 10^{-16}\frac{n_\gamma}{n_B} \qquad (12\text{–}57)$$

where n_γ is the photon density. This does not take neutrino entropy into account, which is comparable to the photon entropy for each type of neutrino (electron-, muon-, or tau-neutrino) present though somewhat smaller: first, because there are only 7/8 neutrino/antineutrino pairs per pair of orthogonally polarized photons; and second, because, as we will see in section 12:14, the present-day neutrino temperature is lower than the photon temperature by an added factor of $(11/4)^{1/3}$.

Equation (12–43) gives the total energy density per fermion of a given spin. Considering neutrinos and antineutrinos to have only one possible spin each, and taking electrons and positrons to have two spin modes each, we find for the era when all these species are in thermal equilibrium with photons, an epoch characterized by temperatures in the range $10^{12} \geq T \geq 10^{10}°$K, that the total energy density in electrons and photons and electron- and muon-neutrinos (tau-neutrinos have not yet been observed) is

$$\rho = 2 \times 2\left(\frac{7}{16}\right)aT^4 + aT^4 + 4\left(\frac{7}{16}\right)aT^4 = \frac{9}{2}aT^4 \qquad (12\text{–}58)$$

so that (12–54) leads to an entropy per unit volume $s = S/V$

$$s = 6aT^4 \qquad (12\text{–}59)$$

at these high temperatures. Since the subsequent evolution of the cosmos can be considered to be adiabatic, the total entropy remains constant, and the entropy per unit volume is inversely proportional to the scale factor cubed,

$$sa^3 = \text{constant} \qquad (12\text{–}60)$$

Here again it is important not to confuse the scale factor a with the radiation constant for which we have used the same symbol in equations (12–53) to (12–59).

2:14 EARLY ELEMENT FORMATION

The microwave background radiation suggests that the universe was once very hot and also attests to a homogeneity and isotropy of the universe which indicates that conditions at any given epoch have always been identical throughout the visible part of the cosmos. A test of this hypothesis lies in the question of whether the chemical composition found in the universe is compatible with these assumptions: If we consider the field equations (12–10) and (12–11) to hold with the cosmological constant and the Riemann curvature constant both set equal to zero, then the rate at which the universe evolves, its temperature at every epoch, and its density all are defined. We can then investigate the reaction chain drawn in Fig. 8.13, for which interaction cross sections and reaction rates are known. That can tell us whether the abundances of the different chemical elements and isotopes observed today could have resulted from processes active when the universe was younger. We do not, of course, know that $k = 0$ and $\Lambda = 0$ actually characterize the cosmos, but these values are plausible for conditions seen today. At times when the universe was denser, the k and Λ terms will have had an even less significant influence, since the density and expansion rate in equation (12–10) will have been far greater than now observed, while k and Λ will have remained unchanged.

PROBLEM 12–5. Convince yourself quantitatively that this last statement is correct.

Let us now return to the chronology begun in section 12:7 and investigate it for the conditions $k = \Lambda = 0$. In this model we start at a time when:

(i) The universe is a little more than 10 milliseconds old (Bo85). The temperature has dropped to below $10^{11}°$K, and neutrons and protons have number densities $n(\mathcal{N})$ and $n(P)$ that reflect thermal equilibrium through a Boltzmann factor corresponding to the difference, $c^2\Delta m$, in their mass-energies.

$$\frac{n(\mathcal{N})}{n(P)} = \exp\left(-\frac{c^2\Delta m}{kT}\right) \qquad (12\text{--}61)$$

This equilibrium is maintained by interactions with ambient electrons, positrons and electron neutrinos as well as antineutrinos, through reactions of the type given by equations (8–107) to (8–109). Just as in equation (12–44), the mass density in neutrinos and antineutrinos is related to the mass density of photons by the ratio 7/8. The neutrino/antineutrino energies are still sufficiently high at this temperature to keep electron–positron pairs in equilibrium abundance through the reactions.

$$e^+ + e^- \Leftrightarrow \nu + \nu \qquad (12\text{--}62)$$

(ii) At $t \sim 0.1$ sec, the temperature is $T \sim 3 \times 10^{10}°$K. The weak interactions now become too slow to compete with the expansion rate, and the neutrinos become decoupled from the matter. The energy at this stage is still of the order of a few Mev per particle, and hence electron–positron pairs persist in equilibrium with photons.

$$e^+ + e^- \Leftrightarrow \gamma + \gamma \qquad (12\text{--}63)$$

(iii) At $t \sim 1$ sec, $T \sim 10^{10}°$K, particle energies are of the order 1 Mev. Reactions of the type (8–107) to (8–109) become too weak to maintain equilibrium between neutron and proton number densities, and the ratio of neutrons to protons effectively freezes out at a value characteristic of $10^{10}°$K. Thereafter, the ratio changes only because the neutrons decay into protons with a half-life of order 10.6 min. The uncertainty in this decay rate is still of order 0.2 min.

(iv) At $t \sim 10$ sec, $T \sim 3 \times 10^9°$K, energies drop below the electron–positron rest mass and these particles annihilate in pairs, heating the radiation and matter remaining behind. No further pairs are produced, since the photon energies now are too low. But the heat generated in annihilation causes a temperature difference between the photon bath and the bath of neutrinos which earlier decoupled from matter and radiation. Henceforth, the neutrino temperature uniformly remains below the photon temperature. If neutrinos have zero rest mass, their contemporary background temperature should be found to be roughly 2°K, while the radiation temperature is known to be just below 3°K. This neutrino bath has never been detected, because we currently do not have the apparatus to even search for it.

12:14

PROBLEM 12–6. Show that the annihilation of electron–positron pairs raises the photon temperature by a factor of $(11/4)^{1/3}$ compared to that of the decoupled neutrino background bath. Note that the electron–positron annihilation is a phase transition here, and therefore takes place as an adiabatic process.

At this stage, deuterons can already be formed, through the reaction

$$\mathcal{N} + P = D + \gamma \tag{12-64}$$

but they are quickly destroyed by photodissociation.

(v) At $t > 10^2$ sec and $T < 10^9$°K this photodissociation becomes less frequent as photon energies decline. At this stage the following reactions all can set in, to produce the stable isotope He^4.

$$
\left.
\begin{aligned}
D + \mathcal{N} &\Leftrightarrow H^3 + \gamma \\
D + D &\Leftrightarrow H^3 + P
\end{aligned}
\right\}
\quad
\left\{
\begin{aligned}
H^3 + P &\Leftrightarrow He^4 + \gamma \\
H^3 + D &\Leftrightarrow He^4 + \mathcal{N}
\end{aligned}
\right.
$$

$$
\left.
\begin{aligned}
D + P &\Leftrightarrow He^3 + \gamma \\
D + D &\Leftrightarrow He^3 + \mathcal{N}
\end{aligned}
\right\}
\quad
\left\{
\begin{aligned}
He^3 + \mathcal{N} &\Leftrightarrow He^4 + \gamma \\
He^3 + D &\Leftrightarrow He^4 + P \\
He^3 + He^3 &\Leftrightarrow He^4 + 2P
\end{aligned}
\right.
\tag{12-65}
$$

It is not possible to form stable elements of mass 5 or 8 at these temperatures and densities, though some traces of beryllium and lithium isotopes of mass 7 can be formed, the latter in quantities that should be measurable.

$$
\begin{aligned}
He^4 + He^3 &\rightarrow Be^7 + \gamma \\
He^4 + H^3 &\rightarrow Li^7 + \gamma
\end{aligned}
\tag{12-66}
$$

Under certain circumstances the beryllium nuclei can absorb positrons and be converted to Li^7.

Figure 12.5 shows the final abundances that can be expected for different ratios of photon-to-baryon densities that measurements might ultimately reveal. The contemporary baryon density is so poorly known that the entire range shown must be taken seriously.

Figure 12.6 then shows the restrictions on the contemporary baryon density that come from observations of different isotopes in what may be considered virgin material not processed by subsequent nuclear reactions in stars. In this figure conservative upper and lower limits for each species are shown in bold lines and somewhat less secure values in dashed lines. With these figures we conclude that the upper bound to the baryon density cannot exceed 6 to 9 × 10^{-31} g cm^{-3}, while the lower bound cannot be below 2–3 × 10^{-31} g cm^{-3}. These figures, however, are also made uncertain by uncertainties in the Hubble constant measured today. Values of Ω inferred lie between 0.01 and 0.2.

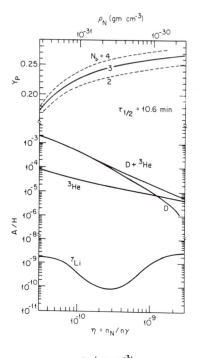

Fig. 12.5 The nuclear abundances predicted by the standard model versus the nucleon-to-photon ratio or the estimated present-day cosmic nucleon mass density (g cm^{-3}). Y_p, the primordial ^{4}He mass fraction, is plotted on an expanded linear scale. The three curves are for $N_v = 2, 3$ or 4 different neutrino types. The neutron half-life is taken to be 10.6 min; the uncertainty corresponds to just ± 0.2 min. The results for D, ^{3}He, and ^{7}Li are plotted as ratios—by number—relative to H. (Bo85.) (Reproduced with permission, from the Annual Review of Astronomy and Astrophysics, Volume 23, © 1985 by Annual Reviews Inc.)

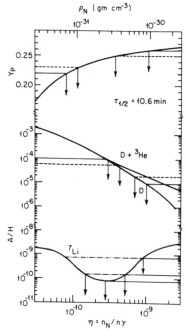

Fig. 12.6 Comparison of predicted and observed abundances. See the caption for Fig. 12.5 for details and the text for discussion. (Bo85.) (Reproduced, with permission, from the Annual Review of Astronomy and Astrophysics, Volume 23, © 1985 by Annual Reviews Inc.)

12:15 GRAVITATIONAL CONDENSATION OF MATTER

To understand the problems surrounding the formation of stars and galaxies we can consider an isolated cloud which initially is neither expanding nor contracting. If it is sufficiently cool, as we will at first assume, it will be free to collapse under its own gravitational attraction. We will consider the density of the cloud to be uniform throughout a sphere of radius r_0. Then the forces on matter at the surface of the sphere will produce an acceleration

$$\ddot{r} = -\frac{GM}{r^2} \tag{12–67}$$

Integration with respect to time yields

$$\frac{\dot{r}^2}{2} = \frac{GM}{r} = \frac{4\pi}{3} r_0 \rho_0 G \left(\frac{r_0}{r} - 1\right) \tag{12–68}$$

where ρ_0 is the initial density.

PROBLEM 12–7. Show that this equation can be integrated a second time, for example through a substitution of a new variable u given by $r = r_0 \sin^2 u$, to give the free-fall collapse time

$$t_{\text{ff}} = \sqrt{\frac{3\pi}{32 G \rho_0}} \tag{12–69}$$

Note that this expression is independent of the size of the cloud. A second point of interest is that the free-fall expression is unaffected by any spherical distribution of matter that lies outside the cloud we have just considered. That can be seen by viewing the potential within a spherical shell of radius R due to a mass distribution whose surface density is $M/4\pi R^2$ everywhere on the sphere.

PROBLEM 12–8. Show that the potential anywhere within that sphere is

$$V = -\frac{MG}{R} \tag{12–70}$$

and, further, show that the potential due to any spherical distribution of matter outside an empty sphere has a value that is constant throughout the sphere.

For any spherically symmetric distribution of matter in spherically symmetric motion, the dynamics within a central sphere always remain unaffected by the distribution outside. This result is valid also in general relativity and has the most wide-ranging consequences. It is named after George Birkhoff, who first showed its generality in what has come to be known as *Birkhoff's theorem* (Bi23).

12:16 PRIMORDIAL MASS CONDENSATIONS

Primordial mass condensations, whether in the form of stars, clusters of stars, galaxies, clusters of galaxies, or in some other form, have consistently presented a puzzle. If we take the density given by equations (12–27) and (12–29) for the recent, matter-dominated era

$$\rho = \frac{3\Omega}{8\pi G \tau^2} = \frac{3\Omega H^2}{8\pi G} \tag{12–71}$$

and substitute this density into the expression just derived for the free-fall time, we obtain an expression for t_a, the age of the universe, or τ, the reciprocal of the Hubble constant in terms of the free-fall time:

$$t_{ff} = \frac{\pi}{2} \frac{\tau}{\Omega^{1/2}} = \frac{\pi}{3} \frac{t_a}{\Omega^{1/2}} \tag{12–72}$$

and since Ω is less than unity in the observed distribution of matter, the free-fall time required to form condensations exceeds the age of the universe at any given epoch. This suggests that gravitational effects alone cannot have led to galaxy formation in recent aeons, at least not starting from a homogenous distribution of matter.

The homogeneity of matter, on the other hand, seems to have been high at the epoch of decoupling of radiation from matter; otherwise, the background radiation observed today would most probably show an imprint due to the existence of primordial condensations that dated back to an era marked by redshift $z = 10^3$. The homogeneity of the background, taken together with this slow free-fall time, makes it difficult to understand how condensations could have formed.

Two comments might still be added. The first is that the free-fall time computed here assumes that the universe was not even expanding at some time $t = 0$. The known expansion of the universe, if anything, makes the problem more worrisome as we will see below. The second factor to note is that there is no particular scale favored by the free fall. Collapse on all scales takes equally long.

12:17 LOCALIZED FORCES AND CONDENSATION

Since gravitational forces alone seem incapable of producing sufficiently rapid condensation, other types of forces have sometimes been sought. We look for a force

12:17

which would have a coupling constant, an effective gravitational-like force constant which would be N times stronger:

$$G_{\text{eff}} = NG. \tag{12-73}$$

We then ask how large the number N has to be to both produce sufficiently rapid collapse and lead to the collapse of massive aggregates.

To see this in more concrete fashion, we can think of a kind of force originally advanced by Lyman Spitzer (Sp41). In a different context he sought a force that might cause interstellar dust grains to attract each other. He noted that if grains were embedded in a strong radiation bath, photons impinging on one dust grain would be absorbed and prevented from reaching a more distant grain lying in the first grain's shadow. Since photons exert a pressure, the first grain would thus experience a force directed toward the second. This would depend on the cross section for absorption of radiation, both of the first and of the second grain, since the second grain would also cast a shadow on the first, for radiation coming from the opposite direction. It would also depend on the radiation density and on the inverse square of the distance, since the flux from a radiation source diminishes as r^{-2}. Finally, since nearby grains in any case tend to absorb radiation, and in that way also shadow more distant grains, the effect is exponentially damped at large distances.

Quite generally we postulate that an attractive force can come about because particles of type i—they might be photons, electrons or some as yet unrecognized form of matter—impinge on hydrogen atoms or ions. These particles should act dissipatively, giving up their energy and effectively sticking to the hydrogen on impact. Assume a cross section σ_i for particles approaching with speed v_i and momentum p_i. Then the force on a hydrogen atom at distance r, embedded in a medium with number density n_i, is

$$F_i = \frac{n_i p_i \sigma_i^2 v_i}{4\pi r^2} e^{-r/L} \tag{12-74}$$

where L is the distance over which an e-fold reduction of particles i is effected through interaction with hydrogen. We see that we can set

$$\frac{n_i p_i v_i \sigma_i^2}{4\pi r^2} = \frac{N m_H^2 G}{r^2} \tag{12-75}$$

in order to meet the requirement (12–73). Hence

$$\sigma_i = \left(\frac{4\pi m_H^2 N G}{n_i p_i v_i}\right)^{1/2} \quad \text{and} \quad G_{\text{eff}} = \frac{n_i p_i v_i \sigma_i^2}{4\pi m_H^2} \tag{12-76}$$

12:17

If we now ask the free-fall collapse to take place in less than the age of the universe, we can write approximately that

$$t_{ff} = \left(\frac{3\pi}{32G_{eff}\rho}\right)^{1/2} < H^{-1}, \qquad N > \frac{\pi^2}{4\Omega} \qquad (12-77)$$

where we have made use of equations (12–71) and (12–73). This leads to the condition for the interaction cross section

$$\sigma_i > \left(\frac{m_H^2 G\pi^3}{n_i p_i v_i \Omega}\right)^{1/2} \sim \left(\frac{m_H^2 G\pi^3}{\Omega \rho_i}\right)^{1/2} \qquad (12-78)$$

where ρ_i is an energy density. If we take Ω to have a value of order 0.03, and ρ_i is in units of erg cm^{-3}, then we see that the cross section has to exceed

$$\sigma_i \geq 1.4 \times 10^{-26}\rho_i^{-1/2} \qquad (12-79)$$

If we consider a collapse just after decoupling, then the radiation temperature is about 4000°K, and the radiation energy density is ~ 2 erg cm^{-3}, so that σ_i must exceed 10^{-26} cm^2. We might think that the Thomson scattering cross section (6–171) would satisfy that criterion, since its magnitude is 6.5×10^{-25} cm^2. But scattering is not a dissipative process; and as long as the radiation is simply scattered, it diffuses throughout a cloud and does not properly shield particles being shadowed by others. For electrons to produce an effective attractive force, they would have to be at a temperature far below that of the radiation, and even then the fact that they mainly scatter rather than absorb would further reduce their effectiveness by a factor that turns out to be order $\gamma(v_i)$ in equation (5–43). Other radiation forces also seem to be ineffective. We cannot invoke a stronger radiation field at an earlier epoch since, as we will show in section 12:18, the isotropy of the background suggests that condensations did not form until after decoupling of matter from radiation, which means that the temperature cannot have exceeded about 4000°K, the temperature characterizing the decoupling era. And while ionization cross sections and resonant absorption cross sections for spectral-line transitions can be far larger than the Thomson scattering cross section, few photons have the energy to excite or even ionize hydrogen at as low a temperature as 4000°K. No force of the type described here appears to exist.

If we now turn to particles that could produce this kind of shadowing force, we can first think of the momentum that could be transferred, say, when electrons combine with protons to form hydrogen, as the primordial gas cools through 4000°K. Quite generally we can then ask whether the effective force produced by these particles is high enough to overcome the kinetic energies of hydrogen atoms and produce collapse. For that to occur we must have a sufficiently high mass enclosed

12:17

in a sphere of radius L (see equation 12–74) so that

$$\frac{4\pi}{3} \frac{n_H m_H L^3 G_{\text{eff}}}{L} > v_H^2, \qquad (12\text{–}80)$$

a condition under which an atom's potential energy exceeds its kinetic energy and it falls into the condensation. We can set

$$L \sim (n_H \sigma_i)^{-1}$$

the distance over which particles i typically are absorbed. Then

$$\frac{4\pi}{3} \frac{m_H G_{\text{eff}}}{n_H \sigma_i^2} > v_H^2 \qquad (12\text{–}81)$$

and substituting for G_{eff} from (12–76) we see that the condition reduces to

$$\frac{n_i p_i v_i}{3 n_H m_H v_H^2} > 1 \qquad (12\text{–}82)$$

This condition is impossible to satisfy unless the temperatures of the particles i are three times the temperature of hydrogen, or unless the particles i have three times the number density of hydrogen. For $p_i v_i$ and $m_H v_H^2$ in equilibrium simply are a measure of some temperature and generally equal $3kT$. Electrons clearly cannot qualify for particles i, since their number density cannot exceed that of hydrogen number densities by appreciable amounts, and they also come into rapid thermal equilibrium with hydrogen.

The conclusion of this section, therefore, is that neither photons nor particles we recognize in the universe right now seem capable of producing both the rapid collapse and the massive condensations needed to account for the formation of stars or galaxies or clusters of galaxies in a standard relativistic universe.

For this reason astrophysicists have chosen to consider other possibilities which include:

(i) Particles that emit no radiation but do exhibit a gravitational force. These particles would raise the value of Ω in equation (12–77) and would therefore permit ordinary gravitation to produce the required collapse. For such particles there also would be no shielding length L over which the force would diminish relative to inverse square law forces.

(ii) Deviations from a standard cosmological model. In Chapter 10 we saw that a Lemaître model (Fig. 10.7) reaches a pause during its expansion, and condensations could presumably occur then. However, such a model does require a finite cosmological constant Λ, and that may already be ruled out by observations.

(iii) Massive inhomogeneities may exist in the universe in the form of sheets or strings which readily scatter background radiation and do not show up against a uniform background, but which can act as condensation sites for the formation of clusters of galaxies.

Much of current cosmology concerns itself with the investigation of such possibilities.

12:18 COLLAPSE DURING THE RADIATION-DOMINATED ERA?

To see how collapse is inhibited in the radiation-dominated era, we consider a homogeneous medium which has suffered a spherically symmetric perturbation favorable to inducing collapse. We imagine regions 1 and 2, Fig. 12.7, separated by a narrow spherical shell devoid of matter or radiation. We know from Birkhoff's theorem that the evolution of the inner sphere will not be affected by the spherically symmetrically distributed mass outside the outer shell. For $k = 0$, $\Lambda = 0$, the field equations in regions 1 and 2, both of which are unperturbed, are

$$\frac{8\pi}{3} G\rho_i = \frac{\dot{a}_i^2}{a_i^2}, \qquad i = 1, 2 \tag{12-83}$$

$$\frac{8\pi G P_i}{c^2} = \frac{8\pi}{3} G\rho_i = \frac{2\ddot{a}_i}{a_i} + \frac{8\pi}{3} G\rho_i, \qquad i = 1, 2 \tag{12-84}$$

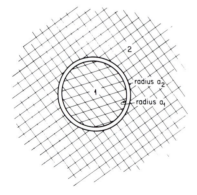

Fig. 12.7   Diagram to illustrate the stability of a spherical cloud embedded in a medium of the same density.

radius a_2

radius a_1

PROBLEM 12–9.   Show that these equations can be integrated to yield a time-dependent evolution of the radii a_i given by

$$a_i^2 = a_{i0}^2 \left(\frac{8\pi G\rho_{i0}}{3}\right)^{1/2} (t - t_0) + a_{i0}^2, \qquad i = 1, 2 \tag{12-85}$$

where the added subscript 0 indicates the conditions at some initial time t_0 at which the perturbation was induced.

Because the mass M enclosed by the two shells is identical,

$$M = \left(\frac{4\pi\rho_{10}a_{10}^3}{3}\right) \tag{12-86}$$

and remains constant throughout. The surface area of each shell grows at a rate proportional to the time and to the square root of the initial raidus. After sufficiently long intervals $(t - t_0)$,

$$\frac{a_1^2(t)}{a_2^2(t)} = \frac{a_{10}^{1/2}\sqrt{2MG}(t - t_0) + a_{10}^2}{a_{20}^{1/2}\sqrt{2MG}(t - t_0) + a_{20}^2} \sim \left(\frac{a_{10}}{a_{20}}\right)^{1/2} \tag{12-87}$$

which is always larger than $(a_{10}/a_{20})^2$, the initial ratio of surface areas. The perturbation is therefore damped out. It does not produce a relative increase in density within the inner region.

We have independent evidence for a lack of galaxy or cluster-sized aggregates forming that early. Observations mentioned in section 12:16 of an extreme homogeneity or absence of clumpiness in the microwave background radiation suggest the following. Figure 12.4 shows that the epoch of recombination, the time when radiation decoupled from matter, still occurred at a time when radiation dominated the mass density of the universe. Suppose that a condensation had formed at that stage and that its density differential with respect to the ambient universe was $\delta\rho$ within a sphere of radius R. Assume also that this condensation had taken place adiabatically. Then equation (4–124) tells us that the temperature of condensation would have risen by an amount of δT with respect to the ambient medium, where

$$\frac{\delta T}{T} = -(\gamma - 1)\frac{\delta V}{V} = (\gamma - 1)\frac{\delta\rho}{\rho} \tag{12-88}$$

Since $\gamma = 4/3$ for radiation and $\gamma = 5/3$ for ionized or for monatomic neutral matter, the *adiabatic* temperature shift with respect to an ambient medium would lie in the range

$$\frac{1}{3}\frac{\delta\rho}{\rho} \leq \frac{\delta T}{T} \leq \frac{2}{3}\frac{\delta\rho}{\rho} \tag{12-89}$$

If this temperature enhancement persisted through decoupling, we should expect to see increased microwave background flux levels from regions of condensation. That deviations in the microwave background are not found at levels of order $\delta T/T \sim o(10^{-5})$ suggests that density enhancements must have been small at the time of decoupling, probably no greater than $\delta\rho/\rho \leq 10^{-4}$. Crudely, we can think of the microwave background as arriving from a distance c/H. If the angular resolution of the measurement is Θ, the sizes of condensations examined for a temperature deviation are $R = \Theta c/H$. For measurements with a beam size measured as one minute of arc, the radiation would appear to emanate from a region $R \sim 1$ Mpc, today, for $H = 75$ km/sec Mpc.

A countereffect must, however, also be considered. Radiation escaping from a massive object suffers a redshift, so that the apparent surface temperature of a condensation at the same temperature as its surroundings appears reduced by an amount

$$\frac{\delta T}{T} \sim -\frac{4\pi}{3}\frac{G}{c^2}\delta\rho\left(\frac{R^3}{R}\right) \tag{12–90}$$

as follows from equation (5–60), which can be used to determine the frequency shift, and from the invariance of v/T discussed in section 5:9. If we now substitute equation (12–12) into (12–90), we obtain the *isothermal shift*

$$\frac{\delta T}{T} = -\frac{\Theta^2}{2}\frac{\delta\rho}{\rho} \tag{12–91}$$

This shift depends not only on the density excess $\delta\rho$, but also on the angular extent of the cloud Θ measured with values of R and H referred to the time at which radiation was emitted. However, since the angle Θ is conserved in an expanding universe, as in Fig. 10.1(b), equation (12–91) holds with the same value for Θ even today. This effect leads to smaller temperature shifts than those encountered in equation (12–89), unless one deals with angles approaching a radian. If the condensation process had somehow been isothermal instead of adiabatic, (12–91) might actually constitute the dominant temperature shift. In that case, a fractional temperature change of order 10^{-5} could correspond to appreciable values of $\delta\rho/\rho$ if observed over sufficiently small angles, $\Theta \leq 10$ arc min.

12:19 DARK MATTER

Though the mass density of today's universe appears too low to permit the formation of condensations, there is powerful evidence that the amount of matter actually present in the universe far exceeds the visible mass.

A first suspicion of the circumstance has been evident for decades. Clusters of galaxies contain too little apparent mass to stay bound. This was already mentioned in sections 3:15 and 4:4.

Second, when the rotation curves of galaxies are studied, remarkably constant rotational velocities in the range 150–300 km sec^{-1} are found for virtually all spirals,

12:19

and these rotational speeds persist out to very large radii. A constant rotational velocity, however, suggests a constant enclosed mass $M(R)$ within a spherical galaxy halo of radius R. Then, by equation (3–44)

$$M(R) = \frac{v^2 R}{G} \qquad (12\text{–}92)$$

implying that the enclosed mass within a spherical distribution is proportional to the radius. This assumption about the distribution can be made partly because the rotational velocities are so constant, and partly because the mass $M(R)$ calculated in this way turns out to be considerably in excess of the mass we can account for in luminous stars or in the observed interstellar dust and gases. The deviation from spherical symmetry could therefore be small because the material in the galaxy's plane is only a minor fraction of the total mass present.

PROBLEM 12–10. A similar result follows from an analysis of the vertical distribution of stars at increasing distance from the Galactic plane. Consider stars at height z above or below the plane and let the plane be pictured as a disk of uniform aereal density σ and radius R. Show that the force acting on unit stellar mass located along the disk axis is (Fig. 12.8)

$$F(z) = -\int_0^R \left(\frac{2\pi r\sigma G}{(r^2 + z^2)}\right)\frac{z}{(r^2 + z^2)^{1/2}} dr = -2\pi\sigma \qquad \text{for } R \gg Z \quad (12\text{–}93)$$

From this, show that for a large disk, the *scale height* is given by

$$h = \frac{v^2}{4\pi\sigma G} \qquad (12\text{–}94)$$

where v is a measure of the vertical component of stellar random velocity in the Galaxy's plane.

Fig. 12.8 Diagram to illustrate the Galaxy's scale height for stars passing through the plane.

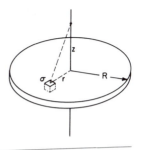

For a galaxy having mass 10^{10} $M_\odot$, radius 10 kpc and hence a uniform aereal density 7×10^{-3} g cm^{-2}, the scale height would be 56 pc for a random velocity component of 10 km sec^{-1}.

When measurements of this type are carried out for the Galaxy, a considerable amount of mass is unaccounted for. In the Galaxy and in other spirals, this mass may be an order of magnitude larger than mass estimates based on the matter detectable by electromagnetic radiation it absorbs or emits. A factor of two may be due to an increase in aereal density over and above those estimates. Another rather larger factor comes from the finding that faint traces of gas in a galaxy's disk, orbiting at far larger radii than normally associated with the visible mass, still show the enclosed mass $M(R)$ rising.

Large quantities of dark matter also are inferred from the distribution of hot gases surrounding some elliptical galaxies. These gaseous coronae appear to be in hydrostatic equilibrium, so equation (8–7) can be applied. Measurement of the X-ray surface brightness of the gas, taken together with the temperature that can be derived from X-ray observations of this hot gas emitting free–free radiation—bremsstrahlung—tell us both the local gas pressure and density in an ideal gas approximation. The derived total masses attracting this gas are estimated to be from 1 to 5×10^{12} $M_\odot$ out to 100 kpc in some of the best-studied cases (Fa85). These elliptical galaxies appear to have a *mass–luminosity ratio* of ~ 100 $M_\odot/L_\odot$. The dark matter clearly is far fainter than mass found in solar-mass stars.

What is the nature of this dark matter? Currently, we do not know. It cannot be largely baryonic; otherwise, the amount of helium formed in the first few minutes of cosmic evolution would have been different. The helium content of the universe is quite sensitive to the total density of baryons at that time. The possibility that neutrinos might be massive, and that they might constitute the missing mass, has been suggested; but for neutrinos to contribute a significant amount of mass, their rest masses would have to be of the order of 10 ev.

PROBLEM 12–11. A supernova exploded in the Large Magellanic Cloud on February 23, 1987. The data hinted that neutrinos arriving early were slightly more energetic than neutrinos arriving a fraction of a second later (Hi87). Suppose that there was a 10-Mev energy difference between early- and late-arriving neutrinos and that the arrival times lapsed by 0.3 sec after traveling a distance of 50 kpc. What is the upper limit to the neutrino mass? Show that this mass is insufficient to provide a critical density, even if three neutrino species and their antineutrinos are considered to be providing a cosmological background.

At any rate, it appears as though progress on understanding the triggering of galaxy formation may depend on the nature of dark matter. Nevertheless, once collapse is initiated, some of the subsequent evolution can be described, at least in general terms. We turn to that next.

12:20 THE JEANS CRITERION

At several places in this book we have encountered similarities in the behavior of gases subjected to Coulomb forces as compared to Newtonian gravitational forces. These similarities arose in the scattering of particles (3:14), in the behavior of ionized gases and assemblies of stars (4:21), as well as in the absorption of radiation in a medium (6:16). Here we will consider the evolution of a disturbance in a neutral cloud in which only gravitational forces are at play and will find that results already derived in section 6:11 can be directly adopted. We recall that we were concerned there with a wave propagating through an ionized medium. By analogy, we here consider a neutral medium in which small density perturbations have arisen. The perturbations $\delta\rho$ can have either a positive or a negative sign. A region of higher density than the average is attracted toward a gravitationally attracting center, while a region of low density is buoyantly repelled. In equation (4–138) we pointed out that gravitational behavior could be derived from electrostatic behavior if we replace the product of charges Q_1Q_2 by $-Gm_1m_2$, where G is the gravitational constant and m_1 and m_2 are interacting masses. In this spirit we turn to section 6:11 and ask how a somewhat perturbed medium acts under its own gravitational influence. We then find that a density perturbation of the form (6–52), $f = f_0\cos(kx \pm \omega t)$, in a medium of mean density ρ obeys a relation of the form of (6–54) which, with the help of (4–138), becomes

$$\omega^2 = k^2c^2 - 4\pi G\rho \qquad (12\text{–}95)$$

Here c is the isothermal speed of sound (9–31) and ω is imaginary for wave numbers k below a critical value

$$k_J = \left(\frac{4\pi G\rho}{c^2}\right)^{1/2} = \frac{2\pi}{\lambda_J}, \qquad \lambda_J^2 = \frac{\pi c^2}{G\rho} \qquad (12\text{–}96)$$

where λ_J is called the *Jeans length*. The Jeans length is the minimum wavelength for a perturbation which can grow in response to gravitational forces in a medium of density ρ. The negative sign of the gravitational term in equation (12–95) implies an exponential growth of the disturbance with an *e*-folding time

$$\tau = \frac{2\pi}{i\omega} = \frac{2\pi}{c(k_J^2 - k^2)^{1/2}}, \qquad |\mathbf{k}| < k_J \qquad (12\text{–}97)$$

The *Jeans mass*, the mass involved in this contraction, is

$$M_J = \frac{4\pi}{3}\rho\lambda_J^3 = \frac{4\pi\rho}{3}\left(\frac{\pi\gamma P}{G\rho^2}\right)^{3/2} \qquad (12\text{–}98)$$

where we have made use of (9–31) in relating the speed of sound to density, pressure P, and ratio of specific heats γ. We note that the Jeans stability criterion assumes

12:20

that the disturbance under consideration grows adiabatically; the gas has no time to radiate or conduct energy away to achieve isothermality throughout the gas. Another, deeper difficulty with the Jeans approach relates to questions that arise if the medium is infinite. For an unperturbed medium there is nothing that shields out gravitational perturbations at a distance and no Debye shielding length corresponding to that found for electrostatic disturbances in a plasma. Though the quantity we call the Jeans mass is usually written in the form of (12–98), it would make somewhat more sense to consider the condensing mass to be more like $[4\pi/3]\rho[\lambda_J/2]^3$, where λ_J is considered the diameter rather than the radius of the collapsing region, and we consider condensation to take place in a region spanned by one wavelength.

We must also be careful not to apply this criterion to the expanding universe without taking some care. We note that expression (8–1) can give us the total potential energy within a sphere of radius $\lambda_J/2$. When this is divided by the total kinetic energy for the gas contained, we obtain

$$\frac{\text{Potential energy}}{\text{Kinetic energy}} \sim \frac{\frac{3}{5}\left(\frac{M_J}{8}\right)^2 G/(\lambda_J/2)}{\left(\frac{M_J}{8}\right)\left(\frac{3c^2}{2\gamma}\right)} \sim \frac{4\pi^2\gamma}{30} \geq 1 \tag{12–99}$$

indicating that the Jeans criterion simply tells us that contraction can occur very roughly when the potential energy due to self-gravitation exceeds the kinetic energy of thermal random motion in the gas. This is especially true if the gas has no internal degrees of freedom—essentially when it is monatomic. In the cosmological case we can only apply the Jeans criterion when the cloud is sufficiently small, so that the dominant velocities over distance λ_J are still thermal velocities, rather than cosmic expansion velocities. If larger clouds are considered, the denominator in equation (12–99) needs to be replaced by the kinetic energy of expansion; but if the expansion velocity is proportional to distance, it will increase linearly with λ_J as λ_J is allowed to grow. Then both numerator and denominator in an expression of the type (12–99) will grow in proportion to λ_J^2, so that unless collapse occurs for random gas velocities exceeding the expansion velocity, no collapse will be expected at all, at least not in this approximation.

All this is just another way of explaining the difficulties we have in accounting for the formation of galaxies in a rapidly expanding universe.

12:21 STAR FORMATION

When we turn our attention to star formation, we can still apply the Jeans criterion, but now the expansion of the universe generally can be neglected because we assume

12:21

that the formation of stars takes place in clouds of gas that somehow have condensed as galaxies have formed. The formation of stars then takes place in substantially denser, usually cooler regions where the density ρ is high and the speed of sound c is low. Instead of talking of the Jeans criterion, we then may also use a related concept, namely that of hydrostatic equilibrium. Here, we turn to equation (8–7) and note that a cloud will contract if the pressure at its center is not sufficiently high to withstand compression from matter gravitationally bearing down on it from larger radial distances. For a cloud of uniform density, the pressure at the center is obtained by integrating (8–1):

$$dP = -\left\{\frac{\rho GM[r]}{r^2}\right\} dr = -\left[\frac{4\pi\rho^2 Gr}{3}\right] dr \qquad (12\text{--}100)$$

and the cloud will collapse unless the central gas pressure $\rho kT/m$ exceeds the value

$$P = \frac{\rho kT}{m} > \frac{\rho GM}{2R} \qquad (12\text{--}101)$$

where M is the mass contained in a spherical cloud of radius R, and m is the atomic or molecular mass characteristic of the gas, the mass of atomic or molecular hydrogen, depending on whether the gas is predominantly in atomic or molecular form. If the temperature is 20°K and the density of a molecular cloud is $\rho \sim 10^{-20}$ g cm^{-3}, we would require $R \sim 8 \times 10^{17}$ cm and $M = 2 \times 10^{34}$ g. However, while the temperatures in molecular clouds often appear to reach such low values and hence would appear to promise that stars of several solar masses could easily form, thermal velocities rarely are the dominant velocities that prevail. At 20°K, the velocities of typical molecules are $[3kT/m]^{1/2}$ which, for $m = 3.2 \times 10^{-24}$ g, are about 0.5 km sec^{-1}, while observed bulk velocities in such clouds usually exceed a few kilometers per second. These random bulk velocities, as well as rotational velocities that could be of the same order, keep the cloud from collapsing.

12:22 TRIGGERED COLLAPSE

In section 1:4 we discussed current views on how stars are formed and mentioned some of the conceptual difficulties involved—the need to shed angular momentum, the required dissipation of magnetic fields, and the loss of energy entailed in gravitational collapse leading to star formation. We now ask: What kind of trigger starts such a chain of events? If the random bulk motions prevent collapse, as discussed in section 12:21, then some process must be found which initiates contraction. For star formation in ordinary spiral galaxies, the distribution of young stars along spiral arms provides one hint. Another comes from luminous galaxies in

12:22

which star formation appears to be particularly active; those galaxies generally also appear disturbed and exhibit high-velocity gas flows. These factors all point to shock compression to trigger the collapse of a cloud. For spiral arms this compression can take place as the arm—the *spiral density wave* which the arm represents—sweeps across a molecular cloud. The compression there is produced by the gravitational potential of the arm, this potential wave having an angular velocity about the center of the galaxy that differs from the orbital angular velocity of the gas clouds and stars. Alternatively, in the neighborhood of young stars, which in any case are associated with the spiral arms, compression can occur at the edge of expanding ionization fronts or at shocks preceding an ionization front into a dark cloud. Finally, in regions where a burst of star formation has already taken place, explosions from supernovae evolving from young stars also may be providing the required shock-compression trigger that leads to the formation of a further generation of stars.

We may turn to the shock compression work of section 9:3.

PROBLEM 12–12.   In equation (9–27) we defined a compression ratio $\Psi = \rho_0/\rho_i$ corresponding to the increased density of gas flowing out of a shock relative to the inflow density. Show that even when extremely high shock velocities are invoked, the compression ratio in an adiabatic shock will not exceed

$$\Psi = \frac{\gamma_0 + 1}{\gamma_0 - 1} \tag{12–102}$$

where γ_0 is the ratio of heat capacities in the outflowing gas. When the flow is not adiabatic and an amount of heat $-Q$ per unit mass is used in exciting or dissociating the gas, show that the compression becomes

$$\Psi = \frac{\gamma_0 \pm \{2(1 - \gamma_0^2)Q/v_i^2 + 1\}^{1/2}}{(2Q/v_i^2 + 1)(\gamma_0 - 1)} \tag{12–103}$$

For molecular hydrogen at very low temperatures the ratio of heat capacities is $\gamma = 5/3$. At temperatures between $100°K$ and $300°K$, as rotationally excited states of the molecules become populated (see section 4:18 and Problem 7–5), γ drops from 5/3 to 7/5 as further internal degrees of freedom add to the heat capacity of the gas. At even higher temperatures, vibrational degrees of freedom become invoked and the ratio of heat capacities can drop toward 9/7. However, for cool post-shock gases, $\gamma = 5/3$, and the compression attained in an adiabatic shock cannot exceed $\Psi = 4$. The compression rises to $\Psi = 6$ as rotational states become increasingly populated. But as vibrational states become excited and dissociation tends to set in, the flow no longer is adiabatic because Q no longer is zero and we then need to use the more general expression (12–103). We can see that the compression then becomes large, provided $-Q \sim v_i^2/2$. This, however, occurs only at rather high shock velocities, at several tens of kilometers per second. All these conditions are appreciably altered in the presence of magnetic fields which change the *jump conditions* (9–19, 9–20 and 9–22). The hydrodynamics then are fairly involved, though the general approach taken above can still lead to useful insights.

12:23 ENERGY DISSIPATION

The compressive shocks just described can trigger collapse, but they do not guarantee permanent compression. Unless a compressed cloud can also dissipate energy, it will rebound elastically. In order for the cloud to remain compressed it must cool itself on a time scale comparable to the compression time. Energy must be radiated away, preferably altogether beyond the borders of the cloud. Often this is a two-step process in which an atom or molecule first is collisionally excited; in a second step it then radiates the excitation energy away. If that process is repeated often enough, the energy drain on the cloud is appreciable and it cools even as it is compressed. For excitation to take place, the translational energy of the gas constituents in a shock must approach or exceed the excitation energy for low-lying atomic or molecular levels. Once that threshold is exceeded, the collisional excitation cross sections for virtually all atoms or molecules found in galactic clouds tend to be of order 10^{-16} cm^2. In that respect there is rather little difference between the various atomic and molecular constituents.

The efficiency with which these constituents radiate, however, varies enormously. Neither atomic nor molecular hydrogen radiates efficiently below 1000°K. At these temperatures, the only available atomic hydrogen transition is the 21-cm, hyperfine transition corresponding to an energy jump of merely 6×10^{-6} ev. Since the Einstein spontaneous decay coefficient for this transition is only $A = 2.87 \times 10^{-15}$ sec^{-1}, the cooling rate through 21-cm emission could maximally be of the order of 1°K in a hundred million years. In contrast, cloud collapse is believed to occur during tens of thousands of years, and involves temperatures in the hundreds of degrees Kelvin. Similarly, molecular hydrogen, a symmetric dipole molecule, can radiate only through quadrupole emission which, as discussed in section 6:13, is an inefficient process. Thus we find that cooling in a compressed molecular cloud depends on impurity constitutents, such as CO and H$_2$O which, though low in abundance, are readily excited through collisions and rapidly radiate that energy away only to be collisionally excited once again to repeat this cycle over and over again. While these molecules are largely excited into rotational states on collision, atomic impurities like oxygen, carbon or singly ionized carbon can be excited to low-lying fine-structure levels within their ground electronic states. A fine-structure transition involves a change in electron-spin angular momentum relative to orbital angular momentum.

To obtain a quantitative estimate of cooling rates, we consider a cloud with density n and an impurity concentration X of a species of atom or molecule with a low-lying level that can be collisionally excited by hydrogen molecules—mass m. The excitation cross section is σ. If the Einstein coefficient for spontaneous emission of a photon with energy ε is A and the gas temperature is T, then the cooling rate per unit volume is

$$L = Xn^2 \left(\frac{3kT}{m}\right)^{1/2} \sigma\varepsilon, \tag{12-104}$$

provided the collisional excitation rate is far slower than the spontaneous emission rate

$$n\left(\frac{3kT}{m}\right)^{1/2} \sigma \ll A \tag{12-105}$$

Otherwise collisional excitation can be followed by de-exciting collisions and radiation becomes less efficient. Equation (12-104) can be rewritten, in terms of frequently encountered cloud parameters, as

$$L = 1.5 \times 10^{-18} \left(\frac{X}{10^{-4}}\right)\left(\frac{n}{10^6 \text{ cm}^{-3}}\right)^2 \left(\frac{T}{70°\text{K}}\right)^{1/2} \left(\frac{\sigma}{10^{-16} \text{ cm}^2}\right)\left(\frac{\varepsilon}{10^{-3} \text{ ev}}\right) \text{erg cm}^{-3} \text{ sec}^{-1} \tag{12-106}$$

which we note is independent of A. In employing this format for writing an equation, we are stating effectively that $X \sim 10^{-4}$, $n \sim 10^6 \text{ cm}^{-3}$, and so on, are typical values for impurity concentrations, hydrogen density and other parameters, and that L scales in proportion to X, to n^2 and so forth. The cooling rate L should still be compared to the heat content per unit volume, H

$$H \sim nkT = 10^{-8} \left(\frac{n}{10^6 \text{ cm}^{-3}}\right)\left(\frac{T}{70°\text{K}}\right) \text{erg cm}^{-3} \tag{12-107}$$

The ratio of these two quantities gives the cooling time

$$t_{\text{cool}} \sim \frac{H}{L} = 6.7 \times 10^9 \left(\frac{10^{-4}}{X}\right)\left(\frac{10^6 \text{ cm}^{-3}}{n}\right)\left(\frac{T}{70\text{K}}\right)^{1/2} \left(\frac{10^{-16} \text{ cm}^2}{\sigma}\right)\left(\frac{10^{-3} \text{ ev}}{\varepsilon}\right) \text{sec} \tag{12-108}$$

This cooling time is roughly 200 y for the parameters assumed and corresponds to the time required by a shock at speed 15 km/sec to cross a distance of 10^{16} cm. This suggests that turbulent motions at supersonic velocities should be rapidly damped and cannot long persist in a cloud. We may still compare the cooling time to the free-fall time for a spherical cloud. This is

$$t_{ff} \sim \left(\frac{3\pi}{32\rho G}\right)^{1/2} \sim 40{,}000 \left(\frac{10^6}{n}\right)^{1/2} y \qquad (12\text{–}109)$$

This tells us that the collapsing cloud can cool itself far more rapidly than free fall. It also gives an explanation for our not having found any protostars thus far. The collapse time is simply too short. The free-fall time is a hundred times shorter than the main sequence lifetime of the most massive, shortest-lived stars. For other stars, it represents an even smaller fraction of the total lifetime. There are very few protostars around at any given epoch. We may expect to have to look far across the Galaxy to find one.

The restriction on the Einstein coefficient is

$$A \gg 10^{-5} \left(\frac{n}{10^6 \text{ cm}^{-3}}\right)\left(\frac{T}{70°K}\right)^{1/2}\left(\frac{\sigma}{10^{-16}\text{ cm}^2}\right) \text{sec}^{-1} \qquad (12\text{–}110)$$

This needs to be compared to the Einstein A values for potential coolants. For CO the Jth rotational state has a coefficient $A_J \sim 1.118 \times 10^{-7} J^3 \text{ sec}^{-1}$. The wavelengths for these transitions lie at $\lambda_J \sim 2600/J$ microns equivalent to a temperature $T_J = hc/\lambda_{Jk} \sim 2.765 J^2 °K$. Atomic oxygen has a fine-structure transition at 63 microns with $A = 9 \times 10^{-5} \text{ sec}^{-1}$. We see that both CO at $J > 7$ and atomic oxygen meet the requirements of (12–110). Water vapor also has a large number of lines with high A values. The relative cooling capabilities of CO and H_2O are shown in Fig. 12.9.

Fig. 12.9 Total cooling as a function of H_2 density for a kinetic temperature of 40°K for typical atomic and molecular abundances expected in clouds (from P.F. Goldsmith and W.D. Langer, Go78).

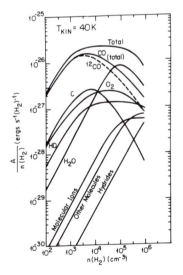

All this assumes that the optical depth of the cloud is low. If the column density rises to the point where self-absorption dominates, radiative cooling by atoms and molecules becomes progressively less efficient. Here a typical value for the Einstein absorption coefficient within the spectral line envelope is $B(v)$ at line frequency v.

$$B(v) = \frac{c^2}{8\pi v^2} \frac{A}{\Delta v} \tag{12-111}$$

where Δv is the Doppler width and the column density cannot exceed

$$N_X = [B(v)]^{-1} \tag{12-112}$$

without appreciable line trapping within the cloud. When the *column density* in the line becomes excessive, atomic and molecular cooling becomes less efficient and dust grains take over the prime cooling role.

12:24 COOLING OF DENSE CLOUDS BY GRAIN RADIATION

Once an interstellar cloud becomes dense enough to absorb spectral line radiation emitted by its principal atomic and molecular constituents, grain emission may begin to dominate cooling. However, grains can cool the gas only as rapidly as atoms or molecules transfer their energy to the grains. If the grains are taken to be roughly spherical with radius a, the rate of heat transfer to a grain is

$$\frac{dQ_{gr}}{dt} = n_2 \pi a^2 \left(\frac{3kT_2}{m_2}\right)^{1/2} (T_2 - T_{gr})\alpha \left(\frac{c_v T_2}{N}\right) \tag{12-113}$$

Here, the expression on the left is the heat dQ_{gr} transferred to a grain in time dt, and we see that this is proportional to the following: n_2, the number of hydrogen molecules per unit volume, assuming H_2 to be the dominant gas constituent; to the grain collision cross section for gas impact, πa^2; to the speed $(3kT_2/m_2)^{1/2}$ with which the molecules of mass m_2 travel at the gas temperature T_2; to the difference in grain and gas temperature $T_2 - T_{gr}$; to an efficiency factor α with which energy is transferred from a molecule to a grain on impact; and, finally, to the energy the molecule has that it could transfer, $c_v T_2/N$. Here we have used the same notation as in section 4:18.

The number density of grains, n_{gr}, is determined by the fraction by mass, X_{gr}, of matter in the form of grains and by the mass density ρ_{gr} of the grain material. Per unit volume in space we then have

$$n_{gr} = \frac{n_2 X_{gr} m_2}{\rho_{gr}(4\pi a^3/3)} = 3.8 \times 10^{-9} \left(\frac{n_2}{10^4 \text{ cm}^{-3}}\right)\left(\frac{X_{gr}}{10^{-3}}\right)\left(\frac{2 \text{ g cm}^{-3}}{\rho_{gr}}\right)\left(\frac{10^{-5} \text{ cm}}{a}\right)^3 \text{ cm}^{-3} \tag{12-114}$$

where the numerator of the intermediate expression gives the mass of grains in unit volume of space, while the denominator is the mass per grain. The expression on the right exhibits representative values for typical parameters characterizing molecular clouds. We can now multiply the expressions for dQ_{gr}/dt and n_{gr} to obtain the cooling rate per unit volume of space, L_{gr}

$$L_{gr} = \frac{3}{4}n_2^2 X_{gr}(3m_2k)^{1/2}(T_2 - T_{gr})\alpha c_v T_2^{3/2}(\mathcal{N}\rho_{gr}a)^{-1}$$

$$= 10^{-22}\left(\frac{n_2}{10^4\ cm^{-3}}\right)^2\left(\frac{X_{gr}}{10^{-3}}\right)\left(\frac{\alpha}{1/3}\right)\left(\frac{2\ g\ cm^{-3}}{\rho_{gr}}\right)\left(\frac{10^{-5}\ cm}{a}\right) \quad (12\text{-}115)$$

$$\times\left(\frac{c_v/\mathcal{N}k}{3/2}\right)\left(\frac{T_2}{50°K}\right)^{3/2}\left(\frac{T_2 - T_{gr}}{30°K}\right)\frac{erg}{cm^3\ sec}$$

This cooling rate must be compared to the radiative cooling rate for grains at temperature T_{gr} radiating with efficiency η.

$$L_{rad} = 4\pi a^2 n_{gr}\sigma T_{gr}^4\eta = \frac{3n_2 X_{gr}m_2\sigma T_{gr}^4\eta}{\rho_{gr}a}$$

$$= 4.4 \times 10^{-21}\left(\frac{n_2}{10^4\ cm^{-3}}\right)\left(\frac{X_{gr}}{10^{-3}}\right)\left(\frac{T_{gr}}{20°K}\right)^4\left(\frac{\eta}{10^{-4}}\right) \quad (12\text{-}116)$$

$$\times\left(\frac{10^{-5}\ cm}{a}\right)\left(\frac{2\ g\ cm^{-3}}{\rho_{gr}}\right)\frac{erg}{cm^3\ sec}$$

Here σ is the Stefan-Boltzmann constant and η has been chosen as 10^{-4}, roughly corresponding to a radiation efficiency comparable to the ratio of grain radius to wavelength, a/λ, for grains radiating at wavelengths just short of one millimeter. Even if this efficiency is an order of magnitude lower, $\eta \sim 10^{-5}$, the grains can still radiate rapidly enough to keep the gas temperature below 50°K; even then we would have $L_{rad} > L_{gr}$, meaning that the rate of radiation by grains is at least as rapid as the rate at which the gas can heat the grains.

As the cooling cloud contracts, the density rises and L_{gr} increases as the square of the density, while L_{rad} only grows linearly with density. With the above parameters we would have equality of grain heating and cooling, $L_{gr} \sim L_{rad}$, at densities $n_2 \sim 5 \times 10^5\ cm^{-3}$; but that number depends quite critically on the difference between gas and grain temperatures. As the density increases, the temperature difference between gas and grains tends to decline. If the grains become hotter they begin emitting at shorter wavelengths where their efficiency $\eta \sim a/\lambda$ increases roughly in proportion to T_{gr}, since a thermally emitting body shifts its peak emission frequency in proportion to T, as evident from (4–71), so that $\eta \propto T_{gr} \propto \lambda^{-1}$. As a result L_{gr} rises roughly in proportion to T_{gr}^5.

Grains play a major role in the cooling of a contracting cloud, radiating away energy gained as gravitational potential energy is released. Their predominance in cooling is due to the broad wavelength range over which they emit and is not greatly affected by their low efficiency in radiating. That low efficiency implies that energy radiated by a given grain in the cloud will not be readily absorbed by another grain and will therefore escape. Per unit volume the opacity is

$$\kappa_{gr} = \pi a^2 n_{gr} \eta = \frac{3 n_2 X_{gr} m_2 \eta}{4 \rho_{gr} a} = 1.2 \times 10^{-22} \text{ cm}^{-1} \qquad (12\text{--}117)$$

for the same parameters used in equation (12–116). Since the opacity can only change in response to an increase in density or radiating efficiency, and since η is unlikely to exceed 10^{-3} at temperatures characteristic of protostellar clouds, n_2 can rise to a value of 10^9 cm^{-3} and still leave the opacity as low as $\sim 10^{-16}$ cm^{-1}. At those densities a sphere of radius 10^{16} cm would just barely have unit opacity, but the total mass encompassed would exceed 1 M$_\odot$. Once the cloud becomes opaque, perhaps at the time its radius is of order of a few hundred astronomical units, it can at best only emit as a blackbody. Also, as long as thermal equilibrium is assumed, rapid contraction would require an increase in temperature to make up for the decreasing surface area available for radiation. Finally, we may want to examine the relationship between the cooling time and the free-fall time. The temperature at the surface of the nebula is given by equation (4–117) in conjunction with equations (4–128) to (4–132). The total potential energy of the cloud is given by equation (8–1). The cooling time is therefore

$$t_{cool} \sim \frac{3 M^2 G}{5 R} \left(\frac{1}{4 \pi R^2 \sigma T^4} \right) = \frac{243 k^4 \mathcal{N}^4 R}{20 \pi M^2 G^4 \sigma} \qquad (12\text{--}118)$$

where $\mathcal{N}$ is Avogadro's number and k is Boltzmann's constant. Taking $\mathcal{N} = m_H^{-1}$, we find the cooling time to be roughly one year. In contrast, the free-fall time is $(3\pi/32 G n_2 m_2)^{1/2}$ as seen from equation (12–69); for a protostellar nebula of constant density that would amount to about 3000 y. The nebula, therefore, can cool itself faster than it would collapse through free fall. We can therefore assume that the collapse proceeds essentially adiabatically, unless excessive angular momentum that cannot be shed, or excessive internal magnetic pressures halt the contraction. Just how magnetic fields or angular momentum are shed is not yet properly understood.

    This is about as much as we can say about the formation of stars, from the view of collapsing cloud dynamics. Another source of information, however, is available from a study of the remnants of earliest phases of the protosolar nebula, remnants from a time when the sun and the planets were just forming. Those remnants are some of the earliest-formed meteorites, or rather meteorite fragments. In order to decipher the message these meteorites have preserved and brought down to us over the aeons, we must first go back to a number of thermodynamic considerations. We do that in section 12:25 and then look at the meteoritic data in 12:26.

12:24

12:25 CONDENSATION IN THE EARLY SOLAR NEBULA

We now ask ourselves how gases in the early solar nebula condensed into solid matter which ultimately went into the formation of planetary bodies. Clearly, the first law of thermodynamics—energy conservation—must hold in this process. We start with equation (4–112)

$$dQ = dU + P \, dV \qquad (12\text{–}119)$$

and note that the second law requires

$$T \, dS \geq dQ \qquad (12\text{–}120)$$

meaning that the entropy either increases or, at best, remains constant in any physical process as long as we deal with the entire system, generally the entire volume within which that process takes place. Equality in this relation holds only when the process can occur reversibly. There are relatively few genuinely reversible processes, but sublimation and condensation is one of them. Typically, a closed vessel kept at constant temperature and pressure near the sublimation point of a substance will see the growth of some crystals at the expense of others while the vapor pressure in the vessel remains constant. No net work is done in this equilibrium state since the pressure and volume remain constant, but heat is transferred from growing crystals to subliming crystals. That heat is the latent heat of evaporation per mole of substance, λ. We, therefore, see the entropy change for a mole of condensing material to be

$$\Delta S = -\frac{\lambda}{T} \qquad (12\text{–}121)$$

The latent heat λ depends on both the pressure and temperature, but the equilibrium vapor pressure rises rapidly near one particular temperature—the sublimation temperature. Below this narrow temperature range one finds the bulk material largely equilibrated in the condensed phase, while above this range it is almost exclusively gaseous.

We now turn to a thermodynamic function termed the *Gibbs free energy* named after the 19th century American thermodynamicist J. Willard Gibbs. It is defined as

$$G \equiv U + PV - TS \qquad (12\text{–}122)$$

This is a function which describes the state of the system without regard to the ways in which the system originated. Such functions can be differentiated exactly and we write

$$dG = dU + P \, dV + V \, dP - S \, dT - T \, dS \qquad (12\text{–}123)$$

Applying equations (12–121) and (12–122) we obtain

$$dG \leq V \, dP - S \, dT \qquad (12\text{–}124)$$

12:25

and we see that for changes occurring at constant temperature and pressure, $dG \leq 0$, and the free energy either decreases or remains constant. A constant free energy requires a reversible process, such as the condensation or sublimation just discussed.

Now, consider two phases designated by v for vapor and s for solid. In equilibrium we will have

$$V_s\, dP - S_s\, dT = V_v\, dP - S_v\, dT \qquad (12\text{--}125)$$

or

$$\frac{dP}{dT} = \frac{S_s - S_v}{V_s - V_v} = \frac{\Delta S}{\Delta V} = \frac{-\lambda}{T\Delta V} \qquad (12\text{--}126)$$

For condensation, $V_s \ll V_v$ and $\Delta V = -V_v = -RT/P$, where R is the gas constant defined in section 4:6. Hence

$$\frac{dP}{dT} = \frac{P\lambda}{RT^2} \qquad (12\text{--}127)$$

from which we obtain the equation already mentioned in Table 9.3

$$\ln P = \frac{-\lambda}{RT} + \text{constant} \qquad (12\text{--}128)$$

This reciprocal relationship between the logarithm of the vapor pressure and the temperature is shown by the straight lines of Fig. 12.10, which shows the rapid drop in vapor pressure for elements of interest in the condensation of the early solar nebula. Table 12.1 also shows the condensation temperatures of a number of pure elements at two different total pressures, respectively 1 and 6.6×10^{-3} atmospheres, in the early solar nebula (1 atm $= 760$ torr $= 1.01 \times 10^6$ dyn cm^{-2}).

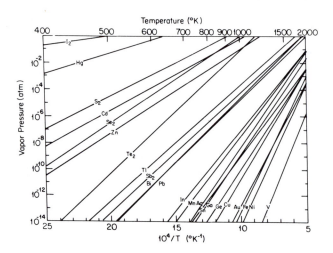

Fig. 12.10 Vapor pressures of elements found in meteorites (Reprinted with permission from Larimer, La67.)

12:25

Table 12.1 Condensation Temperatures for Compounds and Elements in the Early Solar Nebula, at Two Different Total (hydrogen) Gas Pressures, P_T (Reprinted with permission from Larimer, La67).

$P_\tau = 1$ atm		$P_\tau = 6.6 \times 10^{-3}$ atm	
Compound or Element	$T(°K)$	Compound or Element	$T(°K)$
$MgAl_2O_4$	2050	$CaTiO_3$	1740
$CaTiO_3$	2010	$MgAl_2O_4$	1680
Al_2SiO_5	1920	Al_2SiO_5	1650
Ca_2SiO_4	1900	$CaAl_2Si_2O_8$	1620
$CaAl_2Si_2O_8$	1900	Fe	1620
$CaSiO_3$	1860	Ca_2SiO_4	1600
Fe	1790	$CaSiO_3$	1580
$CaMgSi_2O_6$	1770	$CaMgSi_2O_6$	1560
$KAlSi_3O_8$	1720	$KAlSi_3O_8$	1470
Ni	1690	$MgSiO_3$	1470
$MgSiO_3$	1670*	SiO_2	1450
SiO_2	1650	Ni	1440
Mg_2SiO_4	1620*	Mg_2SiO_4	1420
$NaAlSi_3O_8$	1550	$NaAlSi_3O_8$	1320
$MnSiO_3$	1410	$MnSiO_3$	1240
Na_2SiO_3	1350	MnS	1160
K_2SiO_3	1320	Na_2SiO_3	1160
MnS	1300	K_2SiO_3	1120
Cu	1260	Cu	1090
Ge	1150	Ge	970
Au	1100	Au	920
Ga	1015	Ga	880
Sn	940	Zn_2SiO_4	820
Zn_2SiO_4	930	Sn	806
Ag	880	Ag	788
ZnS	790	ZnS	730
FeS	680	FeS	680
Pb	655	Pb	570
CdS	625	CdS	570
Bi	620	$PbCl_2$	535
$PbCl_2$	570	Bi	530
Tl	540	Tl	475
In	400	Fe_3O_4	400
Fe_3O_4	400	In	360
H_2O	260	H_2O	210
Hg	196	Hg	181

* The condensation temperatures for $MgSiO_3$ and Mg_2SiO_4 are somewhat uncertain.

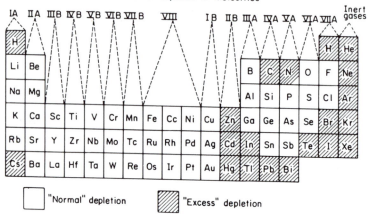

Fig. 12.11 Elements found depleted in most chondritic meteorites, relative to type I carbonaceous chondrites. "Normal depletion" corresponds to depletion by a factor of 0.1–0.5, while "excess depletion" refers to factors of 0.001–0.1. (Reprinted with permission from Larimer, La67.)

We notice first that the elements which are most volatile—those with the highest vapor pressure at a given temperature—tend to be clustered in the right half of the periodic table, as shown in Fig. 12.11.

If the early solar nebula had cooled down slowly, we would expect that the first condensates would have contained only refractory materials, materials that evaporate at high temperatures. As the nebula cooled, more volatile substances would have condensed out. The condensation sequence with declining temperature would then have followed the order indicated in Table 12.1.

PROBLEM 12–13. Consider an early solar nebula consisting largely of H_2 at 2000°K orbiting a central sun. If the nebular mass is 1 $M_\odot$ and the projected density of matter onto a central plane of this rotating disk is constant out to a distance of about 10^{14} cm, roughly 7 AU, show that the expected nebular pressure is $\sim 4 \times 10^{-2}$ atm, well within the range of pressures covered in Table 12.1.

12:26 THE EVIDENCE PROVIDED BY METEORITES

At least some types of meteorites are believed to be remnants of the earliest stages of evolution of the solar nebula. They are thought to have condensed well before the first planets formed. Meteorites generally are classed either as *iron* or as *stony*.

The iron meteorites are metallic and rich in iron. The stony meteorites can taken on different forms. Of particular interest to studies of the formation of the solar system are *chondrites*, stony meteorites containing *chondrules*. Chondrules, in turn, are millimeter-sized silicate spherules that look as though they might have been droplets frozen from a melt. They consist largely of *olivine*, a mineral whose chemical makeup is $(Mg, Fe)_2 SiO_4$, *pyroxine* $(Mg, Fe)SiO_3$, and *plagioclase feldspar*, which is a solid solution of $CaAl_2 Si_2 O_8$ and $NaAlSi_3 O_8$. Table 12.1 shows that all of these minerals condense out at temperatures $\geq 1240°K$ in the pressure range shown. The chondrules are embedded in a *matrix*, a more finely ground mass, generally of the same composition. In the matrix one also finds millimeter-sized particles of nickel-iron with a nickel content ranging from about 5% to 60%. *Troilite*, whose chemical composition is FeS, is also present.

The chondrites can be divided into three groups. *Carbonaceous* chondrites, designated by a letter C, are highly oxidized. In particular, their iron content is always highly oxidized, meaning that each iron atom donates two or three of its outer shell electrons to other elements in the chondritic mineral. Carbonaceous chondrites derive their name from the carbon-rich compounds they contain. In contrast, *enstatites*, or E *chondrites*, are highly reduced, containing iron only in the metallic form or as troilite. Between these two extremes one finds *ordinary*, or O *chondrites*.

The carbonaceous chondrites are subdivided into three classes. Type I is virtually free of chondrules and consists largely of a mineral matrix, while type III consists of 70–80% chondrules with very little matrix in between. Type II is intermediate to these. Chemical analyses show that most chondritic meteorites are quite strongly depleted in the more volatile elements, when compared to Type I. This is because the matrix is richer in volatile elements than are the chondrules.

We now ask why these two chondritic constituents, the chondrules and the matrix, should differ so greatly in their content of volatiles.

The chondrules are thought to have been the first solids to condense out of a solar nebula which began as a high-temperature, gaseous mass. The most refractory materials would have condensed out first, forming the chondrules. As Table 12.1 suggests, highly volatile material, such as bismuth Bi, lead Pb and indium In, would have remained in vapor form and would only have condensed at much lower temperatures. Those chondrites rich in chondrules, therefore, contain largely those refractory constituents which condensed out first.

Actual depletion for ordinary chondrites relative to a cosmic abundance distribution of elements is shown in Fig. 12.12. We see that bismuth and indium are depleted by two to three orders of magnitude.

Two possible condensation sequences showing the nebular temperatures at which different compounds condensed are shown in Fig. 12.13. The actual condensation temperatures depend to some extent on whether the nebula cools slowly or rapidly. The slower cooling assumes that complete diffusion can take place, with the forma-

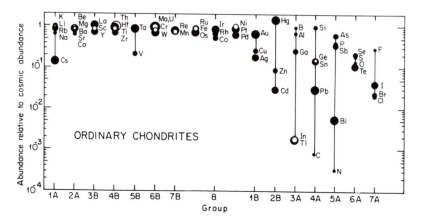

Fig. 12.12 Abundance of the various elements in ordinary chondrites, relative to the cosmic abundance given in Table 1.2 (after Anders, An72).

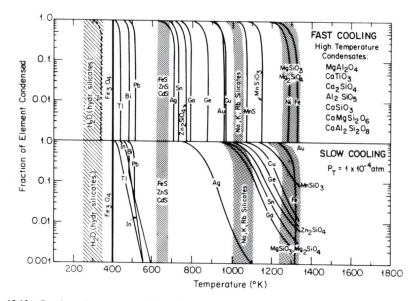

Fig. 12.13 Condensation sequence of gas whose initial composition corresponds to the cosmic abundance of elements. For grains of radius $\sim 10^{-5}$ cm, the upper sequence applies for times of minutes or hours, while the lower sequence corresponds to cooling times of years or centuries. The shaded areas show condensation or chemical transformaton of major constituents (after Anders, An72).

12:26

tion of alloys and solutions to the limit of solubility of all the substances involved. This permits the condensation temperatures for minor elements to be higher, and also widens their condensation ranges.

The last few years have seen a great deal of progress in the analysis of meteorites and in understanding their past histories. A fairly detailed picture of the evolution of the early solar nebula has emerged. Of primary importance is the clear indication that the nebula started at temperatures in excess of perhaps 1800°K, and that it cooled to below 400°K before the planets formed. Evidence for this lower temperature comes from the long chain hydrocarbons in the carbonaceous chondrites which can only form at low temperatures in reactions of the type

$$20\,CO + 41\,H_2 \Leftrightarrow C_{20}H_{42} + 20\,H_2O \qquad (12\text{--}129)$$

through the interaction of residual carbon monoxide, CO, with hydrogen. These reactions, however, only take place in the temperature range between 300°K and 400°K for pressures likely to prevail. At temperatures above 450°K, these gases react differently, giving

$$CO + 3\,H_2 \Leftrightarrow CH_4 + H_2O \qquad (12\text{--}130)$$

The double arrows in these two reactions show that they can go in either direction, but the hydrocarbons at the right are favored toward lower temperatures and the higher CO content seen on the left side of the equations is favored at higher temperatures (Fig. 12.14). CO, of course, is a constituent of interstellar clouds and

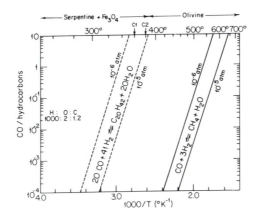

Fig. 12.14 Equilibrium between CO and H_2 at different temperatures and pressures. If equilibrium is maintained on cooling, CO is largely converted into CH_4 before more complex molecules can be formed at lower temperatures. However, this reaction is slow, and some CO may survive to lower temperatures where complex organic molecules can form, particularly if Fe_3O_4 (iron rust) and hydrated silicates like serpentine, $Mg_3Si_2O_5(OH)_4$, form at 380–400°K. Both of these are effective catalysts for the Fischer–Tropsch reaction (see text) (after Anders, An72).

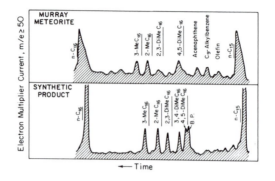

Fig. 12.15 Gas chromatogram of hydrocarbons in the range containing 15–16 carbon atoms. The synthetic product was made by a Fischer–Tropsch reaction (see text) (after Anders, An72).

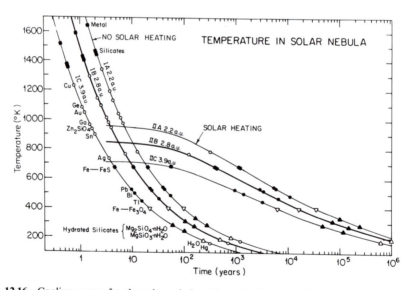

Fig. 12.16 Cooling curves for the solar nebula with and without heating by the highly luminous protosun. Presumably the initial cooling follows curves IA, IB and IC at 2.2, 2.8 and 3.9 AU, respectively, from the sun. Later on, curves IIA, IIB and IIC are more likely to be germane. Accretion temperatures of different types of chondrites are indicated. These are inferred from the condensation temperatures of Pb, Bi, In, Tl and H_2O, as well as the threshold of the $Fe \rightarrow Fe_3O_4$ reaction. A time scale for accretion of chondritic material of 10^4–10^5 y is implied. Just at what distances the various chondrites form is uncertain. (Reprinted with permission from Larimer and Anders, La67a.)

12:26

is expected to be present in the early nebula. The production of hydrocarbons from CO and H_2 proceeds in the laboratory through the Fischer–Tropsch reaction, which is just the reaction (12–129) catalyzed by iron or cobalt in industrial syntheses. That this reaction is responsible for the carbon compounds found in meteorites is indicated by Fig. 12.15, which shows that, among about 10^4 possible hydrocarbon molecules that can be formed using 16 carbon atoms and an arbitrary amount of hydrogen, only six are present in appreciable abundance in the meteorite analyzed. Five of these, all underlined in the figure, are common to both samples. Acenaphthene, not detectable in the synthetic example, has been seen in other products prepared by the Fischer–Tropsch method at higher temperatures.

All this still leaves uncertain the cooling rate of the solar nebula. That depends on whether the nebula is self-shielding, in that the inner parts prevent solar heating of the outer parts, or whether solar heating plays an important role. The cooling rates under these two conditions are shown in Fig. 12.16. Either way, the early solar nebula cooled in a remarkably short time, perhaps 10^4–10^5 y, a mere instant when compared to the sun's lifetime as a main sequence star, $\sim 5 \times 10^9$ y.

12:27 FORMATION OF PRIMITIVE CONDENSATES IN THE EARLY SOLAR NEBULA

Given that the early solar nebula had a very high temperature, and that much of the interstellar dust would have evaporated, we can ask how rapidly new grains would grow as the nebula cooled.

Once condensation temperatures are reached, grain growth is quite rapid. We imagine a seed grain, perhaps an interstellar grain that has survived, a grain that might be entering the cooling nebula from a surrounding region which had not participated in the collapse, or perhaps a seed spontaneously formed from the vapors. Consider that the seed has radius a and is located in the plane of the nebular disk. We know that all the freely orbiting material will pass through the disk twice per orbital period P. At the earth's distance from the sun, that means that material passes through this plane twice yearly. In the answer to Problem 12–13 we saw that the aereal density of this matter was $\sigma = 6.4 \times 10^4$ g cm^{-2}. If a fraction, say $f = 10^{-4}$, of this mass condenses at a particular temperature, the seed will grow at a rate

$$\frac{da}{dt} = \frac{\pi a^2 \sigma f}{4\pi a^2 \rho_{gr}} \left(\frac{2}{P}\right) = \frac{\sigma f}{2\rho_{gr}P} \qquad (12\text{–}131)$$

For a grain density $\rho_{gr} = 3$ g cm^{-3}, we then obtain a growth rate at the earth's distance from the sun of roughly 1 cm/y. However, since there is only $f\sigma \sim 6$ g cm^{-2}

of condensable material in the disk, we see that the sweep-up will only take a few years, even if relatively few seed particles are initially present. Nucleation of seeds is almost inevitable if only because Galactic cosmic rays which always abound can produce nucleation in any supersaturated vapor—as they also do in a Wilson cloud chamber.

12:28 FORMATION OF PLANETESIMALS

Once centimeter-sized particles, or at least the observed millimeter-sized chondrules, form in the protoplanetary nebula, their scale height is quite small. The thermal velocity of a millimeter-sized particle with mass $\sim 10^{-3}$ g and temperature $10^3 °K$ is $(3kT/m)^{1/2} \sim 2 \times 10^{-5}$ cm sec^{-1}, so that the scale height as obtained in Problem 12–10 is effectively less than the chondrule size, even if we use an aereal density due to condensed matter alone. The layer of condensed particles is therefore exceedingly thin, and the velocity of sound in the disk—the speed with which sound is propagated through collisions by grains—is extremely low. The one factor which could make the speed of sound in the disk higher is the circumstance that the grains may be touching and therefore the bulk speed of sound in the material is of importance.

Goldreich and Ward (Go73) have considered the stability of such a disk. They start with a dispersion relation, similar to the Jeans criterion, but applicable to rotating axisymmetric disks in which matter is orbiting along Keplerian trajectories:

$$\omega^2 = k^2 c^2 + \Omega^2 - 2\pi G\sigma k f \qquad (12\text{–}132)$$

where Ω is the frequency of the orbital rotation in radians per second, k is the wave number $k = 2\pi/\lambda$ and λ is the wavelength of the disturbance, while ω is the frequency characterizing the disturbance. When that frequency goes from a real to an imaginary value, instability sets in. That means that the criterion for instability is

$$2\pi G\sigma k f > \Omega^2 \qquad (12\text{–}133)$$

since the speed of sound is essentially negligible. The minimum wavelength which is unstable then becomes

$$\lambda_{min} = 4\pi^2 G\sigma\Omega^{-2}f \qquad (12\text{–}134)$$

The mass contained in such a fragment is $f\sigma\lambda_{min}^2$, so that the minimum condensation mass would be

$$M_{min} = 16\pi^4 G^2\sigma^3\Omega^{-4}f^3 \qquad (12\text{–}135)$$

which has a value of order 10^{18} g, while $\lambda_{min} \sim 4 \times 10^8$ cm. For unit density we obtain *planetesimals* roughly 10 km in size.

We note a number of points. First, if equation (12–132) is multiplied by the square of some wavelength λ and by a local density ρ, we see that the first term on the right is comparable to the internal energy of the gas; the second term corresponds to a rotational kinetic energy; and the third term corresponds to self-gravitational attraction. When the wavelength is sufficiently large so that the self-gravity exceeds the rotational and internal energies combined, instability and contraction can set in.

Second, the maximum wave number $k = 2\pi/\lambda_{min} \sim 10^{-8}$ cm when multiplied by a speed of sound derived from grain velocities is far smaller than Ω which, at the distance of the earth from the sun, has a value $\sim 2 \times 10^{-7}$ rad $\sec^{-1}$. If the speed of sound were higher in the disk, and the first term on the right became dominant over the second, the instability criterion would become

$$\frac{2\pi}{\lambda_{min}} = k_{max} < \frac{2\pi G \sigma f}{c^2}, \qquad \lambda_{min} > \frac{c^2}{G \sigma f} \qquad (12\text{--}136)$$

so that the minimum wavelength would become $\sim 2.5 \times 10^6 c^2$ cm. The argument for the formation of planetesimals through gravitational instabilities arising in a disk, therefore, depends quite crucially on the speed of sound being low.

Third, the speed with which such bodies form is going to be of order $\omega^{-1} \sim \lambda_{min}/2\pi c$, which even for $c \sim 2 \times 10^{-5}$ cm $\sec^{-1}$ and $\lambda_{min} \sim 4 \times 10^8$ cm is only about a hundred thousand years, quite brief compared to the age of the solar system. Again, the speed of sound is quite critical in this estimate.

Once these planetesimals have been formed, further growth of planets may occur through their gravitational accretion into large bodies. Just how that takes place is not understood.

ADDITIONAL PROBLEMS

PROBLEM 12–14. The expansion of the universe causes a freely moving galaxy to slow down relative to its surroundings as it reaches galaxies whose recession velocities equal the galaxy's original velocity with respect to the substratum. Because of this, we believe that the motion of the Galaxy with respect to the microwave background radiation must be caused by gravitational attraction toward mass concentrations that are relatively nearby. We speak of a *density contrast C*, in the universe around us, representing a deviation $\delta\rho$ from some mean density ρ, and having a particular dimension δr. The contrast is then some function $c(\delta\rho/\rho, \rho, \delta r)$. The larger $\delta\rho/\rho$ and the smaller ρ and δr, the greater the contrast. We also speak of *biased galaxy formation*, meaning that galaxies and clusters formed preferably where the contrast was originally large. Such theories postulate such a contrast without necessarily specifying a mechanism that could have produced it.

The 600 km $\sec^{-1}$ velocity of the Galaxy through the microwave background radiation has been decomposed into a velocity of 220 km $\sec^{-1}$ of infall toward the Virgo cluster of our whole local group of galaxies, plus an additional fall of the entire Virgo cluster toward the Hydra-Centaurus complex of galaxy clusters, at

495 km sec^{-1}. If the distance to these two postulated mass concentrations is 16 and 40 Mpc respectively, what kind of density contrast does that imply? For purposes of this problem consider ρ equal to ρ_{crit}—$\Omega = 1$—and consider C to equal $\delta\rho/\rho$ alone, averaged over a volume $4\pi r^3/3$, where r is the distance of the purported source. Assume that each infall started with the moving bodies at rest 10^{10} y ago.

PROBLEM 12–15. It is possible that neutrinos have a rest mass m_ν of the order of a few electron volts. They become nonrelativistic after their decoupling from electrons and positrons, and when their temperature has dropped to $T \sim m_\nu c^2/k$. (a) Show that the minimum radius over which instability can set in is $c^2[3/17.5(4\pi Gk)]^{1/2}T^{-2}$. (b) Also, show that collapse occurs only for radii less than c/H, where $H \sim (T/3°\text{K})^{3/2}H_0$, with H_0 today's Hubble constant. (c) What is the minimum neutrino mass for which instability will set in, and what is the corresponding value of R? Compare m_ν to the value obtained in Problem 12–11. (d) What is the mass contained within the radius R? (e) What is a lower limit to the free-fall time during collapse? In this connection it is worth noting that similar estimates can be made for other types of particles that decouple from the primordial matter/radiation bath. Such particles may be the main constituent of the *dark matter* observed only through its gravitational pull. When the decoupling particles are relativistic at decoupling, we speak of *hot dark matter*. When they are massive and nonrelativistic at decoupling, we speak of *cold dark matter*. Hot dark matter, because of its initially high velocity, tends to produce condensations that are highly extended. Smaller condensations are then expected to form at a later stage in the contracting extended region as it contracts and becomes denser. This is called a *top-down condensation sequence*. For cold dark matter, we expect small condensations to form first and larger ones only later. Together, neutrinos and antineutrinos, are considered to constitute hot dark matter because they are relativistic at decoupling.

ANSWERS TO PROBLEMS

12–1. (a) Since $k = \Lambda = 0$, and $P = \rho/3c^2$, $\ddot{a} = 0$ in equation (12–11). We know that for a relativistic gas $\rho \propto a^{-4}$, so that (12–10) can be written as

$$H = \frac{\dot{a}}{a} = \left(\frac{8\pi G\rho_0}{3}\right)^{1/2}\frac{a_0^2}{a^2}$$

where a_0 and ρ_0 are values chosen for some initial time $t_0 = 0$. On integrating we obtain a world time t_a:

$$a_0^{-2}\left(\frac{3}{8\pi G\rho_0}\right)^{1/2}\int_0^a a\,da \int_0^{t_a} dt = t_a \sim \frac{1}{2H} \qquad \text{for } a_0 \ll a.$$

(b) When the pressure can be taken as negligible, $P = 0$ and

$$\frac{\dot{a}}{2a} = -\frac{\ddot{a}}{\dot{a}}$$

which integrates to $\dot{a} = Aa^{-1/2}$, where

$$H = \frac{\dot{a}}{a} = \left(\frac{8\pi G\rho'_0 a'^3_0}{a^3}\right)^{1/2} \equiv \frac{A}{a^{3/2}}$$

defines the value of the constant of integration A. A second integration then yields, for $t_a \gg t'_0$,

$$t_a \sim t - t'_0 = \frac{2}{3}\left(\frac{1}{8\pi G\rho'_0 a'^3_0}\right)^{1/2}(a^{3/2} - a'^{3/2}_0) \sim \frac{2}{3H} \qquad \text{for } a'_0 \ll a.$$

12–2. From Problem 12–1 and equation 12–21 we see that

$$a(t) \propto t^{1/2} \qquad \text{for a relativistic gas}$$

$$\propto t^{2/3} \qquad \text{for non-relativistic matter}$$

If we insert these relations in equation (12–26) and integrate, we see that the time required to cross distance parameter χ becomes, respectively,

$$\chi \propto (t^{1/2} - t_0^{1/2})$$

and

$$\chi \propto (t^{1/3} - t_0'^{1/3})$$

This shows that when $t \gg t_0$ or t'_0, doubling the parameter χ, or equivalently making a round trip measured by parameter χ, respectively, increases the travel time by factors of 4 and 8. Note that the result also holds when t and t_0 or t'_0 do not differ greatly, because $(t^{1/2} - t_0^{1/2})^2$ and $(t^{1/3} - t_0^{1/3})^3 < (t - t_0)$.

12–3. During inflation, the number density of all particles decreases by $\sim(10^{26})^3 = 10^{78}$. The initial mass density before inflation was 10^{77} g cm^{-3}, so that a volume 10 cm in diameter which defines all matter ultimately to become part of the universe within present horizons would have originated in a volume element 10^{-75} cm^3 before inflation. The mass content of that region would have been 100 g. Even if filled entirely with monopoles of energy 10^{16} Gev $= 1.8 \times 10^{-8}$ g, it could at most have contained 10^{10} monopoles. The number of galaxies within the current cosmic horizon is believed to be 10^{11}.

12–4. Equation 12–53 leads to $dS = (4a/3)d(VT^3)$, which leads to the desired result, though there is an additive constant of integration that needs to be added. However, the requirement that the entropy be proportional to the volume can only be met if that constant is zero.

12–5. The Λ term in equations (12–10) and (12–11) is constant. It cannot greatly exceed today's Hubble constant without violating equation (12–10) for the observed Hubble constant and current mass density. Since the mass density in previous epochs was far larger than today (12–17, 12–19), while Λ remained constant, both the left side of (12–10) and the second term on the right of this equation must have far exceeded the value of Λ at earlier times, when the elements were forming. If equation (12–10) for negligibly small values of Λ is inserted into (12–11), we then also see that the expansion rate of the universe must monotonically have been decelerated at a particularly rapid rate at early time, when a was small and ρ was large. This implies that $\dot{a}$ at early times was far higher than today, so that the influence of a nonzero value $k = \pm 1$ would have been even less significant then than now.

12–6. Equation (12–58) tells us that for each electron and for each positron, as well as for each of their respective spin states, the energy density before annihilation is $7aT^4/16$. On annihilation, which occurs over a relatively narrow temperature range and takes place as a phase transition, this energy density adds to the photon density aT^4, giving a total energy density $(11aT^4/4)$. Before the phase transition the entropy is expressed by a relation of form $S_b = (11/4)(4/3)aT_b^3 V_b$. After the phase transition it simply has the form $S_a = 4aT_a^3 V_a^3/3$, as equation (12–54) shows. Since the entropy does not change in the process,

$$\frac{T_a}{T_b} = \left(\frac{11}{4}\right)^{1/3} \left(\frac{V_b}{V_a}\right)^{1/3}$$

where V_b and V_a are the volumes before the onset of annihilation and after annihilation has run its course. During the transition the universe expands, while the photon temperature remains more or less constant. However, the neutrinos are cooled by the expansion and hence a photon-to-neutrino temperature ratio $(11/4)^{1/3}$ becomes established at annihilation and is maintained forever after.

12–7. Substitute a new variable, u, given by $r = r_0 \sin^2 u$, so that $dr = 2r_0 \sin u \cos u \, du$, which gives the desired result through the integral

$$\int_{\pi/2}^{0} \sin^2 u \, du = -\frac{\pi}{4}$$

12–8. Set the surface mass density on the sphere equal to $\sigma = M/4\pi R^2$ and consider a point at some off-axis distance a. The potential at that point is

$$V = -\int_{0}^{\pi} \frac{2\pi\sigma GR^2 \sin\Theta \, d\Theta}{\sqrt{(R\cos\Theta - a)^2 + R^2 \sin^2\Theta}}$$

where normal polar coordinates have been used. Integration leads to

$$V = -\frac{2\pi\sigma GR}{a}\left(R^2 - 2aR\cos\Theta + a^2\right)^{1/2}\Bigg]_0^\pi = -4\pi\sigma GR$$

Further, since any spherical distribution can be built up from a continuous distribution of spherical shells, the potential anywhere within an empty central sphere will be constant throughout.

12–9.   We see that equation 12–84 gives

$$\frac{\ddot{a}_i}{a_i} = 0$$

while equation (12–19), when substituted into equation (12–83), leads directly to the expression

$$\dot{a}_i a_i = \text{constant} = a_{i0}^2 \sqrt{\frac{8\pi G\rho_{i0}}{3}}$$

whose integration with respect to time yields equation (12–85).

12–10.   The force per unit mass integrated over the whole disk is

$$F(z) = \int_0^R \left(\frac{2\pi r\sigma G}{r^2 + z^2}\right)\left(\frac{z}{(r^2 + z^2)^{1/2}}\right) dr = 2\pi\sigma - \frac{2\pi\sigma z G}{r}$$

where the first term in brackets is the force along the direction to a particular surface element, and the second term in brackets provides the component perpendicular to the plane. For $R \gg z$ this gives the desired result on integration. Hence, for a large disk the potential changes only along a direction perpendicular to the plane and is

$$2\pi\sigma G z$$

This gives a scale height h for random velocity components perpendicular to the plane v,

$$h = \frac{(v^2/2)}{2\pi\sigma G}$$

12–11.   The velocity difference between particles is at most

$$\Delta v = \frac{D}{t_1} - \frac{D}{t_2} = \frac{c^2\Delta t}{D}$$

and

$$\left(1 - \frac{v^2}{c^2}\right)^{1/2} \sim \left(\frac{2\Delta v}{c}\right)^{1/2} \sim \left(\frac{2c\Delta t}{D}\right)^{1/2} \sim 3.5 \times 10^{-7}$$

$$m_v c^2 \sim 10 \text{ Mev}\left(1 - \frac{v^2}{c^2}\right)^{1/2} \sim 3.5 \text{ ev}, \qquad m_v \sim 6 \times 10^{-33} \text{ g}$$

From (4–72) and (12–44), $(n_v + n_{\bar{v}}) \sim 17.5 T_v^3 \sim 140$ cm^{-3}, where from problem 12:6, T_v is chosen $(11/4)^{1/3}$ lower than the microwave background temperature, or roughly $T_v \sim 2°$K. While this would be the actual neutrino temperature only for massless neutrinos—the temperature for massive neutrinos being lower—this, in fact, is the correct effective temperature to be used in estimating n_v. For three neutrino types and their antiparticles that yields 420 cm^{-3} or maximally 3×10^{-30} g cm^{-3}, which is below the critical mass estimated in section 12:6.

12–12. If we set $Q = 0$ in equation (9–24) for an adiabatic flow, and substitute for the inflow parameters v_i, ρ_i and P_i in terms of the corresponding outflow parameters by means of equations (9–19), (9–27) and (9–28), then for extremely high velocities

$$v_i^2 \gg c_i^2 = \frac{\gamma_i P_i}{\rho_i}$$

and the desired relation (12–102) is obtained. A similar procedure leads to (12–103).

12–13. At 2000°K, the random velocity of hydrogen molecules is $(3kT/m)^{1/2} \sim 5$ km sec^{-1}. The orbital velocity of the material at the distance of the earth from the sun is ~ 30 km sec^{-1}. That clearly indicates a high degree of flattening if the aereal density is sufficiently high. That density is $\sigma = M_0(\pi R^2)^{-1} \sim 6.4 \times 10^4$ g cm^{-2}. The scale height (see problem 12–10) then becomes 5×10^{11} cm. This leads to a number density of molecules $n = 1.3 \times 10^{17}$ cm^{-2} and a pressure of 4×10^4 dyn cm^{-2} = 4×10^{-2} atm.

12–14. For an infall velocity $\dot{r}$ after a time t, the acceleration must be $MG/r^2 \sim \dot{r}/t$, so that $\delta\rho \sim 3M/4\pi r^3 \sim 3\dot{r}/4\pi t Gr$. From this, one obtains $\delta\rho/\rho_{\text{crit}} \sim 2\dot{r}/H^2 tr$. Using $H = 2.5 \times 10^{-18}$ sec^{-1}, we then obtain

$$\frac{\delta\rho}{\rho_{\text{crit}}} \sim 0.49 \qquad \text{for 16 Mpc distance}$$

$$\frac{\delta\rho}{\rho_{\text{crit}}} \sim 0.44 \qquad \text{for 40 Mpc distance}$$

12–15. (a) The criterion for collapse is

$$\frac{MG}{R} = \frac{4\pi R^2}{3}[n_v + n_{\bar{v}}]m_v G \geq \frac{kT}{m_v} \sim c^2$$

with $n_v + n_{\bar{v}} = 17.5T^3$ for a thermal neutrino/antineutrino gas. This gives the desired maximum radius.

(b) This criterion simply makes sure that the collapsing material is within a radius less than the distance to the cosmic horizon.

(c) For these two criteria to apply we need $T \sim 3 \times 10^{5\circ}$K, $R \sim 3.8 \times 10^{20}$ cm, $m_v \geq 4.6 \times 10^{-32}$ g ~ 26 ev which is considerably larger than the result obtained in problem 12–11.

(d) The mass of this condensation of neutrinos and antineutrinos is $\sim 10^{49}$ g.

(e) If the matter had started at rest, we would have $t_{ff} = [3\pi/32G\rho]^{1/2} \sim 1.4 \times 10^{10}$ sec for $\rho = 17.5T^3 m_v$. However, the Hubble constant at that time is only $H \sim 8 \times 10^{-11}$ sec^{-1} with the age of the universe at that time, $\sim 6 \times 10^9$ sec. Since the matter has an initially high expansion velocity at the edge of the region, collapse of such a cloud is only marginal even at the relatively high neutrino mass mentioned above. Note, incidentally, that today's neutrino number density would be (see problem 12–6) $\sim 17.5T^3 \sim 140$ for $T \sim 2°$K. This is independent of whether neutrinos are massive or not. For the neutrino's mass just computed we would therefore get a cosmic density $\sim 6 \times 10^{-30}$ g cm^{-3}, close to the critical density.

Epilogue

At crucial points in this book we have been stopped by unsolved problems. Some of the most important questions that remain unanswered are:

(1) Do the laws of physics as we know them apply on the scale of the universe?

(2) Is there a connection between the structure of the universe and the structure of elementary particles?

(3) Does the universe have a beginning and an end in time, and what exactly is time?

(4) How are galaxies born and how do they die?

(5) How are stars formed and how do they die?

(6) What is the origin of cosmic magnetic fields?

(7) Is there a basic preference for matter over antimatter in the universe?

(8) What is the origin of life and do other intelligent civilizations exist?

(9) Are we even asking the right kind of questions?

E

Appendix A

Astronomical Terminology

A:1 INTRODUCTION

When we discover a new type of astronomical entity on a photographic plate of the sky or in a radio-astronomical record, we refer to it as a new *object*. It need not be a star. It might be a galaxy, a planet, or perhaps a cloud of interstellar matter. The word "object" is convenient because it allows us to discuss the entity before its true character is established. Astronomy seeks to provide an accurate description of all natural objects beyond the earth's atmosphere.

From time to time the brightness of an object may change, or its color might become altered, or else it might go through some other kind of transition. We then talk about the occurrence of an *event*. Astrophysics attempts to explain the sequence of events that mark the evolution of astronomical objects.

A great variety of different objects populates the universe. Three of these concern us most immediately in everyday life: The sun that lights our atmosphere during the day and establishes the moderate temperatures needed for the existence of life, the earth that forms our habitat, and the moon that occasionally lights the night sky. Fainter, but far more numerous, are the stars that we can only see after the sun has set.

The objects we detect can be divided into two groups. Many of them are faint, and we would not be able to see them if they were not very close to the sun; others are bright, but at much larger distances. The first group of objects, taken together with the sun, comprise the *solar system*. They form a gravitationally bound group orbiting a common center of mass. Within the solar system the sun itself is of greatest astronomical interest in many ways. It is the one star that we can study in great detail and at close range. Ultimately it may reveal precisely what nuclear processes take place in its center and just how a star derives its energy. Complementing such observations, the study of planets, comets, and meteorites may ultimately reveal the history of the solar system and the origins of life. Both of these are fascinating problems!

A:2 THE SUN

The sun is a star. Stars are luminous bodies whose masses range from about 10^{32} to 10^{35} g. Their *luminosity* in the visual part of the spectrum normally lies in the range between 10^{-4} and 10^4 times the sun's energy outflow. The *surface temperatures* of these stars may range from no more than $\sim 1000°$K to about 50,000°K. Just how we can determine the relative brightness of stars will be seen later in this Appendix. The determination of temperatures is discussed in Chapter 4.

The sun, viewed as a star, has the following features:

(a) Its radius is 6.96×10^{10} cm. Although occasional prominences jut out from the solar surface, its basic shape is spherical. The equatorial radius is only a fractional amount larger than the polar radius: $[(r_{eq} - r_{pol})/r] \simeq 6 \times 10^{-6}$ (Di86).

(b) The sun emits a total flux of 3.9×10^{33} erg sec^{-1}. Nearly half of this radiation is visible, but an appreciable fraction of the power is emitted in the near ultraviolet and near infrared parts of the spectrum. Solar X-ray and radio emission make only very slight contributions to the total luminosity.

(c) The sun's mass is 1.99×10^{33} g.

(d) We recognize three principal layers that make up the sun's atmosphere. They are the photosphere, chromosphere, and corona.

(i). The *photosphere* is the surface layer from which the sun's visible light emanates. It has a temperature of about 6000°K.

(ii). The *chromosphere* is a layer some ten to fifteen thousand kilometers thick. It separates the relatively cool photosphere from the far hotter corona.

(iii). The *corona* extends from 1.03 $R_\odot$, or about 20,000 km above the photosphere, out to at least several solar radii. The outer boundary has not been defined. The corona is not a static structure; instead its outer edge merges continuously into the interplanetary gas that streams outward from the sun at speeds of several hundred kilometers per second. This streaming ionized gas, mainly protons and electrons, is called the *solar wind*. The temperature of the corona is $\sim 1.5 \times 10^{6°}$K.

(e) Sunspots and sunspot groups, cool regions on the solar surface, move with the sun as it rotates, and allow us to determine a 27-day rotation period. This period is only an apparent rotation rate as viewed from the earth which itself

A:2

orbits about the sun. The actual rotation period with respect to the fixed stars is only about twenty-five and a half days at a latitude of 15° and varies slightly with latitude; the solar surface does not rotate as a solid shell. The sun exhibits an 11-year *solar cycle* during which time the number of sunspots increases to a maximum and then declines to a minimum. At minimum the number of spots on the sun may be as low as zero. At maximum the number of individual sunspots or members of a sunspot group may amount to 150. There are special ways of counting to arrive at this *sunspot number* and a continuous record is kept through the collaborative effort of a number of observatories.

The 11-year cycle actually amounts to only half of a longer 22-year cycle that takes into account the polarity and arrangement of magnetic fields in sunspot pairs.

(f) A variety of different events can take place on the sun. Each type has a name of its own. One of the most interesting is a *flare*, a brief burst of light near a sunspot group. Associated with the visible flare is the emission of solar cosmic ray particles, X-rays, ultraviolet radiation, and radio waves. Flares also are associated with the emission of clouds of electrons and protons that constitute a large component added to the normal solar wind. After a day or two, required for the sun-to-earth transit at a speed of $\sim 10^3$ km sec^{-1}, these particles can impinge on the earth's *magnetosphere* (magnetic field and ionosphere), giving rise to *magnetic storms* and *aurorae*. These disturbances tend to corrugate the ionosphere and make it difficult to reflect radio waves smoothly. Since radio communication depends on smooth, continuous ionospheric reflection, reliable radio communication is sometimes disrupted for as long as a day during such *magnetic storms*.

A:3 THE SOLAR SYSTEM

A variety of different objects orbit the sun. Together they make up the *solar system*. The earth is representative of planetary objects. *Planets* are large bodies orbiting the sun. They are seen primarily by reflected sunlight. The majority emit hardly any radiation by themselves. In order of increasing distance from the sun, the planets are Mercury, Venus, Earth, Mars, Jupiter, Saturn, Uranus, Neptune, and Pluto. All the planets orbit the sun in one direction; this direction is called *direct*. Bodies moving in the opposite direction are said to have *retrograde orbits*. Table 1.3 gives some of the more important data about planets. It shows that the different planets are characterized by a wide range of size, surface temperature and chemistry, magnetic field strength, and so on. One of the aims of astrophysics is to understand such differences, perhaps in terms of the history of the solar system.

Besides the nine planets we have listed, there are many more minor planets orbiting the sun. They are sometimes also called *planetoids* or *asteroids*. Most of them travel along paths lying between the orbits of Mars and Jupiter, a region known as the *asteroidal belt*. The largest asteroid is Ceres. Its radius is 350 km. Its mass is about one ten-thousandth that of the earth.

Many of the smaller known asteroids have diameters of the order of a kilometer. These objects number in the thousands and there must be many more orbiting masses that are too small to have been observed. Among these are bodies that might only be a few meters in diameter or smaller. From time to time, some of these approach the earth and survive the journey through the atmosphere. Such an object that actually impacts on the earth's surface is called a *meteorite*. Meteorites are studied with great interest because they are a direct means of learning about the physical and chemical history of at least a small class of extraterrestrial solar system objects.

Even smaller than the meteorites are grains of dust that also circle the sun along orbits similar to those of planets. From time to time a grain of dust may enter the atmosphere. Much of it may burn through heat generated by its penetration into the atmosphere, and the particle becomes luminous through combustion and can be observed as a *meteor*, historically called a *shooting star*.

In contrast to meteoritic material, meteoric matter does not generally reach the earth's surface in recognizable form. However, some fragments do appear to survive and are believed to contribute to a shower of fine dust that continually rains down on the earth. Most of this dust has a micrometeoritic origin. *Micrometeorites* are micron- (10^{-4} cm) or submicron-sized grains of interplanetary origin that drift down through the atmosphere and impinge on the earth's surface. They have a large surface-to-mass ratio and are easily slowed down in the upper atmosphere without becoming excessively hot. Once they have lost speed they slowly drift down through the air. Some of these grains may be formed in the burnup of larger meteors; others may come in unchanged from interplanetary space. Collections of these grains can be made from the arctic snows or deep ocean sediments, far from sources of industrial smoke.

The identification of this extraterrestrial mass is not simple. It is difficult to distinguish cosmic matter from dust generated in, say, volcanic explosions. Nevertheless, some researchers have claimed that the amount of material deposited on the earth each day is of the order of several tons, though the bulk of that is in asteroidal impacts occurring only once in a few millian years. These figures may be compared with dust collected by means of special satellite-borne devices; but the inter-comparison has no real relevance because the total mass carried by the far more abundant dust grains is completely negligible by comparison.

A cloud of micrometeoritic dust exists in the space between the planets and possibly also as a tenuous dust belt about the earth. The dust reflects sunlight

A:3

and gives rise to a glow known as the *zodiacal light*. The zodiacal light can be seen, on very clear days, as a tongue-shaped glow jutting up over the western horizon after sunset or the eastern horizon before sunrise. The glow is concentrated about the *ecliptic*, the plane in which the earth orbits the sun.

We recognize that these planetary and interplanetary objects continually interact. There are indications that planets and their satellites have often collided with huge meteorites, objects that are as large as the asteroids. The surface of both Mars and the moon are pockmarked by what are believed to be impact craters. The earth too shows vestiges of such bombardment; but our atmosphere erodes away and destroys crater outlines in a time of the order of several million years, whereas on the moon erosion times are of the order of billions of years.

We should notice that in talking about planets, asteroids, meteorites, meteors, and micrometeoritic dust grains we are enumerating different-sized members of an otherwise homogeneous group. The major known difference between these objects is their size. Other differences can be directly related to size. For example, it is clear that planets may have atmospheres while micrometeorites do not. But this difference arises because only massive objects can retain a surrounding blanket of gas. The gravitational attraction of small grains just is not strong enough to retain gases at temperatures encountered in interplanetary space. The different names, given to these different-sized objects, have arisen because they were initially discovered by a variety of differing techniques; and although we have known the planets, meteorites, meteors, and other interplanetary objects for a long time, we have just recently come to understand their origin and interrelation.

A set of objects similar to the planets are the *satellites* or *moons*. A satellite orbits about its parent planet and these two objects together orbit about the sun. In physical makeup and size, satellites are not markedly different from planets. The planet Mercury is only four times as massive as our moon. Ganymede, one of Jupiter's satellites, Titan, one of Saturn's satellites, and Triton, one of Neptune's satellites, all are nearly twice as massive as the moon. Titan even has an atmosphere. Many other satellites are less massive; they look very much like asteroids. An extreme of the moon phenomenon is provided by the rings of Saturn, Jupiter and Uranus, consisting of clouds of fine dust—micrometeoritic grains, all orbiting the parent planet like minute interacting moons.

Evidently there are great physical similarities between satellites and planetary objects of comparable size. The main difference lies in the orbital motion of the two classes of objects. It is interesting that an asteroid could be captured by Jupiter to become one of its satellites. The reverse process might also be possible.

The somewhat vague distinction between planets and interplanetary objects is not unique. Differences between stars and planets also are somewhat vague. We talk about *binaries* in which two stars orbit about a common center of gravity.

A:3

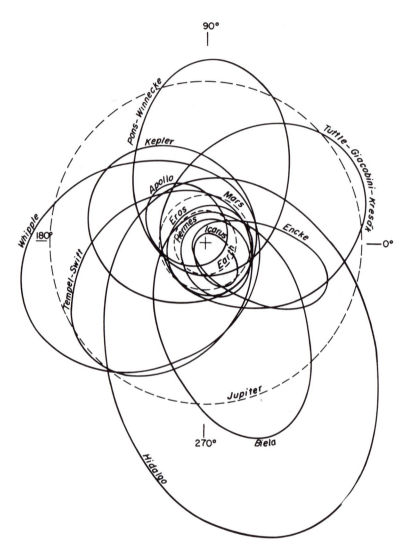

Fig. A.1 Comparison of planetary, asteroidal, and short-period cometary orbits. Although the Earth, Mars, and Jupiter have nearly circular orbits, the orbits of the asteroids, Icarus, Hermes, Eros, Apollo, Kepler, and Hidalgo are appreciably eccentric, as are those of comets Encke, Pons-Winnecke, Tempel-Swift, Whipple, Tuttle-Giacobini-Kresak, and Biela. Comets are named after their discoverers. Many comets and asteroids have aphelion distances near Jupiter's orbit, and Jupiter has a controlling influence on the shape of the orbits and may have "captured" comets from parabolic orbits into short-period orbits.

A:3

Often one of these objects is much smaller than the other, sometimes no more than one thousandth as massive. Jupiter also has about one thousandth the mass of the sun. Perhaps, as some astronomers have suggested, Jupiter should more appropriately be called a star. Clearly, size alone does not provide an appropriate distinction between stars and planets. A more pertinent criterion can be formulated in terms of nuclear processes that go on in stars, chains of events that cannot be triggerred in the interior of less massive planetary objects (Sa70a).

We should still mention one final class of objects belonging to the solar system: the comets. Their orbits are neither strictly planetary, nor are they at all similar to those of satellites. Some comets have elliptic orbits about the sun. Their periods may range from a few years to many hundreds of years. Other comets have nearly parabolic orbits and must be reaching the sun from the far reaches of the solar system. Comets are objects which, on approaching the sun from large distances, disintegrate through solar heating: gases that initially were in a frozen state are evaporated off and dust grains originally held in place by these volatile substances become released. The dust and gas, respectively, are seen in reflected and re-emitted sunlight. They make the comet appear diffuse (Fig. A.2). Comet tails are produced when the gas and dust are repelled from the sun through electron and proton bombardment and by the pressure of sunlight. The dust from a comet tail produces a meteor shower when the earth passes through the remnants of the tail (Wa56).

A:4 STELLAR SYSTEMS AND GALAXIES

Before we turn to a description of individual stars, we should first consider the groupings in which stars occur.

Stars often are assembled in a number of characteristic configurations, and we classify these systems primarily according to their size and appearance. Many stars are single. Others are accompanied by no more than one companion; such pairs are called *binaries*. Depending on their separation and orientation, binary stars can be classified as *visual, spectroscopic* or *eclipsing*. The limit of visual resolution of a binary is given by available optical techniques. Refinements are continually being made, but interferometric techniques now allow us to resolve stars about 0.01 arc seconds apart (Ha67). For smaller separations, we cannot use interferometric techniques. The two stars in such a close pair constitute a spectroscopic binary and have to be resolved indirectly by means of their differing spectra. We sometimes encounter a special but very important type of spectroscopic binary in which the stars orbit about each other roughly in a plane that contains the observer's line of sight. One star may then be seen eclipsing the other and a change in brightness is observed. An eclipse of this kind becomes

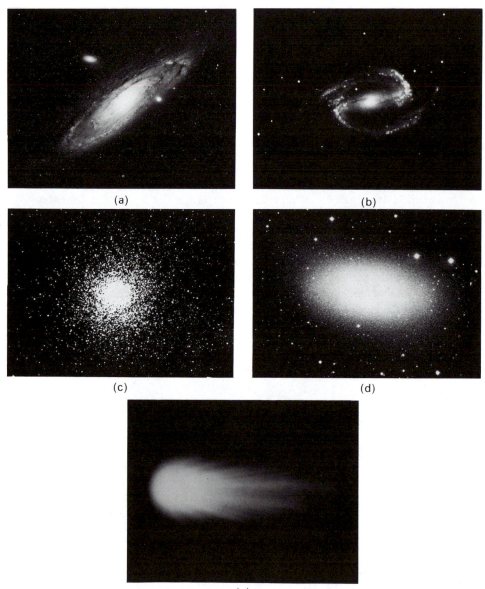

(a)

(b)

(c)

(d)

(e)

Fig. A.2 (*a*) The Andromeda galaxy, NGC 224, Messier 31, a spiral with two smaller companion galaxies, one of which, NGC 205, an elliptical galaxy is shown enlarged (*d*). The barred spiral galaxy (*b*) is NGC 1300. Its spiral classification is SBb. These three pictures were photographed at the Mount Wilson Observatory. The globular cluster (*c*) is Messier 3 (M3), also known as NGC 5272. The comet (*e*) is comet Brooks, and the photograph was taken October 21, 1911. Only the region of the comet around the head is shown. Photographs (*c*) and (*e*) were taken at the Lick Observatory.

A:4

probable only when the two companions are very close together, no more than a few radii apart. We call such systems *eclipsing binaries*. Binaries are important because they provide the only means of determining accurate mass values for stars (other than the sun). How these masses are determined is shown in the discussion of orbital motions (section 3:5).

Close binaries also are important because if one of the two stars begins to expand, as it moves onto the red giant branch of the Hertzsprung-Russell diagram (section A:5g and Fig. 1.5), the surface material may become more strongly attracted to the companion star. If the companion is compact, the infalling material can radiate X-rays on impact. Portions of the giant star, previously in its interior, thereby become visible. This allows us to check for systematic production of the heavy elements in the star and thus also to test the theory of chemical evolution and energy production in stars (section 8:13).

Binary stars are not the only known close configuration. There exist many ternaries consisting of three stars; and higher multiple systems are not uncommon. Perhaps one in every five "stars" shows evidence of being a binary and about one in every 20 "stars" shows evidence of being a higher order system. For stars more massive than the sun, these fractions are considerably higher; single stars occur only in one third of the observed cases. The proportion of observed binaries also differs for stars with differing spectral features.

Sometimes stars form an aggregate of half a dozen or a dozen members. This is called a stellar *group*. There also exist stellar *associations* that are groups of some 30 or so stars mutually receding from one another. Such stars appear to have had a common point of origin in the past. They are thought to have been formed together and to have become separated shortly after formation. By observing the size of the association and the rate at which it is expanding, we can determine how long ago the expansion started and how old the stars must be.

There are two principal groupings called clusters: *galactic clusters* and *globular clusters*. The galactic clusters usually comprise 50 to several hundred stars loosely and amorphously distributed but moving with a common velocity through the surrounding field of stars. In contrast, globular clusters (Fig. A.2) are much larger, containing several hundred thousand stars and having a very striking spherical (globular) appearance. Stars in a cluster appear to have had a common origin. We think they were formed during a relatively short time interval, long ago, and have had a common history.

Clusters are not just populated by single stars. Binaries and higher multiples and groups of stars often form small subsystems in clusters. Normally stars and clusters are members of *galaxies*. These are more or less well defined, characteristically shaped systems containing between 10^8 and 10^{12} stars (Fig. A.2). Some galaxies appear elongated and are called elliptical or E galaxies. Highly elongated ellipticals are classed as E7 galaxies. If no elongation can be detected and the galaxy has a circular appearance, it is called a globular galaxy and is classified as E0. Other numerals, between 0 and 7, indicate increasing apparent elongation. The observed elongation need not correspond directly to the actual elongation of the galaxy because the observer on earth can only see a given galaxy in a fixed projection.

Elliptical galaxies show no particular structure except that they are brightest in the center and appear less dense at the periphery. In contrast, *spiral galaxies* (S galaxies) and *barred spiral galaxies* (SB) show a strong spiral structure. These galaxies are denoted by a symbol O, a, b, or c following the spiral designation to indicate increasing openness of the spiral arms. In this notation, a compact spiral is designated SO and a barred spiral with far-flung spiral arms and quite open structure is designated SBc (see Fig. A.2). (See also Fig. 10.11(a), page 467.)

In comparing galaxies on a photographic plate, one expects nearby galaxies to appear larger than more distant objects; this means that the angular diameter of a regular galaxy can be taken as a rough indicator of its distance. When the spectra of different galaxies are correlated with distance, we find that a few nearby galaxies have blue shifted spectra (Bu71b); but all the more distant galaxies have spectra that are systematically shifted toward the red part of the spectrum. Galaxies at larger apparent distances, as judged from their diameters or brightness, are increasingly red shifted! This correlation is so well established that we now take an observation of a remote galaxy's red shift as a standard indicator of its distance.

Not all galaxies can be described by designations E, S or SB. Some are more randomly shaped. Such galaxies are classified as irregular galaxies, designated by the symbol Ir. Peculiar galaxies of one kind or another are denoted by a letter p following the type designation, for example, E5p.

Of course, galaxies do not contain stars alone. Between the stars there exists interstellar gas and dust. In some spiral galaxies the total mass of dust and gas is comparable to the total stellar mass observed. The exact mass ratios are not known because one is not sure whether all the gas in existence has yet been detected.

Dust clouds can be detected through their extinction, which obscures the view of more distant stars. Moreover, dust absorbs optical and ultraviolet radiation and reemits at long infrared wavelengths. This process is so effective that some galaxies radiate for more strongly in the infrared than in all other spectral ranges combined.

Gas also may be detected in absorption or through emission of radiation. Through spectroscopic studies in the radio, infrared, visible, and ultraviolet parts of the spectrum, many ions, atoms, and molecules have been identified, and the temperature, density, and radial velocity of such gases has been determined.

Galaxies are not the largest aggregates in the universe. There exist many pairs

A:4

and groups of galaxies. Fig. 1.10 shows one such group. Our *Galaxy*, the *Milky Way* to which the sun belongs, is a member of the *Local Group* that contains somewhat more than a dozen galaxies. The Andromeda nebula and the Galaxy are the largest members. The other members of the group have a combined mass about one tenth the mass of the Andromeda Nebula (Table 1.4).

Larger *clusters of galaxies* containing up to several thousand galaxies also exist. Groupings on a larger scale include filamentary structures composed of tenuous chains of galaxies, enormous voids surrounded by denser concentrations of galaxies, and possibly also superclusters—entire groupings of clusters of galaxies. Beyond that scale, no further aggregations are apparent. On the largest scales, the universe can best be described as consisting of randomly grouped aggregates and voids.

The scheme of classification of galaxies leaves a number of borderline cases in doubt. Small E0 galaxies are not appreciably different from the largest globular clusters. Double galaxies sometimes cannot be distinguished from irregular ones; and the distinction between a group or a cluster of galaxies may also be a matter of taste. But the classification is useful nevertheless; it gives handy names to frequently found objects without making any attempt to provide rigorous distinctions.

Crossing the vast spaces between the galaxies are quanta of electromagnetic radiation and highly energetic cosmic ray particles that travel at almost the speed of light. These are the carriers of information that permit us to detect the existence of the distant objects.

There is one overwhelming feature that characterizes the galaxies. Everywhere, at large distances, the spectra of galaxies and clusters of galaxies appear shifted toward the red, long wavelength, end of the spectrum. The farther we look, the greater is the red shift. Most astrophysicists attribute the *red shift* to a high recession velocity. The galaxies appear to be flying apart. The universe expands!

A:5 BRIGHTNESS OF STARS

(a) The Magnitude Scale

One of the first things we notice after a casual look at the sky is that some stars appear brighter than others. We can visually sort them into different brightness groups. In doing this, it becomes apparent that the eye can clearly distinguish the brightness of two objects only if one of them is approximately 2.5 times as bright as the other. The factor of 2.5 can therefore serve as a rough indicator of apparent brightness, or *apparent visual magnitude*, m_v of stars.

Stars of first magnitude, $m_v = 1$, are brighter by a factor of ~ 2.5 than stars of second magnitude, $m_v = 2$, and so on. The magnitude scale extends into the region of negative values; but the sun, moon, Mercury, Venus, Mars, Jupiter and the three stars, Sirius, Canopus, and α Centauri are the only objects bright enough to have apparent visual magnitudes less than zero.

Normally it would be cumbersome to use a factor of 2.5 in computing relative brightnesses of stars of different magnitudes. Since this factor has arisen not because of some feature peculiar to the stars that we study, but is quite arbitrarily dependent on a property of the eye, we are tempted to discard it altogether in favor of a purely decimal system; but a brightness ratio of 10 is not useful for visual purposes. As a result, a compromise that accommodates some of the advantages of each of these systems is in use. We define a magnitude in such a way that stars whose brightness differs by precisely five magnitudes, have a brightness ratio of exactly 100. Since $100^{1/5} = 2.512$, we still have reasonable agreement with what the eye sees, and for computational work we can use standard logarithmic tables to the base 10.

(b) Color

The observed brightness of a star depends on whether it is seen by eye, recorded on a photographic plate, or detected by means of a radio telescope. For different astronomical objects the ratio of radiation emitted in the visible and radio regions of the spectrum varies widely. The spectrum of an object can be roughly described by observing it with a variety of different detectors in several different spectral regions. The apparent magnitudes obtained in these measurements can then be intercompared. Several standard filters and instruments have been developed for this purpose so that we may intercompare data from observatories all over the world. The resulting brightness indicators are listed below:

m_v denotes *visual brightness.*

m_{pg} denotes *photographic brightness.* The photographic plate is more sensitive to blue light than the eye; nowadays this brightness is usually labeled B, for blue. If a photographic plate is to be used to obtain the equivalent of a visual brightness, a special filter has to be used to pass yellow light and reject some of the blue light.

V or m_{pv} denotes photovisual brightness obtained with a photographic plate and the above mentioned yellow transmitting filter. This brightness is generally denoted by V for visual. m_{pg} and m_{pv} are older notations.

U denotes the *ultraviolet brightness* obtained with a particular ultraviolet transmitting filter (Table A.1).

I denotes an *infrared brightness* obtained in the photographic part of the infrared. At longer wavelengths photographic plates no longer are sensitive, but a number of infrared spectral magnitudes have been defined so that results obtained with lead sulfide, indium antimonide, and other infrared detectors might be intercompared by different observers. These magnitudes are labeled *J, K, L, M, N*, and *Q*.

Table A.1 lists the wavelengths at which these magnitudes are determined.

Table A.1 Effective Wavelength for Standard Brightness Measurements.

Symbol	Effective Wavelength	Symbol	Effective Wavelength
U	0.36μ	*K*	2.2μ
B	0.44	*L*	3.4
V	0.55	*M*	5.0
R	0.70	*N*	10.2
I	0.90	*Q*	21
J	1.25		

1μ (micron) = 10^{-6} m = 10^{-4} cm = 10^4 Å (angstrom units)

m_{bol} denotes the total apparent brightness of an object, integrated over all wavelengths. This *bolometric magnitude* is the brightness that would be measured by a bolometer—a detector equally sensitive to radiation energy at all wavelengths.

(c) Color Index

The difference in brightness as measured with differing filters gives an indication of a star's color. The ratio of blue to yellow light received from a star is given by the difference in magnitude—logarithm of the brightness—of the star measured with blue and visual filters. This quantity is known as the *color index*:

$$C = B - V$$

Differences such as $U - B$ also are referred to as color indexes.

The intercomparison of colors involved in producing a reliable color index can only be achieved if we can standardize photographic plates and filters used in the measurements. And even then errors can creep into the intercomparison. For this reason some standard stars have been selected to define a point where the color index is zero. These stars are denoted by the spectral-type symbol A0, (cf A:6).

A:5

(d) Bolometric Correction

Normally the bolometric brightness of a star can only be obtained by indirect means. We may be able to measure the apparent visual brightness; but to estimate the total radiative emission of the star, over its entire spectrum, we must make some assumptions about the surface temperature and emissivity. This estimate is obtained in terms of a factor known as the *bolometric correction*, BC, defined as the difference between the bolometric and visual magnitudes of a star. The bolometric correction always is positive

$$BC = m_v - m_{bol}$$

Estimation of the bolometric correction is not simple. The color index $B - V$ can be used to obtain a rough assessment of the surface temperature. We can then estimate the total radiation output, on the assumption that we understand how electromagnetic radiation is transferred through a star's atmosphere. Generally this assumption is not warranted. In the last decade, stellar brightness values, obtained in the ultraviolet and infrared regions of the spectrum, have not generally agreed with predictions based on earlier theoretical models. Of course, these new data are being incorporated into the theory of stellar atmospheres and more reliable bolometric correction values are becoming available. In the meantime, provisory tables of bolometric corrections for stars with different spectral indices are in use.

(e) Absolute Magnitude

For many purposes we need to know the absolute magnitude rather than the apparent brightness of a star. It is therefore important to convert apparent magnitudes into absolute values. We define the *absolute magnitude* of a star as the apparent magnitude we would measure if the star were placed a distance of 10 pc from an observer. (1 pc $= 3 \times 10^{18}$ cm. See section 2:2, page 54.)

Suppose the distance of a star is r pc. Its brightness diminishes as the square of the distance between star and observer. The apparent magnitude of the star will therefore be greater, by an additive term $\log_{2.5} r^2/r_0^2$, than its absolute magnitude.

$$m = M + \log_{2.5} \frac{r^2}{r_0^2} = M + 5 \log \frac{r}{r_0}$$

where the logarithm is taken to the base 10 when no subscript appears. Since $r_0 = 10$ pc, we have the further relation for the *distance modulus*, μ_0,

$$\mu_0 \equiv m - M = 5 \log r - 5 \tag{A-1}$$

A:5

Thus far no attention has been paid to the extinction of light by interstellar dust. Clearly the apparent magnitude is diminished through extinction and a positive factor A has to be subtracted from the right side of equation A–1 to restore M to its proper value

$$M = m + 5 - 5 \log r - A \qquad (A-2)$$

Obtaining the star's distance, r, often is less difficult than assessing the interstellar extinction A. We discuss this difficulty in section A:6a below.

The detector and filter used in obtaining the apparent magnitude m, in equation A–2, determines the value of the absolute magnitude M. We can therefore use subscripts v, pg, pv, and bol for absolute magnitudes in exactly the same way as for apparent magnitudes.

(f) Luminosity

Once we have obtained the bolometric absolute magnitude of a star, we can obtain its total radiative emission, or luminosity, L, directly in terms of the solar luminosity:

$$\log\left(\frac{L}{L_\odot}\right) = \frac{1}{2.5}\left[M_{\mathrm{bol}_\odot} - M_{\mathrm{bol}}\right] \qquad (A-3)$$

The luminosity of the sun, $L_\odot$, is 3.8×10^{33} erg sec^{-1} and the solar bolometric magnitude, $M_{\mathrm{bol}_\odot}$, is 4.6. The luminosity of stars varies widely. A supernova explosion can be as bright as all the stars in a galaxy for a brief interval of a few days. The brightest stable stars are a hundred thousand times more luminous than the sun. At the other extreme, a white dwarf may be a factor of a thousand times fainter than the sun; and even fainter stars may well exist.

(g) The Hertzsprung-Russell Diagram

One of the most useful diagrams in all astronomy is the Hertzsprung-Russell, H-R, diagram. It presents a comparison of brightness and temperature plotted for any chosen group of stars. We will see in Chapter 2 that the diagram is valuable in estimating the dimensions of galaxies and intergalactic distances; more important, such H-R diagrams, obtained for different stellar age groups, provide the main empirical foundation for the theory of stellar evolution.

H-R diagrams can take many different forms. The color index is an indicator of a star's surface temperature, as shown in Chapter 4. Hence the abscissa sometimes is used to show a star's color index, and we then speak of a *color-magnitude*, instead of an H-R, *diagram*. The ordinate can show either M_v, or M_{bol} or lumi-

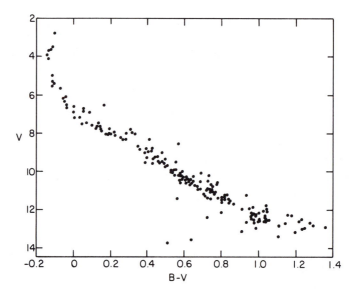

Fig. A.3 Color-magnitude diagram of the Pleiades cluster stars, after correction for interstellar extinction effects. The Pleiades cluster contains some of the most recently formed stars in the Galaxy. (After Mitchell and Johnson, Mi57b).

nosity. When only an intercomparison of stars, all of which are known to be equally distant, is needed, it suffices to plot the apparent magnitude. Figure A.3 shows such a plot of the Pleiades cluster stars. Figure 1.6 plots the characteristics of M3. M3 is an old globular cluster in our Galaxy, while the Pleiades are among the most recently born Galactic stars. This difference is reflected in the difference of the two diagrams.

These two figures, as well as Fig. 1.4, show that stars are found to fall only in a few select areas of the H-R and color-magnitude diagrams. The largest number of stars cluster about a fairly straight line called the *main sequence*. This is particularly clear for the Pleiades cluster. The main sequence runs from the upper left to the lower right end of the diagram, or from bright blue down to faint red stars. To the right and above the main sequence (Fig. 1.5) there lie bright red stars along a track called the *red giant branch*. There also is a *horizontal branch* that joins the far end of the red giant branch to the main sequence. These two branches show up particularly in Fig. 1.6. In the horizontal branch, we find some stars that periodically vary with brightness. Finally, some faint *white dwarf stars* lie below and to the left of the main sequence. The rest of the diagram usually is empty.

A:5

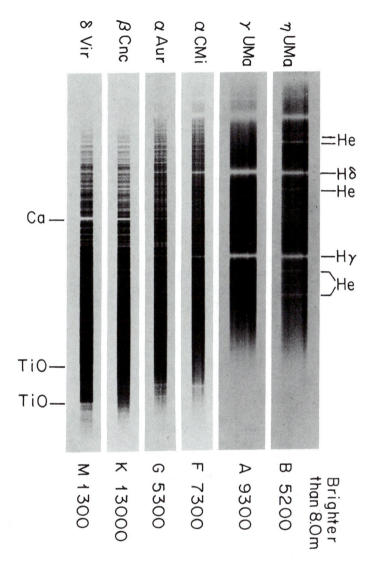

Fig. A.4 Schematic diagram of spectra of typical stars representing different spectral types. The number of stars brighter than the eighth magnitude in each class is listed on the right, next to the star's spectral type. (With the permission of the Yerkes Observatory, University of Chicago.)

A:6 CLASSIFICATION OF STARS

(a) Classification System

The classification of stars is a complex task, primarily because we find many special cases hard to fit into a clean pattern. Currently a "two-dimensional" scheme is widely accepted. One of these "dimensions" is a star's spectrum; the other is its brightness. Each star is therefore assigned a two parameter classification code. Although the object of this section is to describe this code, we should note that the ultimate basis of the classification scheme is an extensive collection of spectra such as those shown in Fig. A.4. Each spectrum is representative of a particular type of star.

Stars are classified primarily according to their spectra, which are related to their color. Although the primary recognition marks are spectral, the sequence of the classification is largely in terms of decreasing stellar surface temperature— that is, an increase in the star's radiation at longer wavelengths. The bluest common stars are labeled O, and increasingly red stars are classed according to the sequence (Table A.2)

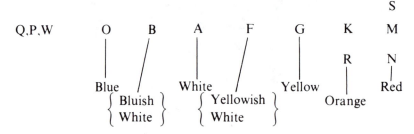

Over 99% of all stars belong to the basic series B, A, F, G, K, and M. Stars with designation O, R, N, and S are comparatively rare and so also are spectral types: Q denoting *novae*—stars that suddenly brighten by many orders of magnitude becoming far brighter than any nonvariable star.

P denotes *planetary nebulae*, hot stars with surrounding envelopes of intensely ionized gases, and W refers to Wolf-Rayet stars, which exhibit broad emission bands of ionized carbon, nitrogen, and helium. These stars appear to consist of a nuclear-processed interior exposed by extreme surface mass loss.

The classes R and N denote stars containing unusually strong molecular bands of diatomic carbon, C_2, and cyanogen, CN. The S stars are characterized by bands of titanium oxide, TiO, and zirconium oxide, ZrO.

Stars classed as W, O, B are sometimes said to be *early types*, while stars of class G, K, M, R, N, S are designated *late types*.

The transition from one spectral class to another proceeds in 10 smaller steps. Each spectral class is subdivided into 10 subclasses denoted by arabic numerals after the letter. A5 lies intermediate between spectral types A0 and A9; and F0 is just slightly redder than A9.

Table A.2 Spectral Classification of Stars.[a]

Type	Main Characteristics	Subtypes	Spectral Criteria	Typical Stars
Q	Nova: sudden brightness increase by 10 to 12 magnitudes			T Pyx Q Cyg
P	Planetary nebula: hot star with intensely ionized gas envelope			NGC 6720 NGC 6853
W	Wolf-Rayet stars		Broad emission of O III to O VI, N III to N V, C II to C IV, and He I and He II.	
O	Hottest stars, continuum strong in UV (O5 to O9)		O II λ 4650 dominates	BD + 35°4013
			He II λ 4686 dominates $\}$ emission	BD + 35°4001
			Lines narrower $\quad\}$ lines	BD + 36°3987
			Absorption lines dominate; only He II, C II in emission	ζ Pup, λ Cep
			Si IV λ 4089 at maximum	29 CMa
			O II λ 4649, He II λ 4686 strong	τ CMa
B	Neutral helium dominates	B0	C III/4650 at maximum	ε Ori
		B1	He I λ 4472 > O II λ 4649	β CMa, β Cen
		B2	He I lines are maximum	δ Ori, α Lup
		B3	He II lines are disappearing	π^4 Ori, α Pav
		B5	Si λ 4128 > He λ 4121	19 Tau, ϕ Vel
		B8	λ 4472 = Mg λ 4481	β Per, δ Gru
		B9	He I λ 4026 just visible	λ Aql, λ Cen
A	Hydrogen lines decreasing from maximum at A0	A0	Balmer lines at maximum	α CMa
		A2	Ca II K = 0.4 Hδ	S CMa, ι Cen
		A3	K = 0.8 Hδ	α PsA, τ^3 Eri
		A5	K > Hδ	β Tri, α Pic
F	Metallic lines becoming noticeable	F0	K = H + Hδ	δ Gem, α Car
		F2	G band becoming noticeable	π Sgr
		F5	G band becoming continuous	α CMi, ρ Pup
		F8	Balmer lines slightly stronger than in sun	β Vir, α For

[a]Compiled mainly from Keenan in *Stars and Stellar Systems*, K.A. Strard (ed.), with permission from the University of Chicago Press (Ke63b) (based on Cannon and Pickering Ca24) and also from Allen (A155). This table, which is based on the Henry Draper classification scheme, is a rough guide to the spectral features of stars. The classification of stars, however, remains an ongoing process and changes occur. (With the permission of Athlone Press of the University of London, second edition © C.W. Allen 1955 and 1963, and with the permission of the University of Chicago Press.)

Table A.2 (*continued*)

Type	Main Characteristics	Subtypes	Spectral Criteria	Typical Stars
G	Solar-type	G0	Ca λ 4227 = Hδ	α Aur, β Hya
	spectra	G5	Fe λ 4325 > Hγ on small-scale plates	κ Gem, α Ret
K	Metallic lines	K0	H and K at maximum strength	α Boo, α Phe
	dominate	K2	Continuum becoming weak in blue	β Cnc, ν Lib
		K5	G band no longer continuous	α Tau
M	TiO bands		TiO bands noticeable	α Ori, α Hya
			Bands conspicuous	ρ Per, γ Cru
			Spectrum fluted by the strong bands	W Cyg, RX Aqr
			Mira variables, Hγ, Hδ	χ Cyg, o Cet
R,N	CN, CO, C$_2$ bands		CN, CO, C$_2$ bands appear instead of TiO. R stars show pronounced H and K lines.	
S	ZrO bands		ZrO bands	R Gem

The classification scheme also allows us to denote a star's luminosity class by placing a Roman numeral after the spectral type. Each of these luminosity classes has a name:

> I — Supergiant
> II — Bright Giant
> III — Normal Giant
> IV — Subgiant
> V — Main Sequence
> — Subdwarf
> — White dwarf

The sun has spectral type G1 V indicating that it is a yellow main sequence star.

Sometimes we find classes I, II, and III collected under the heading "giant" while stars of group V are called "dwarfs." Letters "g" or "d" are placed in front of the spectral class symbol to denote these types. Similarly placed letters "sd" and "w" denote subdwarfs and white dwarf stars. Another classification feature concerns supergiants, which are often separated into two luminosity classes Ia and Ib depending on whether they are bright or faint.

A letter "e" following a spectral classification symbol denotes the presence of emission lines in the star's spectrum. There is one exception to this rule. The combination Oe5 denotes O stars in the range from O5 to O9; it has no further connection with emission.

A letter "p" following the spectral symbol denotes that the star has some form of peculiarity.

A:6

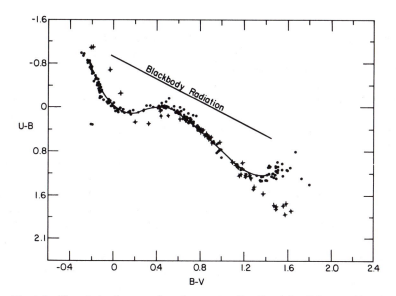

Fig. A.5 The relation between the color systems $U - B$ and $B - V$ for unreddened main-sequence stars (dots) and little reddened supergiants and yellow giants (crossed dots). The line along which blackbody radiators would fall is also shown. (After Johnson and Morgan, Jo53).

The color designation (stellar spectral type) given here is nearly linear in the color index $B - V$. It is not however linear in $U - V$ nor do the $U - V$ values decrease monotonically with increasingly late spectral type. Small differences in color indexes exist for giants and main sequence stars of the same spectral type. This unfortunate difficulty has arisen for historical reasons and should be corrected in future revisions of the spectral classification scheme. We might still see how well stellar colors approach those of a blackbody. The closeness of fit is shown in Fig. A.5, called a color-color diagram.

Four factors are responsible for the rather large deviations from a blackbody. (i) For stars around spectral type A, where the fit to the blackbody spectrum is poorest, absorption by hydrogen atoms in their first excited states produces a deviation. We talk about the *Balmer jump* in connection with the sharp rise in absorption at wavelengths corresponding to the Balmer continuum produced by these excited atoms in the outer atmosphere of a star. (ii) Cool stars have H^- ions in the outer atmospheres. These ions absorb radiation selectively, making the star appear bluer. (iii) The relatively high abundance of metals in population I stars produces a number of absorption lines that change the color of a star, moving it toward the lower right of Fig. 1.5. (iv) Finally, no star looks completely

A:6

black, because its outer layers are not equally opaque at all wavelengths. Light at different wavelengths therefore reaches us from different depths within the star, and these levels are at differing temperatures. The resulting spectrum of starlight therefore corresponds to a mixture of temperatures, rather than to blackbody radiation at one well-defined temperature.

Determination of the spectral type of a star by means of its color index alone would be very difficult, since proper account would have to be taken of the changes in color produced by interstellar dust. Small dust grains tend to absorb and scatter blue light more strongly than red. Light from a distant star therefore appears much redder than when emitted. To discover the true color index of the star a correction has to be introduced for interstellar reddening. However, in order to make this correction, we have to know how much interstellar dust lies along the line of sight to a given star, and to what extent a given quantity of dust changes the color balance. None of this information is normally available. Instead, we have to make use of a circular line of reasoning. We know that nearby stars of any given spectral type exhibit characteristic spectral lines either in absorption or in emission. Since these stars are near, there is little intervening interstellar dust, and their spectra can be taken to be unreddened. We can therefore draw up tables listing the spectral line features of each color class. A distant star can then be classed in terms of its spectral lines rather than its color index and the color index can be used to verify the class assignment. If the color is redder than expected, we have an indication of reddening by interstellar dust. Whether dust is actually present can then be checked—in many instances—by seeing whether other stars in the immediate neighborhood of the given object all are reddened by about the same amount. If they are, we have completed the analysis. The results give the correct spectral identification of stars in the chosen region and, in addition, we are given the extent to which interstellar dust changes the color index. A similar analysis can also be applied to determine the extent to which the overall brightness of the star is diminished through extinction by interstellar dust. This analysis allows us to determine the amount of obscuration in all the spectral ranges for which observations exist.

As already stated, the color and spectrum of a star depends on its surface temperature. Table A.3 gives the effective temperature for representative stars. As discussed in Chapter 4, the effective temperature is measured in terms of the radiant power emitted by the star over unit surface area. Since our information about the extreme ultraviolet emission of stars is still quite incomplete, we may expect that the values given in this table will change over the next few years. The uncertainties for O stars are particularly great and an effective temperature is not listed for them. The provisory nature of the data is emphasized in the table heading.

By analyzing the spectra of stars we can obtain their speed of rotation from the

A:6

Table A.3 Provisory Effective Stellar Temperatures.[a]

Types	Main-Sequence Subgiants		Giants		Supergiants	
	V	IV	III	II	Ib	Ia
			T_e (°K)			
O4	41500		41500?		41000?	
O8	35600		35600		35200	
B0	29500		31000		25700	
B5	15400		14800		13100	
A0	10000		9700		10200	
A3			8500			
F0			7200			
F5	6700	6600	6500	6350	6200	....
G0	6000	5720	5500	5350	5050	....
G5	5520	5150	4800	4650	4500	....
K0	5120	4750	4400	4350	4100	....
K5	4350	....	3700	3600	3500	....
M0	3750	....	3500	3400	3300	....
M2	3350	....	3100	2050	....	....

[a]Adapted from Keenan Ke63b and from Bö81. See also text.

broadening of stellar spectral lines. If the axis of rotation of a star is inclined at an angle i, relative to the line of sight, we then obtain a measure of $v_e \sin i$, where v_e is the equatorial velocity of the star. Only those stars whose spin axes are perpendicular to the line of sight will exhibit the full Doppler broadening due to the rotation of the star; but by analyzing the distribution of line widths, we can statistically determine both the rotational velocity and the distribution function of the angle i (Hu65). As far as we can tell, rotation axes of stars are randomly oriented with respect to the Galaxy's rotation axis. Table A.4 gives some typical values of v_e for different types of stars. Figure 1.9 shows the angular momentum for unit stellar mass for these stars. (See page 33.)

Table A.4 Stellar Rotation for Stars of Luminosity Class III and V (After Allen Al64).

Spectral Type	Mean v_e (km sec^{-1})	
	III	V
O5		190
B0	95	200
B5	120	210
A0	140	190
A5	170	160
F0	130	95
F5	60	25
G0	20	< 12
K,M	< 12	< 12

(b) Variable stars

Two main types of variable stars can be listed. *Extrinsic variables* such as (i) close binary stars whose combined brightness varies because one star can eclipse the other, and (ii) stars in nebulosities that are eclipsed by clouds or that illuminate passing clouds from time to time. These are called *T-Tauri variables*. They are named after the star in which variable features of this kind were first detected.

The second class of variable stars contains *intrinsic variables*—stars whose luminosity actually changes with time. The brightness variations may be repetitive as for periodic variables, erratic as for irregular variables, or the behavior may be semiregular. The distinction is not always clearcut. A brief summary of some characteristics of the pulsating variables is given in Table A.5. These stars are important in the construction of a reliable cosmic distance scale.

Other types of intrinsic variables include exploding stars such as novae, recurrent novae, supernovae, dwarf novae, and shell stars.

The brightness of a nova rises 10 to 12 magnitudes in a few hours. The return to the star's previous low brightness may take no more than a few months, or it may take a century. Both extremes have been observed. The absolute photographic brightness at maximum is about − 7.

Table A.5 Properties of Pulsating Variables.

Type	Range of Period, P	Spectral Type	Mean brightness M_v and variation ΔM_v	Remarks
RR Lyrae (Cluster Variables)	< 1d	A4 to F4	$M_v = 0.6$ $\Delta M_v \sim 1.0$	Found in the halo of the Galaxy
Classical Cepheids	1–50	F to K	$M_v = -2.6$ to -5.3 $M_v, \Delta M_v$ depend on P $\Delta M_v \sim 0.4$ to ~ 1.4	Found in the disk of the Galaxy
W Virginis stars (Type II Cepheids)	> 10	F,G	$M_v =$ one or two mag. less luminous than Class. Ceph. of similar period. $\Delta M_v = 1.2$	Halo population
Mira Stars (Long Period Variables)	100–1000	Red giant	$M_v \sim$ from -2.2 to 0, $\Delta M_v =$ from 3 to 5 for increasing period	Intermediate between disk and halo
Semiregular Variables	40–150	Red giant	$M_v = 0$ to -1 $\Delta M_v \sim 1.6$	Disk population

Recurrent novae brighten by about 7.5 magnitudes at periods of several decades. Their peak brightness is about the same as that of ordinary novae. The brightness decline usually takes 10 to 100 days but sometimes is outside this range.

Supernovae are about 10 magnitudes brighter than novae. The brightness may become as great as that of a whole galaxy. Two types have been recognized. Type I has $M_v = -16$ at maximum. Type II has $M_v = -14$ at maximum and exhibits the spectrum of an ordinary nova.

On exploding, a supernova can thrust about one solar mass of matter into interstellar space at speeds of order 1000 km sec^{-1}. Often these gaseous shells persist as *supernova remnants* for several thousand years. On photographic plates they appear as filamentary arcs surrounding the point of initial explosion.

Dwarf novae brighten by about 4 magnitudes to a maximum absolute brightness of $M_v = +4$ to $+6$. Their spectral type normally is A. Their outbursts are repeated every few weeks.

Shell stars are B stars having bright spectral lines. The stars seem to shed shells. A rise in brightness of one magnitude can occur.

Flare stars sporadically brighten by ~ 1 magnitude over intervals measured in tens of minutes. They then relapse. These stars are yellow or red dwarfs of low luminosity. The flares may well be similar to those seen on the sun, except that they occur on a larger scale. In extreme cases the star brightens a hundred fold.

R Coronae Borealis stars are stars that suddenly dim by as much as eight magnitudes and then within weeks return to their initial brightness. At maximum the spectrum is of class R, rich in carbon.

The variable stars are not very common, but they are interesting for two reasons. First, some of the variable stars have a well-established brightness pattern that allows one to use them as distance indicators (see chapter 2). Second, the intrinsic variables show symptoms of unstable conditions inside a star or on its surface. In that sense the variable stars may provide an important clue to the structure of stars and perhaps to the energy balance, or imbalance, at different stages of stellar evolution.

T-Tauris and novae, which apparently eject material that forms dust, are found to be strong emitters of infrared radiation.

A:7 THE DISTRIBUTION OF STARS IN SPACE AND VELOCITY

We judge the radial velocities of stars by their spectral line shifts; the transverse velocity can be obtained, for nearby stars, from the *proper motion*—the angular velocity across the sky—and from the star's distance, if known. We find that stars of differing spectral type have quite different motions. Stars in the Galactic plane have low relative velocities, while stars that comprise the *Galactic halo* have large velocities, relative to the sun. These latter objects are said to belong to population II, while those orbiting close to the plane are called population I stars. In practice there is no clear-cut discontinuity between these populations (Ku54). This is rather well illustrated by the continuous variation in velocities given in Table A.6. A star's velocity is correlated with the mean height, above the Galactic plane.

The question of real interest is whether the stars were formed and have always traveled in their present orbits. In that case we would expect the trend of velocities to define the sequence in which stars were formed from the interstellar gas. Alternatively, these different velocities may have been acquired in time, as a result of distant encounters among stars (section 3:14). In that case the stars all may have been formed at low velocities, in the galactic plane. We do not know, but we hope that studies of stellar dynamics may clear up this important question!

By noting the distribution of stars in the solar neighborhood, we at least obtain some idea about how many stars of a given kind have been formed in the Galaxy. If we can compute the life span of a star, as outlined in Chapter 8, then we can also judge the rate at which stars are born. For the short-lived stars, such birth rates represent current formation rates; and we can look for observational evidence to corroborate estimates of longevity, once the spatial number density of a given type of star has been established (Fig. A.6). Such studies still are in relatively preliminary stages, because we are not quite sure what the

A:7

Fig. A.6 Present formation rate, ψ, of bright stars per square parsec of area (projected onto the Galactic plane) during 10^{10} y. The mass of stars of different brightness is shown. (After M. Schmidt, Sc63).

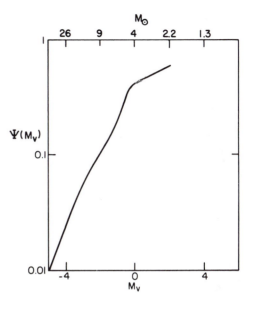

appearance of a star should be at birth, particularly if it still is surrounded by some of the dust from which it has been formed (section 1:4) (Da67).

A:8 PULSARS, RADIO STARS, AND X-RAY SOURCES

(a) Pulsars

With the exception of the Crab Nebula pulsar, the Vela pulsar, and 1509–58, pulsars have thus far been identified only in the radio wavelength region. The first two emit optical light as well, and all three give off X-rays or γ-radiation. Their remarkable feature is the regularity of the pulses, typically spaced anywhere between 1.5 msec and 4 sec apart. For many of these objects the rate is constant to one part in 10^8.

Within each pulse, there are subpulses that march, relative to the overall envelope, with regularities both in their phase and in their changing sense of polarization.

The coherence and pulse rates already tell us that the source is small compared to normal stars. We are dealing with *neutron stars*, stars whose cores consist of closely packed, degenerate neutrons. In such a star the mass of the sun is packed into a volume about 10 km in diameter.

The most widely accepted model for the pulsars has a neutron star rotating with a period equal to the interval between the main pulses. The method of

A:8

Table A.6 Stellar Velocities Relative to the Sun, and Mean Height Above Galactic Plane.[a]

Objects	Velocity[b], v km sec^{-1}	Density, ρ $10^{-3} M_\odot$ pc^{-3}	Height, h pc
Interstellar clouds			
large clouds	8		
small clouds	25		
Early main sequence stars:			
O5–B5	10 ⎱	0.9	50
B8–B9	12 ⎰		60
A0–A9	15	1	115
F0–F9	20	3	190
Late main sequence stars:			
F5–G0	23		
G0–K6	25	12 ⎱	350
K8–M5	32	30 ⎰	
Red giant stars:			
K0–K9	21	0.1	270
M0–M9	23	0.01	
High velocity stars:			
RR Lyrae variables	120	10^{-5}	
Subdwarfs	150	1.5	
Globular clusters	120–180	10^{-3}	

[a]Stellar velocities collected by Spitzer and Schwarzschild from other sources (Sp51a). Densities, ρ, and Heights, h, after Allen (Al64). (With the permission of the Athlone Press of the University of London, second edition © C.W. Allen 1955 and 1964.)

[b]Root mean square value for component of velocity projected onto the Galactic plane.

generating the pulses, however, has not yet been settled. In all the theories we have, the radiation is emitted in a direction tangential to the charged particles moving with the rotating star and, hence, there is a loss of angular momentum and a corresponding slowdown of the star's rotation and of the pulse rate. Careful measurements indeed show this slowdown in a number of pulsars (Go68).

From time to time a discontinuous change in the period can also occur; and this kind of change is not yet understood. There are other remarkable features that remain unexplained: Giant pulses, thousands of times brighter than a normal pulse, but appearing only about once in ten thousand pulses; null pulses, where there is no intensity at all, appearing at times; sudden changes in the pulse structure, with an equally sudden flipping back to the original pulsing mode. All these provide puzzles we must seek to fathom.

A small number of pulsars are associated with known gaseous *supernova remnants.* One is in the constellation Vela. Another is a star in the Crab Nebula, remnant of a supernova seen in 1054 AD. It was identified, more than 25 years before the pulsar's discovery, as the stellar remnant of the supernova. Within the past decade a handful of binary pulsars has been discovered, two compact sources orbiting each other.

The fastest known pulsar emits pulses with a periodicity of only 1.5 msec. For all short-period pulsars there is an observed slow-down of the period by about 10^{-15} sec each second. The slower pulsing pulsars must be appreciably older. The Crab pulsar now pulses every 33 msec. Using the present slowdown rate, we can make a rough linear extrapolation of the Crab pulsar's period, backward in time, and see that this is indeed the object that exploded in 1054 AD.

A particularly interesting feature of pulsars is that they may also be responsible for cosmic ray particles. The charged particles, which give rise to the observed pulsed electromagnetic radiation, are believed to be highly relativistic; and it is

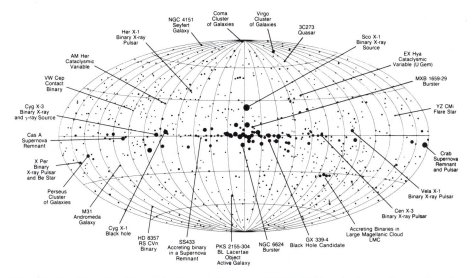

Fig. A.7 The X-ray sky. Map of the X-ray sky known in 1987 plotted in galactic coordinates. Note the concentration toward the galactic center and galactic plane. Bright sources are shown as larger circles. A number of individual sources are identified both by their respective catalogue designations and by source type. (With the permission of Kent S. Wood and the Naval Research Laboratory, as well as of Jay M. Pasachoff and W.B. Saunders, Co.)

A:8

entirely possible that some of the particles generated in pulsars have energies as high as the highest energy *cosmic rays* (section 5:10) observed. In that case, it may well be that cosmic rays are a more local phenomenon than had long been believed. If this hypothesis is true, we might expect an anisotropy in the cosmic ray flux at the highest energies. It is also true that we would expect a different chemical composition in cosmic rays than found in ordinary stellar material, because the material of a neutron star has probably undergone severe nuclear changes during the evolution of the star.

One interesting feature of the pulsars is the arrival time of pulses at different radio frequencies. Although equally spaced, these pulses arrive at slightly different times because the refractive index of the interstellar medium, and of any outer layers of the pulsar, are slightly different for differing radio frequencies. This allows us to compute the number of electrons along the line of sight to the object. We can then make a rough estimate of the distance of the object and its radio brightness (section 6:11).

The pulsars—some 60 of them were known in 1971—are concentrated toward the plane of the Galaxy. Based on arguments of distance and concentration, we are sure that these objects are stars within our own Galaxy. Other galaxies, however, doubtlessly contain pulsars too.

(b) Radio Stars

The first radio star to be discovered was the sun (Re44). Its radio emission is very weak and we detect it clearly only because we are near. For more than a decade after the sun's detection, all discovered radio sources were extragalactic radio galaxies or quasars, or involved nebulosities such as supernovae or ionized hydrogen regions in the Galaxy. However, in recent years, several new classes of radio stars have been observed. Besides pulsars, novae and X-ray emitting stars have also been detected at radio frequencies. Red supergiants, red dwarf flare stars, and a blue dwarf companion to a red supergiant have also been studied. These objects are very faint and only the most sophisticated techniques available to us are suitable for their study (Hj71).

(c) X-Ray Stars

Many extragalactic sources are known to emit X-rays, among than M87, a spherical galaxy that emits radio waves and strikingly exhibits jets of relativistic particles apparently shot out from its center; 3C 273, the brightest quasar; and many other quasars, radio galaxies and Seyfert galaxies. In addition, X-rays have been detected from the direction of several clusters of galaxies. The most readily observed X-ray sources, however, are galactic. Figure A.7 shows a clustering of the sources about

A:8

the galactic plane. These sources are associated with stars and fall into several groups.

(i) The Crab pulsar emits extremely regular pulses with a 0.033-sec period. This is somewhat of an exception; but the gaseous supernova remnant surrounding the Crab is quite typical in being a bright X-ray source.

(ii) Centaurus X-3 is a source that pulses semiregularly. For periods of a day and a half it pulses with a period that slowly increases from 4.84 to 4.87 sec. Then it suddenly drops in intensity in a period of an hour, only to start all over again a half day later. Sources of this kind appear to be close binary stars.

(iii) There are a variety of X-ray sources that seem to have some regularity to their brightness changes that occur on the scale of seconds. But we have not yet been able to define this regularity in all cases.

(iv) Novalike sources that flare up for a month or so and then die away. Several of these have been discovered each year. Some X-ray stars are associated with white dwarf stars, with planetary nebulae, or neutron stars. Some might be associated also with *black holes*, stars in an ultimate state of collapse (section 8:19).

A:9 QUASARS AND QUASISTELLAR OBJECTS (QSO's)

These objects (Ha63, Gr64) are listed separately because their nature is not yet properly understood.

Quasars have many features in common with some types of radio galaxies; in particular the visible spectra bear a strong resemblance to the nuclei of *Seyfert galaxies*, which are spiral galaxies with compact nuclei that emit strongly in the infrared and exhibit highly broadened emission lines from ionized gases. In both the quasars and Seyfert nuclei, we find highly ionized gases with spectra indicating temperatures of the order of 10^5 to $10^{6°}$K and number densities $\sim 10^6$ cm^{-3}. The conditions resemble those found in the solar corona. In the quasars and Seyfert nuclei the spectra of these gases show velocity differences of the order of 1000 or 2000 km/sec, indicating either (a) that gases are being shot out of these objects at high velocity, (b) that they are falling in at high speed, (c) that there is a fast rotation, or (d) that there is a great deal of turbulent motion present. Most likely, a combination of two or three factors is involved.

The quasars show brightness variations on a time scale of months and are, therefore, believed to be less than a light-month $\sim 10^{17}$ cm in diameter. This argument, however, is weak, because it assumes that the brightness changes are coherent, while actually we may be dealing with independent outbursts in different portions of a much larger object in which independent outbursts occur on the time scale of weeks.

A:9

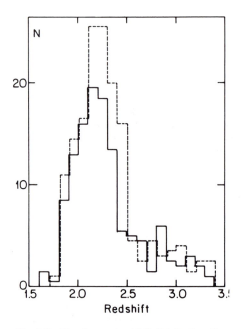

Fig. A.8   The observed redshift distributions for Ly-α emitting quasars obtained in two different surveys. The dashed curve represents quasars found in slitless spectrograph photographic surveys using large telescopes. The solid curve shows the findings with similar techniques using smaller Schmidt telescope surveys. The agreement is very good. N is the number observed (after Vé86).

Very strikingly the quasars also show (Fig. A.8) spectra that are highly red shifted, indicating that they are at extreme distances and hence must be extremely luminous to appear as bright as they do. Only extreme infrared galaxies, whose peak emission occurs at wavelengths of $\sim 100\mu$, are comparably luminous. Some may emit more than 10^{46} erg sec^{-1}—a hundred times more than our galaxy. Since the sizes of these objects are so small, the surface brightness must be some 10 orders of magnitude greater than that of normal galaxies. Extremely high X-ray luminosity also characterizes many quasars. One quasar, 3C273, is also known as a gamma-ray emitting source and many others may emit at these wavelengths as well.

The terms "quasar" and "QSO" are often used interchangeably. However, some astronomers reserve "quasar" for those QSO's that emit strongly at radio frequencies, and use "QSO" to denote both radio quiet and powerful radio emitters—that is, the whole class of these compact objects.

A:10 PHOTONS AND COSMIC RAY PARTICLES

The earth, the solar system, and the Galaxy all are bathed in streams of photons and highly relativistic particles. Inside the Galaxy, photon densities are higher than outside, since starlight and infrared emission give a stronger local illumination; but there is a microwave component with the spectrum of a blackbody at 3°K that seems to be equally strong inside the Galaxy and in the ambient universe (Pe65).

Cosmic ray particles, highly energetic electrons and nucleons, constitute a denser energy bath in the earth's vicinity than starlight and microwave photons combined; but we do not know how the particles are distributed in extragalactic space. Table A.7 shows the energy densities of these components. X-rays and γ-rays, highly energetic photons, have far smaller energy densities than visible and microwave radiation (see also Fig. 12.2).

Table A.7 Energy and Number Densities of Photons and Cosmic Rays.

	Cosmic Ray Particles	Visible Light	Microwaves
Energy density in Galaxy ($\mathrm{ergs\,cm^{-3}}$)	10^{-12}	$\sim 2 \times 10^{-13}$	$\sim 5 \times 10^{-13}$
Extragalactic energy density ($\mathrm{ergs\,cm^{-3}}$)	?	$\sim 2 \times 10^{-14}$	$\sim 5 \times 10^{-13}$
Number density in Galaxy ($\mathrm{cm^{-3}}$)	$\sim 10^{-9}$	$\sim 10^{-1}$	$\sim 10^{3}$
Extragalactic number density ($\mathrm{cm^{-3}}$)	?	$\sim 10^{-2}$	$\sim 10^{3}$

The microwave component is particularly interesting. As far as we can tell it is isotropic, has a blackbody spectrum, and pervades the entire universe. While a blackbody spectrum is normally associated only with opaque emitters, the universe is so tenuous that it hardly seems capable of producing this radiation at the present epoch. Does this mean that an ever-tenuous steady state universe (10:2) could never have produced this background, and that an evolving cosmos could only have generated the radiation, at a time when the universe was far denser and more compact than now? We cannot yet tell.

Appendix B

Astrophysical Constants*

B:1 PHYSICAL CONSTANTS

Velocity of light	$c = 2.998 \times 10^{10}$ cm sec^{-1}
Planck constant	$h = 6.626 \times 10^{-27}$ erg sec
Gravitational constant	$G = 6.67 \times 10^{-8}$ dyn cm^2 g^{-2}
Electron charge	$e = 4.803 \times 10^{-10}$ esu
Mass of electron	$m_e = 9.1096 \times 10^{-28}$ g
Mass of proton	$m_p = 1.6724 \times 10^{-24}$ g
Mass of hydrogen atom	$m_H = 1.6733 \times 10^{-24}$ g
Mass of neutron	$m_N = 1.6747 \times 10^{-24}$ g
Atomic mass unit	$amu = (1/12)\, m_{C^{12}} = 1.661 \times 10^{-24}$ g
Avogadro's number	6.0222×10^{23}
Boltzmann constant	$k = 1.380 \times 10^{-16}$ erg deg^{-1}
Electron volt	$ev = 1.602 \times 10^{-12}$ erg
Radiation density constant	$a = 7.5646 \times 10^{-15}$ erg cm^{-3} deg^{-4}
Stefan-Boltzmann constant	$\sigma = 5.67 \times 10^{-5}$ erg cm^{-2} deg^{-4} sec^{-1}
Rydberg constant	$R_\infty = 2.17992 \times 10^{-11}$ erg

B:2 ASTRONOMICAL CONSTANTS

Year $= 3.156 \times 10^7$ sec

Astronomical unit $AU = 1.496 \times 10^{13}$ cm
(mean sun earth distance)

---

*(Al64), (Ta70).

Parsec, pc $= 3.086 \times 10^{18}$ cm $= 2.06 \times 10^5$ AU $= 3.261$ light years
Solar mass $\hspace{3.5cm} M_\odot = 1.99 \times 10^{33}$ g
Solar radius $\hspace{3.5cm} R_\odot = 6.96 \times 10^{10}$ cm
Solar luminosity $\hspace{3cm} L_\odot = 3.9 \times 10^{33}$ erg sec^{-1}
Star with $M_{bol} = 0$ radiates 3.02×10^{28} watt
Aeon $= 1$ æ $= 10^9$ y

List of References

Articles and books that are cited in the text are referred to by the first two letters of the author's name and the last two digits of the year in which the publication appears. An article by Johnson published in 1928 is designated (Jo28). Where several individuals have co-authored a work, only the first author's name is used. A publication that appeared prior to 1901 carries a numeral designation 00, for example Newton's *Principia* carries the designation (Ne00). An asterisk following a given designation, for example (He67)*, implies that the reference is important for the entire section in which it is cited.

The following abbreviations have been used:

A + A: Astronomy and Astrophysics.
Ann. New York Acad. Sci.: Annals of the New York Academy of Sciences.
Ann. Rev. Astron. and Astrophys.: Annual Reviews of Astronomy and Astrophysics.
A.J.: Astronomical Journal (USA).
Ap.J.: Astrophysical Journal (USA). If the letter L precedes the page number, the article appeared in the affiliated journal Astrophysical Journal Letters. The Astrophysical Journal also publishes a supplement, denoted by "Suppl."
Astrophys. Let.: Astrophysical Letters.
B.A.N.: Bulletin of the Astronomical Society of the Netherlands.
I.A.U.: International Astronomical Union. The organization issues symposium proceedings and a variety of other publications.
J. Roy. Astron. Soc. Can.: Journal of the Royal Astronomical Society of Canada.
J. Phys. USSR: Journal of Physics of the Soviet Union.
MNRAS: Monthly Notices of the Royal Astronomical Society (London).
PASP: Publications of the Astronomical Society of the Pacific.
Phil. Mag.: Philosophical Magazine (London).
Phys. Rev.: Physical Reviews (USA).
Phys. Rev. Let.: Physical Review Letters (USA).
Proc. Nat. Acad. Sci.: Proceedings of the National Academy of Sciences (USA).
Proc. Roy. Soc.: Proceedings of the Royal Society (London).

PASJ: Publications of the Astronomical Society of Japan.

Quart. J. Roy. Astron. Soc.: Quarterly Journal of the Royal Astronomical Society (London).

Rev. Mod. Phys.: Reviews of Modern Physics (USA).

Soviet A.J.: Astronomical Journal (Soviet Union). This journal also appears in English translation.

Z. Astrophys.: Zeitschrift für Astrophysik (Germany).

(Aa60) S. Aarseth, "The Rotation of a Barred Galaxy under Gravitational Forces," *MNRAS, 121*, 525 (1960).

(Aa61) S. Aarseth, "The Rotation of Barred Galaxies under Gravitational Forces II," *MNRAS, 122*, 535 (1961).

(Al55) C.W. Allen, *Astrophysical Quantities*, Athlone Press, London (1955).

(Al63) L.H. Aller, "Astrophysics" in *The Atmospheres of the Sun and Stars,* 2nd ed., Ronald Press, New York (1963).

(Al64) C.W. Allen, *Astrophysical Quantities*, 2nd ed., Athlone Press, London (1964).

(Al68) T. Alväger and M.N. Kreisler, "Quest for Faster than Light Particles," *Phys. Rev., 171*, 1357 (1968).

(Al70) Margo F. Aller and C.R. Cowley, "The Possible Identification of Prometheum in HR46," *Ap.J., 162*, L145 (1970).

(Al70a) W.J. Altenhoff *et al., Astron. and Astrophys. Suppl. 1*, 319 (1970).

(Al82) A. Albrecht, P.J. Steinhardt, M.S. Turner, F. Wilczek, "Reheating an Inflationary Universe", *Phys. Rev. Let. 48*, 1437 (1982).

(An72) E. Anders "Physico-Chemical Processes in the Solar Nebula, as inferred from Metearites" in *l'Origine du Système Solaire* H. Reeves (Ed.), Centre National de la Rechershe Scientifique, Paris (1972).

(Ar65) H. Arp, "A Very Small, Condensed Galaxy," *Ap.J., 142*, 402 (1965).

(Ar70) W.D. Arnett and D.D. Clayton, "Explosive Nucleosynthesis in Stars," *Nature, 227*, 780 (1970).

(Ar71) H. Arp, "Observational Paradoxes in Extragalactic Astronomy," *Science, 174*, 1189 (1971).

(As71) M.E. Ash, I.I. Shapiro, and W.B. Smith, "The System of Planetary Masses," *Science, 174*, 551 (1971).

(Ba71a) E.S. Barghoorn, "The Oldest Fossils," *Scientific American* May (1971), p. 30.

(Ba71b) J.N. Bahcall and S. Frautschi, "The Hadron Barrier in Cosmology and Gravitational Collapse," *Ap.J., 170* L81 (1971).

(Ba72) J.N. Bahcall and R.L. Sears, "Solar Neutrinos," *Ann. Rev. Astron. and Astrophys.* (1972).

(Ba85) J.N. Bahcall, B.T. Cleveland, R. Davis, Jr. and J.K. Rowley, "Chlorine and Gallium Solar Neutrino Experiments," *Ap.J. Let 292*, L79 (1985).

(Be39) H. Bethe, "Energy Production in Stars," *Phys. Rev., 55*, 434 (1939).

(Be71) A. Bers, R. Fox, C.G. Kuper, and S.G. Lipson, "The Impossiblity of Free Tachyons" in *Relativity and Gravitation*, C.G. Kuper and A. Peres (Eds.) Gordon and Breach, New York (1971).

(Be87) C. Beichman "The IRAS View of the Galaxy and the Solar System," *Ann. Rev. Astron. and Astrophys. 25*, 521 (1987).

(Bi23) G.D. Birkhoff, *Relativity and Modern Physics*, Harvard University Press, Cambridge, M.A. (1923).

(Bi87) R.M. Bionta *et al*, "Observation of a Neutrino Burst in Coincidence with Supernova 1987A in the Large Magallanic Cloud," *Phys. Rev. Let. 58*, 1494 (1987).

(Bl61) A. Blaauw, "On the Origin of the O- and B- type Stars with High Velocities, (The 'Run-away' Stars), and Some Related Problems," *B.A.N., 15*, 265 (1961).

(Bo48) H. Bondi and T. Gold, "The Steady State Theory of the Expanding Universe," *MNRAS, 108*, 252 (1948).

(Bo52) H. Bondi, *Cosmology*, Cambridge University Press, Cambridge, England (1952).

(Bo64a) S. Bowyer, E. Byram, T. Chubb, and H. Friedman, "X-ray Sources in the Galaxy," *Nature, 201*, 1307 (1964).

(Bo64b) S. Bowyer, E. Byram, T. Chubb, and H. Friedman, "Lunar Occultation of X-ray Emission from the Crab Nebula," *Science, 146*, 912 (1964).

(Bo71) B.J. Bok, "Observational Evidence for Galactic Spiral Structure" in *Highlights in Astronomy*, DeJager, Ed. I.A.U. (1971).

(Bö81) E. Böhm-Vitense "The Effective Temperature Scale" *Ann. Rev. Astron. and Astrophys. 19*, 295 (1981).

(Bo85) A.M. Boesgaard and G. Steigman, "Big Bang Nucleasynthesis: Theories and Observations" *Ann. Rev. Astron. and Astrophys. 23*, 319 (1985).

(Br68) P.F. Browne, "The Generation of Magnetic Fields by Viscous Forces in a Non-Uniformly Rotating Plasma," *Astrophys. Let., 2*, 217 (1968).

(Br72) A.H. Bridle, M.M. Davis, E.B. Fomalont, and J. Lequeux "Counts of Intense Extragalactic Radio Sources at 1400 MHz," *Nature Physical Science, 235*, 123 (1972).

(Bu57) E.M. Burbidge, G.R. Burbidge, W.A. Fowler and F. Hoyle, "Synthesis of the Elements in Stars" *Rev. Mod. Phys. 29*, 547 (1957).

(Bu60) E.M. Burbidge, G.R. Burbidge, and K.H. Prendergast, "Motions in Barred Spiral Galaxies II. The Rotation of NGC 7479," *Ap.J., 132*, 654 (1960).

(Bu69) G.R. Burbidge and E.M. Burbidge, "Quasistellar Objects—A Progress Report," *Nature, 224*, 21 (1969).

(Bu71a) E.M. Burbidge and W.L.W. Sargent, "Velocity Dispersion and Discrepant Redshifts in Groups of Galaxies," from Pontifical Academy of Science, *Scripta Varia No. 35: Nuclei of Galaxies* (1971).

(Bu71b) E.M. Burbidge and P.M. Hodge, "Is NGC 4569 a Member of the Virgo Cluster?" *Ap.J., 166*, 1 (1971).

(By67) E.T. Byram, T.A. Chubb, and H. Friedman, "Cosmic X-ray Sources, Galactic and Extragalactic," *Science, 152*, 66 (1967).

(Ca24) A.J. Cannon and E.C. Pickering, "The Henry Draper Catalogue," *Ann. Astron. Obs. Harvard College, 91* to *99* (1924).

(Ca66) D. Cattani and C. Sacchi, "A Theory on the Creation of Stellar Magnetic Fields," *Nuovo Cimento, 46B*, 8046 (1966).

(Ca68) A.G.W. Cameron "A New Table of Abundances of the Elements in the Solar System," in *Origin and Distribution of the Elements,* L.H. Arens, Ed., Pergamon Press, Elmsford, New York (1968), p. 125.

(Ca70a) G.R. Carruthers, "Rocket Observations of Interstellar Molecular Hydrogen," *Ap.J., 161*, L81 (1970).

(Ca70b) A. Cavaliere, P. Morrison, F. Pacini, "A Model for the Radiation from the Compact Strong Sources," *Ap.J., 162*, L133 (1970).

(Ch39) S. Chandrasekhar, *An Introduction to the Study of Stellar Structure,* University of Chicago Press, Chicago (1939), Dover, New York (1957).

(Ch43) S. Chandrasekhar, *Principles of Stellar Dynamics*, University of Chicago Press, Chicago (1943), Dover, New York (1960).

(Ch50) S. Chandrasekhar, *Radiative Transfer*, University of Chicago Press, Chicago (1950), Dover, New York (1960).

(Ch58) S. Chandrasekhar, "On the Continuous Absorption Coefficient of the Negative Hydrogen Ion," *Ap.J., 128*, 114 (1958).

(Ch64a) H.Y. Chiu, "Supernovae, Neutrinos and Neutron Stars," *Annals of Physics, 26*, 364 (1964).

(Ch64b) J.H. Christenson, J.W. Cronin, V.L. Fitch, and R. Turlay, "Evidence for the 2π Decay of the K_2^0 Meson," *Phys. Rev. Let., 13*, 138 (1964).

(Ch70) W.Y. Chau, "Gravitational Radiation and the Oblique Rotator Model," *Nature, 228*, 655 (1970).

(Cl68) D.D. Clayton, *Principles of Stellar Evolution and Nucleosynthesis*, McGraw-Hill, New York (1968).

(Co57) T.G. Cowling, *Magnetohydrodynamics*, Interscience, New York (1957).

(Co68) C.C. Counselman III and I.I. Shapiro, "Scientific Uses of Pulsars," *Science, 162*, 352 (1968).

(Co70) R. Cowsik, Y. Pal, and T.N. Rengarajan, "A Search for a Consistent Model for the Electromagnetic Spectrum of the Crab Nebula" *Astrophys. and Space Sci., 6*, 390 (1970).

(Co71a) D.P. Cox and E. Daltabuit, "Radiative Cooling of a Low Density Plasma," *Ap.J., 167*, 113 (1971).

(Co71b) R. Cowsik and P.B. Price, "Origins of Cosmic Rays," *Physics Today,* October (1971), p. 30.

(Co72) J.J. Condon, "Decimetric Spectra of Extragalactic Radio Sources," Doctoral Dissertation in Astronomy, Cornell University (1972).

(Da51) L. Davis and J.L. Greenstein, "The Polarization of Starlight by Aligned Dust Grains," *Ap.J., 114*, 206 (1951).

(Da67) K. Davidson and M. Harwit, "Infrared and Radio Appearance of Cocoon Stars," *Ap.J., 148*, 443 (1967).

(Da68) R. Davis, Jr., D.S. Harmer, K.C. Hoffman, "Search for Neutrinos from the Sun," *Phys. Rev. Let., 20*, 1205 (1968).

(Da69) K. Davidson and Y. Terzian, "Dispersion Measures of Pulsars," *A.J., 74*, 849 (1969).

(Da70a) K. Davidson, "The Development of a Cocoon Star," *Astrophysics and Space Science, 6*, 422 (1970).

(Da71) R. Davis Jr., L.C. Rogers, and V. Radeka, "Report on the Brookhaven Solar Neutrino Experiment," *Bull. Amer. Phys. Soc. Ser. 11, 16*, 631 (1971).

(De68) S.F. Dermott, "On the Origin of Commensurabilities in the Solar System-II-The Orbital Period Relation," *MNRAS, 141*, 363 (1968).

(De73) S.F. Dermott "Bode's Law and the Resonant Structure of the Solar System" *Nature Physical Science 244*, 18 (1973).

(deS17) W. deSitter, "On Einstein's Theory of Gravitation and its Astronomical Consequences," *MNRAS, 78*, 3 (1917).

(Di31) P.A.M. Dirac, "Quantized Singularities in the Electro-magnetic Field," *Proc. Roy. Soc., A, 133*, 60 (1931).

(Di38) P.A.M. Dirac, "A New Basis for Cosmology," *Proc. Roy. Soc., A, 165*, 199 (1938).

(Di67a) R.H. Dicke, "Gravitation and Cosmic Physics," *Am. J. of Physics, 35*, 559 (1967).

(Di69) R.H. Dicke, "The Age of the Galaxy from the Decay of Uranium," *Ap.J., 155*, 123 (1969).

(Di86) R.H. Dicke, J.R. Kuhn and K.J. Libbrecht, "The variable oblateness of the sun: measurements of 1984," *Ap.J. 311*, 1025 (1986).

(Dr62) F.D. Drake, "Intelligent Life in Space," MacMillan, New York (1962).

(Du62) S. Dushman and J.M. Lafferty, "Scientific Foundations of Vacuum Technique," John Wiley, New York (1962).

(Dy60) F.J. Dyson, "Search for Artificial Stellar Sources of Infrared Radiation," *Science, 131*, 1667 (1960).

(Eb55) R. Ebert, "Über die Verdichtung von Hı Gebieten," *Z. Astrophys., 37*, 217 (1955).

(Ed30) A.S. Eddington, "On the Instability of Einstein's Spherical World," *MNRAS, 90*, 668 (1930).

(Eg62) O. Eggen, D. Lynden-Bell, and A. Sandage, "Evidence From the Motions of Old Stars that the Galaxy Collapsed," *Ap.J., 136*, 748 (1962).

(Ei05a) A. Einstein, "On the Electrodynamics of Moving Bodies," *Ann. d. Phys., 17*, 891 (1905), translated and reprinted in *The Principle of Relativity,* A. Sommerfeld, Ed., Dover, New York.

(Ei05b) A. Einstein, "Does the Inertia of a Body Depend Upon Its Energy Content?" *Ann. d. Phys., 18,* 639 (1905) translated and reprinted in *The Principle of Relativity,* A. Sommerfeld, Ed., Dover, New York.

(Ei07) A. Einstein, "On the Influence of Gravitation on the Propagation of Light," *Jahrbuch für Radioakt. und Elektronik, 4,* 1907 (translated in *The Principle of Relativity,* A. Sommerfeld, Ed., Dover, New York.

(Ei16) A. Einstein, "The Principle of General Relativity" *Ann. d. Phys., 49,* 769 (1916), translated and reprinted in *The Principle of Relativity,* A. Sommerfeld, Ed., Dover, New York.

(Ei17) A. Einstein, *Kosmologische Betrachtungen zur Allgemeinen Relativitätstheorie,"* S.B. Preuss. Akad. Wiss, (1917). p. 142.

(Ez65) D. Ezer and A.G.W. Cameron, "The Early Evolution of the Sun," *Icarus, 1,* 422 (1963).

(Fa85) A.C. Fabian "Gas in elliptical galaxies," *Nature, 314,* 130 (1985).

(Fo62) W.A. Fowler, J.L. Greenstein, and F. Hoyle, "Nucleosynthesis During the Early History of the Solar System," *Geophys. J., 6,* 148 (1962).

(Fr22) A. Friedmann, "Über die Krümmung des Raumes," *Z. Phys., 10,* 377 (1922).

(Fr46) J. Frenkel, *Kinetic Theory of Liquids,* Oxford University Press, Oxford, England (1946).

(Fr69) G. Fritz, R.C. Henry, J.F. Meekins, T.A. Chubb, and H. Friedman, "X-ray Pulsars in the Crab Nebula," *Science, 164,* 709 (1969).

(Ga00) Galileo Galilei, *Dialogues Concerning Two New Sciences,* translated by H. Crew and A. de Salvio, Northwestern University Press, (1946).

(Ge70) S.L. Geisel, D.E. Kleinmann, and F.J. Low, "Infrared Emission from Nebulae," *Ap.J., 161,* L101 (1970).

(Gi62) R. Giacconi, H. Gursky, F. Paolini, and B. Rossi, "Evidence for X-rays From Sources Outside the Solar System," *Phys. Rev. Let., 9,* 439 (1962).

(Gi64) V.L. Ginzburg and S.I. Syrovatskii, "The Origin of Cosmic Rays," Pergamon Press, Elmsford, New York (1964).

(Gi69) V.L. Ginzburg, "Elementary Processes for Cosmic Ray Astrophysics," Gordon and Breach, New York (1969).

(Gi72) R. Giacconi, H. Gursky, E. Kellogg, S. Murray, E. Schreier, and H. Tananbaum. "The Uhuru Catalog of X-Ray Sources," *Ap.J. 178,* 281 (1972).

(Go52) T. Gold, "The Alignment of Galactic Dust," *MNRAS, 112,* 215 (1952).

(Go62) T. Gold, "The Arrow of Time," *Am. J. of Phys., 30,* 403 (1962).

(Go63) R.J. Gould and E.E. Salpeter, "The Interstellar Abundance of the Hydrogen Molecule I. Basic Processes," *Ap.J., 138,* 393 (1963).

(Go68) T. Gold, "Rotating Neutron Stars as the Origin of the Pulsating Radio Sources," *Nature, 218,* 731 (1968).

(Go69) T. Gold, "Rotating Neutron Stars and the Nature of Pulsars," *Nature, 221,* 25 (1969).

(Go73) P. Goldreich and W.R. Ward, "The Formation of Planetesimals," *Ap.J. 183,*

1051 (1973).

(Go78) P.F. Goldsmith and W.D. Langer, "Molecular Cooling and Thermal Balance of Dense Interstellar Clouds," *Ap.J., 222,* 881 (1978).

(Gr64) J.L. Greenstein and M. Schmidt, "The Quasi-Stellar Radio Sources 3C48 and 3C273," *Ap.J., 140,* 1 (1964).

(Gr66) K. Greisen, "End to the Cosmic Ray Spectrum," *Phys. Rev. Let., 16,* 748 (1966).

(Gr68) J.M. Greenberg, "Interstellar Grains" in *Nebulae and Interstellar Matter,* B.M. Middlehurst and L.H. Aller, Eds., University of Chicago Press, Chicago (1968).

(Gu54) S.N. Gupta, "Gravitation and Electromagnetism," *Phys. Rev., 96,* 1683 (1954).

(Gu65) J.E. Gunn and B.A. Peterson, "On the Density of Neutral Hydrogen in Intergalactic Space," *Ap.J., 142,* 1633 (1965).

(Gu66) H. Gursky, R. Giacconi, P. Gorenstein, J.R. Waters, M. Oda, H. Bradt, G. Garmire, and B.V. Sreekantan, "A Measurement of the Location of the X-ray Source Sco X-1," *Ap.J., 146,* 310 (1966).

(Gu69) J.W. Gunn and J.P. Ostriker, "Acceleration of High-Energy Cosmic Rays by Pulsars," *Phys. Rev. Let., 22,* 778 (1969).

(Gu71) H. Gursky, E.M. Kellogg, C. Leong, H. Tananbaum, and R. Giacconi, "Detection of X-rays from the Seyfert Galaxies, NGC 1275 and NGC 4151 by the Uhuru Satellite," *Ap.J., 165* L43 (1971).

(Gu81) Alan H. Guth, "Inflationary Universe: A Possible Solution to the Horizon and Flatness Problem," *Phys. Rev. D. 23,* 347 (1981).

(Ha54) R. Hanbury Brown, and R.Q. Twiss, "A New Type of Interferometer for Use in Radio Astronomy," *Phil. Mag., 45,* 663 (1954).

(Ha61) M. Harwit, "Can Gravitational Forces Alone Account for Galaxy Formation in a Steady State Universe?" *MNRAS, 122,* 47; *123,* 257 (1961).

(Ha62a) M. Harwit, "Dust, Radiation Pressure and Star Formation," *Ap.J., 136,* 832 (1962).

(Ha62b) M. Harwit and F. Hoyle, "Plasma Dynamics in Comets II," *Ap.J., 135,* 875 (1962).

(Ha62c) C. Hayashi, R. Hoshi, and D. Sugimoto, "Evolution of the Stars," *Progress of Theor. Physics,* Suppl. 22 (1962).

(Ha63) C. Hazard, M. B. Mackey, and A. J. Shimmins, "Investigation of the Radio Source 3C 273 by the Method of Lunar Occultations," *Nature, 197,* 1037 (1963).

(Ha65) E.R. Harrison, "Olbers' Paradox and the Background Radiation Density in an Isotropic Homogeneous Universe," *MNRAS, 131,* 1 (1965).

(Ha66) C. Hayashi, "Evolution of Protostars," *Ann. Rev. Astron. and Astrophys., 4,* 171 (1966).

(Ha67) R. Hanbury Brown, J. Davis, L.R. Allen, J.M. Rome, "The Stellar Interferometer at Narrabri Observatory-II. The Angular Diameters of 15 stars," *MNRAS, 137*, 393 (1967).

(Ha70) M. Harwit, "Is Magnetic Alignment of Interstellar Dust Really Necessary?," *Nature, 226*, 61 (1970).

(Ha71) D.A. Harper and F.J. Low, "Far Infrared Emission from H_{II} Regions," *Ap.J., 165*, L9 (1971).

(Ha72) M. Harwit, B.T. Soifer, J.R. Houck, and J.L. Pipher, "Why Many Infrared Astronomical Sources Emit at $100 \, \mu m$," *Nature Physical Science, 236*, 103 (1972).

(Ha87) C.G.T. Haslam and J.L. Osborne "The infrared and radio-continuum emission of the galactic disk" *Nature 327*, 211 (1987).

(He50) G. Herzberg, *Molecular Spectra and Molecular Structure I. Spectra of Diatomic Molecules* 2nd ed., Van Nostrand, Princeton, New Jersey, (1950).

(He62) O. Heckmann and E. Schücking, "Relativistic Cosmology," in *Gravitation*, L. Witten, Ed., John Wiley, New York. (1962).

(He67) G. Herzberg, "The Spectra of Hydrogen and Their Role in the Development of Our Understanding of the Structure of Matter and of the Universe," *Trans. Roy. Soc.,* Canada *V,* Ser *IV,* 3 (1967).

(He68a) C.E. Heiles, "Normal OH Emission and Interstellar Dust Clouds," *Ap.J., 151*, 919 (1968).

(He68b) A. Hewish, S.J. Bell, J.D.H. Pilkington, P.F. Scott, and R.A. Collins, "Observation of a Rapidly Pulsating Radio Source," *Nature, 217*, 709 (1968).

(He69) C.E. Heiles, "Temperatures and OH Optical Depths in Dust Clouds," *Ap.J., 157*, 123 (1969).

(Hi71) J.M. Hill, "A Measurement of the Gravitational Deflection of Radio Waves by the Sun," *MNRAS, 153*, 7p (1971).

(Hi84) A.M. Hilas, "The Origin of Ultra-high-energy Cosmic Rays," *Ann. Rev. Astron. and Astrophys. 22*, 425 (1984).

(Hi87) K. Hirata *et al*, "Observation of a Neutrino Burst from the Supernova SN 1987A," *Phys. Rev. Let 58*, 1490 (1987).

(Hj71) R.M. Hjellming and C.M. Wade, "Radio Stars," *Science, 173*, 1087 (1971).

(Ho48) F. Hoyle, "A New Model for the Expanding Universe," *MNRAS, 108*, 372 (1948).

(Ho56) F. Hoyle and A. Sandage, "Second Order Term in the Redshift-Magnitude Relation," *PASP, 68*, 306 (1956).

(Ho57) F. Hoyle, "The Black Cloud," Signet, New York (1957).

(Hö65) B. Höglund, and P.G. Mezger, "Hydrogen Emission Line $n_{110} \rightarrow n_{109}$: Detection at 5009 Megahertz in Galactic H_{II} Regions," *Science, 150*, 339 (1965).

(Ho69) L.M. Hobbs, "Regional Studies of Interstellar Sodium Lines," *Ap.J., 158*, 461 (1969).

(Ho74) G. t'Hooft, "Magnetic Monopoles in Unified Gauge Theories" *Nuclear Physics B79*, 276 (1974).

(Hu29) E. Hubble, "Distance and Radial Velocity Among Extragalactic Nebulae," *Proc. Nat. Acad. Sci., 15*, 168 (1929).

(Hu65) S. Huang, "Rotational Behavior of the Main-Sequence Stars and its Plausible Consequences Concerning Formation of Planetary Systems I and II," *Ap.J., 141*, 985 (1965) and *150*, 229 (1967).

(Ib65) I. Iben, Jr., "Stellar Evolution-I. The Approach to the Main Sequence," *Ap.J., 141*, 993 (1965).

(Ib70) I. Iben, "Globular-Cluster Stars" *Scientific American,* July 1970, p. 27.

(Já50) L. Jánossy, *Cosmic Rays*, 2nd ed., The Clarendon Press, Oxford, England (1950).

(Ja70a) D.A. Jauncey, A.E. Niell, and J.J. Condon, "Improved Spectra of Some Ohio Radio Sources with Unusual Spectra," *Ap.J., 162*, L31 (1970).

(Ja70b) J.C. Jackson, "The Dynamics of Clusters of Galaxies in Universes with Non-Zero Cosmological Constant, and the Virial Theorem Mass Discrepancy," *MNRAS, 148*, 249 (1970).

(Je69) E.B. Jenkins, D.C. Morton, and T.A. Matilsky, "Interstellar Lα Absorption in β^1, δ, and π Scorpii," *Ap.J., 158*, 473 (1969).

(Je70) E.B. Jenkins, "Observations of Interstellar Lyman-α Absorption," *IAU Symposium* #36: *Ultraviolet Stellar Spectra and Related Ground Based Observations*, L. Houziaux and H.E. Butler, Eds., D. Reidel Publ. Co., Holland (1970).

(Jo28) J.B. Johnson, "Thermal Agitation of Electricity in Conductors," *Phys. Rev., 32*, 97 (1928).

(Jo53) H.L. Johnson and W.W. Morgan, "Fundamental Stellar Photometry for Standards of Spectral Type on the Revised System of the Yerkes Spectral Atlas," *Ap.J., 117*, 313 (1953).

(Jo56) H.L. Johnson and A.R. Sandage, "Three-Color Photometry in the Globular Cluster M3," *Ap.J., 124*, 379 (1956).

(Jo67a) H.L. Johnson, "Infrared Stars," *Science, 157*, 635 (1967).

(Jo67b) R.V. Jones and L. Spitzer, Jr., "Magnetic Alignment of Interstellar Grains," *Ap.J., 147*, 943 (1967).

(Ka54) F.D. Kahn, "The Acceleration of Interstellar Clouds," *B.A.N. 12*, 187 (1954).

(Ka59) N.S. Kardashev, "On the Possibility of Detection of Allowed Lines of Atomic Hydrogen in the Radio-Frequency Spectrum," *Astronomicheskii Zhurnal, 36*, 838 (1959), *Soviet A.J., 3*, 813 (1959).

(Ka68) F.D. Kahn, "Problems of Gas Dynamics in Planetary Nebulae," *I.A.U.,* Symposium on Planetary Nebulae, D. Osterbrock, and C.R. O'Dell, Eds., Springer-Verlag, New York (1968).

(Ke63a) R.P. Kerr, "Gravitational Field of a Spinning Mass as an Example of Algebraically Special Metrics," *Phys. Rev. Let., 11*, 237 (1963).

(Ke63b) P.C. Keenan, "Classification of Stellar Spectra" in *Stars and Stellar Systems*, Vol. 3 K.A. Strand, Ed. (1963).

(Ke65) F.J. Kerr and G. Westerhout, "Distribution of Interstellar Hydrogen" in *Stars and Stellar Systems V: Galactic Structure*, A. Blaauw and M. Schmidt, Eds., University of Chicago Press, Chicago (1965).

(Ke68) K.I. Kellermann, I.I.K. Pauliny-Toth, and M.M. Davis, "The Dependence of Radio Source Counts and the Spectral Index Distribution on Frequency," *Astrophys. Let., 2*, 105 (1968).

(Ke69) K.I. Kellermann, I.I.K. Pauliny-Thoth, and P.J.S. Williams, "The Spectra of Radio Sources in The Revised 3C Catalogue," *Ap.J., 157*, 1 (1969).

(Ke70) J.C. Kemp, J.B. Swedlund, J.D. Landstreet, and J.R.P. Angel, "Discovery of Circularly Polarized Light from a White Dwarf," *Ap.J., 161*, L77 (1970).

(Ke71) K.I. Kellermann, D.L. Jauncey, M.H. Cohen, D.B. Shaffer, B.G. Clark, J. Broderick, B. Rönnäng, O.E.H. Rydbeck, L. Matveyenko, I. Moiseyev, V.V. Vitkevitch, B.F.C. Cooper, and R. Batchelor, "High Resolution Observations of Compact Radio Sources at 6 and 18 Centimeters," *Ap.J., 169*, 1 (1971).

(Kl71) O. Klein, "Arguments Concerning Relativity and Cosmology," *Science, 171*, 339 (1971).

(Ko69) M. Kondo, "On the Formation of Condensations in an Expanding Universe," *PASJ, 21*, 54 (1969).

(Kr68) K.S. Krishna Swamy and C.R. O'Dell, "Thermal Emission by Particles in NGC 7027," *Ap.J., 151*, L61 (1968).

(Ku54) B.W. Kukarkin, *Erforschung der Struktur und Entwicklung der Sternsysteme auf der Grundlage des Studiums veränderlicher Sterne*, Akademie Verlag, Berlin (1954).

(Kv70) K. Kvenvolden, J. Lawless, K. Pering, E. Peterson, J. Flores, C. Ponnamperuma, I.R. Kaplan, and C. Moore, "Evidence for Extraterrestrial Amino-acids and Hydrocarbons in the Murchison Meteorite," *Nature, 228*, 923 (1970).

(La51) L. Landau and E. Lifshitz, *The Classical Theory of Fields*, Addison-Wesley, New York (1951).

(La67) J.W. Larimer, "Chemical fractionations in meteorites-I. Condensation of the elements," *Geochimica et Cosmochimica Acta, 31*, 1215 (1967).

(La67a) J.W. Larimer and E. Anders "Chemical fractionations in meteorites-II. Abundance patterns and their interpretation," *Geochimica et Cosmochimica Acta, 31*, 1239 (1967).

(La72) R.B. Larson, "Infall of Matter in Galaxies." *Nature, 236*, 21 (1972).

(La74) K. Lang, *Astrophysical Formulae*, Springer (1974), P47.

(Le31) G. Lemaître, "A Homogeneous Universe of Constant Mass and Increasing

Radius accounting for the Radial Velocity of Extra-Galactic Nebulae," *MNRAS 91*, 483 (1931).

(Le50) G. Lemaître, *The Primeval Atom,* Van Nostrand, Princeton, New Jersey (1950).

(Le56) T.D. Lee and C.N. Yang, "Question of Parity Conservation in Weak Interactions," *Phys. Rev. 104*, 254 (1956).

(Le72) T.J. Lee, "Astrophysics and Vacuum Technology," *Journal of Vacuum Science and Technology,* Jan/Feb (1972).

(Li46) E.M. Lifshitz, "On the Gravitational Stability of the Expanding Universe," *J. Phys. USSR, 10*, 116 (1946).

(Li67) C.C. Lin, "The Dynamics of Disk-Shaped Galaxies." *Ann. Rev. Astron. Astrophys., 5*, 453 (1967).

(Lo04) H.A. Lorentz, "Electromagnetic Phenomena in a System Moving with Any Velocity Less Than Light," *Proceedings of the Acad. Sci.*, Amsterdam *6*, 1904, reprinted in *The Principle of Relativity*, A. Sommerfeld, Ed., Dover, New York.

(Lo78) M. Longair "Cosmological Aspects of Infrared and Millimetre Astronomy," in *Infrared Astronomy*, Eds. G. Setti and G.G. Fazio, Reidel, Dordrecht, Holland (1978).

(Ly63) C. R. Lynds and A. Sandage, "Evidence for an Explosion in the Center of the Galaxy M82," *Ap.J., 137*, 1005 (1963).

(Ma70) D.S. Mathewson and V.L. Ford, "Polarization Observations of 1800 Stars," *Memoirs of the Royal Astronomical Society*, 74, 139 (1970).

(Ma72a) P. M. Mathews and M. Lakshmanan, "On the apparent Visual Forms of Relativistically Moving Objects," *Nuovo Cimento, 12 B, 168* (1972).

(Ma72b) R.N. Manchester, "Pulsar Rotation and Dispersion Measures and the Galactic Magnetic Field," *Ap.J., 172*, 43 (1972).

(Ma79) T. Maihara, N. Oda, and H. Okuda, *Ap.J. Let. 227* L132 (1979).

(McNa65) D. McNally, "On the Distribution of Angular Momentum Among Main Sequence Stars," *Observatory, 85*, 166 (1965).

(Me68) P.G. Mezger, "A New Class of Compact, High Density H II Regions" in *Interstellar Ionized Hydrogen*, Y. Terzian, Ed., Benjamin, Menlo Park, Calif. (1968).

(Me69) P. Meyer, "Cosmic Rays in the Galaxy," in *Ann. Rev. Astron. Astrophys., 7*, 1 (1969).

(Mi08) H. Minkowski, "Space and Time" an address delivered to the German Natural Scientists and Physicians (1908), translated and reprinted in *The Principle of Relativity*, A. Sommerfeld, Ed., Dover, New York.

(Mi57a) S.L. Miller, "The Formation of Organic Compounds on the Primitive Earth," *Ann. New York Acad. Sci., 69*, 260 (1957).

(Mi57b) R.I. Mitchell and H.L. Johnson, "The Color-Magnitude Diagram of the Pleiades Cluster," *Ap.J., 125*, 418 (1957).

(Mi59) S.L. Miller and H.C. Urey, "Organic Compound Synthesis on the Primitive Earth," *Science, 130,* 245 (1959).

(Mo67) D.C. Morton, "Mass Loss from Three OB Supergiants in Orion," *Ap.J., 150,* 535 (1967).

(Mo68) P.M. Morse and K.U. Ingard, *Theoretical Acoustics,* McGraw-Hill, New York (1968).

(Mu87) J. Murthy *et al,* "IUE Observations of Hydrogen and Deuterium in the Local Interstellar Medium" *Ap.J.* 315, 675 (1987).

(Ne00) Isaac Newton, *Mathematical Principles of Natural Philosophy* and *Systems of the World,* revised translation by Florian Cajori, University of California Press, Berkeley (1962).

(Nu84) H. Nussbaumer and W. Schmutz, *A + A 138,* 495–496 (1984).

(Ny28) H. Nyquist, "Thermal Agitation of Electric Charge in Conductors," *Phys. Rev., 32,* 110 (1928).

(Od65) M. Oda, G.W. Garmire, M. Wada, R. Giacconi, H. Gursky, and J.R. Waters, "Angular Sizes of the X-ray Sources in Scorpio and Sagittarius," *Nature, 205,* 554 (1965).

(Oo27a) J.H. Oort, "Observational Evidence Confirming Lindblad's Hypothesis of a Rotation of the Galactic System." *B.A.N., 3,* 275 (1927).

(Oo27b) J.H. Oort, "Investigations Concerning the Rotational Motion of the Galactic System, Together with New Determinations of Secular Parallaxes, Precession and Motion of the Equinox," *B.A.N., 4,* 79 (1927).

(Oo65) J.H. Oort, "Stellar Dynamics" in *Galactic Structure,* A. Blaauw and M. Schmidt, University of Chicago Press, Chicago (1965), p. 455.

(Op39a) J.R. Oppenheimer and G.M. Volkoff, "On Massive Neutron Cores," *Phys. Rev., 55,* 374 (1939).

(Op39b) J.R. Openheimer and H. Snyder, "On Continued Gravitational Contraction," *Phys. Rev., 56,* 455 (1939).

(Op61a) A.I. Oparin, *Life, Its Nature, Origin and Development,* Academic Press, New York (1961).

(Op61b) A.I. Oparin and V.G. Fessenkov, *Life in the Universe,* Foreign Languages Publishing House, Moscow; also Twayne and Co., New York (1961).

(Ov69) O.E. Overseth, "Experiments in Time Reversal," *Scientific American,* October (1969) p. 89.

(Pa68) F. Pacini, "Rotating Neutron Stars, Pulsars and Supernova Remnants," *Nature, 219,* 145 (1968).

(Pe65) A. A. Penzias and R. W. Wilson "A Measurement of Excess Antenna Temperature at 4080 Mc/s," *Ap.J., 142,* 420 (1965).

(Pe69) J.V. Peach, "Brightest Members of Clusters of Galaxies," *Nature, 223,* 1141 (1969).

(Pe71) R. Penrose and R.M. Floyd, "Extraction of Energy from a Black Hole,"

Nature, 229, 177 (1971).

(Pe85) R.C. Peterson, "Radial Velocities of Remote Globular Clusters: Stalking the Missing Mass," *Ap.J. 297*, 309 (1985).

(Pi66) G.C. Pimentel, K.C. Atwood, H. Gaffron, H.K. Hartline, T.H. Jukes, E.C. Pollard, and C. Sagan, "Exotic Biochemistries in Exobiology" in *Biology and the Exploration of Mars*, C.S. Pittendrigh, W. Vishniac, and J.P. Pearman, Eds., Nat. Acad. Sci., NRC (Washington) (1966).

(Po56) S. Pottasch, "A Study of Bright Rims in Diffuse Nebulae," *B.A.N., 13*, 77 (1956).

(Po58) S. Pottasch, "Dynamics of Bright Rims in Diffuse Nebulae," *B.A.N., 14*, 29 (1958)

(Po68) G.G. Pooley and M. Ryle, "The Extension of the Number-Flux Density Relation for Radio Sources to Very Small Flux Densities," *MNRAS, 139*, 515 (1968).

(Po71) K.A. Pounds, "Recent Developments in X-ray Astronomy," *Nature, 229*, 303 (1971).

(Po74) A.M. Polyakov, "Particle Spectrum in Quantum Field Theory," *JETP Letters 20*, 194 (1974).

(Pr61) I. Prigogine, "Thermodynamics of Irreversible Processes," John Wiley, New York (1961).

(Pr68) K.H. Prendergast and G.R. Burbidge, "On the Nature of Some Galactic X-ray Sources," *Ap.J., 151*, L83 (1968).

(Ra71) D.M. Rank, C.H. Townes, and W.J. Welch, "Interstellar Molecules and Dense Clouds," *Science, 174*, 1083 (1971).

(Re44) G. Reber, "Cosmic Static," *Ap.J., 100*, 279 (1944).

(Re68a) V.C. Reddish, "The Evolution of Galaxies," *Quart. J. Roy. Astron. Soc., 9*, 409 (1968).

(Re68b) M.J. Rees, "Proton Synchrotron Emission from Compact Radio Sources," *Astrophys. Let., 2*, 1 (1968).

(Re68c) M.J. Rees and W.L.W. Sargent, "Composition and Origin of Cosmic Rays," *Nature, 219*, 1005 (1968).

(Re70) M.J. Rees and J. Silk, "The Origin of Galaxies," *Scientific American, 222*, (June 1970), p. 26.

(Re71) M.J. Rees, "New Interpretation of Extragalactic Radio Sources," *Nature, 229*, 312 (1971).

(Re87) N. Reid "The stellar mass function at low luminosities," *MNRAS, 225*, 873 (1987).

(Ri56) W. Rindler, "Visual Horizons in World Models," *MNRAS, 116*, 662 (1956).

(Ro33) H.P. Robertson, "Relativistic Cosmology," *Rev. Mod. Phys., 5*, 62 (1933).

(Ro55) H.P. Robertson, "The Theoretical Aspects of the Nebular Red Shift," *PASP, 67*, 82 (1955).

(Ro64a) Bruno Rossi, *Cosmic Rays*, McGraw Hill, New York (1964).

(Ro64b) P.G. Roll, R. Krotkov, and R.H. Dicke, "The Equivalence of Inertial and Passive Gravitational Mass," *Annals of Physics* (USA), *26*, 442 (1964).

(Ro65) F. Rosebury, "Handbook of Electron Tube and Vacuum Techniques," Addison-Wesley, Reading, Mass. (1965).

(Ro68) H.P. Robertson and T.W. Noonan, *Relativity and Cosmology*, Sanders, (1968).

(Ru71a) R. Ruffini and J.A. Wheeler, "Introducing the Black Hole," *Physics Today*, *24*, 30, January 1971.

(Ru71b) M. Ruderman, "Solid Stars," *Scientific American*, February (1971), p. 29.

(Ry68) M. Ryle, "The Counts of Radio Sources" in *Ann. Rev. Astron. and Astrophys.*, *6*, 249 (1968).

(Sa52) E.E. Salpeter, "Nuclear Reactions in Stars Without Hydrogen," *Ap.J.*, *115* 326 (1952).

(Sa55) E.E. Salpeter, "Nuclear Reactions in Stars II. Protons on Light Nuclei," *Phys. Rev.*, *97*, 1237 (1955).

(Sa57) A. Sandage, "Observational Approach to Evolution-II.A Computed Luminosity Function for K0-K2 Stars from $M_v = +5$ to $M_v = -4.5$," *Ap.J.*, *125*, 435 (1957).

(Sa58) A. Sandage, "Current Problems in the Extragalactic Distance Scale," *Ap.J.*, *127*, 513 (1958).

(Sa66a) A.R. Sandage, P. Osmer, R. Giacconi, P. Gorenstein, H. Gursky, J. Waters, H. Bradt, G. Garmire, B.V. Sreekantan, M. Oda, K. Osawa, and J. Jugaku, "On the Optical Identification of Sco X-1" *Ap.J.*, *146*, 316 (1966).

(Sa66b) C. Sagan and R.G. Walker, "The Infrared Detectability of Dyson Civilizations," *Ap.J.*, *144*, 1216 (1966).

(Sa67) E.E. Salpeter, "Stellar Structure Leading up to White Dwarfs and Neutron Stars" in "Relativity Theory and Stellar Structure," Chapter 3 in *Lectures in Applied Mathematics*, Vol. 10., American Mathematical Society (1967).

(Sa68a) D.H. Sadler, "Astronomical Measures of Time," *Quarterly Journal, Roy. Astr. Soc.*, London, *9*, 281 (1968).

(Sa68b) E.E. Salpeter, "Evolution of the Central Stars of Planetary Nebulae: Theory," *IAU Symposium No. 34*, North-Holland, Amsterdam (1968) p. 409.

(Sa69a) E.E. Salpeter and J.N. Bahcall, "On the Masses of Quasi-Stellar Objects," *Ap.J.*, *158*, L 15 (1969).

(Sa69b) E.E. Salpeter, "Neutrinos and Stellar Evolution," from *Yeshiva University Annual Science Conference Proceedings*, Vol. II (1969).

(Sa70a) E.E. Salpeter, "Solid State Astrophysics," in *Methods and Problems of Theoretical Physics* J.E. Bowcock, Ed., North-Holland, Amsterdam (1970).

(Sa70b) C. Sagan, "Life," in *Encyclopedia Britannica* (1970).

(Sc44) E. Schrödinger, *What is Life,* Cambridge University Press, Cambridge England (1944).

(Sc58a) L.I. Schiff, "Sign of the Gravitational Mass of a Positron," *Phys. Rev. Let., 1,* 254 (1958).

(Sc58b) M. Schwarzschild, "Structure and Evolution of the Stars," Princeton University Press, Princeton, New Jersy (1958).

(Sc63) M. Schmidt, "The Rate of Star Formation II. The Rate of Formation of Stars of Different Mass." *Ap.J., 137,* 758 (1963).

(Sc70) M. Schwarzchild, "Stellar Evolution in Globular Clusters," *Quarterly J. Royal Astron. Soc., 11,* 12 (1970).

(Se65) P.A. Seeger, W.A. Fowler, and D.D. Clayton, "Nucleosynthesis of Heavy Elements by Neutron Capture," *Ap.J.* Suppl. No. 97, *11,* 121 (1965).

(Sh60) I.S. Shklovskii, "Cosmic Radiowaves," Harvard University Press, Cambridge, Mass. (1960).

(Sh66) I.S. Shklovskii and C. Sagan, "Intelligent Life in the Universe," Delta, New York (1966).

(Sh68) A.J. Shimmins, J.G. Bolton, and J.V. Wall, "Counts of Radio Sources at 2,700 MHz," *Nature, 217,* 818 (1968).

(Sh71) I.I. Shapiro, W.B. Smith, M.B. Ash, R.P. Ingalls, G.H. Pettengill, "Gravitational Constant: Experimental Bound on Its Time Variation," *Phys. Rev. Let., 26,* 27 (1971).

(Sp51a) L. Spitzer, Jr. and M. Schwarzschild, "The Possible Influence of Interstellar Clouds on Stellar Velocities," *Ap.J., 114,* 394 (1951).

(Sp51b) L. Spitzer, Jr. and J.L. Greenstein, "Continuous Emission from Planetarv Nebulae," *Ap.J., 114,* 407 (1951).

(Sp62) L. Spitzer Jr., *The Physics of Fully Ionized Gases,* Interscience, New York (1962).

(Sp71) H. Spinrad, W.L. Sargent, J.B. Oke, G. Neugebauer, R. Landau, I.R. King, J.E. Gunn, G. Garmire, and N.H. Dieter, "Maffei 1: A New Massive Member of the Local Group?" *Ap.J., 163,* L25 (1971).

(St39) B. Strömgren, "The Physical State of Interstellar Hydrogen," *Ap.J., 89,* 526 (1939).

(St65) B. Strömgren, "Stellar Models for Main-sequence Stars and Subdwarfs," in *Stellar Structure*, L.H. Aller and D.B. McLaughlin, Eds., University of Chicago Press, Chicago (1965).

(St68) F.W. Stecker, "Effect of Photomeson Production by the Universal Radiation Field on High-Energy Cosmic Rays," *Phys. Rev. Let., 23,* 1016 (1968).

(St69) T.P. Stecher, "Interstellar Extinction in the Ultraviolet II," *Ap.J., 157,* L125 (1969).

(St72) L.J. Stief, B. Donn, B. Glicker, E.P. Gentieu, and J.E. Mentall, "Photochemistry and Lifetimes of Interstellar Molecules," *Ap.J., 171,* 21 (1972).

(Ta70) B.N. Taylor, D.N. Langenberg, and W.H. Parker, "The Fundamental Physical Constants," *Scientific American*, October (1970) p. 62.

(Te48) E. Teller, "On the Change of Physical Constants," *Phys. Rev., 73*, 801 (1948).

(Te59) J. Terrell, "Invisibility of the Lorentz Contraction," *Phys. Rev., 116*, 1041 (1959).

(Te72) Y. Terzian, *A Tabulation of Pulsar Observations*, Earth and Terrestrial Sciences, Gordon and Breach, New York (May 1972).

(Th69) D.J. Thouless, "Causality and Tachyons," *Nature, 244*, 506 (1969).

(Th72) R.I. Thompson, "Carbon Stars and The CNO Bi-Cycle," *Ap.J., 172*, 391 (1972).

(To47) C.H. Townes, "Interpretation of Radio Radiation from the Milky Way," *Ap.J., 105*, 235 (1947).

(Un69) A. Unsöld, "Stellar Abundances and the Origin of Elements," *Science, 163*, 1015 (1969).

(vdBe68) S.v.d. Bergh, "Galaxies of the Local Group," *J. of the Royal Astron. Soc. Canad., 62*, 145,219 (1968).

(vdBe72) S. v.d. Bergh, "Search for Faint Companions to M31," *Ap.J., 171*, L31 (1972).

(vdHu57) H.C. van de Hulst, "Light Scattering by Small Particles," John Wiley, New York (1957).

(Va70) R.S. Van Dyck, Jr., C.E. Johnson, and H.A. Shugant, "Radiative Lifetime of Metastable 2^1S_0 State of Helium," *Phys. Rev. Let., 25*, 1403 (1970).

(Va71) L.V. Vallen, "The History and Stability of Atmospheric Oxygen," *Science, 171*, 439 (1971).

(Vé86) P. Véron, "The Cosmic evolution of quasars at high redshifts," *Astron. and Astrophys. 170*, 37 (1986).

(vHo55) S. von Hoerner, "Über die Bahnform der Kugelförmigen Sternhaufen," *Z. Astrophys., 35*, 255 (1955).

(vHo57) S. von Hoerner, "The Internal Structure of Globular Clusters," *Ap.J., 125*, 451 (1957).

(Wa34) A.G. Walker, "Distance in an Expanding Universe," *MNRAS, 94*, 159 (1934).

(Wa56) F.G. Watson, "Between the Planets," rev. ed., Harvard University Press, Cambridge, Mass. (1956).

(Wa67) R.V. Wagoner, "Cosmological Element Production," *Science, 155*, 1369 (1967).

(Wa71) R.V. Wagoner, "Production of Helium in Massive Objects," in *Highlights of Astronomy*, C. de Jager, Ed., I.A.U. (1971).

(We62) S. Weinberg, "The Neutrino Problem in Cosmology," *Nuovo Cimento, 25*, 15 (1962).

(We68) V. Weidemann, "White Dwarfs" in *Ann. Rev. Astron. and Astrophys.,* 6, 351 (1968).

(We70) J. Weber, "Anisotropy and Polarization in the Gravitational-Radiation Experiments," *Phys. Rev. Let., 25,* 180 (1970).

(We71) A.S. Webster and M.S. Longair, "The Diffusion of Relativistic Electrons from Infrared Sources and Their X-ray Emission," *MNRAS, 151,* 261 (1971).

(We77) S. Weinberg, *The First Three Minutes,* Basic Books, New York, 1977.

(Wh64) F.L. Whipple, "The History of the Solar System," *Proc. Nat. Acad. Sci., 52,* 565 (1964).

(Wi57) O.C. Wilson and M.K.V. Bappu, "H and K Emission in Late-Type Stars: Dependence of Line Width on Luminosity and Related Topics," *Ap.J., 125,* 661 (1957).

(Wi58) D.H. Wilkinson, "Do the 'Constants of Nature' Change with Time?" *Phil. Mag.,* Ser 8:3, 582 (1958).

(Wo57) L. Woltjer, "The Crab Nebula," *B.A.N., 14,* 39 (1957).

(Za54) H. Zanstra, "A simple Approximate Formula for the Recombination Coefficient of Hydrogen," *Observatory, 74,* 66 (1954).

Index